ARCTIC

SEARCHING EXPEDITION:

A

JOURNAL OF A BOAT-VOYAGE THROUGH RUPERT'S LAND AND THE ARCTIC SEA,

IN SEARCH OF THE DISCOVERY SHIPS UNDER COMMAND OF

SIR JOHN FRANKLIN.

WITH AN APPENDIX ON THE PHYSICAL GEOGRAPHY OF NORTH AMERICA.

BY SIR JOHN RICHARDSON, C.B., F.R.S.,

INSPECTOR OF NAVAL HOSPITALS AND FLEETS,
ETC., ETC., ETC.

NEW YORK:
HARPER & BROTHERS, PUBLISHERS,
82 CLIFF STREET.
1852.

CONTENTS.

CHAPTER I.

CHAPTER II.

CHAPTER III.

CHAPTER IV.

PAGE

CHAPTER XII.

ON THE KUTCHIN OR LOUCHEUX.

CHAPTER XIII.

OF THE 'TINNE OR CHEPEWYANS.

CHAPTER XIV.

EYTHINYUWUK, OR CREES AND CHIPPEWAYS.

CHAPTER XV.

OCCURRENCES IN WINTER.

CHAPTER XVI.

APPENDIX.

No. I.—PHYSICAL GEOGRAPHY.

No. II.—CLIMATOLOGY.

No. III.—THE GEOGRAPHICAL DISTRIBUTION OF PLANTS NORTH OF THE 49TH PARALLEL OF LATITUDE.

PAGE

No. IV.—LIST OF INSECTS.

No. V.—VOCABULARIES.

WOODCUTS AND DIAGRAMS.

NOTICE.

In the Indian names which occur in the following narrative, *u* is to be sounded like *oo*, in "moon;" *yu* as in "yule," or like "you;" and *i* as in "ravine."

ARCTIC SEARCHING EXPEDITION.

CHAPTER I.

Route assigned to the Expedition under Command of Sir John Franklin.—Names of the Officers.—Erebus and Terror.—Date of its Sailing.—Last Letters.—Sir John Franklin's Last Official Letter.—Last Sight of the Expedition.—Sir John Ross proposes a Search.—Discussion of various Opinions offered respecting the Fate of the Expedition.—Plans of Search adopted.—Main Objects of the Overland Searching Expedition.—Instructions from the Admiralty.

HER MAJESTY'S government having deemed it expedient that a further attempt should be made for the accomplishment of a northwest passage by sea from the Atlantic to the Pacific, the "Erebus" and "Terror" were fitted out for that service, and placed under the command of Captain Sir John Franklin, K.C.H. He was directed by the Admiralty instructions, dated on the 5th of May, 1845, to proceed with all dispatch to Lancaster Sound, and, passing through it, to push on to the westward, in the latitude of 74½°, without loss of time or *stopping to examine any openings to the northward*, until he reached the longitude of Cape Walker, which is situated in about 98° west. He was to use every effort to penetrate to the *southward* and *westward* of that point, and to pursue as direct a course for Beering's Straits as circumstances might permit. He was cautioned not to attempt to pass by the western extremity of Melville Island, until he had ascertained that a permanent barrier of ice or other obstacle closed the prescribed route. In the event of not being able to penetrate to the westward, he was to enter Wellington Sound in his second summer.

He was further directed to transmit accounts of his proceedings to the Admiralty, by means of the natives and the Hudson's Bay Company, should opportunities offer; and also, after passing the 65th meridian, to throw overboard daily a copper cylinder, containing a paper stating the ship's position. It was also understood that he would cause piles of stones or signal-posts to be erected on conspicuous headlands at convenient times, though the instructions do not contain a clause to that effect.*

The following officers joined the expedition :

EREBUS.	TERROR.
Captain, Sir John Franklin, Kt. K.C.H.	*Captain*, Francis R. M. Crozier.
Commander, James Fitzjames.	*Lieutenant*, Edward Little.
Lieutenant, Graham Gore.	*Lieutenant*, George H. Hodgson.
Lieut., H. P. D. Le Vesconte.	*Lieutenant*, John Irving.
Lieut., James W. Fairholme.	*Ice-Master*, Thomas Blanky.
Ice-Master, James Read.	*Surgeon*, John S. Peddie.
Surgeon, Stephen S. Stanley.	*Assist.-Surgeon*, A. M'Donald.
Paymaster, C. H. Osmer.	*Sec. Master*, Gillies A. Maclean.
Assist.-Surg., H. D. S. Goodsir.	*Clerk-in-Charge*, Edward J. H. Helpman.
Sec. Master, Henry F. Collins.	

And the conjoined crews of the two ships amounted to 130 souls.

The "Erebus," originally built for a bomb-vessel, and therefore strongly framed, was of 370 tons measurement, and had been fortified, in 1839, after the most approved plan, by an extra or double exterior planking and diagonal bracing within, for Sir James C. Ross's Antarctic voyage, from which she returned in 1843. Having been carefully examined and refitted for Sir John Franklin, she was considered to be as strongly prepared to resist the pressure of the ice as the resources of science, and the utmost care of Mr. Rice, the skillful master-shipwright who superintended the preparations, could insure. The "Terror," of 340 tons, was also constructed for a bomb-vessel, and had the bluff form, capacious hold, and strong framework of that class of war vessels. When commanded by Captain Sir George Back, on his voyage to Repulse Bay in 1836–7, she had been beset for more than eleven months in drifting floes of ice, and exposed to every variety of assault and pressure to which a vessel was liable in such a dangerous position. In this severe and lengthened trial, the "Terror" had been often pressed more or less out of the water, or thrown over on one side, and had, in consequence thereof, sustained some

* The instructions are published at length in a parliamentary Blue Book, and all known particulars respecting the expedition have been communicated from time to time to the public by the same channel. The above abstract mentions the leading points which would direct the course of the expedition.

damage, particularly in the stern post. All defects, however, were made good in 1839, when she sailed for the Antarctic Seas, under the command of Captain Crozier, the second officer of Sir James C. Ross's expedition. She was again examined, and made as strong as ever, before Captain Crozier took the command of her a second time in 1845.

The best plans that former experience could suggest for ventilating and warming the ships in the winter were adopted, and full supplies of every requisite for Arctic navigation were provided, including an ample stock of warm bedding, clothing, and provisions, with a proportion of preserved meats and pemican.

The expedition sailed from England on the 19th of May, 1845, and, early in July, had reached Whalefish Islands, near Disco, on the Greenland coast of Davis's Straits, where, having found a convenient port, the transport which accompanied it was cleared and sent home to England, bringing the last letters that have been received from the officers or crew. The following extract of a letter, from Lieutenant Fairholme, of the "Erebus," will serve to show the cheerful anticipation of success which prevailed throughout the party, and the happy terms on which they were with each other:

"We have anchored in a narrow channel between two of the islands, protected on all sides by land, and in as convenient a place for our purpose as could possibly be found. Here we are with the transport lashed alongside, transferring most actively all her stores to the two ships. I hope that this operation will be completed by to-morrow night, in which case Wednesday will be devoted to swinging the ships for local attraction, and I suppose Thursday will see us under way with our heads to the northward. We have had the observatory up here, on a small rock on which Parry formerly observed, and have got a very satisfactory set of magnetic and other observations. Of our prospects we know little more than when we left England, but look forward with anxiety to our reaching 72°, where it seems we are likely to meet the first obstruction, if any exists. On board we are as comfortable as it is possible to be. I need hardly tell you how much we are all delighted with our captain. He has, I am sure, won not only the respect but the love of every person on board by his amiable manner and kindness to all; and his influence is always employed for some good purpose both among the officers and men. He has been most successful in his selection of officers, and a more agreeable set could hardly be found. Sir John is in much better health than when we left England, and really looks ten years younger. He takes an active part in every thing that goes on, and his long experience in such services as this makes him a most valuable adviser. *July 10th.*—The transport is just reported clear, so I hope that we may be able to swing the ships to-morrow and get away on Saturday. We are very much crowded; in fact, not an inch of stowage has been lost, and the decks are still covered with casks, &c. Our supply of coals has encroached seriously on the ship's stowage; but as we consume both this and provisions as we go, the evil will be continually lessening."

Letters from most of the other officers, written in a similarly buoyant and hopeful spirit, were received in England at the same time with the above. An extract of a letter from Sir John Franklin himself to Lieutenant Colonel Sabine deserves to be quoted, as expressing his own opinion of his resources, and also his intention of remaining out should he fail after a second winter in finding an outlet to the southwestward from Barrow's Strait. The letter is dated from Whalefish Islands, on the 9th of July, 1845, and, after noticing that the "Erebus" and "Terror" had on board provisions, fuel, clothing, and stores for three years complete, from that date, adds, "I hope my dear wife and daughter will not be over-anxious if we should not return by the time they have fixed upon; and I must beg of you to give them the benefit of your advice and experience when that time arrives, for you know well that, without success in our object, even after the *second winter*, we should wish to try some other channel if the state of our provisions and the health of the crews justify it."

The following is the last official letter written by Sir John Franklin to the Admiralty.

"Her Majesty's Ship 'Erebus,'
Whalefish Islands, July 12, 1845

"Sir—I have the honor to acquaint you, for the information of the Lords Commissioners of the Admiralty, that Her Majesty's ships 'Erebus' and 'Terror,' with the transport, arrived at this anchorage on the 4th instant, having had a passage of one month from Stromness. The transport was immediately taken alongside this ship, that she might be more readily cleared; and we have been constantly employed at that operation till last evening, the delay having been caused not so much in getting the stores transferred to either of the ships, as in making the best stowage of them below, as well as on the upper deck. The ships are now complete with supplies of every kind for three years: they are, therefore, very deep; but happily we have no reason to expect much sea as we proceed further.

"The magnetic instruments were landed the same morning; so also were the other instruments requisite for ascertaining the position of the observatory; and it is satisfactory to find that the results of the observations for latitude and longitude accord very nearly with those assigned to the same place by Sir Edward Parry. Those for dip and variation are equally satisfactory, which were made by Captain Crozier with the instruments belonging to the 'Terror,' and by Commander James with those of the 'Erebus.'

"The ships are now being swung, for the purpose of ascertaining the dip and deviation of the needle on board, as was done at Greenhithe; which I trust will be completed this afternoon, and I hope to be able to sail in the night.

"The governor and principal persons are at this time absent from Disco; so that I have not been able to receive any communication from head quarters as to the state of the ice to the north. I have, however, learned from a Danish carpenter in charge of the Esquimaux at these islands, that, though the winter was severe, the spring was not later than usual, nor was the ice later in breaking away hereabout. He supposes, also, that it is now loose as far as 74°, and as far as Lancaster Sound, without much obstruction.

"The transport will sail for England this day. I shall instruct the agent, Lieutenant Griffiths, to proceed to Deptford, and report his arrival to the Secretary of the Admiralty. I have much satisfaction in bearing my testimony to the careful and zealous manner in which Lieutenant Griffiths has performed the service intrusted to him, and would beg to recommend him as an officer who appears to have seen much service, to the favorable consideration of their Lordships.

"It is unnecessary to assure their Lordships of the energy and zeal of Captain Crozier, Commander Fitzjames, and of the officers and men with whom I have the happiness of being employed on this service.

"I have, &c.,
"JOHN FRANKLIN,
"Captain.

"The Right Hon. H. L. Cary, M.P.
"&c. &c. &c."

The two ships were seen on the 26th of the same month (July) in latitude, 74° 48′ N., longitude 66° 13′ W., moored to an iceberg, waiting for a favorable opportunity of entering or rounding the "middle ice" and crossing to Lancaster Sound, distant in a direct westerly line from their position about 220 geographical miles. On that day a boat from the discovery ships, manned by seven officers, one of whom was Commander Fitzjames, boarded the "Prince of Wales," whaler, Captain Dannett. They were all in high spirits and invited Captain Dannett to dine with Sir

John Franklin on the following day, which had he done, he would doubtless have been the bearer of letters for England, but a favorable breeze springing up he separated from them. The ice was then heavy but loose, and the officers expressed good hopes of soon accomplishing the enterprise. Captain Dannett was favored with very fine weather during the three following weeks, and thought that the expedition must have made good progress. This was the last sight that was obtained of Franklin's ships.

In January, 1847, a year and a half after the above date, Captain Sir John Ross addressed a letter to the Admiralty, wherein he stated his conviction that the discovery ships were frozen up at the *western end of Melville Island*, from whence their return would be forever prevented by the accumulation of ice behind them, and volunteered his services to carry relief to the crews. Sir John also laid statements of his apprehensions before the Royal and Geographical Societies, and, the public attention being thereby roused, several writers in the newspapers and other periodicals published their sentiments on the subject, a variety of plans of relief were suggested, and many volunteers came forward to execute them.

The Lords Commissioners of the Admiralty, though judging that the second winter was too early a period of Sir John Franklin's absence to give rise to well founded apprehensions for his safety, lost no time in calling for the opinions of several naval officers who were well acquainted with Arctic navigation, and in concerting plans of relief, to be carried out when the proper time should arrive.

A brief review of the replies most worthy of notice may help the reader to form a judgment of the plans that were eventually adopted by the Admiralty for the discovery and relief of the absent voyagers. It is convenient to consider first the notions of those who believe that Sir John Franklin never entered Lancaster Sound, either because the ships met with some fatal disaster in Baffin's Bay, and went down with the entire loss of both crews, or that Sir John endeavored to fulfill the purposes of the expedition by taking some other route than the one exclusively marked out for him by his instructions. That the ships were not suddenly wrecked by a storm, or overwhelmed by the pressure of the ice, may be concluded from facts gathered from the records of the Davis's Straits whale-fishery, by which we learn that of the many vessels which have been crushed in the ice, in the course of several centuries, the whole or greater part of the crews have almost always escaped with their boats. It is, therefore, scarcely possible to believe that two vessels so strongly fortified as the "Erebus" and "Terror," and found by previous trials to be capable of sustaining so enormous a pressure, should both of them have been so suddenly crushed as to allow no time for active officers and men, disciplined and prepared for emergencies of the kind, to get out their boats. And having done so they would have had little difficulty in reaching one of the many whalers, that were occupied in the pursuit of fish in those seas for six weeks after the discovery ships were last seen. Moreover, had the ships been wrecked, some fragments of their spars or hulls would have been found floating by the whalers, or being cast on the eastern or western shores of the bay, would have been reported by the Greenlanders or Eskimos. Neither are any severe storms recorded as having occurred then or there, nor did any unusual calamity befall the fishing vessels that year.

With respect to Sir John Franklin having chosen to enter Jones's or Smith's Sounds in preference to Lancaster Sound, his known habit of strict adherence to his instructions is a sufficient answer, and the extract quoted above from his letter to Lieutenant Colonel Sabine, which gives his latest thoughts on the subject, plainly says that such a course would not be pursued until a *second winter* had proved the impracticability of the route laid down for him. This point is mooted, because Mr. Hamilton, surgeon in Orkney, states that Sir John, when dining with him on the last day that he passed in Great Britain, mentioned his determination of trying Jones's Sound. But Sir John's communication to Colonel Sabine shows that this could be meant to refer only to the contingency of a full trial by Lancaster Sound proving fruitless. Supposing that, contrary to all former experience, he had found the mouth of Lancaster Sound so barred by ice as to preclude his entrance, then, after waiting till he had become convinced that it would remain closed for the season, he might have tried to find a way, by Jones's Sound, into Wellington Sound; but in such a case, we may hold it as certain that he would have erected conspicuous cairns, and deposited memoranda of his past proceedings and future intentions, at the entrance of Lancaster Sound.

Taking it, then, for granted that the expedition entered Lan caster Sound, the most probable conjecture respecting the direction in which it advanced is that Sir John, literally following his instructions, did not stop to examine any openings either to the northward or southward of Barrow's Strait, but continued to push on to the westward until he reached Cape Walker, in longitude 98°, when he inclined to the southwest, and steered as directly as he could for Beering's Straits. But even supposing that the state of the ice permitted him to take the desired route, and to turn to the southwestward by the first opening beyond the 98th meridian, we are ignorant of the exact position of that opening, the tract between Cape Walker and Banks's Land being totally unknown. That a passage to the southward does exist in that space, and terminates between Victoria and Wollaston Lands in Coronation Gulf, is inferred from the observed setting of the flood tide. There is, it is true, an uncertainty in our endeavors to determine the directions of the tides in these narrow seas, where the currents are influenced by prevailing winds; but Mr. Thomas Simpson, who was an acute observer, remarked that the flood tide brought much ice into Coronation Gulf round the west end of Victoria Land, and facts collected on three visits which I have made to that gulf lead me to concur with him. Entirely in accordance with this opinion is the fact noted by Sir Edward Parry, that the flood tide came from the north between Cornwallis and the neighboring islands, and that the ice was continually setting round the west end of Melville Island and passing onward to the southeast.

These observations, while they point to an opening to the eastward of Banks's Land, may be adduced as an argument against the existence of a passage directly to the westward be tween it and Melville Island; and, though they are not con clusive, they are supported by another remark of Sir Edward Parry's that he thought there was some peculiar obstruction immediately to the west of that island, which produced a permanent barrier of ice.

But wherever the opening which we presume to exist may be situated, the channels among the islands are probably not direct, and may be intricate. Vessels, therefore, having pushed into one of them, would be exposed to the ice closing in behind and barring all regress. Sir John Ross, whose opinions are first recorded in the parliamentary Blue Book, believes that "Sir John Franklin put his ships into the drift ice at the western end of Melville Island," and that, "if not totally lost, they must have been carried by the ice, which is known to drift to the southward, on land (Banks's Land) seen at a great distance in that direction, and from which the accumulation of ice behind them will," says he, "as in my own case, forever prevent the return of the ships."

Sir W. Edward Parry is of opinion that Sir John Franklin would endeavor "to get to the southward and westward before he approached the southwestern extremity of Melville Island, that is, between the 100th and 110th degree of longitude:" "how far they may have penetrated to the southward between those meridians, must be a matter of speculation, depending on the state of the ice and the existence of land in a space hitherto blank in our maps." "Be this as it may, I (Sir W. E. Parry) consider it not improbable, as suggested by Dr. King, that an attempt will be made by them to fall back on the western coast of North Somerset, wherever that may be found, as being the nearest point affording a hope of communication, either with whalers or with ships sent expressly in search of the expedition."

Sir James C. Ross says: "It is far more probable, however, that Sir John Franklin, in obedience to his instructions, would endeavor to push the ships to the south and west as soon as they passed Cape Walker; and the consequence of such a measure, owing to the known prevalence of westerly winds, and the drift of the main body of the ice, would be their inevitable embarrassment; and if he persevered in that direction, which he probably would do, I have no hesitation in stating my conviction, that he would never be able to extricate his ships, and would ultimately be obliged to abandon them. It is, therefore, in latitude 73° N. and longitude 135° W. that we may expect to find them involved in the ice, or shut up in some harbor."

The opinions here quoted are contingent on the supposition, that Sir John Franklin found the state of the ice to be such that he could take the routes in question; but the several officers quoted admit that, in the event of no opening through the ice in a westerly or southwesterly direction being found, Sir John would attempt Wellington Sound, or any other northern opening that was accessible. Commander Fitzjames, in a letter dated January, 1845, says: "The northwest passage is certainly to be gone through by Barrow's Straits, but whether south or north of Parry's Group remains to be proved. I am for going far north, edging northwest till in longitude 140° W., if possible." Mr. John Barrow, to whom this letter was addressed, appends to it the following memorandum: "Captain Fitzjames was much inclined to try the passage to the northward of Parry's Islands, and he would, no doubt, endeavor to persuade Sir John Franklin to pursue that course, if they failed to get to the southward."

My own opinion, submitted to the Admiralty in compliance with their commands, was substantially the same with that of Sir James Clark Ross, though formed independently; and I further suggested that, in the event of accident to the ships, or their abandonment in the ice, the members of the expedition would make either for Lancaster Sound to meet the whalers, or Mackenzie River, to seek relief at the Hudson's Bay posts, as they judged either of these places most easy of attainment.*

After deliberately weighing these and other suggestions, and fully considering the numerous plans submitted to them, the Admiralty determined that, if no intelligence of the missing ships arrived by the close of autumn, 1847, they would send out three several searching expeditions—one to Lancaster Sound, another down the Mackenzie River, and the third to Beering's Straits.

The object of the first, and the most important of the three, was to follow up the route supposed to have been pursued by Sir John Franklin; and, by searching diligently for any signal-posts he might have erected, to trace him out, and carry the required relief to his exhausted crews. Sir James Clark Ross was appointed to the command of this expedition, consisting of the "Enterprise" and "Investigator;" and, as his plan of proceeding bears upon my own instructions, I give it at length:

"As vessels destined to follow the track of the expedition must, necessarily, encounter the same difficulties, and be liable to the same severe

---

[* Since the publication of the English edition of this work, the return of the American Arctic Expedition (Oct., 1851), has brought intelligence proving that Sir John Franklin's expedition was at Beechy Cape, at the Entrance of Wellington Sound, from January 1, to April 3, 1846, at least. The graves of three members of his party, bearing these dates, were discovered at that spot.—AM. PUB.]

pressure from the great body of the ice they must pass through in their way to Lancaster Sound, it is desirable that two ships, of not less than 500 tons, be purchased for this service, and fortified and equipped, in every respect as were the 'Erebus' and 'Terror,' for the Antartic Seas.

"Each ship should, in addition, be supplied with a small vessel or launch of about 20 tons, which she could hoist in, to be fitted with a steam-engine and boiler of ten-horse power, for a purpose to be hereafter noticed.

"The ships should sail at the end of April next, and proceed to Lancaster Sound, with as little delay as possible, carefully searching both shores of that extensive inlet, and of Barrow's Strait, and then progress to the westward.

"Should the period at which they arrive in Barrow's Strait admit of it, Wellington Channel should next be examined, and the coast between Cape Clarence and Cape Walker explored, either in the ships or by boats, as may at the time appear most advisable. As this coast has been generally found encumbered with ice, it is not desirable that both ships should proceed so far along it as to hazard their getting beset there and shut up for the winter; but in the event of finding a convenient harbor near Garnier Bay or Cape Rennell, it would be a good position in which to secure one of the ships for the winter.

"From this position the coast line might be explored, as far as it extends to the westward, by detached parties early in the spring, as well as the western coast of Boothia, a considerable distance to the southward, and at a more advanced period of the season the whole distance to Cape Nicolai might be completed.

"A second party might be sent to the southwest as far as practicable, and a third to the northwest, or in any other direction deemed advisable at the time.

"As soon as the formation of water along the coast, between the land and main body of the ice admitted, the small steam-launch should be dispatched into Lancaster Sound, to communicate with the whale-ships at the usual time of their arrival in those regions, by which means information of the safety or return of Sir John Franklin might be conveyed to the ships before their liberation from their winter quarters, as well as any further instructions the Lords Commissioners might be pleased to send for their future guidance.

"The easternmost vessel having been safely secured in winter quarters, the other ship should proceed alone to the westward, and endeavor to reach Winter Harbor in Melville Island, or some convenient port in Banks's Land, in which to pass the winter.

"From this point, also, parties should be dispatched early in spring, before the breaking up of the ice. The first should trace the western coast of Banks's Land, and proceeding to Cape Bathurst, or some other conspicuous point of the continent, previously agreed on with Sir John Richardson, reach the Hudson's Bay Company's settlement of Fort Good Hope on the Mackenzie, whence they may travel southward by the usual route of the traders to York Factory, and thence to England.

"The second party should explore the eastern shore of Banks's Land, and, making for Cape Krusenstern, communicate with Sir John Richardson's party on its descending the Coppermine River, and either assist him in completing the examination of Wollaston and Victoria Land, or return to England by any route he should direct.

"These two parties would pass over that space in which most probably the ships have become involved (if at all), and would, therefore, have the best chance of communicating to Sir John Franklin information of the measures that have been adopted for his relief, and of directing him to the best point to proceed, if he should consider it necessary to abandon his ships.

"Other parties may be dispatched, as might appear desirable to the commander of the expedition, according to circumstances; but the steam-launches should certainly be employed to keep up the communication between the ships, to transmit such information for the guidance of each other as might be necessary for the safety and success of the undertaking.

(Signed) "JAMES C. ROSS,
"Captain, R.N.

"Athenæum, 2 December, 1847."

By a subsequent arrangement between Sir James Ross and myself, under the sanction of the Admiralty, I undertook to deposit pemican at Fort Good Hope and Point Separation on the Mackenzie, and Capes Bathurst, Parry, Krusenstern, and Hearne, on the sea-coast, for the use of Sir James Ross's detached parties.

The Beering's Straits expedition was composed of the "Herald," Captain Kellet, then employed in surveying the Pacific coasts of America, and the "Plover," Commander Moore. The vessels were expected to arrive in Beering's Straits about the 1st of July, 1848, and were directed to "proceed along the American coast as far as possible, consistent with the certainty of preventing the ships being beset by the ice." A harbor was to be sought for the "Plover" within the Straits, to which that vessel was to be conducted; and two whale-boats were to go on to the eastward in search of the missing voyagers, and to communicate, if possible, with the Mackenzie River party. The "Plover" was fitted out in the Thames in December, 1847; but having been found to leak when she went to sea, was compelled to put into Plymouth for repair, and did not finally leave England until February, 1848. This tardy departure, conjoined with her dull sailing, prevented her from passing Beering's Straits at all in 1848; but she wintered near Cape Tschukotskoi, on the Asiatic coast, just outside of the Straits.

The "Herald" visited Kotzebue Sound, repassed the Straits before the arrival of the "Plover," and returned to winter in South America, with the intention of going northward again next season.

The main object of the searching party intrusted to my charge, was to trace the coast between the Mackenzie and Coppermine Rivers, and the shores of Victoria and Wollaston Lands lying opposite to Cape Krusenstern. In a preceding page I have adduced reasons for believing that there is a passage to the northward between these lands; and if so, its position makes it the most direct route from the continent to the unknown tract interposed between Cape Walker and Banks's Land, into which Sir John Franklin was expressly ordered to carry his ships. Should he have done so, and his egress by the way he entered be barred by the ice closing in behind him as already suggested, there remained a probability that the annual progression of the ice southward would eventually carry the ships into Coronation Gulf, or, if abandoned before that event, their crews were to be sought for on their way to the continent.

At the time when Sir John Franklin left England, two other openings from the north into the sea washing the continental shores were supposed to exist. The most westerly of these is between Boothia and Victoria Land, and it was part of Sir James Ross's plan to examine the whole western side of Boothia and North Somerset by one of his steam-barges.

The other supposed entrance was by Regent's Inlet. Dease and Simpson had left only a small space unsurveyed between that inlet and the sea, which was known to afford in good seasons a passage all the way to Beering's Straits; and this might have recommended the route by Regent's Inlet for trial. But, exclusive of its being absolutely prohibited by Sir John Franklin's instructions, Sir Edward Parry and Sir James Ross, on whose opinions Sir John placed deservedly the greatest reliance, were decidedly averse to his attemping a passage in that direction; and it was known that Sir John Franklin had resolved on trying all the other openings before he entered Regent's Inlet, which was to be his last resource. It fortunately happened before any of the searching expeditions were finally organized, that the non-existence of a passage through that inlet was fully ascertained.

Mr. John Rae, a Chief Trader in the service of the Hudson's Bay Company, left Fort Churchill in the beginning of the summer of 1846, with two boats, for the express purpose of completing the survey of Regent's Inlet. He arrived in Repulse Bay in the month of August of that year, and immediately crossed an isthmus, forty-three miles wide, to the inlet, taking one boat with him. Finding that the season was too far advanced for him to complete the survey that year, he determined, with a boldness and confidence in his own resources that has never been surpassed, to winter in Repulse Bay, and to finish his survey of Regent's Inlet on the ice next spring; so that he might be able to return to Churchill and York Factory by open water in the summer of 1847. He therefore recrossed the isthmus again with his boat, and set about collecting provisions and fuel for a ten months' winter. To one less experienced and hardy, the desolate shores of Repulse Bay would have forbidden such an attempt. They yielded neither drift-wood nor shrubby plants of any kind; but Mr. Rae employed part of his men to gather the withered stems of the *Andromeda tetragona,* a small herbaceous plant which grew in abundance on the rocks, and to pile it in cocks like hay: others he set to build a house of stone and earth, large enough to shelter his party, amounting in all to sixteen; while he himself and his Eskimo interpreter were occupied in killing deer for winter consumption. He succeeded in laying up a sufficient stock of venison, and kept his people in health and strength for next year's operations, though not in comfort, for the chimney was so badly constructed for ventilation, that when the fire was lighted it was necessary to open the door, and thus to reduce the temperature of the apartment, nearly to that of the external air. The fire was, therefore, used as seldom as possible, and only for cooking or melting snow to drink. In the spring he completed the survey of Prince Regent's Inlet on foot, thereby proving that no passage existed through it, and confirming the Eskimo report, first made to Sir Edward Parry, and afterward to Sir John Ross. A party of Eskimo, who resided near Mr. Rae in the winter, informed him, through his interpreter, that they had not seen Franklin's ships, thereby excluding the Gulf of Boothia from the list of places to be searched.

Having thus mentioned the opinions most worthy of note, respecting the quarters in which search was to be made, the plans of search adopted by the Admiralty after duly weighing a great variety of suggestions, and the extent of coast and parts of the Arctic Sea embraced in the three expeditions of the summer of 1848, I subjoin the instructions I received from the Admiralty:

Instructions to Sir John Richardson, M.D., 16th March, 1848. By the Commissioners for executing the office of Lord High Admiral, &c.

"Whereas, we think fit that you should be employed on an overland expedition in search of Her Majesty's ships 'Erebus' and 'Terror,' under the command of Captain Sir John Franklin, which ships are engaged in a voyage of discovery in the Arctic Seas, you are hereby required and directed to take under your orders Mr. Rae, who has been selected to accompany you, and to leave England on the 25th instant, by the mail steamer for Halifax in Nova Scotia, and New York; and on your arrival at the latter place, you are to proceed immediately to Montreal, for the purpose of conferring with Sir George Simpson, Governor of the Hudson's Bay Company's Settlements, and making arrangements with him for your future supplies and communications.

"You should next travel to Penetanguishene, on Lake Huron, and from thence, by a steamer, which sails on the 1st and 15th of every month of open water, to Saut Ste. Marie, at the foot of Lake Superior, and there embark in a canoe, which, with its crew, will have been provided for you, by that time, by Sir George Simpson.

"Following the usual canoe route by Fort William, Rainy Lake, the Lake of the Woods, Lake Winipeg, and the Saskatchewan River, it is hoped that you will overtake the boats now under charge of Mr. Bell, in July, 1848, somewhere near Isle à la Crosse, or perhaps the Methy Portage.

"You will then send the canoe with its crew back to Canada, and having stowed the four boats for their sea voyage, you will go on as rapidly as you can to the mouth of the Mackenzie; leaving Mr. Bell to follow with the heavier laden barge, to turn off at Great Bear Lake, and erect your winter residence at Fort Confidence, establish fisheries, and send out hunters.

"Making a moderate allowance for unavoidable detention by ice, thick fogs, and storms, the examination of the coast between the Mackenzie and the Coppermine Rivers will probably occupy 30 days; but you can not calculate to be able to keep the sea later than the 15th of September, for, from the beginning of that month, the young ice covers the sea almost every night, and very greatly impedes the boats, until the day is well advanced.

"If you reach the sea in the first week of August, it is hoped you will be able to make the complete voyage to the Coppermine River, and also to coast a considerable part of the western and southern shores of Wollaston Land, and to ascend the Coppermine to some convenient point, where the boats can be left with the provisions ready for the next year's voyage; and you will instruct Mr. Bell to send two hunters to the banks of the river to provide food for the party on the route to Fort Confidence, and thus spare you any further consumption of the pemican reserved for the following summer.

"As it may happen, however, from your late arrival on the coast, or subsequent unexpected detentions, that you can not with safety attempt to reach the Coppermine, you have our full permission in such a case to return to Fort Good Hope, on the Mackenzie, there to deposit two of the boats, with all the sea stores, and to proceed with the other two boats, and the whole of the crews, to winter quarters on Great Bear Lake.

"And you have also our permission to deviate from the line of route along the coast, should you receive accounts from the Eskimos, which may appear credible, of the crews of the 'Erebus' and 'Terror,' or some part of them, being in some other direction.

"For the purpose of more widely extending your search, you are at liberty to leave Mr. Rae and a party of volunteers to winter on the coast, if, by the establishment of a sufficient fishery, or by killing a number of deer or musk oxen, you may be able to lay up provisions enough for them until you can rejoin them next summer.

"As you have been informed by Captain Sir James Ross, of Her Majesty's ship 'Enterprise,' who is about to be employed on a similar search in another direction, of the probable directions in which the parties he will send out toward the continent will travel, you are to leave a deposit of pemican for their use at the following points—namely, Point Separation, Cape Bathurst, Cape Parry, and Cape Krusenstern; and as Sir James Ross is desirous that some pemican should be stored at Fort Good Hope, for the use of a party which he purposes sending thither in the spring of 1849, you are to make the necessary arrangements with Sir George Simpson for that purpose, as his directions to that effect must be sent early enough to meet the Company's brigade of Mackenzie River boats at Methy Portage, in July, 1848.

"Should it appear necessary to continue the search a second summer (1849), and should the boats have been housed on the Coppermine, you are to descend that river, on the breaking up of the ice in June, 1849, and to examine the passages between Wollaston and Banks's and Victoria Lands, so as to cross the routes of some of Sir James C. Ross's detached parties, and to return to Great Bear Lake in September, 1849, and withdraw the whole party from thence to winter on Great Slave Lake, which would be as far south as you will have a prospect of traveling before the close of the river navigation.

"Should you have found it necessary to return to the Mackenzie (September, 1848), instead of pushing on to the Coppermine, the search in the summer of 1849 would, of course, have to be commenced from the former river again; but should circumstances render it practicable and desirable to send some of the party down the Coppermine with one or two boats, you are at liberty to do so.

"A passage for yourself and Mr. Rae will be provided in the 'America,' British and North American mail-steamer, which sails from Liverpool on the 25th of March, and you will receive a letter of credit on Her Majesty's Consul at New York for the amount of the expense of your journey from New York to Saut Ste. Marie, and the carriage of the instruments, &c.

"And in the event of intelligence of the 'Erebus' and 'Terror' reaching England after your departure, a communication will be made to the Hudson's Bay Company to ascertain the most expeditious route to forward your recall.

"We consider it scarcely necessary to furnish you with any instructions contingent on a successful search after the above-mentioned expedition, or any parties belonging to it. The circumstances of the case, and your own local knowledge and experience, will best point out the means to be adopted for the speedy transmission to this country of intelligence to the above effect, as well as of aiding and directing in the return of any such parties to England.

"We are only anxious that the search so laudably undertaken by you and your colleagues should not be unnecessarily or hazardously prolonged; and while we are confident that no pains or labor will be spared in the execution of this service, we fear lest the zeal and anxiety of the party so employed may carry them further than would be otherwise prudent.

"It is on this account you are to understand that your search is not to be prolonged after the winter of 1849, and which will be passed on the Great Slave Lake; but that, at the earliest practical moment after the breaking up of the weather in the spring of 1850, you will take such steps for the return of the party under your orders to England as circumstances may render expedient.

"It must be supposed that the instructions now afforded you can scarcely meet every contingency that may arise out of a service of the above description; but reposing, as we do, the utmost confidence in your discretion and judgment, you are not only at liberty to deviate from any point of them that may seem at variance with the objects of the expedition, but you are further empowered to take such other steps as shall be desirable at the time, and which are not provided for in these orders.

"Given under our hands, 16th March, 1848.

(Signed) "AUCKLAND.
"J. W. D. DUNDAS.

"To Sir John Richardson, M.D., &c.
"By command, &c.
(Signed) "W. A. B. Hamilton."

CHAPTER II.

Overland Searching Expedition.—Routes through the Interior.—Hudson's Bay Ships.—Pemican.—Boats.—Boat Party leaves England.—Arrives at Winter Quarters.—Volunteers.—Mr. John Rae appointed to the Expedition.—The Author and Mr. Rae sail from England.—Land at New York.—Proceed to Montreal and La Chine.—Canoe-Men.—Saut Ste. Marie.—Voyage to the North.—Reach Cumberland House.

THE preceding pages contain an exposition of the objects of the expedition, with a general outline of the course to be pursued after leaving the Mackenzie; but as that great river can be attained only by a long and laborious lake and river navigation, it is proper that I should introduce the narrative by a brief account of that first stage of our overland journey. There are two routes to the Mackenzie, one of which, traced at an early period by the Canadian fur companies, passes through Lakes Huron and Superior, the Kamenistikwoya, or Dog River, the Lake of the Woods, Rainy Lake, Lake Winipeg, Cedar Lake, the Saskatchewan River, Beaver and Half-moon Lakes, Churchill or English River, Isle à la Crosse, Buffalo and Methy Lakes to the Methy Portage, and the Clear-water or Little Athabasca River, one of the affluents of the Mackenzie. From thence there is a continuous water-course to the sea, through the Elk or Athabasca River, Athabasca Lake, Slave River and Lake, and the Mackenzie proper.

The length of this interior navigation from Montreal to the Arctic Sea is, in round numbers, four thousand four hundred miles, of which sixteen hundred miles are performed on the Mackenzie and its affluents, from Methy Portage northward, and in which the only interruptions to boat navigation are a few cascades and rapids in Clear-water and Slave Rivers.

During the existence of the Northwest, X-Y, and other fur companies trading from Canada, supplies were conveyed to their northern posts by the way of the Ottawa River and great Canada lakes; but they reached the distant establishments on the Mackenzie only in the second summer, having been deposited in the first year at the dépôt on Rainy River. Owing to the shallowness of the streams, and badness of the portage roads over the heights between Lake Superior and Rainy Lake, the transport of goods requires to be performed in canoes, with much manual labor, and is, consequently, very expensive. On this account the Hudson's Bay Company, who are now the sole possessors of the northern fur trade, no longer take their trading goods from Cànada, but send them by the shorter and cheaper way of Hudson's Bay; though they still employ two or three canoes on the Lake Superior route, to accommodate the Governor in his annual journeys from his residence at La Chine to Norway House, and for the transport of newly-hired servants to the interior, or for bringing down officers coming out on furlough, and men whose period of service has expired. No repairs having of late years been made on the portage roads, they have very much deteriorated, and are truly execrable.

The distance between York Factory, in Hudson's Bay, and Norway House, situated near the northeast corner of Lake Winipeg, does not much exceed three hundred miles; and as the navigation, though much interrupted by rapids and cascades, admits, in the majority of seasons, of boats carrying a cargo of between fifty and sixty hundred-weight, it offers a much more economical approach to the interior of the fur countries than the other; since one of these boats may be managed by the same crew that is required for a canoe carrying only twenty hundred-weight. The Hudson's Bay ships are generally two in number; one of them being employed in taking supplies to Moose Factory, at the bottom of James's Bay, and the other to York Factory, in latitude 57° N., longitude 92½° W., on the west coast of Hudson's Bay. They sail annually from the Thames on the first Saturday in June, and, after touching at the Orkneys, to receive laborers for the Company's service, proceed on their voyage to Hudson's Straits. The York Factory ship has dropped her anchor at the mouth of Hayes River as early as the 5th of August, and as late as the beginning of September. A tardy arrival is very inconvenient, both in respect of forwarding goods into the interior, and also with regard to the return of the ship to England, there being in such a case scarcely time for the embarkation of the cargo of furs and the passage of Hudson's Straits before the winter sets in.

This brief notice of the modes of communication with Rupert's

Land—for so the possessions of the Hudson's Bay Company are named—is given, to explain some parts of the plan of the expedition, and particularly to show why the stores and men were sent out by ships which sailed in June, 1847, although the expediency of searching expeditions was not considered by the Admiralty to be established until the last of the whalers came in at the close of that season, without bringing tidings of the discovery ships. It was arranged that in that case, the officers were to leave England early in 1848, and, traveling as rapidly as they could through the United States and Canada, were to overtake the party conveying the stores in the vicinity of Methy Portage.

In April, 1847, I had the advantage of a personal interview with Sir George Simpson, Governor-in-chief of Rupert's Land, who was then on a visit to England, and of concerting with him the measures necessary for the future progress of the expedition; and I may state here that he entered warmly into the projects for the relief of his old acquaintance Sir John Franklin; and from him I received the kindest personal attention, and that support which his thorough knowledge of the resources of the country and his position as Governor enabled him so effectively to bestow. He informed me that the stock of provisions at the various posts in the Hudson's Bay territories was unusually low, through the failure of the bison hunts on the Saskatchewan, and that it would be necessary to carry out pemican from this country, adequate not only to the ulterior purposes of the voyage in the Arctic Sea, but also to the support of the party during the interior navigation in 1847 and 1848. I, therefore, obtained authority from the Admiralty to manufacture, forthwith, the requisite quantity of that kind of food in Clarence Yard; and as I shall have frequent occasion to allude to it in the subsequent narrative, it may be well to describe in this place the mode of its preparation.

The round or buttock of beef of the best quality, having been cut into thin steaks, from which the fat and membranous parts were pared away, was dried in a malt kiln over an oak fire, until its moisture was entirely dissipated, and the fibre of the meat became friable. It was then ground in a malt mill, when it resembled finely grated meat. Being next mixed with nearly an equal weight of beef-suet or lard, the preparation of plain pemican was complete; but to render it more agreeable to the unaccustomed palate, a proportion of the best Zante currants was added to part of it, and part was sweetened with sugar. Both these kinds were much approved of in the sequel by the consumers, but more especially that to which the sugar had been added. After the ingredients were well incorporated by stirring, they were transferred to tin canisters, capable of containing 85 lbs. each ; and, having been firmly rammed down and allowed to contract further by cooling, the air was completely expelled and excluded by filling the canister to the brim with melted lard, through a small hole left in the end, which was then covered with a piece of tin, and soldered up. Finally, the canister was painted and lettered according to its contents. The total quantity of pemican thus made was 17,424 lbs., at a cost of 1*s.* $7\frac{1}{4}$*d.* a pound.* But the expense was somewhat greater than it would otherwise have been from the inexperience of the laborers, who required to be trained, and from the necessity of buying meat in the London market at a rate above the contract price, occasioned by the bullocks slaughtered by the contractor for the naval force at Portsmouth being inadequate to the supply of the required number of rounds. Various temporary expedients were also resorted to in drying part of the meat, the malt kiln and the whole Clarence Yard establishment being at that time fully occupied night and day in preparing flour and biscuit for the relief of the famishing population of Ireland. By the suggestions of Messrs. Davis and

* Particulars of the estimated expense of pemican, manufactured in the Royal Clarence Victualing Yard, in Midsummer quarter, 1847 :

	£.	*s.*	*d.*	£.	*s.*	*d.*
Fresh beef 35,651 lbs. at $6\frac{3}{4}$*d.* per lb.	979	10	1			
Lard . . . . 7,549 " at 88*s.* per cwt.	296	11	4			
Currants . 1,008 " at 84*s.* per cwt.	37	16	0			
Sugar . . . 280 " at 31*s.* 2*d.* per cwt.	3	17	11			
				1,317	15	4
Oak slab 46 fms. at 22*s.* 6*d.* per load	47	5	0			
Hire for laborers	59	8	8			
Hire of kiln and cartage	8	1	0			
				114	14	8
				1,432	10	0
Deduct for scraps of fat sold				35	18	1
				1,396	11	11

Quantity of pemican manufactured 17,424 lbs. ; average cost per lb. 1*s.* $7\frac{1}{4}$*d.*

Grant, the intelligent chief officers of the Victualling Yard, and their constant personal superintendence, every difficulty was obviated.

As the meat in drying loses more than three-fourths of its original weight, the quantity required was considerable, being 35,651 lbs.;* and the sudden abstraction of more than one thousand rounds of beef from Leadenhall Market occasioned speculation among the dealers, and a rise in the price of a penny per pound, with an equally sudden fall when the extra demand was found to be very temporary.

The natives dry their venison by exposing the thin slices to the heat of the sun, on a stage, under which a small fire is kept, more for the purpose of driving away the flies by the smoke than for promoting exsiccation; and then they pound it between two stones on a bison hide. In this process the pounded meat is contaminated by a greater or smaller admixture of hair and other impurities. The fat, which is generally the suet of the bison, is added by the traders, who purchase it separately from the natives, and they complete the process by sewing up the pemican in a bag of undressed hide with the hairy side outward. Each of these bags weighs 90 lbs., and obtains from the Canadian voyagers the designation of "un taureau." A superior pemican is produced by mixing finely powdered meat, sifted from impurities, with marrow fat, and the dried fruit of the Amelanchier.

By order of the Admiralty, four boats were built; two of them in Portsmouth Dock Yard, and two in Camper's Yard at Gosport. These boats, to fit them for river navigation, were required to be of as small a draught of water as was consistent with the power of carrying a cargo of at least two tons; to have the head and stern equally sharp, like a whale-boat, that they might be steered with a sweep oar when running rapids; and to be of as light a weight as possible, for more easy transportation across the numerous portages on the route, and especially the formidable one between Methy Lake and Clear-water River. They were also to be as good sea-boats as a compliance with the other requisites would allow. It is manifest that the invention of a form of boat possessing such various and in some respects antagonistic qualities would task the skill of the constructor, and I felt much indebted to William Rice, Esq., Assistant Master Builder of Portsmouth

* By drying this was reduced to about 8000 lbs.

Yard, for the care and skill with which he worked out a successful result. The Company's boats, or barges, as they term them, are generally about 36 feet long from stem to stern-post, 8 feet wide, stoutly framed and planked, and are capable of carrying seventy packages of 90 lbs. each, with a crew of eight men. The thickness of the planks of these boats is such that they sustain with little injury a severe blow against a rock, to which they are much exposed in descending the rapids; but their weight being proportionally great, they are transported with much labor across the ordinary portages, and it is necessary to avoid this operation altogether at Methy Portage by keeping a relay of boats at each terminus. Moreover, these boats resemble the London river barges in the great rake of the stem and stern, by which they are better fitted for the descent of a rapid, but from the flatness of their floors they are leewardly and bad sea-boats.

Two of the expedition boats measured 30 feet from the fore part of the stem to the after part of the stern-post, 6 feet in breadth of beam, and 2 feet 10 inches in depth; and each of them weighed 6½ cwt., or, including fittings, masts, sails, oars, boat-hook, anchor, lockers, and tools, half a ton. The other two boats measured 28 feet in length, 5 feet 6 inches in width, 2 feet 8 inches in depth; and weighed 5¼ cwt., or, with the movable fittings and equipment, 9 cwt. They were all clinker-built of well-seasoned Norway fir planks $\frac{5}{16}$ of an inch thick; ashen floors placed 9 inches apart; stem, stern-posts, and knees of English oak; and gunwales of rock-elm. To admit of their stowing the requisite cargo, they were necessarily very flat-floored, but screws and bolts were fitted to the kelson, by which a false keel might be readily bolted on before they reached the Arctic Sea, so as to render them more weatherly. The larger boats when quite empty drew 7¼ inches of water, and, when loaded with two tons but without a crew, 14¼ inches. They were constructed of two sizes, that the smaller might stow within the larger ones during the passage across the Atlantic.

For the voyage on the Arctic Sea, a crew of five men to each boat was considered sufficient, but for river navigation a bowman and steersman experienced in the art of running rapids were required in addition. Five seamen and fifteen sappers and miners were selected in the month of May, for the expedition, from a number of volunteers. They were all men of good physical powers, and, with one exception, bore excellent characters in their respective services. The solitary exception was one of the sappers and miners who had repeatedly appeared on the defaulters' list for drunkenness, but as he was reported to be in other respects a good and willing workman, and I knew that he would have no means of obtaining intoxicating drinks in Rupert's Land, I yielded to his request that I would allow him an opportunity of retrieving his character. Few seamen were employed, since I knew from experience that as a class they march badly, particularly when carrying a load, and the bulk of the party was composed of sappers and miners, because that corps contains a large proportion of intelligent artisans. Of the men selected, six were joiners or sawyers, and four were blacksmiths, armorers, or engineers, who could be useful for repairing the boats, working up iron, constructing the buildings of our winter residence, or making the furniture.

Every thing was ready before the appointed day; and the boats and stores, having been sent round from Portsmouth to the Thames, were embarked with the expedition men on board the "Prince of Wales" and "Westminster," bound to York Factory, the exigences of the Hudson's Bay trade of that year requiring two ships to go to that port. The stores consisted of 198 canisters of pemican, each weighing 85 lbs., 10 bags of flour, amounting in all to 8 cwt., 5 bags of sugar, weighing 4^{1} cwt., 2 of tea, weighing 88 lbs., 3 of chocolate, weighing 2 cwt., 10 sides of bacon, amounting to $4\frac{1}{2}$ cwt., and 6 cwt. of biscuit; also 400 rounds of ball cartridge, 90 lbs. of small shot, and 120 lbs. of fine powder in four boat magazines. In the arm-chests and lockers of the boats, there were stowed a musket fitted with a percussion lock for each man, with a serrated bayonet that could be used as a saw; also a complete double set of tools for making or repairing a boat, a tent for each boat's crew, towing-lines, anchors, and one seine net.

Each man was provided with a Flushing jacket and trowsers, a stout blue Guernsey frock, a waterproof over-coat, and a pair of leggings. Instructions were also given that they should be furnished in winter with such moccasins and leather coats as the nature of their employment should render necessary. Could the expedition have depended on procuring supplies of provision at the Company's posts during their progress through the interior, and a sufficient quantity of pemican at one of the northern dépôts for the sea voyage, the boats would have been lightly laden, and a quick advance into the interior might have been anticipated. But such not being the case, it was necessary to employ one of the Company's barges to assist in the transport; and Governor Sir George Simpson undertook to provide one, and to engage a proper crew in Rupert's Land, together with bowmen and steersmen for the expedition boats. He also agreed to select from the Company's stores a complete assortment of nets and other necessaries for the use of the party in the winter of 1847–8.*

The Company's ships sailed from the Thames on the 15th of June, 1847, and, being much delayed by ice in Hudson's Straits, had a long passage; so that the "Prince of Wales" did not cross the bar of Hayes River till the 25th of August, nor the "Westminster" until five days later; and the 8th of September arrived before the expedition stores were landed. Sir George Simpson, on his annual visit to the Company's dépôt at Norway House, had engaged a guide or river pilot, with the requisite number of bowmen, steersmen, and fishermen, and placed the whole under the superintendence of Mr. John Bell, chief trader, who, having resided many years on the Mackenzie, was intimately acquainted with the natives inhabiting that part of the country. Notwithstanding the high wages offered, being much in advance of the rate ordinarily paid by the Company, and though none of these men were required to extend their services beyond the winter quarters of the party in 1848, there was a scarcity of volunteers; and several of the steersmen, that were, from the necessity of the case, engaged, were men of little experience. None of them were acquainted with the neighborhood of Great Bear Lake, and they all anticipated with more or less apprehension a season of extreme hardship in that northern region. Mr. Bell's party consisted of twenty Europeans, a guide, and sixteen Company's voyagers, together with the wives† of three of the latter, and two children; making in all, with himself and two of his own children, forty-five individuals, embarked in five boats. Had the ships arrived early, there was a possibility of the party reaching Isle à la Crosse before the navigation closed, which, in that district, may be expected to

* See Appendix.

† It is desirable to have two or three females at every post in the interior for washing, making, and mending the people's clothes and moccassins, netting snow-shoes, making and repairing fishing nets, and other services of a similar nature.

occur about the 20th of October. But the very late date at which the stores were disembarked precluded such a hope; and the extreme dryness of the season, and consequent lowness of the rivers between York Factory and Lake Winipeg, obliged Mr. Bell to leave a quantity of the pemican and some other packages at York Factory, that he might reduce the draught of his boats.

These facts were communicated to me on the return of the Hudson's Bay ships to England in October; and in February, 1848, I heard by letters forwarded through Canada, that Mr. Bell and his party had, from the causes specified, made slow progress; that the boats had been often stranded and broken in the shallow waters, causing frequent detention for repairs; and that the party was overtaken by winter in Cedar Lake. Mr. Bell forthwith housed the boats, constructed a store-house for the goods, left several men to take care of them, and such of the women and children as were unable to travel over the snow. This being done, he set out with the bulk of the party for Cumberland House, and reached it on the eighth day after leaving Cedar Lake. His first care was to establish a fishery, which he did on Beaver Lake, two days' walk further north; and having sent a division of the men thither, the others were distributed to the several winter employments of cutting firewood, driving sledges with meat or fish, and such-like occupations. The unforeseen stoppage of the boats occasioned a large consumption of the pemican destined for the sea voyage, but was attended by no other bad consequences, and the deficiency was amply made up in spring through the exertions of the gentlemen in charge of the Company's provision posts on the Saskatchewan; so that Mr. Bell, when he resumed his voyage northward in the summer of 1848, was enabled to take with him as much of that kind of food as his boats could stow.

While the body of the party was thus passing the winter at Cumberland House and its vicinity, I was almost daily receiving letters from officers of various ranks in the army and navy, and from civilians of different stations in life, expressing an ardent desire for employment in the expedition. It may interest the reader to know that among the applicants, there were two clergymen, one justice of peace for a Welsh county, several country gentlemen, and some scientific foreigners, all evidently imbued with a generous love of enterprise, and a humane desire to be the means of carrying relief to a large body of their fellow creatures. But as long as there remained a hope of the return of the discovery ships in the autumn of 1847, it was not thought necessary to take any steps for the appointment of a second officer to the party which I was to command. In November, however, when the last whalers from Davis's Straits had come in, I suggested to the late Lord Auckland, then the First Lord of the Admiralty, that Mr. John Rae, chief trader of the Hudson's Bay Company, was fully qualified for the peculiar nature of the service on which we were to be employed. He had resided upward of fifteen years in Prince Rupert's Land, was thoroughly versed in all the methods of developing and turning to advantage the natural products of the country, a skillful hunter, expert in expedients for tempering the severity of the climate, an accurate observer with the sextant and other instruments usually employed to determine the latitude and longitude, or the variations and dip of the magnetic needle, and had just brought to a successful conclusion, under circumstances of very unusual privation, an expedition of discovery fitted out by the Hudson's Bay Company, for the purpose of exploring the limits of Regent's Inlet. Lord Auckland highly approved of my suggestion, and Mr. Rae was appointed, with the assent of the Governor and Committee of the Hudson's Bay Company.

Mr. Rae and I left Liverpool on the 25th of March, 1848, in the North American mail steam-packet "Hibernia," and landed at New York on the morning of the 10th of April. In addition to our personal baggage, we took with us a few very portable astronomical instruments required for determining our positions; and four pocket chronometers, one of them being the property of Mr. Frodsham, which had been used on the several expeditions of Sir W. E. Parry and Sir John Ross, and which he wished to lend gratuitously for service in the present enterprise. We had also a few meteorological instruments, and some others for determining questions in magnetism, that shall be more particularly described hereafter, when their employment comes to be mentioned. An ample supply of paper for botanical purposes, a quantity of stationery, a small selection of books, a medicine chest, a canteen, a compendious cooking apparatus, and a few tins of pemican, completed our baggage, which weighed in the aggregate, above 4000 lbs.

Mr. Barclay, the British consul, assisted with much kindness in expediting our departure from New York. An order from the

United States Treasury directed that our baggage should not be inspected by the custom-house agents, and it was without delay consigned to the care of Messrs. Wells and Co., forwarders, who contracted to send it to Buffalo, by railroad, and from thence to Detroit and Saut Sainte Marie, by the first steamboat, which was advertised to sail from Detroit on the 21st of April. Immediately on landing, the chronometers were placed in the hands of Mr. Blount, of Water-street, that he might ascertain their rate by comparison with the astronomical clock in the observatory. For this service Mr. Blount would receive no remuneration, but, on the contrary, said that he was glad of the opportunity it afforded him of showing his sense of the courtesy he had experienced from the hydrographer of the British Admiralty.

We received the chronometers next day, and embarked in the evening on board the "Empire," for Albany and Troy, with the view of proceeding, by way of Lake Champlain, to Montreal, where the canoe-men engaged for us by Sir George Simpson were ordered to rendezvous.

We waited one day at Whitehall, for the complete disruption of the ice on Lake Champlain*, and did not reach Montreal till the fourth day after leaving New York. Sir George Simpson received us, with his usual kindness and hospitality, at his residence in La Chine, and expedited our arrangements by all the means in his power; but two days were spent in collecting the voyagers† who were engaged as our canoe-men. Four of them, with the levity of their class, were absent at the time finally fixed for our departure, thereby, in terms of their agreements, incurring fines, which were afterward levied by the Hudson's Bay Company.

The steamers commenced running on the St. Lawrence on the 18th of April; we embarked on the 19th, reached Buffalo on the 21st, Detroit on the 23d, and Saut Ste. Marie, at the outlet of Lake Superior, on the 29th, where we again found ourselves in advance of the season, the lake being covered with drift ice.‡

* The ice broke up on Lake Champlain on the 13th of April. On the previous day a steamer was prevented from reaching Whitehall by drift ice filling a narrow passage of the lake.

† The Canadian term "voyageurs" is usually employed to designate these men, as that is the language in which they are addressed; but there seems to be no reason why they should not be called "voyagers," or "canoe-men," in an English work.

‡ In the instructions, the route by Penetanguishene is specified for the

At the Hudson's Bay House, the residence of Chief Factor Ballenden, we found two "north canoes," made ready for us, by direction of Sir George Simpson, and, having engaged four additional men to supply the place of an equal number who had failed to appear at La Chine, our crews now consisted of

First Canoe.

Thomas Karahonton (*dit* Gros Thomas), an Iroquois guide.

Laxard Tacanajazè	Iroquois.
Thomas Nahanajazè	"
François Monegon	"
Thomas Anackera	"
Sauveur St. Martin	Canadian.
Thomas Cadrant	Half-breed.
Joseph Dinduvant	"

Second Canoe.

Charlot Arahota	Iroquois.
Louis Taranta	"
Ignace Atawackon	"
Ignace Sataskatchi	"
Apoquash	Chippeway.
Miskiash	"
Piquatchiash (Peter)	"

Two days were occupied in re-packing our baggage, instruments, and provisions, in cases weighing 90 lbs. each (being the established size for the portages); in which, and in all other matters connected with our equipment and comfort, we experienced great assistance and personal kindness from Mr. Ballenden. On the 2d of May, 1851, we quitted his hospitable roof, but it was the 4th before the ice on the lake broke up, and permitted us to pass the portal of the lake formed by Gros Cap and Point Iroquois.

We accomplished the navigation of the lake on the 12th by arriving at Fort William, attained the summit of the water-shed which separates the St. Lawrence and Winipeg valleys on the 18th*, the mouth of the River Winipeg on the 29th, Norway House, near the efflux of Nelson River, on the 5th of June, and Cumberland House, on the Saskatchewan, on the 13th; our

expedition to take; but the steamer from that port to Saut Ste. Marie was not advertised to start for three weeks later than our time.

* Dog Lake, near the summit of this water-shed, broke up only on the eve of our arrival; an Indian whom we met on the Kamenistikwoya, which flows from it, having crossed it on the preceding day over the ice.

passage through Lake Winipeg having been much delayed by ice, from which we did not disengage ourselves till the 9th.

We learned at Cumberland House, that Mr. Bell had given the boats a thorough repair at Cedar Lake in the spring, had brought them and the stores up on the first opening of the Saskatchewan, and was now a fortnight in advance of us on his way to Methy Portage. The bulk of his party had been maintained at Beaver Lake on fish, but some having wintered in Cedar Lake, to look after the stores, and the fishery there having failed, there had been an unavoidable consumption of the pemican destined for the sea-voyage. The provision posts on the upper part of the Saskatchewan had fortunately been able to replace what was consumed, and Mr. Bell had started from Cumberland House with his boats fully laden.

He had left two men of the English party behind, who were unequal to the labors of the voyage; one of them, because of an injury received in the hand early in the spring, and the other owing to a recurrence of pains in the bones, with which he had formerly been afflicted. After carefully examining these men, I decided upon sending them to York Factory by the first conveyance which offered, that they might return to England in September, in the Hudson's Bay annual ship.

Having thus briefly touched on the line of route pursued by us in a journey of two thousand eight hundred and eighty statute miles, from New York to the wintering place of the boat-party,* I shall detail the events of the remainder of the voyage in form of a daily journal. To have given a full account of the country traveled through between New York and the Saskatchewan, would have swelled the work to an inconvenient size; and I must, therefore, refer the reader, who wishes to have a physical description of that part of the continent, to Sir Charles Lyell's accounts of his recent visits to the United States, to Professor

New York to La Chine	500	miles.
La Chine to Buffalo	372	"
Buffalo to Detroit	230	"
Detroit to Saut Ste. Marie	400	"
Saut Ste. Marie to Fort William	370	"
Fort William to Cumberland House (Franklin's second journey)	1,018	"
	2,880	"

Agassiz's description of Lake Superior, and to Major Long's voyage to the St. Peter's, Red River, and River Winipeg. The Appendix to the present work also contains a summary of the physical geography of North America, wherein the lake basins of the St. Lawrence and Winipeg or Saskatchewan are particularly noticed. This may be consulted by the reader before he enters upon the narrative of the voyage, and I shall give in this place a few remarks, by way of preface to the botanical and geological notices which follow in the journal.

On the bluff granitic promontories and bold acclivities which form the northern shore of Lake Superior, the forest is composed of the white spruce, balsam fir, Weymouth pine, American larch, and canoe birch, with, near the edge of the lake and on the banks of streams, that pleasant intermixture of mountain maple and dogwood* which imparts such a varied and rich gradation of orange and red tints to the autumnal landscape. Other trees exist, but not in sufficient numbers to give a character to the scenery. Oaks are scarce, and beech disappears to the south of the lake. The American yew, which does not rise into a tree like its European namesake, is the common underwood of the more fertile spots, where it grows under the shade to the height of three or four feet, in slender bush-like twigs. On the low sandstone islands deciduous trees, such as the poplars and maples, abound, with the nine-bark spiræa, cockspur thorns, willows, plums, cherries, and mountain-ash.† When we entered the lake on the 4th of May, large accumulations of drift snow on the beaches showed the lateness of the season; none of the deciduous trees had as yet budded: and the precocious catkins of a silvery willow (*Salix candida*), with the humble flowers of a few Saxifrages and Uvulariæ, gave the only promises of spring.

In various parts of the lake, the gorges lying between the jutting bluffs of granite or slate are filled with deposits of sand rising in four or five successive terraces to the height of more than a hundred feet above the present surface of the water. Mr. Logan has measured some of the most remarkable, and Professor Agassiz

* *Abies alba*, *Abies balsamea*, *Pinus strobus*, *Larix americana*, *Betula papyracea*, *Acer montanum*, and *Cornus alba*.

† *Populus tremuloides* et *balsamifera*; *Acer*; *Spirea opulifolia*; *Cratægus crus-galli*, *punctata*, *glandulosa*, et *coccinea*; *Prunus americana*; *Cerasus pumila*, *nigra*, *pennsylvanica*, *virginiana*, et *serotina*; *Pyrus americana*.

devotes an interesting chapter to the discussion of their origin; in which he comes to the conclusion that they were formed by the waters of the lake itself, and have been raised, at various intervals, from the beach to their present levels, by the agency of the innumerable trap dikes which cross the rocks in many directions.

Near Cape Choyyè, on the south side of Michipicoten Bay, a small gorge between two points of granite is filled, to the height of twenty-five feet above the water, with rolled stones and pebbles. These rounded stones vary in size from that of a hogshead to a hen's egg, and form a steeply shelving beach, with a flat terraced summit, the larger boulders being next the water, and the smaller pebbles highest up. As the cove is sheltered from high waves, the terrace could not be thrown up by the waters of the lake standing at their present height; nor can it be owing to the pressure of ice, since that would not graduate the pebbles.

At Michipicoten River we had a curious illustration of the agency of frost, on the outlet of the stream. During the summer, when the waters are low, the waves of the lake throw a sandy bar across the mouth of the river. In winter this bar freezes into a solid rock and closes the channel, but as the spring advances the stream acts upon it and cuts a passage. At the time of our visit, on May 7th, the river was in flood, and the bar remained hard, but was cleft by a narrow channel with precipitous sides like sandstone cliffs, and a cascade one foot high existed. This fall, which was five or six feet high when the river broke, would, we were told, entirely disappear in a few days.

The north coast of Michipicoten Bay is the boldest and most rugged of the shores of the lake, and apparently the least capable of cultivation. It rises to the height of about eight hundred feet, and for twenty-five miles comes so precipitously down to the water that there is no safe landing for a boat. On much of the crags the forest was destroyed by fire, many years ago, and with it the soil, presenting a scene of desolation and barrenness not exceeded on the frozen confines of the Arctic Sea. The few dwarf trees that cling to the crevices of the rocks, struggling, as it were, between life and death, add to the dreariness of the prospect rather than relieve it, and wreaths of drift snow lining many of the recesses, at the time when we passed, though it was in the second week of the glorious month of May, gave a most unfavorable impression of the land and its climate. Professor Agassiz has pointed out the sub-arctic character of the vegetation of Lake Superior, by a lengthened comparison with the sub-alpine tracts of Switzerland; but this is due to the nature of the soil, rather than to the elevation or northern position of the district; for as we advance to the north at an equal elevation above the sea, but more to the westward, so as to enter on silurian or newer deposits in the vicinity of the Lake of the Woods and Rainy River, we find *cacti* and forests having a more southern aspect.

The ascent to the summit of the water-shed between Lakes Superior and Winipeg, by the Kamenistikwoya River, is made by about forty portages, in which the whole or part of a canoe's lading is carried on the men's shoulders; and a greater number occur in the descent to the Winipeg. The summit of the water-shed is an uneven, swampy, granitic country, so much intersected in every direction by lakes that the water surface considerably exceeds that of the dry land. Its mean elevation above Lake Superior is about eight hundred feet, and the granite knolls and sand-banks, which vary its surface, do not rise more than one hundred and fifty or two hundred feet beyond that general level, though their altitude above the river valleys which surround them is occasionally greater, giving the district a hilly aspect. The highest of these eminences does not overtop Thunder Mountain and some other basalt-capped promontories on Lake Superior, and had not the silurian strata, which, judging by the patches which remain, once covered the gneiss and granitic rocks nearly to their summits been removed, the country would have been almost level, and would have formed part of the rolling eastern slope of the continent, above whose plain the highest of the hills on Lake Superior scarcely rises. The summit of this water-shed of the St. Lawrence basin, commencing toward the Labrador coast, runs south 52° west, or about southwest half-west, at the distance of rather more than two hundred miles from the water-course, until it comes opposite to that elbow of the line of the great lakes which Lake Erie forms; it then takes a north 51° west course, or about northwest half-west, toward the northeast end of Lake Winipeg, and onward from thence in the same direction to Coronation Gulf of the Arctic Sea. The angle at which the two arms of this extensive water-shed (but no where mountain ridge) meet between Lakes Huron and Ontario is within half a point of a right one, and the character of the surface is every where the same, bearing, in the ramifications and conjunctions of its narrow valleys filled with water, no distant resemblance to the fiords of the Norway coast. Such a preponderance of fresh water coupled with the tardy melting of the ice in spring, makes a late summer, and augments the severity of the climate beyond that which is due to the northern position of the district.

Though the whole tract is most unfavorable for agriculture, much of the scenery abounds in picturesque beauty. Of this we have an instance in the Thousand Islands Lake, which forms the funnel-shaped outlet of Lake Ontario. At this place the pyrogenous rocks, denuded of newer deposits, cross the river to form a junction with the lofty highlands of the northern counties of New York. The round-backed, wooded hummocks of granite which constitute the more than thousand islets of this expanse of water, are grouped into long vistas, which are alternately disclosed and shut in as we glide smoothly and rapidly among them, in one of the powerful steamers, that carry on the passenger traffic of the lakes. The inferior fertility of this granite belt has deferred the sweeping operations of the settler's ax; the few farm-steadings scattered along the shore enhance the beauty of the forest; and the eye of the traveler finds a pleasant relief in contemplating the scenery, after having dwelt on the monotonous succession of treeless clearances lower down the river. Sooner or later, however, the shores of the Lake of the Thousand Isles will be studded with the summer retreats of the wealthy citizens of the adjacent States, and the incongruities of taste will mar the fair face of nature.

On the summit of the canoe-route between Lakes Superior and Winipeg, a sheet of water bearing the analogous appellation of Thousand Lakes, is also studded with knolls of granite, forming islets; but low mural precipices are more common there; and there is, moreover, an intermixture of accumulations of sand, such as are commonly found on the summit of the water-shed, along its whole range. The general scenery of this lake is similar to that of the Thousand Islands; but though the elevation above the sea does not exceed fourteen hundred feet, the voyagers say that frosts occur on its shores almost every morning throughout the summer.

Silurian strata occur on both flanks of both arms of the watershed above spoken of, to a greater or smaller extent throughout their whole length.* When we descend to Lake Winipeg we come upon epidôtic slates, conglomerates, sandstones, and trap rocks, similar to those which occur on the northern acclivity of the Lake Superior basin; and after passing the straits of Lake Winipeg, we have the granite rocks on the east shore, and silurian rocks (chiefly bird's-eye limestone) on the west and north, the basin of the lake being mostly excavated in the limestone. The two formations approach nearest to each other at the straits in question, where the limestone, sandstone, epidotic slates, green quartz-rock, greenstone, gneiss, and granite occur in the close neighborhood of each other.

The eastern coast-line of Lake Winipeg is, in general, swampy, with granite knolls rising through the soil, but not to such a height as to render the scenery hilly. The pine-forest skirts the shore at the distance of two or three miles, covering gently-rising lands, and the breadth of continuous lake-surface seems to be in process of diminution, in the following way. A bank of sand is first drifted up, in the line of a chain of rocks which may happen to lie across the mouth of an inlet or deep bay. Carices, balsam-poplars, and willows speedily take root therein, and the basin which lies behind, cut off from the parent lake, is gradually converted into a marsh by the luxuriant growth of aquatic plants. The sweet gale next appears on its borders, and drift-wood, much of it rotten and comminuted, is thrown up on the exterior bank, together with some roots and stems of larger trees. The first spring storm covers these with sand, and, in a few weeks, the vigorous vegetation of a short but active summer binds the whole together by a network of the roots of bents and willows. Quantities of drift-sand pass before the high winds into the swamp behind, and, weighing down the flags and willow-branches, prepare a fit soil for succeeding crops. During the winter of this climate, all remains fixed as the summer left it, and as the next season is far advanced before the bank thaws, little of it washes back into the water, but, on the contrary, every gale blowing from the lake brings a fresh

* A *Pentamerus* very like *P. Knightii*, was gathered by Dr. Bigsby on the Lake of the Woods, and presented by him to the British Museum. He probably found it in some of the western arms of the lake, the islands in the more easterly part being mostly granite.

supply of sand from the shoals which are continually forming along the shore. The floods raised by melted snows cut narrow channels through the frozen beach, by which the ponds behind are drained of their superfluous waters. As the soil gradually acquires depth, the balsam-poplars and aspens overpower the willows, which, however, continue to form a line of demarkation between the lake and the encroaching forest.

Considerable sheets of water are also cut off on the northwest side of the lake, where the bird's-eye limestone forms the whole of the coast. Very recently this corner was deeply indented by narrow, branching bays, whose outer points were limestone cliffs. Under the action of frost, the thin horizontal beds of this stone split up, crevices are formed perpendicularly, large blocks are detached, and the cliff is rapidly overthrown, soon becoming masked by its own ruins. In a season or two the slabs break into small fragments, which are tossed up by the waves across the neck of the bay into the form of narrow ridge-like beaches, from twenty to thirty feet high. Mud and vegetable matter gradually fill up the pieces of water thus secluded; a willow swamp is formed; and when the ground is somewhat consolidated, the willows are replaced by a grove of aspens.* Near the First and Second Rocky Points,† the various stages of this process may be inspected, from the rich alluvial flat covered with trees and bounded by cliffs that once overhung the water, to the pond recently cut off by a naked barrier of limestone, pebbles, and slabs, discharging its spring floods into the lake, by a narrow though rapid stream. In some exposed places the pressure of the ice, or power of the waves in heavy gales, has forced the limestone fragments into the woods, and heaped them round the stems of trees, some of which are dying a lingering death; while others, that have been dead for many years, testify to their former vitality, and the mode in which they have perished, by their upright stems, crowned by the decorticated and lichen-covered branches which protrude from the stony bank. The analogy between the entombment of living

* The fact of the formation of these detached ponds, marshes, and alluvial flats, points either to a gradual elevation of the district, or to an enlargement of the outlet of the lake, producing a subsidence of its waters.

† The strata at these points contain many gigantic orthoceratites, some of which have been described by Mr. Stokes in the Geological Transactions.

trees, in their erect position, to the stems of *sigillariæ*, which rise through different layers in the coal-measures, is obvious.*

The action of the ice in pushing boulders into the woods was observed at an earlier period of our voyage, and is noticed in the following terms in my journal: "In the first part of our course through Rainy Lake we followed a rocky channel, which was in many places shallow, and varied in breadth from a mile, down to a few yards. Some long arms stretch out to the right and left of the route, and particularly one to the eastward, into which a fork of Sturgeon River is said to enter. There is considerable current in these narrows. The first expanse of water we traversed is six miles across, and the second is fully wider. They are connected by a rocky channel, on whose shores many boulders are curiously piled up eight or ten feet above the rocks on which they rest. Other boulders lie in lines among the trees near the shore. They have been thrust up, many of them very recently, by the pressure of the ice, since the channel is too narrow for the wind to raise waves powerful enough to move such stones."

The granite and gneiss which form the east shore of Lake Winipeg, strike off at its northeast corner, and, passing to the north of Moose Lake, go on to Beaver Lake, where the canoe-route again touches upon them. At some distance to the westward of them the Saskatchewan, which is the principal feeder of Lake Winipeg, flows through a flat limestone country, which is full of lakes, the reticulating branches of the river, and mud-banks; it has in fact all the characters of a delta, though the divisions of the stream unite into one channel before entering the lake. This flat district extends nearly to the forks of the river, above which the prairie lands commence. Pine Island Lake, Muddy Lake, Cross Lake, and Cedar Lake, where the boats were arrested by ice in 1848, are dilatations of the Saskatchewan, and when the water rises a very few feet, the whole district is flooded; which commonly occurs on the snow melting in spring. Some way to the south lies an eminence of considerable height, named by the

* If one of the spruce firs included in the limestone debris, had its top broken off, and a layer of mud were deposited over all, we should have the counterpart of a sketch of Sir Henry de la Beche's Manual (p. 407). The thick and fleshy rhizomata of the *Calla palustris*, marked with the cicatrices of fallen leaves, and which are abundant in these waters, bear no very distant resemblance to *stigmariæ*.

Crees *Wapŭskëow-watchi,** and by the Canadians *Basquiau*. It separates Winepegoos Lake, and Red-Deer Lake and River from the bed of the Saskatchewan. I am ignorant of its geological structure, not having visited it.

With respect to the forests: The white or sweet cedar (*Cupressus thyoides*) disappears on the south side of Rainy Lake, within the American boundary line. The Weymouth pine, various maples, cockspur thorns, and the fern-leaved Comptonia, reach the southern slope of the Winipeg basin. Oaks extend to the islands and narrows of that lake. The elm, ash, arbor vitæ, and ash-leaved maple terminate on the banks of the Saskatchewan. The "wild rice," or *Folle avoine*† of the voyagers and traders, grows abundantly in the district between Lakes Superior and Winipeg. This grain resembles rice in its qualities, but has a sweeter taste. Though small, it swells much in cooking, and is nourishing, but its black husk renders it uninviting in its natural state. In favorable seasons it affords sustenance to a populous tribe of Indians, but the supply is uncertain, depending greatly on the height of the waters. In harvest time the natives row their canoes among the grass, and, bending its ears over the gunwale, thresh out the grain, which separates readily. They then lay it by for use in neatly-woven rush baskets. This grass finds its northern limit on Lake Winipeg, and it is common in the western waters of the more northern of the United States; but how far south it extends, I have not been able to learn. Strachey, in his "Historie of Travaille in Virginia," speaks of a "graine called *Nattowine*, which groweth as bents do in meadowes. The seeds are not much unlike rice, though much smaller; these they use for a deyntie bread, buttered with deere's suet." (p. 118). It is possible that he may refer to a smaller species (*H. fluitans*) of the same genus, which is known to abound in Georgia; but the seed of that could scarcely be collected in sufficient quantity. The hop-plant (*Humulus lupulus*) reaches the south end of Lake Winipeg, and, according to Mr. Simpson, yields flowers plentifully in the Red River colony. We observed it in the autumn of 1849, growing luxuriantly on the banks of the Kamenistikwoya, and connecting the lower branches of the trees

* *Wapŭs*, strait; *Ke-ow*, woods; *Watchi*, hill: the signification being, "a pass through woods on a hill."

† *Zizania aquatica* L., or *Hydropyrum esculentum* of Link.

with elegant festoons of fragrant flowers. An opinion prevailed among the traders that Lord Selkirk introduced it into this neighborhood when he took possession of the Northwest Company's post of Fort William, upward of thirty years ago; but the plant is indigenous to America, and grows abundantly in the Raton Pass, lying on the 37th parallel, at the height of eight thousand feet above the sea, as well as in many localities of the northern States. Throughout the canoe-route from Lake Superior to Lake Winipeg, no district shows such fertility as the banks of Rainy River. In autumn, especially, the various maples, oaks, sumachs, ampelopsis, cornel bushes, and other trees and shrubs whose leaves before they fall assume glowing tints of orange and red, render the woodland views equal, if not superior, to the finest that I have seen elsewhere on the American continent, from Florida northward. Nor are showy asters *helianthi*, *lophanthi*, *gentianeæ*, *physostegiæ*, *irides*, and many other gay flowers, wanting to complete the adornment of its banks.

From Saut Ste. Marie to the Saskatchewan, and the banks of Churchill River, the native inhabitants term themselves *In-nin-yu-wuk* or *Ey-thinyu-wuk*, and are members of a nation which formerly extended southward to the Delaware. That part of this widely spread people which occupies the north side of Lake Huron, the whole border of Lake Superior, and the country between it and the south end of Lake Winipeg, call themselves *Ochipewa*, written also *Ojibbeway*, or Chippeway;* and the more northerly division, who name themselves *Nathè-wywithin-yu*, are the Crees of the traders, and Knistenaux of French writers In a subsequent chapter I shall speak more particularly of the place which this people hold among the aboriginal nations. At present, I wish merely to point out some of the circumstances which have tended to work out a difference in the moral character of these two tribes, essentially the same people in language and manners. The Crees have now for more than twenty-six years been under the undivided control and paternal government of the Hudson's Bay Company, and are wholly dependent on them for ammunition, European clothing, and other things which have become necessaries. No spirituous liquors are distributed to them, and schoolmasters and missionaries are encouraged and aided by

* They are the *Sauteurs* or *Saulteaux* of the Canadians, and *Sotoos* of the fur traders.

the Company, to introduce among them the elements of religion and civilization. One village has been established near the dépôt at Norway House, and another at the Pas on the Saskatchewan, each having a church, and school-house, and a considerable space of cultivated ground. The conduct of the people is quiet and inoffensive; war is unknown in the Cree district; and the Company's officers find little difficulty in hiring the young men as occasional laborers.

The case is otherwise with the Chippeways, who live within the Company's territories. The vicinity of the rival United States Fur Company's establishments; the vigorous competition which is carried on between them and the Hudson's Bay Company, in prosecution of which spirituous liquors are dispensed by both parties liberally to the natives; and the abundance of *Folle avoine* on Rainy River and the River Winipeg, with the plentiful supply of sturgeon obtained from these waters, rendering the natives independent of either party, have had a demoralizing effect, and neither Protestant nor Roman Catholic missionaries have been able to make any impression upon them. One party of these Indians, from whom we purchased a supply of sturgeon on Rainy River, are briefly characterized in my notes, made on the spot, as being "fat, saucy, dirty, and odorous." A Roman Catholic church, erected some years ago on the banks of the Winipeg, has been abandoned, with the clearing around it, on account of the want of success of the priest in his endeavors to convert the natives; and neither the Hudson's Bay Company nor the United States people have been able to extinguish the deadly feud existing between the Chippeways and Sioux, nor to restrain their war parties.

Very recently the Chippeways of Lake Superior, through some oversight in the Canadian government in not making arrangements with them at the proper time, organized a war party against the mining village of Mica Bay, containing more than a hundred male inhabitants. In passing through Lake Superior we were pleased with the flourishing appearance of this village, containing many nicely white-washed houses, grouped in terraces on the steep bank of the lake. The mines were worked by a company, under a grant from the Canadian legislature, who, at the same period, made many other similar grants of mining localities on the lake, without previously purchasing the Indian rights. As the game is nearly extinct on the borders of the lake, the natives subsist chiefly by the fisheries; and the vicinity of the mining establishments was likely to be beneficial to them rather than injurious, by providing a market for their fish. But when they beheld party after party of white men crowding to their lands, eager to take possession of their lots by erecting buildings, and inquisitively examining every cliff, they acquired exaggerated ideas of the value of their rocks. For two summers they descended in large bodies to Saut Ste. Marie, expecting payment, and, being disappointed, thought they were trifled with. They determined, therefore, in council, to bring matters to a crisis by expelling the aggressors, and, in the autumn of 1849, made a descent upon Mica Bay, and drove away the miners and their families. To repel this attack a regiment was ordered up from Canada, at an expense which would have paid the Indians again and again: but a small part of the force only reached Mica Bay, to find the Chippeways gone; the rest were driven back to Saut Ste. Marie by stormy weather, not without very severe suffering, leading, I have been informed, to loss of life.

CHAPTER III.

Pine Island Lake.—Silurian Strata.—Sturgeon River.—Progress of Spring.—Beaver Lake.—Isle à la Crosse Brigade.—Ridge River.—Native Schoolmaster and his Family.—Two kinds of Sturgeon.—Native Medicines.—Bald Eagles.—Pelicans.—Black-bellied and Cayenne Terns.—Cranes.—Frog Portage.—Missinipi or Churchill River.—Its Lake-like Character.—Poisonous Plants and Native Medicines.—Athabasca Brigade.—Sand-fly Lake.—The Country changes its Aspect.—Bull-dog Fly.—Isle à la Crosse Lake.—Its Altitude above the Sea.—Length of the Missinipi.—Isle à la Crosse Fort.—Roman Catholic Mission.—Deep River.—Canada Lynx.—Buffalo Lake.—Methy River and Lake.—Murrain among the Horses.—Burbot or La Loche.—A Mink.—Methy Portage.—Join Mr. Bell and his Party.

We left Cumberland House at 4 A.M., on the 14th of June, but had not passed above three miles through Pine Island Lake, before we were compelled to seek shelter on a small island by a violent thunder storm, bringing with it torrents of rain. The rain moderating after a few hours, we resumed our voyage; but the high wind continuing and raising the waves, our progress was slow, and the day's voyage did not exceed twenty-two miles. In the part of the lake where we encamped the limestone (silurian*) rises, in successive outcrops, to the height of thirty feet above the water, the strike of the beds being about southwest by west, and northeast by east, or at right angles to the general line of direction of the gneiss and granite formation, which lies to the eastward. Many boulders of granite and trap rocks are scattered

* Some fragments of large *Orthocerata*, and a specimen of *Receptaculites neptunii*, point to the bird's-eye and Trenton limestones as occurring in this neighborhood. Mr. Woodward says of the latter specimen, "The only wood-cut in the New York State Surveys at all resembling your engine-turned fossil, is a very rude representation of part of a circular disk, with *radiating* and *concentric* (not engine) turned lines. It is called *Uphanteria chemungensis*, and is supposed to be a marine plant (p. 183, Vanuxem). A fossil much like yours is figured by De France in the *Dictionnaire des Sciences Naturelles* under the name of *Receptaculites neptuni*, from Chimay, in the Pays Bas. This is certainly the same genus. De Blainville also describes it in his *Actinologie* at the end of the corals, but offers no opinion respecting its affinities. I should compare it with *Eschadites Konigi* of Murchison's upper silurian, but *that* was originally spherical and hollow."

over the surface of the ground, far beyond the reach of any modern means of transport.

Thunder and heavy rain detained us in our encampment the whole of the following day; but some improvement in the weather taking place at midnight, we embarked, and at one in the morning of the 16th entered Sturgeon River, named by the voyagers, on account of its many bad rapids, "*La Rivière Maligne.*" We made two portages, and an hour after noon reached Beaver Lake. The entire bed of the river consists of limestone, sometimes lying in nearly horizontal layers more or less fissured; in other places broken up into large loose slabs, tilted up and riding on each other. Boulders of granite occur in various parts of the river, some of them of considerable magnitude, and rising high out of the water. In the lower part of the river the banks are sandy, a considerable deposit of dry, light soil overlies the limestone, and vegetation is early and vigorous.

When we left Lake Superior, in the middle of May, the deciduous trees gave little promise of life; and, in ascending the Kamenistikwoya, we were glad to let the eye dwell upon the groves of aspens which skirt the streams in that undulating and rocky district, and which, when well massed, gave a pleasing variety to the wintry aspect of the landscape—the silvery hue of their leafless branches and young stems contrasting well with the sombre green of the spruce fir, which forms the bulk of the forest. On the 27th of May, while ascending Church Reach of Rainy River, we had been cheered by the lovely yellowish hue of the aspens just unfolding their young leaves; but the ice, lingering on Lake Winipeg, when we crossed it, had kept down the temperature, spring had not yet assumed its sway, and the trees were leafless. Now, the season seemed to be striding onward rapidly, and the tender foliage was trembling on all sides in the bright sunshine. It was in a patch of burnt woods in this vicinity that, in the year 1820, I discovered the beautiful *Eutoca Franklinii*, now so common an ornament of our gardens.

Constantly, since the 1st of June, the song of the *Fringilla leucophrys* has been heard day and night, and so loudly, in the stillness of the latter season, as to deprive us at first of rest. It whistles the first bar of "Oh dear, what can the matter be!" in a clear tone, as if played on a piccolo fife; and, though the distinctness of the notes rendered them at first very pleasing, yet, as they haunted us up to the Arctic circle, and were loudest at mid night, we came to wish occasionally that the cheerful little songster would time his serenade better. It is a curious illustration of the indifference of the native population to almost every animal that does not yield food or fur, or otherwise contribute to their comfort or discomfort, that none of the Iroquois or Chippeways of our company knew the bird by sight, and they all declared boldly that no one ever saw it. We were, however, enabled, after a little trouble, to identify the songster, his song, and breeding-place. The nest is framed of grass, and placed on the ground under shelter of some small inequality; the eggs, five in number, are grayish, or purplish-white, thickly spotted with brown; and the male hides himself in a neighboring bush while he serenades his mate.

At the outlet of Beaver Lake, and at several succeeding points on both sides of the canoe-route, the thin, slaty limestone forms cliffs, thirty or forty feet high; but about the middle of the lake, there is a small island of greenstone. Beyond this we again touched upon the granite rocks which we had left at the north-east corner of Lake Winipeg, bearing from this place about east 82° south.

At the entrance of Ridge River we met Mr. M'Kenzie, Jun., in charge of a brigade of boats, carrying out the furs of the Isle à la Crosse district, and were glad to obtain from him tidings of Mr. Bell, who was advancing prosperously, though he had been stopped for three days by ice, on the lake which we had just crossed. The Missinipi, or Churchill River, Mr. M'Kenzie told us, did not open till the 6th of the present month, though in common years it seldom continues frozen beyond the 1st.

Soon after parting with this gentleman, we met the school-master of Lac La Ronge district, who, with his wife and four children, were on their way to pass the summer with the Rev. Mr. Hunter, Episcopal clergyman at the Pas. Both husband and wife are half-breeds, and both are lively, active, and intelligent. The family party were traveling in a small canoe, which the husband paddled on the water, and carried over the portages with their light luggage. For their subsistence, they depended on such fish and wild-fowl as they could kill on the route; and the lady was very grateful for a small supply of tea, sugar, and flour which we gave her. The young ones bore the assaults of the musquitoes with a stoical indifference, as an inevitable evil, that had belonged to every summer of their lives, and from which no part of the world, as far as they knew, was exempt. At the Ridge Portage, where we encamped for the night, the rock is gneiss, resembling mica-slate, owing to the quantity of mica that enters into its composition.

On the 17th, we came early to a long and strong rapid, bearing the same appellation with the preceding portage, and which is said to be the highest point to which sturgeon ascend in this river; and it is most probably the northern limit of the range of that fish, on the east side of the Rocky Mountains. It is situated in about 54½ degrees of north latitude. We noticed two species of this fish in the Saskatchewan River system. One of these is described in the *Fauna Boreali-Americana* under the name of *Accipenser rupertianus*, and has a tapering, acute snout. It seldom exceeds ten or fifteen pounds in weight. The other is the *Namèyu* of the Crees, and has not been hitherto described. It very commonly weighs ninety pounds, and attains the weight of one hundred and thirty. Its snout is short and blunt, being only one third of the length of the entire head; its nasal barbels are short, its shields small and remote, and the ventral rows are absent. Its caudal is less oblique than that of the smaller kind, the upper lobe being proportionally shorter. This species ascends the Winipeg River as high as the outlet of Rainy Lake: and the smaller kind is occasionally, though rarely, taken also in that locality, but, in general, it seems to be unable to surmount the cascade at the outlet of the Lake of the Woods. The rocks here are granite, and a mountain-green chlorite slate, similar to that which occurs so abundantly on the north side of the Lake Superior basin; the latter, under the action of the weather, forms a tenacious clayey soil. A hornblende-slate occupies the bed of the river, and rises, on each bank, into rounded knolls and low cliffs. The inequalities of the country here, as well as its vegetation, are very similar to that on the Kamenistikwoya, where the same formation exists.

The woods, being now in full but still tender foliage, were beautiful. The graceful birch, in particular, attracted attention by its white stem, light green spray, and pendent, golden catkins. Willows of a darker foliage lined the river bank; and the background was covered with dark green pines, intermixed with patches of lively aspen, and here and there a tapering larch, gay with its minute tufts of crimson flowers, and young pale green leaves. The balsam poplar, with a silvery foliage though an ungainly stem, and the dank elder, disputed the stand at intervals with the willows; among which the purple twigs of the dog-wood contributed effectively to add variety and harmony to the colors of spring.

The *Actæa alba* grows abundantly here; it is called by the Canadians *le racine d'ours*, and by the Crees, *musqua-mitsu-in* (bears' food). A decoction of its roots and of the tops of the spruce fir is used as a drink in stomachic complaints. The *Acorus calamus* is another of the indigenous plants that enter into the native pharmacopœia, and is used as a remedy in colic. About the size of a small pea of the root, dried before the fire or in the sun, is a dose for an adult, and the pain is said to be removed soon after it is masticated and swallowed. When administered to children, the root is rasped, and the filings swallowed in a glass of water, or of weak tea with sugar. A drop of the juice of the recent root is dropped into inflamed eyes, and the remedy is said to be an effectual though a painful one. I have never seen it tried. The Cree name of this plant is *watchŭskè mitsu-in*, or "that which the musk-rat eats."

At breakfast-time we crossed the Carp Portage, where there is a shelving cascade over granite rocks. The gray sucking carp (*Catastomus hudsonius*) was busy spawning in the eddies, and our voyagers killed several with poles. Two miles above the portage there are some steeply rounded sandy knolls clothed with spruce trees, being the second or high bank of the river, which is elevated above all floods of the present epoch. In some places granite rocks show through sand, heaped round their base. The frequent occurrence of accumulations of sand in this granite and gneiss district, near the water-sheds of contiguous river systems, has been already noticed. In the course of the forenoon we passed the Birch lightening-place (*Demi-charge du bouleau*), where a slaty sienite or greenstone occurs, the beds being inclined to the east-northeast at an angle of 45°; and an hour afterward we crossed the Birch Portage, five hundred and forty paces long. The rocks there are porphyritic granite, portions of which are in thin beds, and are therefore to be entitled gneiss.

The river has the character peculiar to the district, that is, it is formed of branching lake-like expansions without perceptible current, connected by falls or rapids occasioning portages, or by narrow straits through which there flows a strong stream. At four in the afternoon we crossed the Island Portage, where the rock is a fine-grained, laminated granite or gneiss, containing nodules or crystals of quartz, which do not decay so fast as the rest of the stone, and consequently project from its surface: the layers are contorted. In 1825, which was a season of flood, this islet was under water, and our canoes ascended among the bushes.

Two hours later we passed the Pine Portage (*Portage des Epinettes*), and entered Half-Moon Lake (*Lac Mi-rond**). At this portage the rocks are granite, greenstone, and black basalt, or hornblende-rock, containing a few scales of mica, and a very few garnets. The length of the portage is two hundred yards. At our encampment on a small island in Half-Moon Lake the gneiss lay in vertical layers, having a north and south strike. A few garnets were scattered through the stone. This piece of water, and Pelican and Woody Lakes, which adjoin it, are full of fish, and they are consequently haunted by large bodies of pelicans, and several pairs of white-headed eagles (*Haliæëtus albicilla*). This fishing eagle abounds in the watery districts of Rupert's Land; and a nest may be looked for within every twenty or thirty miles. Each pair of birds seems to appropriate a certain range of country on which they suffer no intruders of their own species to encroach; but the nest of the osprey is often placed at no great distance from that of the eagle, which has no disinclination to avail itself of the greater activity of the smaller bird, though of itself it is by no means a bad fisher. The eagle may be known from afar, as it sits in a peculiarly erect position, motionless, on the dead top of a lofty fir, overhanging some rapid abounding in fish. Not unfrequently a raven looks quietly on from a neighboring tree, hoping that some crumb may escape from the claws of the tyrant of the waters. Some of our voyagers had the curiosity to visit an eagle's nest, which was built, on the cleft summit of a balsam poplar, of sticks, many of them as thick as a man's wrist It contained two young birds, well-fledged, with a good store of fish, in a very odoriferous condition. While the men were climbing the tree the female parent hovered close round, and threatened an attack on the invaders; but the male, who is

* Called by mistake *Lac Heron* in Franklin's Overland Journeys.

of much smaller size, kept aloof, making circles high in the air. The heads and tails of both were white.

The pelican, as it assembles in flocks, and is very voracious, destroys still larger quantities of fish than the eagle. It is the *Pelicanus trachyrhynchus* of systematic ornithologists, and ranges as far north as Great Slave Lake, in latitude 60°–61° North. These birds generally choose a rapid for the scene of their exploits, and, commencing at the upper end, suffer themselves to float down with the current, fishing as they go with great success, particularly in the eddies. When satiated, and with full pouches, they stand on a rock or boulder which rises out of the water, and air themselves, keeping their half-bent wings raised from their sides, after the manner of vultures and other gross feeders. Their pouches are frequently so crammed with fish that they can not rise into the air until they have relieved themselves from the load, and on the unexpected approach of a canoe they stoop down, and, drawing the bill between their legs, turn out the fish. They seem to be unable to accomplish this feat when swimming, so that then they are easily overtaken, and may be caught alive, or killed with the blow of a paddle. If they are near the beach when danger threatens, they will land to get rid of the fish more quickly. They fly heavily, and generally low, in small flocks of from eight to twenty individuals, marshaled, not in the cunei-form order of wild geese, but in a line abreast, or slightly *en echellon;* and their snow-white plumage with black-tipped wings, combined with their great size, gives them an imposing appearance. Exceeding the fishing eagle and the swan in bulk, they are the largest birds in the country. Their eggs are deposited on rocky islets among strong rapids, where they can not be easily approached by man or beasts of prey. The species is named from a ridge or crest which rises from the middle line of the upper mandible of the male; sometimes from its whole length, when it is generally uneven; and sometimes from a short part only, when it is semicircular and smooth-edged.

The black-bellied tern (*Hydrochelidon nigra*) is also abundant on these waters, and ranges northward to the upper part of the Mackenzie. And the Cayenne tern (*Sterna cayana*) is common in this quarter and onward to beyond the Arctic Circle; but notwithstanding Mr. Rae's expertness as a fowler, and eagerness to procure me a specimen, the extreme wariness of the bird frustrated all his endeavors until this day, when he brought one down, and gave me an opportunity of examining it, which I was glad to do, since from want of a northern specimen the bird was not noticed in the *Fauna Boreali-Americana*. Mr. Audubon mentions the great difficulty of shooting this bird, and he succeeded in doing so only by employing several boats to approach its haunts in different directions. Albert, our Eskimo interpreter, told me that it does not visit Hudson's Bay.

I was also indebted to Mr. Rae's gun for specimens of the brown crane (*Grus canadensis*). Mr. Audubon, who is so competent an authority on all questions relating to American birds, and whose recent death all lovers of natural history deplore, was of opinion, in common with many other ornithologists, that the brown crane is merely the young of the large white crane (*Grus americana*) ; but, though I concede that the young of the latter are *gray*, I think that the *brown* species is distinct ; first, because it is generally of larger dimensions than the white bird, and secondly, because it breeds on the lower parts of the Mackenzie and near the Arctic coasts, where the *Grus americana* is unknown. As far as I could ascertain, the latter bird does not go much further north than Great Slave Lake. At *Fort aux Liards*, on the River of the Mountains, large flocks of cranes pass continually to the westward, from the 17th to the 20th of September ; the gray and white birds being in different bands, and the former of smaller size, like young birds. Very rarely during the summer a flock of white cranes passes over Fort Simpson, in latitude 62° N. The brown cranes, on the other hand, which frequent the banks of the Mackenzie from Fort Norman, in latitude 65° N., down to the sea-coast, are generally in pairs. They are in the habit of dancing round each other very gracefully on the sand-banks of the river.*

June 18th.—About three hours after embarking we came to the Pelican lightening-place (*Demi-charge de chetauque*), and by breakfast-time we had crossed the three portages of Woody Lake. A micaceous gneiss or mica-slate rock prevails at these portages. A family of Cree Indians, who were encamped on one of the many islands which adorn the scenery of Woody Lake, exchanged

* Much of this information I received from Murdoch MacPherson, Esq., who, during twenty years' residence on the Mackenzie, became thoroughly acquainted with all its feathered and ferine inhabitants.

fish for tobacco, and enabled us to vary the diet of our voyagers, an indulgence which pleases them greatly; for, though they generally prefer pork or pemican to fish, they relish the latter occasionally. At five we crossed the Frog Portage, or *Portage de Traité* of the Canadians, and encamped on the banks of the Missinipi or Churchill River, in the immediate vicinity of a small outpost of the Hudson's Bay Company.*

No change of formation takes place in passing from the Saskatchewan River system to that of the Missinipi. The Frog Portage is low, and Churchill River, in seasons of flood, sometimes overflows it, and discharges some of its superfluous waters into the Woody Lake.† The general level of the country for some distance, or down to the lower end of Half-Moon Lake varies little; but in this and in Pelican Lake there are a few conical eminences, which rise several hundred feet above the water. We did not approach them sufficiently near for examination.

Frog Portage is the most northerly point of the Saskatchewan basin, and lies in 55° 26′ N. latitude, 103° 20′ W. longitude. Below it there is a remarkable parallelism in the courses of Churchill and Saskatchewan rivers, both streams inclining to the northeast in their passage through the "intermediate primitive range," the district from whence they receive lateral supplies being at the same time very greatly narrowed. Several other considerable streams run near them, and parallel to them, but do not originate so far to the westward. In their widely spreading upper branches, and their restricted trunks, they resemble trees. As they are not separated high on the prairie slopes by an elevated water-shed, they may be considered, in reasoning upon the direction of the force which excavated their basins, as one great system, having an eastern direction and outlet, interposed between the Missouri and Mackenzie, which discharge themselves respectively into southern and northern seas.

The Churchill River is the boundary between the Chepewyans

* The Cree term *Missinipi* signifies "much water," and is analogous to that of *Mississippi*, which means "great river;" *nipi* being water, and *sipi* river. The Canadians call it English River, because on it the early fur-traders from Canada encountered the Hudson's Bay Company's people ascending from their principal dépôt at Fort Churchill.

† About forty years ago, in a season remembered especially for the land-floods, a gentleman was drowned on the Frog Portage, by his canoe oversetting against a tree, as he was passing from Churchill River.

and Crees, but a few of the latter frequent its borders, resorting to Lac la Ronge and Isle à la Crosse posts, along with the Chepewyans, for their supplies.

On June the 19th, a fog detained us at our encampment to a later hour than usual; when being unwilling to lose all the morning, we went some distance in the thick weather under the guidance of the post-master, who was acquainted with every rock in the neighborhood. As the sun rose higher the atmosphere cleared, and we ascended the Great Rapid by its southern channel, making a portage part of the way, and poling up the remainder. A recent grave with its wooden cross, marked the burial-place of one of the Hudson's Bay Company's servants, who was drowned here last year. His body was thrown out a little below the rapid.

We next crossed the Rapid lightening-place, and afterward mounted four several rapids, connected with the Barrel Portage. In the afternoon the Island Portage was made, where the river, being pent in for a short space between high, even, rocky banks, is there only five or six hundred yards wide, and has a strong current, requiring much exertion from the canoe-men in paddling round the headlands. Elsewhere, except at the rapids during this day's voyage, Churchill River has more the character of a lake. In the evening we crossed the portage of the Rapid River, one hundred and sixty paces long, which has its name from a tributary stream on which the Hudson's Bay Company have a post, that is visible from the canoe-route. Afterward we passed the lightening-place of the Rapid River, and encamped five or six miles further on, at half-past eight o'clock.

Our Iroquois, being tired with the day's journey, and longing for a fair wind to ease their arms, frequently, in the course of the afternoon, scattered a little water from the blades of their paddles as an offering to *La Vieille*, who presides over the winds. The Canadian voyagers, ever ready to adopt the Indian superstitions, often resort to the same practice, though it is probable that they give only partial credence to it. Formerly the English shipmen, on their way to the White Sea, landed regularly in Lapland to purchase a wind from the witches residing near North Cape; and the rudeness and fears of Frobisher's companions in plucking off the boots or trowsers of a poor old Eskimo woman on the Labrador coast, to see if her feet were cloven, will be remembered by readers of Arctic voyages.

Thoughout the day's voyage, the primitive formation continued. In several places we observed micaceous slate, traversed by large veins of granite, and alternating with beds of the same, also gneiss in thick beds, with its layers much contorted. Below the Great Rapid there are many bluff granite rocks, and some precipices thirty or forty feet high, the higher knolls rising probably from two to three hundred feet above the water. At the Great Rapid a greenstone-slate stained with iron occurs. At the Barrel Portage, a mile or two further on, where the river makes a sharp bend, beds of chlorite-slate occupy its channel for two miles, having a northeast and southwest strike, and a southerly dip of 60° or 70°. Beds of greenstone-slate are interleaved with it. Above the Island Portage a sienite occurs which contains an imbedded mineral; and at the Rapid River Portage, mica-slate, passing into gneiss, prevails, the beds having a southwest and northeast strike. The granite veins here have a general direction nearly coincident with that of the beds, but they are waved up and down. In the vicinity of the veins the layers of slate are much contorted, following the curvatures of the veins closely. At the lightening-place of the Rapid River, there is a fine precipice of granite fifty feet high, which is traversed obliquely from top to bottom by two magnificent veins of flesh-colored porphyry-granite. Five miles further on there are precipices of granite one hundred and fifty feet high.

The country in this neighborhood is hilly, and a few miles back from the river the summits of the eminences appear to the eye to rise four hundred or perhaps five hundred feet above the river. The resemblance of the whole district to Winipeg River is perfect, and the general aspect of the country is much like that of the north shores of Lake Superior, though the water basin is not so deeply excavated.

An hour and a half after starting on the morning of the 20th, we crossed the Mountain Portage, one hundred paces long, where the rock is horneblende-slate. At Little Rock Portage, a short way further on, the thin slaty beds have a northeast and southwest strike. Above this, a dilatation of the river, named Otter Lake, leads to the Otter Portage, of three hundred paces, made over mica-slate. The beach there is strewed with fragments of a crystalline augitic greenstone, showing that that rock is not far distant.

From a party of Chepewyans who were encamped on the Otter Lake, we procured a quantity of a small white root, about the thickness of a goose quill, which had an agreeable nutty flavor. I ascertained that it was the root of the *Sium lineare*. The poisonous roots of *Cicuta virosa*, *maculata*, and *bulbifera*, are often mistaken for the edible one, and have proved fatal to several laborers in the Company's service. The natives distinguish the proper kind by the last year's stem, which has the rays of its umbel ribbed or angled, while the *Cicutæ* have round and smooth flower-stalks. When the plant has put out its leaves by which it is most easily identified, the roots lose their crispness and become woody. The edible root is named *ūskotask* by the Crees, and *queue de rat* by the Canadians. The poisonous kinds are called *manito-skatask*, and by the voyagers *carrotte de Moreau*, after a man who died from eating them.

The *Heuchera Richardsonii*, which abounds on the rocks of this river, is one of the native medicines, its astringent root being chewed and applied as a vulnerary to wounds and sores. Its Cree name is *pichē quaow-utchēpi*. The leaves of the *Ledum palustre* are also chewed and applied to burns, which are said to heal rapidly under its influence. The cake of chewed leaves is left adhering to the sore until it falls off.

In the course of the forenoon we ascended four rapids, occasioning short portages, then the Great Devil's Portage, of fourteen hundred paces; and in the afternoon several other rapids were passed, among which were the Steep Bank, Little Rock, and Trout portages. At the Steep Bank Portage (*Portage des Ecores*), which is one hundred and sixty paces long, gneiss and mica-slate occur interleaved irregularly with each other, and intersected in every direction by reticulating quartz veins; the prevailing rock in the neighborhood being gneiss, and the hills low and barren.

June 21*st*.—Soon after starting we crossed the Thicket Portage (*Portage des Haliers*) of three hunded and sixty paces, and entered Black-bear Islands Lake, a very irregular piece of water, intersected by long promontories and clusters of islands. After four hours' paddling therein we came to a rapid, considered by the guide as the middle of the lake; in three hours more we came to another strong rapid, and after another three hours to the Broken-Canoe Portage, which is at the upper end of this dilatation of the river. Granite is the prevailing rock in the lake, and one of the small islands consists of large balls of that stone, piled on each other like cannon shot in an arsenal. They might be taken for boulders were they not heaped up in a conical form and all of one kind of stone; and they have obviously received their present form by the softer parts of the rock having crumbled and fallen away. At Thicket Portage and the lower end of the lake, the granite is associated with greenstone slate; and at Broken-Canoe Portage, above the lake, a laminated stone exists, whose vertical layers are about an inch think, and have a north and south strike, being parallel to the direction of the ridges of the rock. This stone is composed of flesh-colored quartz, with thin layers of duck-green chlorite, and no felspar. It ought perhaps to be considered as a variety of gneiss.

Later in the afternoon we came to the Birch and Pin Portages, on the last of which we encamped. The granite rocks here are covered by a high bank of sand and gravel, filled with boulders.

June 22d.—Embarking early, we passed through Sand-fly Lake, and afterward Serpent Lake, in which we met the Athabasca brigade of boats, under charge of Chief Trader Armitinger. This gentleman informed us that he met Mr. Bell with our boats on the 19th, on which day they would arrive at Isle à la Crosse. The aspect of the country changes suddenly on entering these lakes. The rising grounds have a more even outline, and one long low range rises over another, as the country recedes from the borders of the water, where it is generally low and swampy. The trees near the water are almost exclusively birch and balsam-poplar, or aspen; the spruce firs occupying the distant elevations, which are generally long round-backed hills, with a few short conical bluffs. Serpent Lake is named from the occurrence on its shores of a small snake.* I was not able to learn that this or any other snake had been detected further to the north. Having passed a high sand-bank on the north side of Serpent Lake, six miles further on, we entered the Snake River, within the mouth of which there is a bank of loam, sand, and rolled stones, thirty-five feet high. The bed of the stream is lined with these stones, and its width is about equal to that of Rainy River. The rocky points, as seen from the canoe, appeared to be of granite. All the boulders that I examined were of a dull brownish-

* *Coluber* or *Tropidinotus sirtalis.*

red, striped or laminated granite, which, on a cursory inspection, might be mistaken for sandstone. Boulders of the same kind occur at the Snake Rapid, where they are intermixed with a few pieces of hornblende rock.

June 23d.—The musquitoes were exceedingly numerous and troublesome during the night and this morning. Our route lay through Sandy Lake and Grassy River, where the country retains the same general aspect that it has on Sand-fly and Serpent Lakes, and where the prevailing rock is a brownish-red, fine-grained sienite, resembling a sandstone. The same rock abounds in Knee Lake, where, however, we saw, for the first time since leaving the south end of Lake Winipeg, fragments of white quartzose sandstone; but did not find the stone *in situ*. The sienite, when traced, is found to pass into hornblendic granite, by the addition of scales of mica to some parts of the same beds. The high banks of Knee Rapid consist of sandy loam crammed with boulders.

The *Tabanus*, named by the voyagers "Bull-dog," has been common for two days. The current notion is, that this fly cuts a piece of flesh from his victim, and at first sight there seems to be truth in the opinion. The fly alights on the hands or face so gently that if not seen he is scarcely felt until he makes his wound, which produces a stinging sensation as if the skin had been touched by a live coal. The hand is quickly raised toward the spot, and the insect flies off. A drop of blood, oozing from the puncture, gives it the appearance of a gaping wound, and the fly is supposed to have carried off a morsel of flesh. In fact, the *Tabanus*, inserts a five-bladed lancet, makes a perforation like a leech-bite, and, introducing his flexible proboscis, proceeds to suck the blood. He is, however, seldom suffered to remain at his repast; unless, as in our case, he be allowed to do so, that his mode of proceeding may be inspected. These *Tabani* are troublesome only toward noon and in a bright sun, when the heat beats down the musquitoes.*

In the afternoon we passed through Primeau's Lake, having previously ascended three strong and bad rapids. At the middle turn of the lake a moderately high, long, and nearly level-topped

* Of the five lancets with which the *Tabanus* wounds his prey, two are broader than the others. They are inclosed in a black, hairy sheath, whose extremity folds back. The palpi are conico-cylindrical and tubular.

hill closes the transverse vista. The channel between the eastern and western portions of the lake winds among extensive sandy flats, covered with bents, and in some places there was a rich crop of grass not in flower, but seemingly a *Poa*. In the evening we encamped at the "Portage of the Exhausted," on the river between Isle à la Crosse and Primeau Lakes. The rock here, and on the two lakes below it, is the brownish-red slaty sienite already mentioned; it has much resemblance to a rock on Lakes Huron and Superior, which seemed there to be associated with a conglomerate. The brownish color belongs to the felspar; a vitreous quartz also enters into its composition, and a little hornblende. It is rather easily frangible, and has a flat, somewhat slaty fracture.

Two hours after embarking on the 24th, we passed the Angle Rapid (*Rapide de l'Equerre*), and subsequently the Noisy (*Rapide Sonante*), and Saginaw Rapids, and entered the small Saginaw Lake, which we crossed in half an hour. At various points we had cursory glances, in passing, of granite forming low rocks. The crooked Rapid, a mile and a half long, conducted us to Isle à la Crosse Lake. In traversing twenty-three geographical miles of this lake, we disturbed many bands of pelicans, which were swimming on the water, or seated on rocky shoals, in flocks numbering forty or fifty birds. On the shores there are fragments of a white quartzose sandstone, but I noticed no limestone. The country consists of gravelly plains, having a coarse, sandy soil, and numerous imbedded boulder stones. Shoals formed by accumulations of boulders, are common in the lake, and in various places close pavements of these stones are surmounted by sandy cliffs twenty or thirty feet high. The bulk of the boulders belongs to the brownish glassy sienite mentioned in a preceding page.

The funnel-shaped arm named Deep River (*La Rivière Creuse*) meets the northern point of the lake at an acute angle, inclosing between it and Clear Lake a triangular peninsula. Beaver River, the principal feeder of the lake, flows from Green Lake, which lies directly to the southward, near the valley of the Saskatchewan in the 54th parallel of latitude. The winter path from Isle à la Crosse to Carlton House ascends this river to its great bend, whence it leads to the Saskatchewan plains, through an undulating country, but without any marked acclivity. I consider it probable, therefore, that Isle à la Crosse Lake and

Carlton House do not differ more than two hundred feet from each other in their height above the sea. The altitude of the latter I have judged to be about eleven hundred feet; and Captain Lefroy, from his experiments with the boiling water thermometer, assigns an elevation of thirteen hundred feet to the former.

Churchill River, disregarding its flexures, has a course to the sea from Isle à la Crosse Fort, of five hundred and twenty-five geographical miles, and the length of the Saskatchewan below Carlton House is six hundred and thirty miles. The general descent of the eastern slope of the continent to Hudson's Bay from these two localities may be reckoned at a little more than two feet a mile. Further to the westward, in the vicinity of Fort George, near the 110th meridian, the upper branches of the Beaver River rise from the very banks of the Saskatchewan.

On Beaver River the strata are limestone, and a line drawn from the north side of Lake Winipeg, to the south side of Isle à la Crosse Lake, runs about north 58° west, and touches upon the northern edge of the limestone in Beaver Lake. That line may therefore, be considered as representing the general direction of the junction of the limestone with the primitive rocks in this district of the country. Judging from relative geographical position and mineralogical resemblances, the north part of Isle à la Crosse Lake belongs to a similar sandstone deposit with that which skirts the primitive rocks on Lake Superior—a peculiar looking sienite being connected with the sandstone in both localities. From its order of occurrence the limestone of Beaver River is probably silurian. My observations were too limited and cursory to carry conviction, even to my own mind, on these points; the circumstances attending the several journeys I have made through these countries having prevented me from obtaining better evidence. In a voyage with ulterior objects through so wide an extent of territory, and with so short a traveling season, every hour is of importance, and whoever has charge of a party must show that he thinks so, otherwise his men can not be induced to keep up their exertions for sixteen hours a day, which is the usual period of labor in summer traveling. Of this time an hour's halt is allowed for breakfast, and half an hour for dinner. We did not reach Isle à la Crosse Fort till half-past nine in the evening, and then learned that Mr. Bell with the boats was four days in advance of us.

June 25th.—A strong gale blowing this morning detained us at the post, and the day being Sunday, our voyagers went to mass at the Roman Catholic chapel, distant about a mile from the fort. This mission was established in 1846 under charge of Monsieur La Flêche, who has been very successful in gaining the confidence of the Indians, and gathering a considerable number into a village round the church. In the course of the day I received a visit from Monsieur La Flêche and his colleague Monsieur Taschè. They are both intelligent, well-informed men, and devoted to the task of instructing the Indians; but the revolution in France having cut off the funds the mission obtained from that country, its progress was likely to be impeded. They spoke thankfully of the assistance and countenance they received from the gentlemen of the Hudson's Bay Company. The character they gave the Chepewyans for honesty, docility, aptness to receive instruction, and attention to the precepts of their teachers, was one of almost unqualified praise, and formed, as they stated, a strong contrast to that of the volatile Crees. They have already taught many of their pupils here to read and write a stenographic syllabic character, first used by the late Reverend Mr. Evans, a Wesleyan missionary, formerly resident at Norway House, but which Monsieur La Flêche has adapted to the Chepewyan language. On asking this gentleman his opinion of the affinity between the Cree and Chepewyan tongues, both of which he spoke fluently, he told me that the grammatical structure of the Chepewyan was different, the words short, and the sounds dissimilar, bearing little resemblance to the soft, flowing compounds of the Cree language.

As there is generally some difficulty in making an early start from a fort, we moved in the evening to the point of the bay, that we might be ready to take advantage of the first favorable moment for proceeding on our voyage.

June 26th.—We embarked before 3 A.M., but a strong headwind blowing, we could proceed only by creeping along-shore under shelter of the projecting points. For some days past the water has been covered with the pollen of the spruce fir, and today we observed that it was thickly spread with the downy seeds of a willow. The banks of Deep River, which forms the discharge of Buffalo and Clear Lakes, consist of gravel and sand containing large boulders, principally of trap and primitive rocks.

The eminences rise from fifteen to forty feet above the river, and the land-streams have cut ravines into the loose soil, the whole being well covered with the ordinary trees of the country. This low land extends to Primeau Lake on the one side, and Buffalo Lake on the other. The beach, especially toward the openings of Cross and Buffalo Lakes, is strewed with fragments of quartzose sandstone, mixed with some pieces of light-red freestone, and many boulders of earthy greenstone, chlorite-slate, porphyritic greenstone slate, and gneiss. Neither mica-slate nor limestone were observed among them, and no rocks *in situ*. Many of the bays have sandy beaches. The Deep River has little current, except where it issues from the lakes.

In the morning a Canada lynx was observed swimming across a strait, where the distance from shore to shore exceeded a mile. We gave chase, and killed it easily. This animal is often seen in the water, and apparently it travels more in the summer than any other beast of prey in this country. We put ashore to sup at seven in the evening, at a point in Buffalo Lake, where we found evidences of the boat party having slept there a night or two previously. Being desirous of overtaking them without delay, we immediately resumed our voyage, but were caught in the middle of the lake by a violent thunder-storm, accompanied by strong gusts of wind. The voyagers were alarmed, and pulled vigorously for the eastern shore, on which we landed soon after eleven. The shores of Buffalo Lake are generally low; but, on the west side, there is an eminence named Grizzle Bear Hill, which is conspicuous at a considerable distance. It probably extends in a northwest direction toward the plateau of Methy Portage and Clear-water River. The valley to the east is occupied by Methy, Buffalo, and Clear Lakes, the last of which is said to have extensive arms.

Embarking at daylight on the 27th, we crossed the remainder of the lake, being about fourteen miles, and entered the Methy River, which we found to our satisfaction higher than usual; as in so shallow a stream the navigation is very tedious in dry seasons. The watermarks on the trees skirting the river showed that the water had fallen at least five feet, since the spring floods. The musquitoes are more numerous in seasons of high water, and this year was no exception to the general rule.

At the Rapid of the Tomb (*La Cimetière*) several pitch or

red pines (*Pinus resinosa*) grow intermixed with black spruces, one of them being a good-sized tree. This is the most northerly situation in which I saw this pine, and the voyagers believe that it does not grow higher than the River Winipeg.

An Indian, who has built a house at the mouth of the river, keeps fifteen or twenty horses, which he lets to the Company's men on Methy Portage, the charge being "a skin," or four shillings, for carrying over a piece of goods or furs weighing ninety pounds. From him we received the very unpleasant intelligence, that not only had his horses died of murrain last autumn, but that all the Company's stock employed on the portage had likewise perished. This calamity foreboded a detention of seven or eight days longer on the portage than we expected, and a consequent reduction of the limited time we had calculated upon for our sea-voyage. I had used every exertion to reach the sea-coast some days before the appointed time, expecting to be able to examine Wollaston's Land this season;—this hope was now almost extinguished. Another stock of horses had been ordered from the Saskatchewan, but they were not likely to arrive till the summer was well advanced.

Methy River flows through a low, swampy country, of which a large portion is a peat moss. Some sandy banks occur here and there, and boulders are scattered over the surface, and line the bed of the stream. We encamped on the driest spot we could find, and had to sustain the unintermitting attacks of myriads of musquitoes all night.

The Methy River, Lake, and Portage, are named from the Cree designation of the Burbot (*Lota maculosa*), (*La Loche* of the Canadians), which abounds in these waters, and often supplies a poor and watery food to voyagers whose provisions are exhausted. Though the fish is less prized than any other in the country, its roe is one of the best, and, with a small addition of flour, makes a palatable and very nourishing bread.

Four hours' paddling brought us, early on the 28th, to the head of the river, and two hours more enabled us to cross to the eastern side of Methy Lake, where we were compelled to put ashore by a strong headwind. A female mink (*Vison lutreola*) was killed as it was crossing a bay of the lake. It had eight swollen teats, and its udder contained milk; so that probably its death insured that of a young progeny also. The feet of this little amphibious animal are webbed for half the length of its toes. It is the *Shakwèshew* or *Atjakashew* of the Crees, the "Mink" of the fur-traders, and the *Foutereau* of the Canadians.

In the evening, the wind having decreased, we paddled under shelter of the western shore to the upper end of the lake, and entered the small creek which leads to the portage.

Mr. Bell was encamped at the landing-place, having arrived on the previous day, which he had spent in preparing and distributing the loads, and the party had advanced one stage of different lengths, according to the carrying powers of the individuals, which were very unequal. On visiting the men, I found two of the sappers and miners lame from the fatigue of crossing the numerous carrying-places on Churchill River, and unfit for any labor on this long portage. Several others appeared feeble; and, judging from the first day's work of the party, I could not estimate the time that would be occupied, should they receive no help in transporting the boats and stores, at less than a fortnight, which would leave us with little prospect of completing our sea-voyage this season. In the equal distribution of the baggage each man had five pieces of ninety pounds' weight each, exclusive of his own bedding and clothing, and of the boats, with their masts, sails, oars, anchors, &c., which could not be transported in fewer than two journeys of the whole party. The Canadian voyagers carry two pieces of the standard weight of ninety pounds at each trip on long portages such as this, and, in shorter ones, often a greater load. Several of our Europeans carried only one piece at a time, and had, consequently, to make five trips with their share of the baggage, besides two with the boats; hence they were unable to make good the ordinary day's journey of two miles, being, at seven trips with the return, twenty-six miles of walking, fourteen of them with a load. The practiced voyager, on the contrary, by carrying greater loads, can reduce the walking by one-third, and some of them by fully one-half.*

* In 1825 Sir John Franklin ascertained the position of the first resting-place, after leaving Methy Lake, to be in latitude 56° 36′ 30″ N., longitude 109° 52′ 54″ W. By carefully pacing the distance from thence to Methy Lake, I found it to be 1790 yards, on a south 43° 25′ east bearing, giving 22″ difference of latitude, and 58″ of longitude. Hence the east end of the portage lies in latitude 56° 36′ 08″ N., longitude 109° 51′ 56″ W.

The usual encampment by the tomb on the south side of the Little Lake

By their agreements, our canoe-men were at liberty to return as soon as we overtook the boats; and, in that case, the additional pieces we had brought, would, of course, be added to the baggage of the boat party; but I engaged them to assist us during the time that we were occupied on the portage, for an increase of wages of four shillings, York currency, per diem, each...

June 29th.—Our canoe-men were early astir this morning, and, before breakfast time, had carried all the cargo of the canoes to the banks of a small lake, being two-thirds of the whole portage, or 16,724 paces: the entire distance from Methy Lake to Clear-water River is 24,593 paces.

By observations with the aneroid and Delcros' barometers, I ascertained that the Little Lake was elevated twenty-two feet above Methy Lake; that the highest part of the pathway between the Little Lake and the Clear-water River rises above the latter six hundred and fifty-six feet, but, above Methy Lake, only sixty-six feet. The Cockscomb, or the crest of the precipitous brow which overlooks the magnificent valley of the Clear-water, is twenty-two feet lower than the summit of the path, or six hundred and thirty-four feet above the last-named river. The port-

is in latitude 56° 40′ 17″ N., longitude 109° 57′ 54″ W., and the north end of the path on the banks of the Clear-water River is in 56° 42′ 51″ N., 109° 59′ 08″ W. The direct distance from one end of the portage to the other is therefore only 7½ geographical miles in a north 27° west course; while the paces, reduced to yards at the rate of 23 feet to every 10 paces (which I found after several trials to be the average), are 18,855, or 10·7 statute miles.

I subjoin the voyagers' names for the several resting-places on the portage, premising, however, that the halting-places vary both in number and position with the loads and strength of the carriers, and that the names are often transposed.

Methy Lake (*Lac la Loche*).		
Thence to *Petit Vieux*	2557	paces.
" *Fontaine du Sable*	3171	"
" *La Vieille*	4591	"
" *Bon Homme ou de Cyprès*	3167	"
" *Petit Lac*	3238	"
" *De Cyprès ou La Vieux*	4302	"
" *La Crête*	1283	"
" *Descente de la Crête*	1984	"
" *La Prairie*	300	"
	24,593	"

age-road is, in fact, nearly level; the inequalities being of small account as far as to the sudden descent of the Cockscomb. In the sandy soil there are many fragments of sandstone, a few of limestone, and scattered boulders of granites, sienites, and greenstones. The deposit of sand is about six hundred feet deep, and most probably incloses solid beds of sandstone. It is based on a (Devonian ?) limestone, which lines the whole bed of the Clearwater River, till its junction with the Elk River, as I shall hereafter mention.*

Captain Lefroy assigns fifteen hundred feet as the elevation of the surface of Methy Lake above the sea,† and, from various esti-

* As the Cockscomb is under the level of the brow of the valley, the depth of sand may be more than 600 feet at its highest points.

† The exact height assigned by Captain Lefroy to Methy Lake is 1540 feet, which I have reduced in the text to the even number of 1500, as agreeing better with my own estimates. If this be nearly correct, Captain Lefroy gives too small an altitude to Isle à la Crosse Lake, since the route from thence to the portage is chiefly lake-way; and the Methy River can not have a descent of 240 feet, which his altitudes would assign to it.

In the year 1848, I made several observations with the aneroid on Methy Portage to ascertain its levels, but they were neither so carefully made nor so extensive as they would have been, had I been less anxiously and constantly employed about the transport of the goods and boat. The error in this case is not, however, likely to be many feet, as the portage is evidently very nearly level as far as the Cockscomb. The height of the latter was ascertained on July 27, 1849, by Delcros' barometer, the observations being as follows:

	Hour. A.M.	Delcros' barom. Millimr.	+0·34 cor. for general error.	Red. to Eng. inches.	Red to temp. 32°	Att. Therm. Centr.	Att. Therm. Fah.	Det. Th.
	h. m.							
Six feet above Clearwater River.........	4 0	72·719	72·753	28·644	28·606	6·4	43·5	40·8
Two feet above Cockscomb...............	4 46	71·079	71·113	27·998	27·944	10·2	50·4	50·9
Six feet above Clearwater River.........	5 20	72·740	72·774	28·652	28·591	11·4	52 5	51·0

These furnish two sets for calculation,

The first giving a height of.................. 640 feet.
And the second of......................... 632 "
The aneroid barometer in 1848, gave.......... 631 "

Mean.......... 634 "

Sir Alexander Mackenzie estimated this declivity at 1000 feet, Lieutenant Hood at 900 feet, both judging merely from the eye and time employed in its descent.

mates of the rate of descent of Mackenzie River and its feeders, I am inclined, independent of his calculations, to consider the Clear-water River at Methy Portage, to be nine hundred feet above the sea, which accords well with his conclusions; since the difference of level between Methy Lake and Clear-water River being five hundred and ninety feet by my barometrical observations, the latter would be nine hundred and fifty feet above the sea by his data.

On the 3d of July, the whole of the baggage and the boats were brought to the banks of the Little Lake; and on the 6th, every thing having been carried over to Clear-water River on the preceding evening, we descended from the Cockscomb, where we had remained encamped for two days, that we might avoid the musquitoes which infested the low grounds. While the boats were loading, we took leave of our canoe-men, who returned to Canada, and at half-past eight, A.M., we pushed off.

The portage occupied nine days from the time of Mr. Bell's arrival; but, with the assistance of horses, we could have passed it easily in three, and saved nearly a week of summer weather, most important for our future operations, besides husbanding the strength of the men. The transport of the four boats, being made on the men's shoulders, employed two days and a half of our time.

CHAPTER IV.

Clear-water River.—Valley of the Washakummow.—Portages.—Limestone Cliffs. —Shale.—Elk, or Athabasca River.—Wapiti.—Devonian Strata.—Geological structure of the banks of the River.—Athabasca Lake, or Lake of the Hills.—Meet Mr. M'Pherson with the Mackenzie River Brigade.—Send home Letters. —L'Esperance's Brigade.—Fort Chepewyan.—Height of Lake Athabasca above the Sea.—Rocks.—Plumbago.—Forest Scenery.—Slave River.—Reindeer Islands.—Portages.—Native Remedies.—Separate from Mr. Bell and his Party.

It is probable that the sands of this district and the adjacent limestones, belong to the Erie division of the New York system of rocks, considered by the United States geologists to be an upper member of the silurian system, but, by various English naturalists, to be rather part of the Devonian, or of the carboniferous series.

The valley of the Clear-water River, or Washakummow, as it is termed by the Crees, is not excelled, or indeed equaled, by any that I have seen in America for beauty; and the reader may obtain a correct notion of its general character by turning to an engraving in the narrative of Sir John Franklin's second Overland Journey, executed from a drawing of Sir George Back's. The view from the Cockscomb extends thirty or forty miles, and discloses, in beautiful perspective, a succession of steep, well-wooded ridges descending on each side from the lofty brows of the valley to the borders of the clear stream which meanders along the bottom. Cliffs of light-colored sand occasionally show themselves, and near the water limestone rocks are almost every where discoverable. The *Pinus banksiana* occupies most of the dry sandy levels; the white spruce, balsam fir, larch, poplar, and birch are also abundant; and, among the shrubs, the *Amelanchier*, several cherries, the silver-foliaged *Eleagnus argentea*, and rusty-leaved *Hippophäe canadensis* are the most conspicuous.

At the portage, the immediate borders of the stream are formed of alluvial sand; but six or seven miles below, limestone in thin, slaty beds crops out on both sides of the river, and, to the left, forms cliffs twenty feet high. A short way further down an isolated pillar of limestone in the same thin layers rises out of the water; and soon after passing it, we come to the White Mud Portage (*Portage de Terre blanche*), of six hundred and seventy paces, where the stream flows over beds of an impure siliceous limestone, in some parts meriting the appellation of a calcareous sandstone, and, for the most part, having a yellowish-gray color. On the portage, and on the neighboring islands and flats, the limestone stands up in mural precipices and thin partitions, like the walls of a ruined city; and the beholder can not help believing that the rock once formed a barrier at this strait, when the upper part of the river must have been one long lake. The steep sandy slopes, as they project from the high sides of the valley, appear as if they had not only been sculptured by torrents of melted snow pouring down from the plateau above in more recent times, but that they had been previously subject to the currents and eddies of a lake. If such was the case, we must admit that other barriers further down were also then or subsequently carried away, as the sides of the valley retain their peculiar forms nearly to the junction of the stream with the Elk River. I have been informed that the country extending from the high bank of the river toward Athabasca Lake is a wooded, sandy plain, abounding in bison and other game.

In the evening we encamped on the Pine Portage (*Portage des Pins*), which is one thousand paces long. The name would indicate that the *Pinus resinosa* grows there; but, if so, I did not observe it, the chief tree near the path being the *Pinus banksiana*, named *Cyprès* by the voyagers. A very dwarf cherry grows at the same place; it resembles a decumbent willow, and is probably the *Cerasus pumila* of Michaux. This is the most northern locality in which it, and the *Hudsonia ericoides*, which was flowering freely at this time, were observed. The *Lonicera parviflora* was also showing a profusion of fragrant, rich, yellow flowers, tinged with red on the ends of the petals, especially before they expand; and on this day we gathered ripe strawberries for the first time in the season.

July 7th.—The Pine Portage was completed in the morning, and an hour later we crossed the Bigstone Portage of six hundred paces. Afterward we passed the Nurse Portage (*Portage de Bonne*), of two thousand six hundred and ten paces; the Cascade Portage, of one thousand three hundred and eighty; and encamped on the Portage of the Woods, two thousand three hundred and fifty paces long; where two of our boats were broken. At this place, and on many other parts of the river, smooth granite boulders line the beach. The strata *in situ* are limestone covered by thick beds of sand.

July 8th.—The boats having been repaired early in the morning, we embarked at half-past six, and at eight came to a sulphureous spring, which issues from the limestone on the bank of the river. Its channel is lined with a snow-white incrustation, the taste of the water is moderately saline and sulphureous, and, from its coolness, rather agreeable than otherwise: it had a slight odor of sulphureted hydrogen. Here I obtained specimens of a terrebratulite (*T. reticularis*).

In the afternoon we passed the mouth of an affluent named the *Pembina*, from the occurrence of the *Viburnum edule* on its banks. I did not observe the fruit of this bush further north than Winipeg River, but I was assured that it grew in various localities up to the Clear-water, beyond which it has not been detected. It is distinguished as a species from the very common cranberry tree, or mooseberry (*Mongsöa meena* of the Crees), by the obtuse sinuses of its leaves; and its fruit has an orange color, is less acid, more fleshy, and more agreeable to the taste. There is a rapid in Clear-water River just above the Pembina, where a section of the north bank is exposed; and I regretted that I had not leisure to examine it. As seen from the boat in passing, it appeared to be formed of sandstone at the base, then of sand, and high up of shale or sandstone in thin layers. Three miles further down a cliff on the south side, about twenty feet high, is composed of an impure limestone, in very thin layers, capped by a more compact, cream-yellow limestone. The sun was intensely hot this day, and, dreading the musquitoes, we avoided the bushy banks of the river, and encamped on an open sand-flat, but did not thereby gain immunity, for we were assailed by myriads during the whole night, a heavy rain having driven them into the tents. The species that now infested us had a light brown color. Each kind remains in force a fortnight or three weeks, and is succeeded by another more bitter than itself.

The Dog-bane and Indian hemp (*Apocynum androsæmifolium* and *hypericifolium*) grow luxuriantly on the sandy banks of this river. They abound in a milky juice, which, when applied to the skin, produces a troublesome eruption. The voyagers, by lying down incautiously among these plants at night, or walking among them with naked legs, often suffer from the irritation, which resembles flea-bites; hence they designate the plant *herb à la puce.* The second-named species grows more robustly and erectly than the other, and furnishes the natives living on the coast of the Pacific with hemp, out of which they form strong and durable fishing nets.

July 9th.—Three miles below our last night's encampment we entered the Elk or Athabasca River, a majestic stream, between a quarter and half a mile wide, with a considerable current, but without rapids.

The lower point of the bank of the Clear-water, where it loses itself in the Elk River, is formed of limestone strata, covered by a thick deposit of bituminous shale, which is probably to be referred to the Marcellus shale of the United States geologists.* The shelving cliff of this shale is one hundred and fifty feet high or upward, and is capped by sand or diluvium. The high cliffs extend for two or three miles up the Clear-water River, above which the sandy slopes for the most part conceal the strata, except at the water's edge, where the limestone crops out. Much of this limestone has a concretionary structure, and easily breaks down. Other beds are more compact.

The same kind of limestone forms the banks of Elk or Athabasca River for thirty-six miles downward, to the site of Berens' Fort, now abandoned. The beds vary in structure, the concretionary form rather prevailing, though some layers are more homogeneous, and others are stained with bitumen. The strata for the most part lie evenly, and have a slight dip, but in several places they are undulated, and in one or two localities dislocated, though I did not observe any dykes or intruding masses of trap rock.

Among the organic remains obtained from the beds of limestone at the water's edge, were *Producti, Spirifers,* an *Orthis* resembling *resupinata, Terebratula reticularis,* and a *Pleurotomaria,* which, in the opinion of Mr. Woodward of the British Museum, who kindly examined the specimens, are characteristic of Devonian strata. In the following season, Mr. Rae picked up from the beach of Clear-water River a fine *Rhynchonella,* which

* See Appendix for a classification of the rocks of the New York system. The Marcellus shale belongs to the Erie division.

retained chestnut-colored bands on the shell. The occurrence of colors in fossil shells of so ancient an epoch is very rare. The specimen has been deposited in the British Museum. In one of the cliffs not far below the Clear-water River, the indurated arenaceous beds resting on the limestone contain pretty thick layers of lignite, much impregnated with bitumen, which has been ascertained by Mr. Bowerbank to be of coniferous origin, though he could not determine the genus of the wood.

Fourteen or fifteen miles below the junction of the Clear-water with the Elk River there are copious springs on the right bank. They rise from the summit of an eminence among the fragments of a ruined shale bank, which they have wholly incrusted with tufa. This incrustation, analyzed for me by Dr. Fife in 1823, was found to be composed principally of sulphate of lime with a slight admixture of sulphate of magnesia and muriate of soda, and with sulphur and iron. Below this there is a fine section of a bituminous cliff from one hundred and twenty to one hundred and thirty feet high, resting on limestone whose beds are undulated in two directions. The limestone is immediately covered by a thin stratum of a yellowish-white earth, which, from the fineness of its grain, appears at first sight to be a marl or clay. It does not, however, effervesce with acids, is harsh and meagre, and, when examined with the microscope, is seen to be chiefly composed of minute fragments of translucent quartz, with a grayish basis in form of an impalpable powder. This seam follows the undulations of the limestone; but the beds of the superincumbent bituminous shale, or rather of sand charged with slaggy mineral pitch are horizontal.

About thirty miles below the Clear-water River, the limestone beds are covered by a bituminous deposit upward of one hundred feet thick, whose lower member is a conglomerate, having an earthy basis much stained with iron and colored by bitumen. Many small grains and angular fragments of transparent and translucent quartz compose a large part of the conglomerate, which also contains water-worn pebbles of white, green, and otherwise colored quartz, from a minute size up to that of a hen's egg, or larger. Pieces of greenstone, and nodules of clay-ironstone, also enter into the composition of this rock, which, in some places, is rather friable, in others, possesses much hardness and tenacity. Some of the beds above this stone are nearly plastic, from the quantity of mineral pitch they contain. Roots of living trees and herbaceous plants push themselves deep into beds highly impregnated with bitumen; and the forest where that mineral is most abundant does not suffer in its growth.

The shale banks are discontinued for a space in the neighborhood of Berens' House, where thin beds of limestone come to the surface, and form cliffs twenty or thirty feet high at the water's edge.

Further down the river still, or about three miles below the Red River, where there was once a trading establishment, now remembered as *La vieux Fort de la Rivière Rouge*, a copious spring of mineral pitch issues from a crevice in a cliff composed of sand and bitumen. It lies a few hundred yards back from the river in the middle of a thick wood. Several small birds were found suffocated in the pitch.

Soon after passing this spot, we saw right ahead, but on the left bank of the river, a ridge of land named the "Bark Mountain," looking blue in the distance, being fully sixty miles off. From its name, I conclude that the canoe birch abounds on it. It is the length of a spring day's march, or about thirty miles, distant from Fort Chepewyan; and bison, moose deer, and other game, are said to resort to it in numbers.

At the deserted post named *Pierre au Calumet*, cream-colored and white limestone cliffs are covered by thick beds of bituminous sand. Below this there is a bituminous cliff, in the middle of which lies a thick bed of the same white earth which I had seen higher up the river in contact with the limestone, and following the undulations of its surface.

A few miles further on, the cliffs for some distance are sandy, and the different beds contain variable quantities of bitumen. Some of the lower layers were so full of that mineral as to soften in the hand, while the upper strata, containing less, were so cemented by iron as to form a firm dark-brown sandstone of much hardness. The cliff is, in most places, capped by sand containing boulders of limestone. One very bituminous bed, carefully examined with the microscope, was found to consist, in addition to the bitumen, of small grains of transparent quartz, unmixed with other rock, but inclosing a few minute fragments of the pearly lining of a shell. A similar bed in another locality contained, besides the quartz, many scales of mica. The whole country for many miles is so full of bitumen that it flows readily into a pit dug a few feet below the surface.

In no place did I observe the limestone alternating with these sandy bituminous beds, but in several localities it is itself highly bituminous, contains shells filled with that mineral, and when struck yields the odor of *Stinkstein.* It is probable that the whole belongs to the same formation, but I do not possess evidènce of the facts to satisfy a geologist.

The rate of our descent of the Elk River must this day have exceeded six geographical miles an hour, indicating a strong current. This river, named also the Athabasca or *Rivière la Biche,* rises in the parallel of 47½° north latitude, near the foot of Mount Brown, a peak of the Rocky Mountains, having a height of sixteen thousand feet above the sea. Its course in a straight line to the influx of Clear-water River is three hundred miles; but the river course, including its windings, must be more than one-third greater. The elevation of its sources is probably seven or eight thousand feet. Lesser Slave Lake, situated about midway between its origin and the junction above mentioned, lies, according to Captain Lefroy, eighteen hundred feet above the sea. Some of the feeders of the Oregon spring from very near the head of the Athabasca, and many tributaries of the Saskatchewan arise not far to the southward. It is the most southern branch of the Mackenzie; and as it originates further from the mouth of that great river than any other affluent, it may be considered as its source. It flows partly through prairie lands, and its Canadian appellation of *Rivière la Biche* indicates that the American red-deer, or Wapiti, frequents its banks. Its English name of Elk River, having reference to the moose deer, is a mistranslation of the Canadian one, and is also inappropriate as a distinctive epithet, though the moose grazes on its banks, as well as on the Mackenzie, down to the sea. The *Wapiti* is not known on Slave River or Lake, but further to the west it ranges as far north as the east branch of the River of the Mountains near the 59th parallel, where Mr. Murdoch M'Pherson informs me that he has partaken of its flesh. From the Saskatchewan and Lesser Slave Lake the country can be traversed by horsemen who are sufficiently acquainted with the district to avoid the deep ravines through which the streams flow. By this route a band of horses were brought to Methy Portage in August, 1848, though they were too much exhausted by their journey to be of service. In 1849 a fine body of upward of forty horses came to the portage from Lesser Slave Lake, early in the season and in good condition.

July 10*th*.—Our voyage this morning was impeded by a strong head-wind, followed by heavy rain, which compelled us to put ashore for four or five hours. We were able to resume our route at 10 A.M., and at noon we came to high sandy banks named *Les Ecores*, resembling the sandy deposits on the Clear-water River. These continue down to the alluvial delta formed by the four or five branches into which the river splits before entering the Athabasca Lake, or Lake of the Hills.

At 5, P.M., we arrived at the head of this delta, and, passing down the main channel, held on our way till 8 o'clock, when we landed to cook supper, and then re-embarked to drift with the current during the night, the crews, with the exception of the steersmen, going to sleep in the boats.

July 11*th*.—We entered Athabasca Lake at three in the morning, but found, to our mortification, that two of the boats, through the inattention of the steersmen, had taken a more easterly branch of the river in the night, which would delay their arrival at Fort Chepewyan for some hours, and consequently be the means of detaining us for that time.

Immediately on emerging from the river we saw the Mackenzie River brigade of boats crossing the lake toward the entrance of the Embarras River, lying four or five miles to the westward of the branch we had descended. On our firing guns and hoisting the sails and ensigns, we were perceived by the officer in charge of the brigade, Chief Factor Murdoch M'Pherson, who waited till we joined him. From this gentleman I received much useful intelligence of the measures he had taken for supplying the expedition with provisions during our winter residence in Fort Confidence, at the north end of Great Bear Lake, and also a list of all the provisions and stores remaining at Fort Simpson, the Company's chief post and dépôt on the Mackenzie; and I have pleasure in acknowledging here, that I am indebted to him for much valuable assistance, as well as for very many acts of personal kindness.

To him we committed the last letters that we could send to our families and friends in Europe this year. I had sent dispatches to the Admiralty from Methy Portage, not being sure that we should meet the Mackenzie River brigade, which is the latest that goes out. It can seldom cross Great Slave Lake before the end of June, and from twenty to twenty-four days are required for the passage of loaded boats from thence to Methy Portage. There the Mackenzie River party are met by a brigade from York Factory, which brings up goods for next year's supply of the northern posts, and takes back the furs brought from the Mackenzie. There is just time in common seasons for that brigade to descend to York Factory before the annual ship sails from thence for England, about the middle of September, or in backward seasons a week or two later; and afterward to return to the colony at Red River, where the crews reside, and from whence they come annually in the spring on this special service. For many years the Methy Portage brigade has been conducted by a guide named L'Esperance, and on that account it is known by the name of L'Esperance's brigade.

After the return of the Mackenzie River boats to Fort Simpson, the winter's supply of goods has to be sent to the outposts; but as some of these are at a distance of four or five weeks' traveling, the parties carrying them are not unfrequently arrested by frost, far from their destination, and the posts suffer severely—sometimes to the length of actual starvation and loss of life; an instance of which occurred before I left the country.

We reached Fort Chepewyan at half-past 7 A.M., but the two boats that strayed from us did not arrive till the afternoon, and the chief artisans being in the missing boats, the intention I had of giving them a complete repair here, and putting on false keels, was frustrated. Their leaks were, however, stopped, and some planks replaced, which detained us till 11 A.M., on July the 12th, when we left the fort.

The height of Lake Athabasca above the sea is estimated by Captain Lefroy at six hundred feet.* Its basin offers another instance of the softer strata having been swept away at the line of their junction with the primitive rocks; and a reference to the map will show that there must have been an evident connection between the cause of this excavation and that of Wollaston and

* Eight months of observations with the boiling-water thermometer by this officer, give an elevation of 468 feet, excluding two observations on which he could not rely. This being, however, in his opinion too low, he assigns the altitude mentioned in the text, after a review of his entire body of observations in various parts of the country, and checking one by another.

Deer's Lakes, belonging to the Missinipi River system.* Wollaston Lake is said to supply a river at one end, which falls into Athabasca Lake, and one at the other, which joins the Missinipi, which, if correct, is not a common occurrence in hydrography, though one or two instances of the kind, in seasons of flood, have been alluded to in the preceding pages.

Much of the country in the immediate vicinity of Chepewyan is composed of rounded knolls of granite, nearly destitute of soil, and many of them smooth and polished. These rocks extend along the north shore of the lake; and the eminences rise in the interior in a confused manner, one over the other, to the height of four or five or perhaps six hundred feet above the water. They also form many islands at the west end of the lake and in front of the fort. Between this end of the lake and the mouth of Peace River there lies a muddy expanse of water, named Lake Mamawee; and, in times of flood, the waters of Peace River flow by this channel into Athabasca Lake, rendering its usually transparent waters very turbid. A short way to the eastward of the fort a gray gneiss rock is associated with reddish granite, and its beds are much contorted, and are traversed by veins of vitreous quartz. Still further off in that direction a cliff of chlorite slate occurs. Plumbago of excellent quality has been found on the shores of this lake, and I have been informed that at its eastern extremity, named the *Fon du Lac*, there is much sandstone—the resemblance to the succession of strata on Lake Superior being maintained here also. Granite rocks, generally forming rounded knolls, prevail in Stony River, by which name the discharge of Athabasca Lake is known, and on whose banks we encamped on the evening of the 12th.

Soon after starting on the morning of the 13th, we passed the mouth of Peace River, or Unjugah, which is the largest branch of the Mackenzie, since it brings down more water than either the Athabasca River or River of the Mountains. When it is flooded it overcomes the stream of Stony River, and carries its muddy waters into Lake Athabasca, meeting there another rush of waters coming through Lake Mamawee; but at other times there is a strong current in Stony River, and at one point a dangerous rapid, where a gentleman of the Northwest Company was drowned many years ago. A delta, intersected by several chan-

* See Appendix.

nels, exists at the junction of Peace River with Athabasca Lake and its outlet. The source of Finlay's branch of this river is nearly in the same parallel with its mouth, but in its course the trunk of the river makes a great curve to the southward, and its southern tributaries rise in the same mountains from which Frazer River issues on the west side of the Rocky Mountains, the upper waters of the Peace River coming in fact through a gap in the chain which forms one of the passes leading to the Pacific coast. Captain Lefroy, who has traveled through this district, makes the following remarks upon its elevation. "The next series of observations was made in the elevated region at the base of the Rocky Mountains, between Peace River and the Saskatchewan, a district remarkable for its gradual and regular ascent, preserving throughout much of the character of a plain country. From Lake Athabasca to Dunvegan, a distance of about six hundred and fifty miles" (250 geographical miles in a straight line) "there occurs but one inconsiderable fall, and a few rapids; the bed of the Peace River preserves a nearly uniform inclination, in which it rises three hundred and ten feet. The stream is, however, more rapid above Fort Vermilion than below it. The depth of the bed of the river below the surrounding country increases with great uniformity as we ascend the river. A defile, very similar to that called the Ramparts on Mackenzie's River, but on a finer scale and with far more picturesque features, occurs about eight miles above the river Cadotte, in long. 117°; and here the river has cut a passage through cliffs of alternating sandstone and limestone to a bed of shale, through which it flows, at a depth of two hundred feet (by estimation) below their summit. The general elevation of the country, however, still continues to increase, and at Dunvegan, it is six hundred feet above the bed of the stream; yet even at this point, except on approaching the deep gorges through which the tributaries of Peace River join its waters, there is little indication of an elevated country; the Rocky Mountains are not visible, and no range of hills meets the eye. A rough trigonometrical measurement gave five hundred and thirty-eight feet as the elevation of Gros Cap, a bold hill behind Dunvegan, above the bed of the river; and the ground was estimated to rise behind Gros Cap, by a gradual ascent, until it attains the general level." (*Lefroy*, l. c.) The elevation above the sea, that this intelligent officer assigns to the country about Dunvegan is sixteen hundred feet, and the region in which the sources of the river occur is probably four times as high.

The oaks, the elms, the ashes, the Weymouth pine, and pitch pine, which reach the Sackatchewan basin, are wanting here, and the balsam-fir is rare; but as these trees form no prominent feature of the landscape in the former quarter, no marked change in the woodland scenery takes place in any part of the Mackenzie River district until we approach the shores of the Arctic Sea. The white spruce continues to be the predominating tree in dry soils whether rich or poor; the Banksian pine occupies a few sandy spots; the black spruce skirts the marshes; and the balsam-poplar and aspen fringe the streams; the latter also springs up in places where the white spruce has been destroyed by fire. The canoe-birch becomes less abundant, is found chiefly in rocky districts, and is very scarce north of the arctic circle. It still, however, attains a good size in the sheltered valleys of the Rocky Mountains, up to the 65th parallel. Willows, dwarf birches, alders, roses, brambles, gooseberries, white cornel, and mooseberry, form the underwood on the margins of the forest; but there is no substitute for the heath, gorse, and broom, which render the English wild grounds so gay. On the barren lands, indeed, the heath has representatives in the Lapland rhododendron, the *Azalea*, *Kalmia*, and *Andromeda tetragona*, but these are almost buried among the *Cornicularia* and *Cetraria nivalis* of the dry spots, or the *Cetraria islandica* and mosses of the moister places, and scarcely enrich the colors of the distant hills.

The granite knolls show themselves at frequent intervals on the banks of Slave River, which is the appellation of the stream formed by the junction of the Peace and Stony Rivers; and in several places, ledges of the rock crossing the river form rapids. One of these is named the Lightening Place of the Hummock, because it occurs at the beginning of a reach two miles long, which is terminated by a sandy bluff on the right bank, twenty or thirty feet high, and covered with Banksian pine. This *Bute*, as it is termed by the Canadian voyagers, is about thirty miles from Fort Chepewyan, and opposite to it there is a limestone cliff, constructed of thin undulated layers. The lower beds of the limestone have a compact structure, a flat conchoidal fracture, and a yellowish gray color. Some of the upper beds contain ineral pitch in fissures, and shells, which Mr. Sowerby in 1827, ascertained to be *Spirifer acuta*, and several new *Terebratulæ* one of them resembling *T. resupinata;* associated with them a *Cirrus* and some crinoidal remains occur. Not far above this cliff, a vitreous reddish-colored sienite protrudes; and half a mile or so below it, the stream passes between rounded hummocks of granite, one of which forms an island, the water-course evidently following the line of dislocation of the strata. The clustered nests of large colonies of the republican swallow (*Hirundo fulva*) adhere to the ledges of the limestone cliffs, and the bank swallow (*Hirundo riparia*) has pierced innumerable holes in the sandy brows.

A small tributary enters the river from the left, behind an island, lying a short way below the *Bute*, and another comes in from the right, beneath which the brown vitreous sienite re-appears, forming a flatly-rounded eminence. Within a mile of this pyrogenous rock, another limestone cliff occurs on the left bank, at the commencement of a pathway which leads over prairie-lands, or through spruce-fir woods, marshes, and by small lakes, to the Salt River, to be hereafter noticed.

A mile and a half below this are the three Rocky Islands (*Isles des Pierres*), which is perhaps the best locality on the river for studying the connection of the limestone with the pyrogenous rocks; and I regretted that I could devote no time to this purpose. The beds of limestone, as seen in passing rapidly along these islands, appeared of various thickness, some being thin and shaly, and almost all more or less undulated, saddle-formed or contorted. On the borders of a channel between two of the islands, a conglomerate is interleaved with sienite; and in the vicinity there are beds of a brownish, finely crystalline limestone, having a conchoidal fracture, the fragments being sharply angular. The conglomerate varies considerably in its texture in different parts of the same bed. It contains, in general, a large proportion of small rounded grains of translucent or milky quartz, with angular fragments of various sizes of vitreous quartz, chlorite-slate, and calc-spar, imbedded in a powdery or friable white basis, which does not effervesce with acids; the whole forming a tough stone. In some beds the quartz grains predominate, so as to render the rock a coarse sandstone; but in other parts, these grains appear to have been fused into a bluish quartz rock, the original granular structure being only faintly discernible, and to be detected chiefly in spots, where some of the powdery basis remains unchanged. In one bed angular fragments of greenstone incrusted with calc-spar occur. The sienite contains grains of hornblende and quartz in about equal quantities, imbedded in a snow-white powdery basis, which appears to be disintegrated felspar.

A mile below the Stony Islands we passed the smaller Balsam Fir Island, below which there is a pretty little *bute* on the left, where the purplish-colored rock that protruded appeared to us in passing to be amygdaloid or porphyritic trap rock. Some miles further down we entered among the rather high and rocky cluster of the Reindeer Islands (*Isles de Carrèbœuf*) by a channel having a north-northwest direction. The rocks here appeared to us as we shot past them, to be principally trap, associated with gneiss, or perhaps chlorite-slate. A point on the main shore, on which I landed in 1820, is composed of felspar and quartz, and is probably a variety of granite.

A short way further down the Great Fir Balsam Island (*La grand Isle des Epinettes*), which is a mile across and three or four long, has a triangular form, and divides the river into two channels. We descended the easternmost, or right-hand one, which is the most direct, and has a high and sandy eastern bank.

Below this a bend of the river is filled with many rocky islands. occasioning numerous rapids and cascades, and seven or eight portages. The river expands here to the width of a mile and a half or two miles; its bed is every where rocky, and the rocks are apparently all primitive; but as the boat-route lies wholly through the eastern channels, we had no opportunity of inspecting the opposite shore closely. The islands are well wooded, and the scenery picturesque. Some of the narrower channels, which would be convenient for the descent of boats, are blocked up by immense rafts of drift timber, which have been accumulating for many years, and which could not be set free without very great and long-continued labor. Large flocks of pelicans have made their nests on the more inaccessible rocks rising from the brows of the cascades. In the evening we ran down the Dog Rapid after lightening the boats, and afterward descended a second rapid, and then encamped on a smooth granite rock early in the evening, there not being time to complete the Chest Portage before dark.

Embarking at 3 A.M. on the 22d of July, we descended a narrow channel to the Chest Portage (*Portage de Cassette*), where our five boats were hauled over a pathway of four hundred and sixty-five paces, and their cargoes carried. A rocky chasm at this place, being one of the numerous channels through which the water flows, incloses a perpendicular cascade upward of twenty feet high; beneath which an isolated column of rock divides the current into two branches, which eddy with great force into the niches and recesses of the stony walls. Huge angular blocks obstruct the water-course, and drift trees, entangled among them, partially denuded of their branches, and wholly of their bark, point in all directions. The overhanging woods almost seclude this gloomy ravine from the sun; and it presents such an aspect of wildness and ruin as rarely occurs even in this country. In one part of the portage road a bed of gneiss is flanked on each side by masses of granite. A labyrinth of passages among granite rocks exists below the portage, many of them entirely choked up with drift timber. In passing rapidly through one of them we grazed a point composed of a crumbling red and gray porphyritic rock, perhaps an amygdaloid; many cubical and irregularly angular fragments had fallen from it.

At the Island Portage, which immediately followed, the cargo is carried only in the ascent of the river. Our boats descended the fall with their entire load. We next crossed the Raft Portage (*Portage d'Embarras*), which occupied us three hours. At the Little Rock Portage, which follows, the rock is composed of felspar, quartz, and chlorite, being the *protogine* of Jurine. It differs from the slaty rock observed near the Reindeer Islands, in not being stratified. At the Burnt Portage, the next in order, the rock, which is a porphyritic granite, acquires a polished glistening surface. There is a cascade here of fifteen or twenty feet. The succeeding portage, named the "Mountain," from the steep bank down which the boats are lowered, is shorter than the others, being only one hundred and seventy paces across. The rock at this place is a red, compact, shining or vitreous-looking granitic porphyry, much fissured, and breaking, by the action of the frost, into cubical or rhomboidal blocks, sometimes of great size. The principal fissures are generally, but not always, parallel to each other, and may be traced for seventy or eighty yards without a break, in a transverse direction to that of the eminences and projecting tongues of the rock. Their course is northeast by north, and southwest by south; and they are, for the most part, four or five feet apart. The minor cracks meet the chief ones at various oblique angles, and sometimes cross them, but not generally. At another denuded point of rock, the wider cracks crossed each other, one set running east-southeast and west-northwest. The recesses left by the blocks which fall away retain their sharp-cornered rectangular shape. A layer of hornblende-slate or basalt shows itself at one spot.

The launching-place for the boats here is both steep and rugged; and a brigade seldom passes without some of the boats being broken. One of ours was injured; but, being soon repaired, we left the portage by six in the evening, and encamped for the night at the south end of the Pelican Portage, which is seven hundred paces long.

The power of the sun, this day, in a cloudless sky, was so great, that Mr. Rae and I were glad to take shelter in the water while the crews were engaged on the portages. The irritability of the human frame is either greater in these northern latitudes, or the sun, notwithstanding its obliquity, acts more powerfully upon it than near the equator; for I have never felt its direct rays so oppressive within the tropics as I have experienced them to be on some occasions in the high latitudes. The luxury of bathing at such times is not without alloy; for, if you choose the mid-day, you are assailed in the water by the *Tabani*, who draw blood in an instant with their formidable lancets; and if you select the morning or evening, then clouds of thirsty musquitoes, hovering around, fasten on the first part that emerges. Leeches also infest the still waters, and are prompt in their aggressions.

The *Geum strictum* grows plentifully on these portages, and is used by the natives for the purpose of increasing the growth of their hair. They dry the flowers in the sun, powder them, and mix them with bear's grease. The *Eleagnus argenta*, which is also abundant on the banks, is named by the Chepewyans *Tàp-pah*, or gray berry. It is the bear-berry of the Crees, and the stinking willow of the traders; so called, because its bark has a disagreeable smell.

July 15th.—The portage was completed, and breakfast prepared and eaten, in five hours and a half. At the lower end of the path, a sienitc rock, composed of crystallized quartz, aurora-red felspar, and greenish-black hornblende, yields large cubical blocks of a handsome stone. One of the small boats was overset in lowering it down a narrow channel, and the oars, a coil of rope, and the boat-lockers were swept away by the current. A boat's anchor, and some clothes belonging to two of the crew, were in one of the lockers.

An hour before noon we had crossed the Portage of the Drowned (*Portage des Noyes*), where granite is the prevailing rock. This being the last of the portages, three of the small boats brought from England were stowed with pemican for the sea-voyage; and Mr. Bell was left to follow with the large boat and the fourth small boat, containing the stores for house-building, nets, ammunition, and other supplies for winter use. He would have accompanied us; but his men had to make oars in place of those which had been lost; an employment which was likely to occupy them for two or three hours.

CHAPTER V.

Pyrogenous Rocks.—Rate of the Current of Slave River.—Salt River and Springs. —Geese.—Great Slave Lake.—Domestic Cattle.—Deadman's Islands.—Horn Mountain.—Hay River.—Alluvial Lignite beds.—Mackenzie's River.—Marcellus Shale.—Fort Simpson.—River of the Mountains.—Rocky Mountains.—Spurs.—Animals.—Affluents of the Mackenzie.—Cheta-ut-tinne.

No primitive rocks were seen on the route down the Mackenzie, on this voyage, after leaving the Portage of the Drowned ; but in 1820, when we crossed Great Slave Lake, near the 113th meridian, we traced the western boundary of these rocks, from near the mouth of Slave River, northward by the Rein-deer Islands to the north side of the lake, and continued to travel within their limits up to Point Lake in the 66th parallel. The western edge of the formation was afterward found at the northeast and eastern arms of Great Bear Lake.

The district intervening between the granite at the Portage of the Drowned and the Salt River is flat, with sandy terraces and slopes rising from the river to the height of from twenty to eighty feet, there being in some places two in others three or more such terraces, while in others the river has made a section of the sandy deposit, and formed a high and steep cliff. The valley of the river, deflected to the westward by the rocks of the portages, passes here through the more level (upper ?) silurian strata.

At Gravel Point (*Pointe de Gravoir*), ten miles from the portages, a bed of concretionary or brecciated limestone protrudes from under a sand-bank forty feet high, and two miles higher up a cliff of cream-colored and brownish limestone stands on the right bank. The country on both sides of the river there appears to be a plain, which has a general level of about fifty feet above the bed of the stream.

Just before arriving off the mouth of Salt River, we picked up one of the boat's lockers containing the anchor, which had been carried away fifteen miles higher up nearly eight hours before, so that it drifted about two miles an hour, including the time it might have been detained in eddies.

In 1820, I ascended the very tortuous Salt River, for twenty miles, for the purpcse of visiting the salt springs, which give it its name. Seven or eight copious springs issue from the base of a long even ridge, some hundreds of feet high, and, spreading their waters over a clayey plain, deposit much pure common salt in large cubical crystals. The *mother water*, flowing off in small rivulets into the Salt River, communicates to it a very bitter taste; but before the united streams join the Slave River, the accession of various fresh-water rivulets dilutes the water so much that it remains only slightly brackish. A few slabs of grayish compact gypsum protrude from the side of the ridge above mentioned, and a pure white gypsum is said to be found at Peace Point on Peace River, distant about sixty or seventy miles in a south-southwest direction, whence we may conjecture that this formation extends so far. From the circumstance that the few fossils gathered from the limestone on Slave River are silurian, I venture to conjecture that these springs may belong to the Onondaga salt group of the Helderberg division of the New York system. The Athabasca and Mackenzie River districts are supplied from hence with abundance of good salt. We obtained some bags of this useful article from Beaulieu, who was guide and hunter to Sir John Franklin on his second overland journey, and who has built a house at the mouth of Salt River. This is a well-chosen locality for his residence: his sons procure abundance of deer and bison meat on the salt plains, which these animals frequent in numbers, from their predilection for that mineral; and Slave River yields plenty of good fish at certain seasons. It is the most southern locality to which the *Inconnu* or *Salmo Mackenzii* comes, on this side of the Rocky Mountains, as it is unable to ascend the cascades in the Slave River. The *Coregoni* are the staple fish of the lakes here, as they are elsewhere throughout the country: and there are also pike, burbot, and excellent trout. A limestone cave in the neighborhood, which was too distant for us to visit, supplies Beaulieu with ice all the summer, and he gave us a lump to cool water for drinking, which was extremely grateful. The ammunition and tobacco with which I repaid these civilities were no less acceptable to him. Indeed, I believe that he turns his residence on the boat-route to good account, as few parties pass without giving him a call.

After a short halt, we resumed our voyage until 7 P.M., when we landed to cook supper, after which we re-embarked to eat it: and, having lashed the boats together, drifted down the stream all night, one man being appointed to steer.

July 16th.—Though we lay down in the best manner we could in the boats during the night, the continuous assaults of the musquitoes deprived every one of rest, and rendered us all so feverish, that we were glad when daybreak called the crews to the oars, and the boats acquired motion through the water, by which we obtained some relief.

The sandy banks of the river show sections in many places upward of twenty feet high; and, in almost all, the sand is distinctly stratified; the layers being of different colors, and often having clayey or loamy seams interposed. The whole of the banks, from Salt River downward to Slave Lake, appear to be alluvial; and many small lakes existing behind them communicate with the river by narrow channels. In ordinary seasons at this date, vast numbers of Canada geese moult in the district, and are followed by their young brood not yet fully fledged, which fall a ready prey to the natives or voyagers descending the river. In 1825 I could have filled a boat with these delicate young birds. This year, owing to the high waters, the greater part of the broods had retreated to the lakes, where grass could be more easily procured, and we obtained only a few. The natives observe, that, besides the old birds which rear young, and moult when their offspring are obtaining their plumage, there are a considerable number who do not breed, but keep in small bands, and are called "barren geese." Of these we saw some flocks; but they were not easily approached without a greater loss of time than we could spare.

We kept at the oars all day, except when we landed to breakfast, or to cook supper, and, after sunset, resumed the plan of drifting, with very little better success, as far as sleep was concerned, than on the preceding night. During the day the sun's rays felt intensely warm; and the puffs of northerly wind blew as hot as if they had passed over the deserts of Arabia. At midnight a strong contrary wind springing up, compelled us to anchor until half-past 2 A.M., on the 17th, when we again took to the oars, and entered Great Slave Lake at 7 in the morning.

Like the Athabasca River, the Slave River joins its lake through a delta of low, well-wooded, alluvial islands, by many channels, having a spread of more than twenty miles. Near the easternmost, which is named John's River (*Rivière à Jean*), is Stony Island, a naked mass of granite, rising fifty or sixty feet above the water; and beyond that, to the eastward, the banks of the lake are wholly primitive. In the vicinity of the westernmost channel of the delta, and from thence to the efflux of the Mackenzie, the whole southern shore of the lake is limestone, associated with a bituminous shale, and belonging, as well as can be ascertained from its fossils, to the Erie division of the New York system, which includes the Marcellus shales, and is referred by English geologists to the carboniferous series. In the small channel which divides Moose-deer Island from the point of the bay on which the present Fort Resolution stands, many boulders of porphyritic and common granites, greenstone, and limestone occur; also large angular blocks, not worn or rounded, of a conglomerate of granite, chert, and hornblende rock cemented by a basis of ironstone.

We reached Fort Resolution at 10 A.M.; and having received some supplies of fish, and two or three deals for repairing the boats, we resumed our voyage, after a halt at the fort of one hour. Domestic cattle have been introduced at this place, and at the posts generally throughout the country, even up to Peel's River and Fort Good Hope, within the Arctic circle. At this season the musquitoes prevent them from feeding, except when urged by extreme hunger; and fires are made for their accommodation near the forts, to which they crowd, and, lying to leeward amidst the smoke, ruminate at their ease. Smoke is the only remedy against these venomous insects; and at this time of the year, when the heat renders a free circulation of air in the houses essential, the rooms are made comfortable by nailing bunting over the windows, and burning turf or rotten wood in a pan on the threshold of the door. At no place on our route were the musquitoes in denser clouds than this day at Fort Resolution; and we gladly left them behind as we launched into the lake with a favorable breeze. We had not gone above two miles, when we saw Mr. Bell and his boats issuing from behind Moose-deer Island, and steering for the house; but time was too precious for us to wait for his coming up. Our route lay through a small group of islands lying five or six miles off the bay into which Buffalo Creek falls. In these islands, a bituminous limestone crops out in thin horizontal layers near the water's edge; but, except in a few places, its beds are concealed under a beach composed of fragments of the same stone, partly rolled and worn, partly with recently broken edges. The islands are most of them low; and the stony beach rising above their centres incloses marshy spots traversed by ridges of sand and gravel, more or less wooded.

At 5 P.M. the wind which had been increasing all the afternoon, rose to a high gale, and we put into a good boat harbor at Deadman's Island and encamped. This spot received its name from a massacre committed by a war party of Beaver Indians, who surprised a body of Dog-ribs encamped there, and destroyed them all. Thirty years ago many of the bones of the victims were to be seen, but they have now disappeared. The influence of the Hudson's Bay Company has put a stop to these war excursions, and tribes formerly the most hostile to each other now meet in amity at the trading posts.

This lake is a breeding station of the *Sterna cayana*. The Arctic tern also hatches on its shores, depositing its eggs among the gravel on the beach. The leaves of the gooseberry bushes had been stripped off by a black-banded caterpillar, and it was evident that the fruit would fail this season. The ice having parted from the shore little more than a fortnight, vegetation was backward. A strong gale, bringing on a keen frost, blew all night, and effectually quelled the musquitoes, so that, though we could not but regret the detention, we all enjoyed some hours of sound repose. But in the morning of July 17th, during one of the squalls of a thunderstorm accompanied by heavy rain, the tent pegs drew from the sandy soil on which we had encamped, and the dripping canvas falling upon us put an end to our rest. We were miserably cold and wet before this mischance was remedied. High winds and a rolling sea kept us stationary all day, and our carpenters took advantage of the delay to secure the thwarts of two of the boats which had given way on the portages.

July 18th.—The gale did not abate so as to allow us to embark until 4 P.M., when we resumed our voyage, and at 9 encamped again in a small boat harbor at Burnt Point. This coast of the lake generally is flat and shelving, and secure landing-places for boats are very scarce. Though we did not discover limestone *in situ* here, the beach is formed of fragments of that stone of very various size, mixed with some bituminous shale, and a few granite boulders. This point is about thirty-five geographical miles from Fort Resolution. In a bay a little to the westward several sulphureous streams issue from a limestone containing corals. The channels of these streams are encrusted with a similar tufa to that observed on Clear-water and Athabasca Rivers, and the organic remains that have been examined indicate the formations to be of the same geological epoch.*

July 19th.—Embarking at three, we passed the mouth of Buffalo Lake River, and after five hours' pulling, put into Canoe or Sandy River to cook breakfast. From this place a rising ground on the north side of the lake is distinctly visible, the distance being about thirty miles or more. The Horn Mountain, an even ridge more to the westward, appeared also in the extreme distance, being at least sixty miles off.

Hay River enters the lake at the distance of eleven or twelve miles from Canoe River. It is formed of two branches, the westernmost of which rises from Hay Lake and the other one originates not far from the banks of Peace River, and flows past Fort Vermilion. Hay River Fort, now abandoned, stood at the junction of the two. On the eastern branch, the country is an agreeable mixture of prairie and woodland, and this is the limit of those vast prairies which extend from New Mexico. Below the forks of Hay River the country is covered with a forest intersected by swamps.

The range of the Wapiti is nearly coincident with the boundaries of these prairies. The bison, though inhabiting the prairies in vast bands, frequents also the wooded country, and once, I be-

* The fragments of black and bituminous shale which strew the beaches of these islands, and which evidently have not traveled far, contain a "pteropodous shell (*Theca*) apparently the *Tentaculites fissurella* of Hall, a *Chonetes*, the *Strophenema setigera* of Hall, and *Avicula lævis* of the same author; at least they are undistinguishable from his figures of these fossils in the Marcellus shale, which according to him is upper silurian, but is probably somewhat newer, and what we call Devonian. Two corals in the associated bituminous limestone are characteristic of the same epoch, namely a *Strombodes* of Hall, having its cysts filled with bitumen, and a *Favosites* very like the common *F. polymorpha* of the Plymouth marbles. I have not identified any of the *Terebratulæ* from Great Slave Lake, but they are certainly either Devonian or carboniferous, and not silurian. There is nothing like a secondary fosil in the collection."—*Woodward* in lit.

lieve, almost all parts of it down to the coasts of the Atlantic; but it had not until lately crossed the Rocky Mountain range, nor is it now known on the Pacific slope, except in a very few places. Its most northern limit is the Horn Mountain mentioned above. The musk-ox does not come to the south of the Arctic circle.

In the evening we pitched our tents on a small island, being one of a group known as Desmarais's Fishery, and from a party of Indians encamped on a neighboring point we obtained a supply of fish.

The whole south shore of the lake, westward from Slave River, is low and level, and is lined in parts, and especially between Hay River and Desmarais's Islands, by rafts of drift-timber, which, pressed on shore by northerly winds, become water-logged and covered with sand mixed with comminuted wood and decaying grasses. As this buried forest accumulates, willows and balsam poplars spring up from its surface, and bind all together by their roots. The swamps that extend backward from the lake appear to have originated very much in this way; and had the locality been one where there was much drift-sand, the erect trees might in parts have been swallowed up and killed by sand-drifts, having their roots in the subjacent lignite or shale. In Slave Lake, however, the sand does not act so conspicuous a part as in Lake Winipeg. The recent deposits conceal the limestone strata, except at a very few places, but the numerous fragments which line the beach show that a bituminous limestone, associated with a black shale having a resinous streak, and a thin marly slate, must exist in the neighborhood. They are referrible, as has been mentioned above, to the Marcellus shale. At the Stony Point, between Hay River and Desmarais's Fishery, fragments of these rocks from the beach, on which some very large boulders of gneiss, sienite, and greenstone also lie. At Desmarais's Fishery I observed the same kind of beach, with the addition of blocks of basalt, of a dull-red sandstone, a coarse conglomerate composed of rounded pieces of sandstone cemented by a basis of red clay strongly impregnated with iron. The limestone fragments contained bivalves and corals. On the 6th of July, in the following year, the whole bay that we had traversed in this day's voyage was filled with ice, not yet parted from the shore; and the lake is scarcely ever navigable in this quarter before the beginning of the month,* so that we were only a fortnight later than we could have hoped to cross the lake, had the boats advanced even to Isle à la Crosse the first season, as they might have done under a very favorable combination of circumstances. But this fortnight, by enabling the expedition to be at the mouth of the Mackenzie on the disruption of the ice of the Arctic Sea, would have been of the very greatest advantage. In fact, a fortnight is no contemptible portion of the six weeks during which the Arctic Sea is navigable for boats. The ice on this lake is sometimes eleven feet thick; at Fort Resolution, and at Big Island, which lies across the western outlet of the lake, it varies from five to seven feet.

July 20th.—This morning we crossed from Desmarais's Fishery obliquely to the north side of the lake, through an archipelago of islets and along the south side of Big Island. There is more or less current in the passages, and from the general shallowness of the water, it is probable that the limestone strata come near the surface, but they are concealed by gravel and boulders. To the south of this traverse, on a strait two miles wide, which separates the site from Big Island, stood formerly Fort George. The limestone beds are said to crop out in its neighborhood.

During the whole summer, in the eddies between the islands of this part of the lake, multitudes of fish may be taken with hooks and by nets, such as trout, white-fish, pike, sucking-carp, and *inconnu.* In spring and autumn wild-fowl may be procured in abundance at several places in the neighborhood, which are their accustomed passes; and the fishery on the north side of Big Island seems to be inexhaustible in the winter. With good fishermen and a proper supply of nets, a large body of men may be wintered here in safety and plenty, and it was to this place that I contemplated conducting any of the crews of the Discovery ships that we might be so happy as to find. To it, also, I purpose to send a large portion of my party in the winter of 1848–9. In no other part of the Hudson's Bay Company's territories, that I am acquainted with, can so many people be maintained, with so much certainty, on the resources of the country. A body of good native hunters, well supplied with ammunition, could not fail to bring from the Horn Mountain an agreeable variety of diet in

* Dean and Simpson made their way through it on the 24th of June; the ice, however, was still adhering to the shore at some points.

form of reindeer and bison meat, and in some seasons the American hare may be snared in great numbers.

After we had rowed about thirty-four or thirty-five miles from our encampment of the preceding night, the funnel-shaped entrance of the river had contracted to a width of about two miles, and the current, as it washed the boulders of the beach, made a bubbling noise, like that of a strong rapid; and not long afterward we shot a rapid, the river having still further narrowed. The barking crow (*Corvus americanus*) is not seen to the northward of this place. In the *Fauna Boreali-Americana*, I have stated that it does not range beyond the 55th parallel; but more correct information, received on the present voyage, enables me to carry its northern limit on to the 61st. It becomes rare before it ceases altogether to be seen, and we have not noticed it in flocks since leaving the Saskatchewan. In its gregarious habits on the latter river it resembles the European rook, but differs from that bird in the care with which it conceals its nest. In the evening we landed to cook supper, and afterward re-embarked to drift with the stream. At midnight, having come to the Little Lake, where there is no current, we could no longer drive; we therefore anchored under a small sandy island, and at 4 A.M. on

July 21st, resumed our voyage. Four hours afterward, we landed at the outlet of the lake to cook breakfast. The morning was close and hazy, with distant thunder; and at 10 A.M. the storm approaching us, we were driven to take shelter for a time under the bank of the river. When the squall abated, we continued our voyage, notwithstanding that the rain fell throughout the day; and during the night we again drifted with the stream, the crews sleeping in the boats.

We made sail on the 22d, at a quarter before 3 A.M., with a fair wind, which soon afterward chopped round against us, and increased to a fresh breeze. At an early hour we passed the mouth of Trout River; and after breakfast descended the westerly reach below the site of the old fort. An hour later we passed the River La Câche, and in an hour and a half more came to Hare-skin River. The rate at which we passed the land must have been at least seven geographical miles an hour; but the distances in this part of Sir John Franklin's chart are too great, and Fort Simpson, which was laid down by him from dead reckoning, is placed twenty miles too far north.

The river having, through the increase of the wind, become too rough for the use of oars, we worked down under sail, and made good progress, arriving at Fort Simpson at five in the afternoon. The position of this place, as ascertained by Mr. Thomas Simpson in 1836, is in latitude 61° 51′ 25″ N.; and longitude deduced from lunar distances, 121° 51′ 15″ W.*

Between Desmarais's Fishery, on Slave Lake, and Fort Simpson, the direct distance is about one hundred and fifty-five geographical miles. In the wider parts of the river the coast is shelving, and not easily approached, in boats, from the shallowness of the water; but in the narrower places the beach is steep, and the channel is full of boulders. In a few spots where sections of the strata are visible, a bituminous shale, containing many fragments of the small pteropodous shell *Tentaculites fissurella*, indicates the formation to be the same with that on the Athabasca River and Slave Lake, which has been said above to be probably the Marcellus shale. Between the old fort and Hare-skin River, the basis of the bank is formed of a grayish green slate-clay, which, under the influence of the weather, breaks into scales like wacké, and at last forms a tenacious clay. The whole banks of the river seem to belong to a shale formation; but from the want of induration of the beds, they have crumbled into a slope more or less steep, and the capping of sand, clay, and boulders has fallen down and covered the declivity. On the south, a long even rising ground, named the Trout Mountain, which runs parallel to the river at a distance of from ten to twenty miles, is visible at intervals the whole way; and a similar but higher range, named the Horn Mountain, exists on the north.

Of the composition of these eminences, I have no information; but I suspect, from the evenness of their outlines and their relative position, that they are escarpments of the sandstone and shale of

* From this it appears that, by some means, an error of twenty miles of latitude had crept into the reckoning of Sir George Back and Lieutenant Kendall in 1825, between the old fort, in long. 120°, where the latitude was obtained by these officers, and Fort Simpson; but, on the other hand, they assigned too little departure, so that the mistake was in the courses as much as in the distance. And in correcting the chart, to give Fort Simpson its proper geographical position, a corresponding alteration must be made in the course and length of the river between that fort and the great bend below it, where the latitude and longitude were again ascertained by the observations of Back and Kendall.

the Erie group, remaining after the excavation of the valley of the river, such as has been already noticed as existing in the Clear-water and Elk Rivers, and as we shall afterward have occasion to mention, when describing the northwest side of Great Bear Lake.

The bank of the river at Fort Simpson is precipitous, and about thirty feet high; but the river sometimes flows over it in the spring floods, occasioned by accumulations of drift ice. It is composed of sand and loam, and the beach is lined with boulders of granite, greenstone, limestone, and sandstone.

Barley is usually sown here from the 20th to the 25th of May, and is expected to be ripe on the 20th of August, after an interval of ninety-two days. In some seasons it has ripened on the 15th. Oats, which take longer time, do not thrive quite so well, and wheat does not come to maturity. Potatoes yield well, and no disease has as yet affected them, though the early frosts sometimes hurt the crop. Barley, in favorable seasons, gives a good return at Fort Norman, which is further down the river; and potatoes and various garden vegetables are also raised there. The 65th parallel of latitude may, therefore, be considered as the northern limit of the *Cerealia* in this meridian; for though in good seasons, and in warm, sheltered spots, a little barley might possibly be reared at Fort Good Hope, the attempts hitherto made there have failed. In Siberia it is said that none of the corn tribe are found north of 60°. But in Norway barley is reported to be cultivated, in certain districts, under the 70th parallel. It takes three months, usually, to ripen on the Mackenzie, and on our arrival at Fort Simpson we found it in full ear, having been sown seventy-five days previously. In October, 1836, a pit sunk by Mr. M'Pherson, in a heavy mixture of sand and clay, to the depth of 16 feet 10 inches, revealed 10 feet 7 inches of thawed soil on the surface, and 6 feet 3 inches of a permanently frozen layer, beneath which the ground was not frozen.

A number of milch cows are kept at Fort Simpson, and one or two fat oxen are killed annually. Hay for the winter provender of the stock is made about one hundred miles up the river, where there are good meadows or marshes, and whence it is rafted down in boats. We met the haymakers, being three men, some hours before we reached the fort, on their way to cut the grass, which is a bent that grows in water. The hay will be brought down in September.

The fort stands on an island at the junction of the River of the Mountains (*Rivière aux Liards*) with the Mackenzie. This large tributary originates in the recesses of the Rocky Mountains, by many small streams which, uniting, form two branches. Both branches rise to the westward of the higher peaks, and afford another of the many instances of streams of magnitude crossing the chain. By Dease's River, which is the westernmost affluent of the north branch, boats pass through the mountains, and gain, after much trying and perilous navigation, and some portages, the Pelly and Lewis, at the junction of which the Company have a post named Pelly Banks. Native traders travel thither twice in the season from Lynn Canal, situated to the north of the island of Sitka, on the 59th parallel. This inlet is frequented by the Hudson's Bay Company's steamers, and, in this present summer of 1848, Mr. Todd, captain of one of these steamers, forwarded letters and newspapers to Mr. Campbell, the officer in charge at Pelly Banks. One of the newspapers, published at Honolulu, which was sent on to Fort Simpson, was transmitted by Mr. M'Pherson to Fort Confidence in the winter, and gave us the first intelligence of the origin of the gold hunt in California, and of the migration within a few days of two thousand men from Oregon, and of most of the Company's servants at Fort Vancouver, on that exciting pursuit. Such unexpected channels does commerce open for the conveyance of intelligence, and had previous arrangements been made, we might, by the route across the Andes at Panama, the Atlantic steamers to California, or the Sandwich Islands, and this northern way back again across the Rocky Mountain ridge, have had much more recent intelligence of our friends in Europe than we were destined to receive during our long winter residence on Great Bear Lake.

The Lewis flows from a large sheet of water, lying within the English boundary, but named the Russian Lake, because Mr. Roderick Campbell, who was the first officer of the Hudson's Bay Company who visited it, met there a party of Russian traders. The influence of these rivals in trade is supposed to have caused the attack made by the natives on Mr. Campbell's post in the winter of 1839, which resulted in the loss of three of his party by famine, and the narrow escape of the remainder from the same fate, as related in the narrative of Dease and Simpson's voyage (p. 173). Mr. Campbell, undaunted by this calamity, renewed his journeys in the same direction, and, in consequence of an agreement that had then been made between the Hudson's Bay and Russian Fur Companies, with less hazard. His first post, named after himself, was on the Pelly, and at the supposed distance from Fort Halkett, on the River of the Mountains, of three hundred miles, by the winter route, which is usually as direct as the nature of the country will admit. From Campbell's post to the Forks or junction of the Lewis and Pelly, where the present fort is situated, the distance is reckoned at two hundred and forty miles on a southwest course. To retrace this length of way, the crew of a light canoe are said to consume twelve days on the tracking line, being at the rate of twenty miles a day, which is generally considered as but an indifferent day's work against the current. It is probable, however, that the river is very tortuous, and that there are many impediments in a stream flowing through so mountainous a region. Of these two branches, the Lewis is the westernmost, and the river formed by their junction, which retains the name of Pelly, falls into the Pacific. By observations made by Mr. Campbell on the temperature of boiling water at Pelly Banks, the height of that post above the sea has been estimated at 1314 feet.

After the union of its two arms, the River of the Mountains flows for a considerable breadth of longitude on the 59th parallel, and near the middle of this part, at the influx of Smith's River, Fort Halkett stands. Fort Liard is situated lower down, after the river has made a sharp turn to the north, in its course toward the Mackenzie, which it joins at Fort Simpson. Though this post is more elevated than Fort Simpson, by at least one hundred and fifty feet, and is only two degrees of latitude to the southward, its climate is said to be very superior, and its vegetable productions of better growth and quality. Barley and oats yield good crops, and in favorable seasons wheat ripens well. This place, then, or the 60th parallel, may be considered as the northern limit of the economical culture of wheat.

It has been already mentioned that the *Wapiti* or *Wawaskeeshoo* of the Crees, the representative of the European red deer, does not range to the north of the River of the Mountains, and the same stream marks the northern limit of the American magpie, Say's grouse, and the white crane (*Grus americana*).

Mr. M'Pherson had most kindly set aside for me a cask of excellent corned beef, cured at the fort, and some bags of very fine potatoes raised at Fort Liard, with several other things which he knew would be serviceable at our winter residence. I left them in store, for Mr. Bell to embark when he came up, together with such supplies of iron-work and dried meat as the dépôt could furnish, and to convey them to our future winter residence, on Great Bear Lake. The boats were hauled up, their bottoms payed over with boiling mineral pitch, and such other repairs made as were necessary. I had intended to give them additional false keels at this place, to render them safer and more weatherly at sea, and, with this view, had long bolts and screws prepared at Portsmouth dockyard, to fit plates sunk in the keels; but the bolts were unluckily left behind at Cumberland House, Mr. Bell not being aware of the purpose for which they were designed, and we could not spare time to make others. All our preparations having been made on the 23d, we left the fort on the 24th, at 5 A.M., and three hours afterward had the first sight of the Rocky Mountains. In nine hours we were exactly opposite the end of the first range, where the Mackenzie, seemingly to avoid the barrier formed by the mountains, makes a sudden flexure from a northwest course to a north-northeast one.

Here I must interrupt the narrative for a little, to give some

ROCKY MOUNTAINS AT THE BEND OF THE RIVER.

account of the geological structure of the country through which the Mackenzie flows.

When the mountains are first seen, in descending the river, they present an assemblage of conical peaks, rising apparently about two thousand feet above the valley ; and it is not until we come opposite to the end of the first mountain, that we observe them to be disposed in parallel ridges, having a direction of about south-southwest and north-northeast ;* which makes an angle of rather more than forty-five degrees with the axis of the great chain, from which they project like spurs The circumstance of the valleys pervading the chain transversely, though with more or less of ascent, explains the reason of the principal rivers on both the eastern and western slopes having their sources beyond the axis of the range, and flowing through it. From some passages in Dr. Hooker's letters, I infer that the Himalayas have a similar configuration.

As the successive spurs and the valleys between them open out to the voyager who descends the river, he observes that the eastern faces of the ridges rise abruptly like a wall, while their western flanks are more shelving. This is not, however, uniformly the case, as in some of the ridges lofty escarpments occur also on their western sides.

The height of the almost precipitous cliff of the first mountain at the bend of the river appeared to the eye, from a distance of seven or eight miles, to be eight or nine hundred feet, though the width of the base of the hill did not exceed a mile. Further back, the summit of the ridge terminated by this mountain was judged to be between two thousand and two thousand eight hundred feet high. The heights here mentioned were estimated solely by the eye, and as in this climate heights and distances are very deceptive they must be considered as very rough approximations. No trees could be detected on the summits when examined with the telescope, but the lower hills, and the slopes to the height of a thousand feet, were well wooded.

The first range re-appears on the east side of the river, and is seen at intervals running in the direction of M'Vicar's Bay of

* I have never had leisure to ascertain the true course of these ranges within six or seven degrees, but from the bearings I have taken several times in passing, I suppose that south 20° west, and north 20° east, is very near their direction.

Great Bear Lake, whose basin interposes between its termination and the granite and gneiss that skirt the eastern arms of that lake.

At the bend of the Mackenzie, the valley which interposes between the first and second ridges does not appear to exceed five miles in width, but it was seen too obliquely to enable us to form a correct judgment. The river flows through this valley for upward of fifty miles, when, making a small bend to the westward, it escapes across the ridge. Thus far the second ridge* runs on the west bank of the river, showing a bold precipitous craggy side at intervals, some parts being concealed from the voyager by the intervening swelling grounds which form the floor of the valley. Where the river cuts it, a high island of limestone stands in mid channel, and on the east bank, a round-topped hill, named the "Rock by the River's Side" † (*Roche qui trempe à l'eau*), rises precipitously from the water's edge to the height of five or six hundred feet or more. The base of this hill scarcely exceeds a mile in diameter, and most of the ridges seem to be of similar breadth. From the Rock by the River's Side the ridge continues, but with interruptions, onward in the same direction to the elevated promontory of Great Bear Lake, named *Sas-choh etha* (Great Bear Hill), which stands between Keith's and M'Vicar's Bays.

The other spurs, which succeed these down to the delta of the river, rise in like manner like rugged walls from the surrounding low, undulating country, the stream escaping through them by successive gaps. Many of the escarpments, when seen from a distance reflecting the rays of the sun, look as bright and white as chalk cliffs: and but for information which I have gleaned from voyagers who have crossed them, I should have been in doubt whether they were not formed of that material or of white sand, instead of being hard limestone.

At this date only a few patches of snow remained in the hollows, having a northern exposure; but in the following year they were entirely covered with snow until late in June, and for some weeks after all the low country had become quite bare. Both the first and second ridges are distinctly stratified at the bend of the river, and seemingly capped with trap. Where they and the

* Partly seen on the right-hand side of the wood-cut, p. 107

† See Wood-cut, p. 113.

succeeding ridges are cut by the river, limestone is the chief rock that is visible; but I have had no opportunity of examining the principal cliffs, and have made but a very cursory inspection of any. The spurs which reach the Mackenzie consist, perhaps, wholly of limestone. Sandstone exists in their vicinity, but I believe it is a newer deposit, belonging to that which forms the floors of the valleys, and rests unconformably on the tilted beds of the ridges. No organic remains were detected in any of the highly inclined beds, but gypsum and chert are of frequent occurrence.

Traders who have crossed from the Atlantic to the Pacific slopes of the continent say that there are fourteen or fifteen ranges of hills, and that when they are viewed from the summit of a peak, the mountain tops appear to be crowded together in great confusion, like a sea of conical billows. My informants could not tell me whether granite, clay-slate, or trap rocks entered into their composition or not; but it is probable that such is the case, as we know it to be in more southern latitudes. I received specimens of semi-opal, plumbago, and specular iron, gathered on one of the ridges. The more westerly ranges have obtained from the traders the name of the Peak Mountains.

On the Mackenzie, a shaly formation makes the chief part of the banks, and also much of the undulating valleys between the elevated spurs. It is based on horizontal beds of limestone, and in some places of sandstone, which abut against the inclined strata of the lofty wall-like ridges, or rest partially on their edges. Covering the shaly beds, there exists in many places a deposit of sand, sometimes cohering so as to form a friable sandstone; and where a good section of the bank occurs, a capping of gravel and boulders, of various thickness, is seen crowning the whole. The shale crumbles readily, and often takes fire spontaneously, occasioning the ruin of the bank, so that it is only by the encroachments of the river carrying away the debris that the true structure is revealed. The boulders that have dropped from above pave the beach in many places as closely and regularly as if it were a work of art, the passage of ice over them driving them firmly and evenly into the bed of tenacious clay which the shale in breaking down produces.

I have no evidence whereby the geological age of the shale may be certainly deduced, but am inclined to consider it as belonging to the epoch of the Marcellus deposit, on account of its exact lithological resemblance to the bituminous beds of Athabasca River, and the occurrence of the *Tentaculites fissurella* in the fragments which line the beach at the west end of Great Slave Lake.* The difficulty of deciding upon the age of the beds through which the river flows is increased by the occurrence among them of a tertiary lignite formation, which also takes fire spontaneously. This general account of the rocks of the Mackenzie is here introduced to facilitate the subsequent descriptions of such points as I landed upon.

With respect to some of the more remarkable quadrupeds that inhabit the Rocky mountains, I may state that the mountain sheep, or big-horn as it is named (*Ovis montana*), frequents the higher peaks down to the delta of the Mackenzie. The Slave Indian appellation of this fine animal is *Sass-sei-yeuneh*, or "Foolish Bear." It keeps to the craggy summits, and can scarcely be approached by the hunter who ascends toward it from below; but should he once get above it, he can come near it easily. Its flesh is said to be equal to well flavored mutton, but its coat resembles that of the reindeer, and is not woolly. The goat-antelope (*Antilocapra americana*), which is covered with a fine long-stapled wool, has its northern limit on the River of the Mountains. Its flesh is much inferior to that of the mountain sheep. Reindeer, of a much larger size and darker color than the "Barren-ground variety," frequent the mountain valleys; and moose deer, extending their range nearly to the Arctic Sea, through the wooded districts only, feed on the banks of the rivers where willows grow. Neither musk-oxen nor bison inhabit this part of the Rocky Mountains; the latter, as has been mentioned, having their northern limit on the Horn Mountain; while the former keep within the Arctic circle, and to the east of the Mackenzie. The little *Pika*, or tail-less hare, occupies the grassy eminences, and lays up a stock of hay for winter use. Say's grouse (*Tetrao Sayi*), named *Ti-choh*, i. e. "big grouse," has not been killed further north than the Nòhhanè Bute; the pin-tailed

* In 1826, Mr. Sowerby referred some fossils which I obtained from the limestone beds of the Mackenzie, to the Oxford oolite and cornbrash. These, which were mostly terebratulites, are not now within my reach, but should his opinion be confirmed by further specimens from the same quarter, they would indicate that the bituminous shale of the Mackenzie belongs to the lias.

grouse goes as far down as the delta ; and the *Tetrao canadensis* lives in the marshy parts of the forest up to Peel's River, and is named *Ti;* while the willow and white-tailed ptarmigans bear the designation of *Kasbah* or *Kampbah,* in the Slave or Chepewyan tongue. The last named is exclusively an Alpine species. The American magpie has not been seen to the north of the River of the Mountains, and is rare even there.

Many large streams join the Mackenzie below Fort Simpson. One, which the Nòhhanè Indians are accustomed to descend, flows down the valley between the first and second mountain ridges, and joins the Mackenzie at its great bend. It is designated from these people, but it must not be confounded with the stream of the same name, which issues also from the hunting-grounds of the Nòhhanès, but falls into the River of the Mountains.

The Willow Lake River enters the Mackenzie a little below the bend, from the right bank. It is ascended by the Marten Lake Indians as far as it is navigable for their canoes, and then a march of four hours, or of from ten to fifteen miles, takes them to Marten Lake.

Another river of considerable size comes in on the left bank, which is named the *Bekka-tess* by the *Dahadinnès* who frequent its banks, and *La Rivière de Gravoir* by the voyagers. It joins the Mackenzie in latitude 64½° N., and is said to issue from a large lake, situated on the summit, or even on the western side, of the Rocky Mountain range. The impediments to its navigation have prevented it from being used as a channel for the Company's trade ; and it has been as yet only partially explored, though it has been thought that a route might be discovered through it to the banks of the Yukon. The *Dahadinnès* speak a dialect of the Chepewyan tongue. Mr. M'Kenzie, the gentleman in charge of Fort Norman, to which these people resort, informed me that their correct designation in their own language is *Cheta-ut-tinné* or *'Dtchetata-ut-tinne,* which, being also the national name of the Beaver and Strong-bow or Mountain Indians, points them out as members of the same nation.

CHAPTER VI.

Rock by the River's side.—Shale Formation.—Fort Norman.—Tertiary Coal Formation.—Lignite Beds.—Fossil Leaves.—Edible Clay.—Spontaneous combustion of the River Bank.—Hill at Bear Lake River.—Hill at the Rapid on that River.—Forest.—Plants.—Birds.

We drifted with the stream all night, and in the morning of July the 25th, a thick fog preventing us from pulling, we continued to drift, trusting that the current would carry us clear of shoals and low islands. The sky cleared at breakfast-time, and by noon we were abreast of the "Rock by the River's Side." In some places, where there are islands, the river is two or three miles wide; in others, it does not appear more than a mile, or a mile and a half. The small island, which lies in the channel just above the Rock by the River's Side, is composed of blackish

ROCK BY THE RIVER'S SIDE.

gray compact limestone, dipping to the south-half-east, at an angle of about twenty degrees; the upper bed, which is thinner and more slaty than the others, being composed of irregularly oblong distinct concretions. On the upper or south side of the Rock by the River's Side the stone is a bituminous limestone, yielding the smell of *stinkstein* when struck; the precipitous face of the rock appears to be the same kind of limestone. Immediately below the Rock, for the distance of half a mile, limestone similar to that of the small island occurs in gently inclined beds.

In the body of this high bluff the beds are nearly vertical; and, as well as I could judge from the view obtained in descending the stream, they were disposed as if the axis of the ridge had been the direction of the elevating force, the beds inclining toward the summit from both sides. In some parts there seemed to be inclined beds lying non-conformably over the ends of the nearly vertical ones, but I could not be certain, without closer examination, that what I saw was not merely oblique sections of the edges of the lower beds.

A thermal spring, much resembling sea-water in its saline contents, issues from the front of the cliff, and the fissure from whence it flows is incrusted with crystallized gypsum.* Shale beds abut against the lower side of the rock, covering the limestone beds above mentioned; but they are in a great measure concealed by the shelving debris of the bank. Contiguous to the upper or south side of the Rock there are sloping banks of gravel, capped by a vertical wall of friable sandstone. And three miles higher up the stream, there are two river terraces, more complete than any I noticed elsewhere on the Mackenzie, though in many places a high and low bank can be traced. These terraces are composed of fine sand: and the slope between them is so steep as to require to be ascended on all fours. Both terraces are very regular in their outlines, and are covered with well grown *Pinus banksiana*. The uppermost is about two hundred and fifty feet above the river. From this terrace, the Rock by the River's Side is clearly seen to be part of a chain, which is crossed there by the river, as has been already mentioned. This is not so evident from the channel of the stream. The high sand-banks continue almost without a break for twenty miles further up, and in some places they are seen to rest upon a gray shale. At one place where there is a good section, it was perceived that the surface of the

* Dr. Davy, who kindly analyzed some water from this spring, ascertained that the chief saline ingredient was sulphate of magnesia.

shale on which the sand reposed was uneven, and much indented also by pot-holes and projecting tongues; the gravel and sand descending into the pits, and the points of shale rising among the sand. The similarity of these shale and sand-cliffs to those at the junction of the Clear-water and Elk Rivers is very great; but the shale generally is not so bituminous as at the latter locality. The surface of the country above is strewed with gravel and boulders, and in the decay of the bank these fall down and line the channel of the river. When the water is high, as it is in the spring, little flat beach is to be seen; but in the autumn, the pavement of boulders to which I have already alluded is exposed. Among these, above the Rock by the River's Side, I observed a considerable number of granites, some gneiss, many sienites, basalts, and greenstones; also felspar rock, felspar porphyries, Lydian-stones, quartz rock, and limestones of various kinds, with quartzose sandstones, white, red, and spotted.

I have been disposed to give a more full abstract of the notes I made in descending and ascending this part of the river, because, in following its oblique course of more than fifty miles, from the first ridge of the Rocky Mountains at the bend, to the second at the Rock by the River's Side, all the various strata of the valley are seen, and, if properly examined, there is little reason to doubt that a key to the geological formations of the entire length of the Mackenzie might be obtained.

On the left bank, six miles below the Rock by the River's Side, beds of shale appear, having a slight dip to the southward; and the ridge, which is prolonged on that side from the rock above-named in a north-northeast direction, appears very rugged, with irregularly serrated summits, the crest being apparently extremely narrow. The country between the ridges seems to be pretty even, except where it is cut by rivulets; and the high bank of the river is level, though in places it looks hummocky or hilly, because of the gullies which intersect it.

In the evening we landed to cook supper at the mouth of Black-water River, which issues from a lake of the same name lying on the eastern bank; and, embarking again to drift during the night, passed a bend of ninety degrees, which the river makes to the westward, and which is known to the voyagers by the appellation of "The Angle" (*L'équerre*). It marks the passage of the river through another range, of which a high hill on the eastern bank, named Clark's Hill, is the most conspicuous part The ridge continued from this hill crosses Bear Lake River in the middle of its course, and there forms a rapid.

A short way below the "Angle," the Red Rock River, named also *Rivière des Grosses Roches*, flows in from the west. It looks wide at its mouth, but is not a large stream Fifteen miles further down, the Gravel or Dahadinnè River, already mentioned, flowing also from the mountains on the left, comes in below the site of an old fort. We were opposite to this when we resumed our oars on the morning of the 26th at four o'clock, and soon afterward, passing a sandy promontory on the left hand, named the "Crumbling Beaver" (*Castor qui déboule*), we arrived at Fort Norman. Obtaining here a bottle of milk as a grateful addition to our breakfast, we landed two hours later to prepare that meal, and at noon reached the mouth of Bear Lake River. Between Fort Norman and this river a tertiary coal formation occurs, which deserves particular notice.

The *coal*, when recently extracted from the beds, is massive, and most generally shows the woody structure distinctly, the beds appearing to be composed of pretty large trunks of trees lying horizontally, and having their woody fibres and layers much twisted and contorted, similar to the white spruce now growing in exposed situations in the same latitude. Specimens of this coal, examined by Mr. Bowerbank, were pronounced by him to be decidedly of coniferous origin, and the structure of the wood to be more like that of *Pinus* than *Araucaria;* but on this latter point he was not so certain. It is probable that the examination of a greater variety of specimens would detect several kinds of wood in the coal, as a bed of fossil leaves connected with the formation reveals the existence at the time of various dicotyledonous trees, probably *Acerineæ*, and of one which I am inclined to consider as belonging to the yew tribe. To these I shall refer again.

When exposed for even a short time to the atmosphere, the coal splits into rhomboidal fragments, which again separate into thin layers, so that it is difficult to preserve a piece large enough to show the woody structure in perfection. Much of it falls eventually into a coarse powder; and if exposed to the action of moist air in the mass it takes fire, and burns with a fetid smell, and little smoke or flame, leaving a brownish-red ash, not one-tenth of the original bulk of coal taken from the purer beds, for some contain much more earthy matter.

Different beds, and even different parts of the same bed, when traced to the distance of a few hundred yards, present examples of "fibrous brown-coal," "earth-coal," "conchoidal brown-coal," and "trapezoidal brown-coal." Some beds have the external characters of "compact bitumen;" but they generally exhibit in the cross fracture concentric layers, although from their jet-like composition the nature of the woody fibres can not be detected by the microscope. Some pieces have a strong resemblance to charcoal in structure, color, and lustre. Very frequently the coal may be named a "bituminous slate," of which it has many of the lithological characters, but on examination with a lens it is seen to be composed of comminuted woody matter, mixed with clay and small imbedded fragments resembling charred wood. Crystals of selenite occur in this slate, and also minute portions of resin, or perhaps of amber. When this shaly coal is burnt, it leaves light, whitish-colored ashes. The shape of the stems and branches of the trees is best preserved when they contain siliceous matter or iron-stone; and in this case, the bark of the tree is often highly bituminized, and falls off from the specimen.

From the readiness with which the coal takes fire spontaneously, the beds are destroyed as they become exposed to the atmosphere; and the bank is constantly tumbling down, so that it is only when the debris have been washed away by the river, that good sections are exposed. The beds were on fire near Bear River, when Sir Alexander Mackenzie discovered them, in 1785, and the smoke, with flame visible by night, has been present in some part or other of the formation ever since.

From one to four beds of coal are exposed above the water level on the banks of the river, the thickest of which exceeds three yards, and was visible a short way above Bear River in the autumn only—the Mackenzie being then seven or eight feet below its spring level.

Interstratified with the coal beds, there are layers of *gravel*, which occasionally, through the intermixture of clay more or less iron-shot, acquire tenacity enough to form vertical cliffs, but more often are very crumbly. The pebbles composing the gravel vary in size from that of a pea to that of an orange, and are formed of Lydian-stone, flinty slate, white quartz, quartose sandstone, and conglomerate, clay-stone, and slate-clay. The gravel is sometimes seamed by thin layers of fine sand, and its beds vary in thickness up to thirty or forty feet.

In place of the gravel, a friable *sandstone* is often interposed between the coal-beds or rests upon them. It is fine-grained, often dark from the dissemination of bituminous matter, and has so little tenacity, that in many places it is excavated by the sand-martens. Being porous, it fills with water, and is frozen into a compact, hard rock, for most of the year; but becomes moist, and breaks down under the influence of the hot rays of the sun in spring.

Potter's clay, of a gray or brown color, alternates with the beds already named, in layers varying from one foot to forty or more in thickness. This clay is often highly bituminous, and is penetrated by ramifications of carbonaceous matter, resembling the roots of vegetables. About ten miles above Great Bear River, a layer of this material, lying immediately over a bed of coal which was on fire, has been baked so as to resemble a fine yellowish-colored *biscuit porcelain*. In a part of this, I found numerous impressions of leaves, most of them dicotyledonous, but one of them apparently coniferous, and belonging, probably, to the yew genus. The existence on many of the leaves of the latter plant of little round bodies like the fructification of ferns, invested the specimens with much interest. The clay had unfortunately cracked so much under the influence of the heat to which it had been subjected, that I could not obtain entire specimens of the larger dicotyledonous leaves, but in the general character of their venation they resemble the *Acerineæ*. Some portions of the clay were semi-vitrified, and so hard as to receive no impression from a file: and I gathered pieces of this kind, composed of blue semi-vitrified layers, alternating with others of a rich buff color. All the indurated clay, containing leaves, splits easily into thin layers, in every one of which there were impressions, so that the various kinds of leaves must have been deposited thickly above one another at this place. The fossiliferous clay is covered by one hundred and forty feet of sand and sandstone, and by some thin layers of conglomerate.

A *pipe-clay* is very generally associated with the coal beds, and is frequently found in contact with the lignite. It exists in beds varying in thickness from six inches to a foot, and is generally of a yellowish-white color, but in some places has a light lake-red tint. It is smooth, without grittiness, and when masticated has a flavor somewhat like the kernel of a hazel nut. When newly dug from its bed, it is plastic, but in drying becomes rather meagre and adheres to the tongue : its streak is less glistening than that of the ordinary English pipe-clay. As the natives eat this earth in times of scarcity, and suppose that thereby they prolong their lives, I requested Dr. Davy and the late Dr. Prout to examine it, but neither of these able chemists could detect any nutritious matter in it. Neither have I been able with the microscope to discover in it the remains of any infusorial animals.* Mr. Nuttall speaks of a similar substance under the name of *pink-clay*, which he observed in the lignite deposits on the Arkansas. It is known generally among the residents at the fur posts on the Mackenzie by the appellation of " white-mud," and is used for whitewashing houses, and also, when soap is scarce, for washing clothes.

In one place in the vicinity of the burned cliff where the leaves were found, several beds of *porcelain-earth* occur from two to three yards thick, and apparently replacing the sandstone of other parts of the formation. It has a whitish color, and at first sight looks like chalk, but some of its beds have a grayish hue from the quantity of carbonaceous matter disseminated in them. Its texture is fine-granular ; it adheres slightly to the tongue, yields readily to the nail, is meagre, and soils the fingers slightly. Besides the coaly matter, it contains, also, a few minute scales of mica, and some of quartz. It is not plastic, and becomes more friable when moistened with water; neither does it effervesce with acids. This lignite formation extends from the Rocky Mountain spur of which Clark's Hill forms a portion, to the spur of which the hill on the lower side of the mouth of Great Bear River is a prominent point, being directly across the valley about twenty-five miles, but considerably further by the course of the Mackenzie. The depth to which the deposit descends below the bed of the Mackenzie was not ascertained, but the height from the surface of the water to the top of the bank varies from ninety to one hundred and fifty feet. Ten or twelve feet or more of the crest of the bank consists of diluvial gravel with boulders, and the

* Baron Humboldt mentions a tribe of Indians residing on the Orinoco, who eat large quantities of clay when food is scarce.

soil is generally peaty to the depth of a foot or two. The beds are usually almost horizontal or have a very moderate inclination, but in some few places they dip very considerably, and in the second reach of the Mackenzie above Great Bear River a bed of stone passes obliquely from the top to the bottom of the clay bank. By the destruction of the coal beds the cliff falls down, the slope is covered with the gravel and boulders, and the latter pave the channel of the river also. The strong current of the river varies its direction from time to time, and as the deposition or removal of alluvial islands expose or protect the banks, the debris of the ruined cliffs accumulates or is carried away. This constant waste of the bank would proceed much more rapidly, were it not that the ground is still frozen hard when barriers of ice, during the high spring floods, often raise the river thirty feet above its ordinary level. Then the frozen earth resists the action of the water as a rock would do, and the surface yields only in proportion as it thaws, which is slowly, since the water loaded with ice is kept down to the freezing point.

I observed that the bank of the river was generally higher than the land behind it, by at least the thickness of the diluvial capping, and sometimes by a part of the sand or clay of the tertiary beds, and that the narrow elevated bank extended in the same form along the principal affluents, a marked instance of which occurs on the south side of Great Bear River. In consequence of this configuration of the surface, the spring floods of melting snow accumulate, and at length make their escape through gullies, contributing further to the ruin of the bank, and giving it a broken and hilly outline when seen from the river. Landslips are of common occurrence, and are occasioned by pressure of water collecting in fissures produced by the partial subsidence of the cliff.*

* Similar tertiary coal formations occur on the flanks of the Rocky Mountains; the most southerly one of which I have any account, being in the Raton Pass, in latitude 37° 15′ N., longitude 104° 35′ W., and upward of seven thousand feet above the level of the sea. Leaves of dicotyledonous trees, obtained in these beds by Lieutenant Abert in 1847, are figured in Colonel Emory's report to Congress (pp. 522, 547). Nuttall observed lignite beds associated with the pink-colored pipe-clay on the Arkansas, somewhere near the 48th parallel. Sir Alexander M'Kenzie states that a narrow strip of marshy, boggy, and uneven ground, producing coal and bitumen, runs along the eastern base of the Rocky Mountains, and he specifies latitude 52° N., longitude $112\frac{1}{2}$° W., on the southern

The Mackenzie traverses the basin in which the tertiary coal is deposited very obliquely, and the Great Bear Lake River cuts it more directly across.

The hill on the north side of the last named river rises about six or seven hundred feet above the water, every where steeply, and in some places precipitously. It is, as has been stated, part of one of the spurs of the Rocky Mountains, a gap in which furnishes a channel for the passage of the Mackenzie. Its base, measured directly across, is about three-quarters of a mile. The Great Bear River flows between its south flank and the tertiary coal beds described above ; but on its north flank horizontal beds of limestone and bituminous shale appear again. The strata of the hill itself are highly inclined upward on both its flanks toward its axis, and some are vertical. I did not procure organic remains from any of the upheaved beds forming these ridges or spurs, whereby their age might be determined, but they are evidently

branch of the Saskatchewan, and latitude 56° N., longitude 116° W. (Edgecoal Creek) in the Peace River, as places where coal beds are exposed. Mr. Drummond procured me specimens of coal with its associated rocks at Edmonton (latitude 53° 45′ N., longitude 113° 20′ W.) on the north branch of the Saskatchewan, and, consequently, between the places mentioned by Sir Alexander Mackenzie. According to Mr. Drummond the coal was in beds varying in thickness from six inches to two feet, and interstratified with clay and sandstone. The examples he selected were precisely similar to the slaty and conchoidal varieties which are found at the mouth of Great Bear River, and the resemblance between the sandstone of the two localities is equally close. He also found a black tertiary pitch coal which breaks into small conchoidal and cubical fragments, which Mr. Small, a clerk of the Hudson's Bay Company, who gave me the first information of these beds, likened well to Spanish liquorice. At Edmonton the more slaty coal-beds pass gradually into a thin, slaty, friable sandstone, which is much impregnated with carbonaceous matter, and contains fragments of fibrous lignite. Hand specimens of this can not be distinguished from others gathered from the shale cliffs on the Athabasca River. Highly bituminized shale, considerably indurated, exists in the vicinity of the coal at Edmonton, and clay-ironstones occur in the clay beds.

Chief Factor Alexander Stewart told me that beds of coal are on fire on the Smoking River, which is a southern affluent of the Peace River, and crosses the 56th parallel of latitude, and also that others exist on the borders of Lesser Slave Lake, that lies between Smoking River and Edmonton. There are coal beds on fire, also, at the present time near Dunvegan on the main stream of the Peace River. All these places are near the base of the Rocky Mountains, or the spurs issuing from that chain, and their altitude above the sea varies from 1800 to 2000 feet and upward. The beds at Great Bear River are probably not above 250 feet above the sea level.

older than the limestone and shale formation which abuts against them or covers their edges, and are, very probably, judging from the scarcity of fossils, of the protozoic epoch.

HILL AT THE RAPID ON BEAR LAKE RIVER.

The Hill at the Rapid, twenty-four miles higher up Bear Lake River, is very similar to the one just noticed, and its beds have the same anticlinal arrangement. It is, as has been already stated, a member of the same spur with Clark's Hill, and from its summit the ridge may be seen extending through a comparatively level country toward the west end of Smith's Bay in Great Bear Lake. The floor of the valley lying between it and the spur at the mouth of the river is well wooded, but is much intersected by lakes, marshes, and considerable streams, some of which fall into the Mackenzie, and others into Bear Lake River. Immediately to the westward of the Hill at the Rapid, but separated from it by a rivulet, there are horizontal beds of friable sandstone, and beyond them a thick deposit of bituminous shale, which extends northward into the high promontory of the Scented Grass Hill, that divides Smith's Bay from Keith's Bay in Great Bear Lake. The excavation of the body of the lake terminates the shale formation in this direction, but more to the westward it can be traced onward to the Arctic Sea.*

* Various detailed accounts of some of the tertiary coal beds, and of the elevated spurs which cross Bear Lake River, are contained in the Geo-

As has been already said, the general aspect of the forest does not alter in the descent of the Mackenzie. The white spruce continues to be the chief tree. In this quarter it attains a girth of four or five feet, and a height of about sixty in a growth of from two to three hundred years, as shown by the annual layers of wood. One tree, cut down in a sheltered valley near Clark's Hill, measured the unusual length of one hundred and twenty-two feet, but was comparatively slender. Most of the timber is twisted, particularly where the trees grow in exposed situations. The Banksian pine was not traced to the north of Great Bear Lake River; but the black spruce, in a stunted form, is found on the borders of swamps as far as the woods extend. The dog-wood, silvery oleaster (*Elæagnus argentea*), *Shepherdia*, and *Amelanchier* grow on banks that in Europe would be covered with gorse and broom, and the southern *Salix candida* is replaced by the more luxuriant and much handsomer *Salix speciosa*, which is the prince of the willow family. The *Hedysarum Mackenzii* and *boreale* flower freely among the boulders that cover the clayey beaches; while the showy yellow flowers and handsome foliage of the *Dryas Drummondii* cover the limestone debris, which give shelter also to the *Androsace Chamæjasmi*. In the heart of the spruce-fir forests, the curious and beautiful *Calypso borealis* lurks, along with some very fine, large, one-flowered, ladies' slippers (*Cypripedia*). There is, in fact, notwithstanding the near neighborhood of the Arctic circle, no want of flowering plants to engage the attention of a student of nature;

logical Appendix to Franklin's Second Overland Journey; and the maps on a large scale, given in that work, may be consulted with advantage by any one who wishes to become well acquainted with the topography of the country, or to trace the course of the ridges here described in the text.

The limestone which forms the body of the hill at the mouth of Great Bear Lake River is blackish-gray, full of sparry veins, or brownish-gray and bituminous, associated with calcareous breccia. On the northern flank of the hill, abutting against the vertical beds, there are layers of bituminous shale, some of which effervesce with acids, while others approach in hardness to flinty slate. Underlying the shale, horizontal beds of limestone are exposed for some miles along the Mackenzie, and from them there issue springs of saline sulphureous waters and mineral pitch.

The horizontal sandstone beds, above the Hill at the Rapid, of the same river, contain fossils, some of which were considered by Mr. Sowerby to belong to the same age with the English oolitic limestones; but they require re-examination, and then we may learn whether the very extensive bituminous formation belongs to the Marcellus shale or to the lias beds.

and many of the feathered inhabitants of the district recall to the traveler or resident fur-trader pictures of southern domestic abodes. The cheerful and familiar *Sylvia æstiva* is one of the earliest arrivals in spring, coming in company with the well-known American robin (*Turdus migratorius*) and the purple and rusty grakles. A little later, the varied thrush makes its appearance from the shores of the Pacific. The white-bellied swallow (*Hirundo bicolor*) breeds, at Fort Norman, in holes of rotten trees; and the *Sialia arctica*, a representative of the bluebird so common in the United States, enlivens the banks of the Mackenzie, coming, however, not from the Atlantic coasts, but from the opposite side of the Rocky Mountain range. On the Mackenzie, there is an intermingling of the floras of both coasts, as well as of the migratory feathered tribes, the Rocky Mountain range not proving a barrier to either.

One of the birds which we traced up to its breeding-places on Bear Lake River, but not to the sea-coast, is the pretty little Bonapartean gull (*Xema Bonapartii*). This species arrives very early in the season, before the ground is denuded of snow, and seeks its food in the first pools of water which form on the borders of Great Bear Lake, and wherein it finds multitudes of minute crustacean animals and larvæ of insects. It flies in flocks, and builds its nests in a colony resembling a rookery, seven or eight on a tree; the nests being framed of sticks, laid flatly. Its voice and mode of flying are like those of a tern; and, like that bird, it rushes fiercely at the head of any one who intrudes on its haunts, screaming loudly. It has, moreover, the strange practice, considering the form of its feet, of perching on posts and trees; and it may be often seen standing gracefully on a summit of a small spruce fir.

The insectivorous habits of this bird, and its gentle, familiar manners, contrast strongly with the predaceous pursuits and voraciousness of the short-billed gull (*Larus brachyrhynchus* of the *Fauna Boreali-Americana*). If a goose was wounded by our sportsmen, these powerful gulls directly assailed it, and soon totally devoured it, with the exception of the larger bones. In the spring of 1849, when Mr. Bell and I were encamped at the head of Bear Lake River, waiting for the disruption of the ice, the gulls robbed us of many geese, leaving nothing but well-picked skeletons. Mr. Bell who was the chief sportsman on this occasion, and spent the day in traversing the half-thawed marshes in quest of game, hung the birds, as he shot them, to the branch of a tree, or deposited them on a rock; but, on collecting the produce of his chase in the evening, he found that the gulls had left him little besides the bones to carry. If by chance a goose, when shot, fell into the river, a gull speedily took his stand on the carcase, and proceed to tear out the entrails, and devour the flesh, as he floated with it down the current. Even the raven kept aloof, when a gull had taken possession of a bird.

The harlequin duck (*Clangula histrionica*) also frequents Bear Lake River; but is comparatively rare in other districts, and is not easy of approach. It congregates in small flocks, which, lighting at the head of a rapid, suffer themselves to glide down with the stream, fishing in the eddies as they go. A sportsman, by secreting himself among the bushes on the strand, conveniently near to an eddy, may, if he has patience to wait, be sure of obtaining a shot. In this way I procured specimens. The osprey and white-headed eagle both build their nests on the banks of Bear Lake River, and the golden-winged woodpecker migrates thus far north, and perhaps further, though it did not come under our observation in a higher latitude.

A small frog (*Bufo americanus*) is common in every pond, and Mr. Bell informed me that he had seen it on Peel River, which is the most northern locality I can name for any American reptile.* A frog resembling it, but perhaps of a different species, abounds on the Saskatchewan, and its cry of love in early spring so much resembles the quack of a duck, that while yet a novice in the sounds of the country, it led me more than once to beat round a small lake in quest of ducks that I thought were marvelously well concealed among the grass.

On Bear Lake River, the frogs make the marshes vocal about the beginning of June. Throughout Rupert's Land, they come abroad immediately after the snow has melted. In the swampy district between Lake Superior and Rainy Lake, they are particularly noisy. While we were descending the Savannah River on the 20th of May, we were exposed to the incessant noise of one called by the voyagers *le crapaud*,† whose cry has an evident

* See note, p. 126.

† This is probably the *Bufo americanus*, also. Mr. Gray of the British Museum, who examined my specimens, found old and young examples of

affinity with the *brekekex* of Asia Minor, and closely resembles the braying sound of a watchman's rattle; but a hundred of the latter, sprung in a circle, would not have equaled the voices of the frogs that we heard at one time. A smaller species, called *la grenouille*, inhabit the same places, and has a shrill, less unpleasing note than the other, yet which was, nevertheless, tiresome from its monotony.

As a contribution to what is known of the geographical distribution of reptiles, on the east side of the Rocky Mountains, frogs may be set down as attaining the 68th parallel of latitude; snakes, as reaching the 56th; and tortoises, as disappearing beyond the 51st, at the south end of Lake Winipeg. There the *Emys geographica* of Le Sueur, named *asatè* by the Chippeways, occurs; and also one with a flexible neck, called by the same people *miskinnah*, which is probably the snapping turtle.*

B. americanus from Lake Winipeg, and young ones from Great Bear Lake. There were also many specimens of *Rana sylvatica* (*la grenouille*) from the former locality; some of *Hyla versicolor* of Le Conte, or *H. verrucosa* of Daudin; and a solitary individual of a *Hylodes*, which he thinks may be new. It resembles, he says, " *H. maculatus* of Agassiz, but differs in color. The back is gray, with three cylindrical dark bands, interrupted and diverging from each other on the hind part of the back. The side of the face has a black streak, which is continued over the base of the fore-arm, and along the side of the body, gradually descending toward the belly. The toes are free and cylindrical, that is, scarcely tapering, and truncate at the end." (*I. E. Gray* in let.)

* By the same post which brought me a proof of this sheet, I had a letter from Mr. Murray, dated on the River Yukon, in which he informs me that a "frog" and "a grass snake" had been killed near his encampment, and that another snake had been killed on the north bend of the Porcupine River, far within the Arctic circle.

CHAPTER VII.

Peregrine Falcon.—The Rapid.—Ramparts.—Hare Indians.—Fort Good Hope.—Hares.—Kutchin.—Their Contests with the Eskimos.—A Fatal Dance.—A Hare Indian devoured by a Brown Bear.—Vegetation.—Narrows.—Richardson Chain of Hills.—Fort Separation.—Cache of Pemican and Memorandum.—Alluvial Delta.—Yukon River.—Reindeer Hills.—M'Gillivray Island.—Harrison Island. Termination of the Forest.—Sacred Island.—Richard's Island.—Point Encounter.

We continued to descend the river until 7 in the evening, when we encamped for the night, as I did not consider it to be safe to drift here, there not being one person in the boats who had ever been in this river before but myself, and I could not trust to my recollections of the best channels after the lapse of so many years since my former visit.

About twelve or fourteen miles below the influx of Great Bear River, the channel of the Mackenzie approaches the spur on its eastern bank, and flows parallel to it for some distance. At the spot where we encamped the beach was formed of displaced bituminous shale with imbedded granite boulders, both evidently derived from the ruined bank, a section of which showed layers of gravel consisting of rolled pieces of shale and a few limestone pebbles, alternating with sand and coarser rolled pieces of limestone. This seemed to be a tertiary deposit formed out of the subjacent beds, but not by the river flowing at its present level.

In the course of the day's voyage we noticed a peregrine falcon's nest, placed on the cliff of a sandstone rock. This falcon is not rare throughout the Mackenzie, where it preys on the passenger pigeons and smaller birds. Mr. M'Pherson related to me one of its feats, which he witnessed some years previously as he was ascending the river. A white owl (*Styx nyctea*), in flying over a cliff, seized and carried off an unfledged peregrine in its claws, and, crossing to the opposite beach, lighted to devour it. The parent bird followed, screaming loudly, and, stooping with extreme rapidity, killed the owl by a single blow, after which it flew quickly back to its nest. On coming to the spot, Mr. M'Pherson picked up the owl, but, though he examined it narrowly, he could not detect in what part the death blow had been received; nor could he, from the distance, perceive whether the peregrine struck it with wing or claws.

July 27th.—Embarking at 3 this morning, we continued our voyage down the river, and for upward of twenty miles pursued a course nearly parallel to the spur which the Mackenzie crosses at the influx of Great Bear River. In latitude 65° 32′ N., longitude 127 W., we were opposite to a magnificent cliff in this ridge, only two or three miles inland, apparently about four hundred feet high, and some miles in length. The escarpment faces directly southward, is remarkably white, and the layers composing it are nearly horizontal, but with some undulation. The heights of the peaks appeared to me to be about eight hundred feet above the water. The beach is composed of fragments of bituminous shale with pieces of lignite; and five or six miles further down, there is a good section of the shale beds interstratified with dark colored sandstone.

At the "Rapid" the Mackenzie crosses another spur, making three elbows in its passage through it. The channel of the river there is formed of limestone, and is shallow, producing, when the water is low, a considerable fall on the east side, and a shelving rapid on the west. At the elbow of the river, above the rapid, one of the hills, which rises steeply from the water's edge on the east bank, is composed of limestone beds, wrapping over one another like the coats of an onion, and curving, at the place where this structure was most distinctly seen, at a spherical angle of 65°, or thereabouts. These inclined beds are capped and covered on the flanks by strata of sandstone, which breaks down readily and forms a steep talus of pale-red sand. A cliff of the upper and more compact sandstone overhangs the crumbling layers beneath it.

Another eminence of the same spur, which rises from the rapid a few miles lower down, shows the same conical elevation with curved concentric beds. In one spot there is a fault, with dislocation of the beds. On both flanks of these inclined beds there are layers of aluminous shale interstratified with limestone and sandstone. Where these shale beds rest on the inclined rocks, they are also inclined, but they rapidly assume the horizontal position as they recede from the hill.

In the earlier part of the summer, a steamboat could ascend the rapid without difficulty; and this great river might be navigated by vessels of considerable burden, from the Portage of the Drowned in Slave River, down to its junction with the sea, being a navigation of from twelve to thirteen hundred miles.

In a dilatation of the river, about ten miles below the rapid, bituminous shale lies horizontally in the hollows of undulated beds of limestone. Having cooked supper at this spot, we embarked to drift for the remainder of the night.

At 5 in the morning of the 28th, we were at the commencement of the Ramparts, where the river is hemmed in to the width of from four hundred to eight hundred yards, and has a strong current. This is the "second rapid" of Mackenzie, who states that it is fifty fathoms deep; but in obtaining such soundings, his lead must have fallen into a crevice, or have been carried down the channel of the stream by the strength of the current; for gentlemen of the Hudson's Bay Company, who are well acquainted with the locality, informed me that a bed of stone crosses the stream, and at the close of the summer, when the river is at the lowest, produces a fall, except on the east side, where there is a channel that boats can ascend by towing. In the dilatation of the river above the Rampart defile, there are some fine examples of sandstone cliffs, which have decayed so as to form caves, pillars, embrasures, and other architectural forms. The beds have slate clay partings and seams of clay ironstone. Associated with them there is a marly stone, containing corallines, referred by Mr. Sowerby to *Amplexus;* and covering the sandstone in many places, and alternating with the upper beds, there is a deposit of bituminous shale.

In making its way through the defile, the river bends suddenly to the east-northeast, and, as the dip of the beds forming the cliffs on each side is in the contrary direction, the strata rise into sight in succession as we descend the river. The cliffs have been denuded of the covering of shale which exists higher up the stream, but the limestone of which they are chiefly formed is stained with bitumen, either in patches or whole layers. At the upper end of the defile, a fine granular, foliated limestone is interleaved with beds containing madrepores, and parted by seams of carbonaceous matter. Near the middle of the defile the limestone contains the *Terebratula sphæroidalis* (or a nearly allied species) which is a fossil of the inferior oolite, also some *Producti* and the coralline named *Amplexus*. Several seams of black shale, about eighteen or twenty inches thick, exist among the

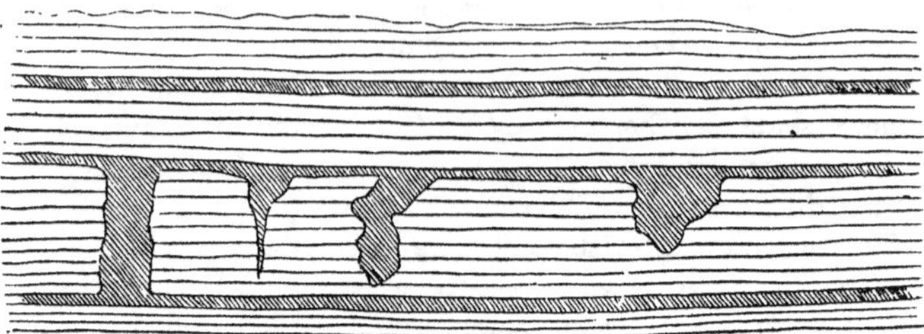

SECTION OF THE RAMPARTS.

limestones, and rise with them in succession above the level of the water; but there are not more than two or three of these seams in the height of the cliff at any one part. The shale is of various degrees of hardness, and passes into a brownish black flinty slate. The dip of the beds is not uniform throughout the defile, being more or less undulated, and for some way the layers are horizontal. In places here and there, the limestone beds are excavated into deep pot-holes filled with shale, resembling the gravel pits which dip into chalk beds.

On the top of the Rampart Cliffs we found a large body of Hare Indians encamped. This is a common summer haunt of these people, who resort thither to avail themselves of the productive fishery which exists above the defile. At this time, owing to the river not having subsided so rapidly as usual, they were taking only a small number of fish, and, consequently, were complaining of want of food. This people, and most of the tribes who live the whole year on the immediate banks of the Mackenzie, depend greatly for subsistence on the hare (*Lepus americanus*). Of these animals they kill incredible numbers; but every six or seven years, from some cause, the hares disappear suddenly throughout the whole country; so that not one can be found either dead or alive. In the following year a few re-appear; and in three years they are as numerous as before. The Canadian lynx migrates when the hares, on which it chiefly preys, become scarce. The musk-rat is subject to periodical murrains, when great numbers lie dead in their nests; but the dead hares are not found, whence we may conjecture that when their numbers become excessive they disappear by migration. I could not learn, however, that the Indians had ever seen them traveling in large bands.

The Hare Indians are a tribe of the Tinnè or Chepewyan nation, and speak a language differing only as a provincial dialect. They are, like the rest of the nation, a timid race, and live in continual dread of the Eskimos, whom they suppose not only to be very warlike and ferocious, but also endowed with great conjuring powers, by which they can compass the death of an enemy at a distance. The possession of fire-arms does not embolden the Tinnè to risk an open encounter with the Eskimo bowmen ; and unless when they are assembled in large numbers, as we found them at the Ramparts, they seldom pitch a tent on the banks of the river, but skulk under the branches of a tree, cut down so as to appear to have fallen naturally from the brow of the cliff; and they do not venture to make a smoke, or rear any object that can be seen from a distance. On the first appearance of a canoe or boat, they hide themselves, with their wives and children, in the woods, until they have reconnoitred, and ascertained the character of the object of their fears. More than once in our descent of the river, when we had landed to cook breakfast or supper, and were not at all aware of the vicinity of natives, a family would crawl from their hiding-places, and come to our fire. They always pleaded want of food ; and as their wretched appearance spoke strongly of their necessities, they invariably shared our meals ; but not unfrequently they sold us a fish or two before we parted ; being probably what they had reserved for their next meal, if we had not furnished them with one. We never found them with abundance of food ; for, in times of plenty, they do not think it necessary to lay up a stock, but let the future provide for itself.

It is supposed that formerly the Eskimos were in the habit of ascending the river to the Ramparts, to collect fragments of flinty slate for lance and arrow-points ; but they have been only once so far up, since the trading-posts were established. An old Indian, who was alive within a few years, told Mr. Bell that on that occasion he was wounded by an arrow ; but that he succeeded in escaping to the top of the cliff, from whence he killed two Eskimos with his fowling-piece.

As we passed the encampment, the Indians rushed down to the river's side, and, launching their canoes, accompanied us to Fort Good Hope, which now stands near its earliest site, a short way below the defile. At the time of Sir John Franklin's descent of the river in 1825 and 1826, the post stood about one hundred miles further down; but it was removed to its present position in 1836, after the destruction of the former establishment by an overflow of the river. The flood, carrying with it large masses of ice, rose thirty feet; and, mowing down the forest timber, swept onward to the fort, which it filled with water, thereby destroying a quantity of valuable furs. Mr. Bell, who was the resident officer at the time, escaped with the other inmates in a boat to the centre of the island; and shortly afterward, the dam of ice giving way, the flood subsided as rapidly as it had risen, leaving the buildings still standing, though much injured. A few turnips, radishes, and some other culinary vegetables, grow at Fort Good Hope in a warm corner, under shelter of the stockades; but none of the Cerealia are cultivated there, nor do potatoes repay the labor of planting. Mr. M'Beath, who had charge of the post, supplied us with some reindeer venison, which he had kept fresh in his ice-cellar, dug under the floor of his hall. This gentleman informed us that no rain had fallen this season in his vicinity, except two very slight showers on one day: there had been no thunder-showers. From him we learnt also that a rumor of guns having been heard on the coast of the Arctic Sea, and supposed to have been fired from the Discovery ships, originated in a story brought by the Kutchin or Loucheux to Peel's River Fort, but that the officer in charge placed no reliance upon it. He also gave us the unpleasant intelligence of three Eskimos having been killed in Peel's River last summer. A large body of that nation, having ascended the Peel River, it was surmised, with hostile intentions, were fired upon by the Kutchin, and three of them killed, upon which they retreated.

The Kutchin and Eskimos of the estuary of the Mackenzie, meet often for purposes of trade, and make truces with each other, but they are mutually suspicious, and their intercourse often ends in bloodshed. The Kutchin have the advantage of fire-arms, but the Eskimos are brave and resolute, and come annually to Separation Point, at the head of the delta, for the purposes of barter. Most of the Kutchin speak the Eskimo language, and from them the latter people have become aware of the existence of a post on the Peel. It is probable, therefore, that the Eskimos had a purpose of opening a trade directly with the white people;

but this, being so obviously contrary to the interest of the Kutchin, was likely to meet with all the opposition they could offer, and hence their firing on the Eskimos without parley. The Kutchin give a very bad character of their neighbors for treachery, and throw on them the whole blame of their mutual quarrels; but the faults are certainly not confined to one side; and, doubtless, were an intercourse once fairly established between the Eskimos and the Company's posts, it might be kept up as peaceably here as it is with the same people elsewhere.

In the course of Mr. Bell's residence on Peel's River, an event occurred in the history of these people, which, in its principal feature, bore no small resemblance to the skirmish between the parties of Joab and Abner, related in the second chapter of the Second Book of Kings. A party of Kutchin having met a number of Eskimos, their demeanor to one another was friendly, and the young men of each nation rose up to dance. The Eskimos, however, being accustomed to carry their knives concealed in their wide sleeves, did so on this occasion, and, grasping them suddenly, on a preconcerted signal in the midst of the dance, thrust them at their Kutchin companions, by which three of the latter fell mortally wounded. A mêlée ensued, in which several were slain on both sides. This is the story told by the Kutchin survivors, but the Eskimos would, perhaps, give a different color to the matter were they the narrators. It is to be hoped that in a few years, the interference of the traders will put an end to these disastrous conflicts, which have long ceased in other parts of the fur countries.*

* An unexpected and cruel massacre of a party of Eskimos, has been reported to the Admiralty by Commander Pullen, since this and some of the following sheets were set in type. This sad occurrence is rendered more lamentable, from a Canadian in the employment of the Hudson's Bay Company, having been a prime actor in the affair. It appears that in the spring of 1850, two men belonging to Peel River Fort had landed on Point Separation, on which a body of Kutchin were at the time encamped. Soon afterward a small number of Eskimos approached in their kaiyaks. The Canadian would have fired upon them instantly, but was restrained by his companion, who did all that he could to prevent the bloodshed that ensued The leading Eskimo called to the Kutchin to lay down their guns; and, to show his own peaceful intentions, he fired his arrows into the sand, and then showed his empty quiver. His signs of amity were replied to by the Canadian firing upon him, and the Kutchin following his example, the party was destroyed. I fear that our endeavors to establish friendly relations with the Eskimos in the estuary of the Mackenzie, may have lured these poor people to the bold advance among their enemies which ended so fatally for them.

By Mr. Bell, I was also informed of the melancholy death of an Indian in the vicinity of Fort Good Hope. This poor man, having set several snares for bears, went to visit them alone. The event showed that he had found a large bear, caught by the head and leg, and endeavored to kill it with arrows, several of which he shot into the neck of the animal. He seems to have been afraid to approach near enough to give full effect to his weapons, and the enraged bear, having broken the snare flew upon him and tore him in pieces. The man's son, a youth of about sixteen years of age, becoming alarmed by the lengthened absence of his father, took his gun, and went in quest of him, following his track. On approaching the scene of the tragedy, the bear hastened to attack him also, but was shot by the lad as he was rushing at him. The boy found his father torn limb from limb, and mostly eaten, except the head, which remained entire. The bear, whose carcass was seen by Mr. Bell, was a brown one, and of great size. Fragments of the snare remained about his neck and leg.

These brown bears are very powerful; and the same gentleman who told me the above story informed me that on the Porcupine River, to the west of the Peel, he saw the footmarks of a large one, which, having seized a moose-deer in the river, had dragged it about a quarter of a mile along the sandy banks, and afterward devoured it all, but part of the hind-quarters. The bones were crushed and broken by the animal's teeth, and, from their size and hardness, Mr. Bell judged the moose to have been upward of a year old, when it would weigh as much as an ox of the same age. The species of these northern brown bears is as yet undetermined. They greatly resemble the *Ursus arctos* of the old continent, if they are not actually the same; and are stronger and more carniverous than the black bears (*Ursus americanus*), which also frequent the Mackenzie. The grisly bears (*Ursus ferox*) reach the same latitudes, but do not generally descend from the mountains.

After a halt of little more than two hours with Mr. M'Beath, we resumed our voyage down the river, and, rowing until suppertime, the crews retired to rest in the boats, which were suffered to drift with the current all night under the guidance of a steersman. On the morning of the 29th, the fog was so dense that for some hours we allowed the boats to follow the current, being afraid to row, lest we should run aground. At night we encamped not far from the Old Fort. The shale, sandstone, and limestone beds, continue throughout the space intervening between the former and present sites of Fort Good Hope. In some places the friable sandstones, yielding readily to the torrents of water which flow over the brow of the cliff in spring, were cut into deep ravines at regular distances, producing conical, truncated eminences, like shot-piles. In others, beds of bituminous shale, one hundred and twenty feet high, existed, interleaved with two or three beds of limestone, and in several places the shale banks were crowned with a thick deposit of sand, which rose above the level of the country behind. This peculiar arrangement, which has been already mentioned as occurring not only on the Mackenzie, but extending also some way up many of its affluents, is conspicuous in the reach immediately above the old site of Fort Good Hope, and has the aspect of ridges of sand left in these situations on the subsidence of waters, that have swept over the neighboring country. These banks rise much beyond any floods of the present day, some of them being fully two hundred feet above the river. In this neighborhood, the drift timber showed that the spring accumulations, at the disruption of the ice, occasionally raises the river at least forty feet. Here, as well as higher up, there is generally a capping of diluvium, with boulders, which roll down and line the beach. Among these, sandstones predominate; but there are many of a beautiful porphyritic granite, and others of sienite, hornblendic rocks, greenstone, &c. No clay-slate nor mica-slate boulders were observed.

Vegetation here preserves the same general character that it has higher up the river. *Salix speciosa* continues to grow twenty feet high in favorable localities; the humbler *Salix myrsinites* skirts stony rivulets; and the *Salix longifolia* covers the flooded sandbanks, and arrests the mud. The *Hedysarum boreale* furnishes long flexible roots, which taste sweet like the liquorice, and are much eaten in the spring by the natives, but become woody and lose their juiciness and crispness as the season advances. The root of the hoary, decumbent, and less elegant, but larger-flowered *Hedysarum Mackenzii* is poisonous, and nearly killed an old Indian woman at Fort Simpson, who had mistaken it for that of the preceding species. Fortunately, it proved emetic; and her stomach having rejected all that she had swallowed, she was restored to health, though her recovery was for some time doubtful.* On the beach I observed a patch of parsley (*Apium petroselinum*) in flower; probably having sprung from seed scattered by a party going to Peel River, as I met with the plant in no other quarter.

July 30th.—In this day's voyage we saw many small parties of Kutchin, seemingly all in want of provisions, owing to the high water spoiling their fishery. From one man, however, we purchased a fine white-fish (*Coregonus*), weighing nearly eight pounds. These families are the most easterly of the Kutchin; and, far from exhibiting the manly conduct and personal cleanliness for which their nation is noted on the banks of the Porcupine and Yukon, have much of the abject demeanor of their neighbors, the Hare Indians. Their jackets differ from those of the Chepewyans in being peaked, after the manner of those of the Eskimos. From their being able to remain in the close vicinity of the latter people, it is evident that they possess more courage than the Hare Indians.

In the morning we passed an affluent thirty or forty yards wide, coming in from the eastward, which is probably the stream mentioned by Sir Alexander Mackenzie as one on whose banks Indians and Eskimos collect flints. These flints are doubtless either chert from the limestone beds, or flinty slate, which exists plentifully in some parts of the shale formation.

Early in the morning of the 31st, we ran through the "Narrows," a defile similar to that of the Ramparts, and in passing which the river makes a similar sharp elbow. The cliffs are composed of sandstone, in some places horizontal, in others dipping to the south by east at a small angle. The stone is of various textures; some of it having a conchoidal fracture, and containing much calcareous matter. The basis is earthy, and the coarsest stone is composed of small, rounded, and also sharply angular grains of opake, white, green, or blue quartz, with grains of Lydian-stone and coal; the basis being also tinged with coaly matter. Other beds pass into a kind of wacké, or shale, which breaks down quickly into very small angular fragments. This shale is often incrusted with alum in powder, and it is sometimes

* There must have been some mistake in the information which I furnished to Sir William J. Hooker, respecting these two plants, as the *H. Mackenzii* is said in the *Flora Boreali-Americana* to have the edible root.

stained with iron, and contains spheroidal nodules of clay-ironstone. The cliffs vary in height from fifty to one hundred and fifty feet, and the capping of clay and loam with boulders is thin. The shale formation extends along the banks of Peel's River; and Mr. Bell informed me that he had procured crystallized alum from some beds in that quarter. The Mackenzie is about one mile and a half wide in these straits. The current sets at the rate of three miles an hour, and we got soundings with eight fathoms in mid-channel. By a mutual understanding between the Eskimos and Kutchin, the Red River, which falls into the Narrows, is considered as the boundary between the two nations.

On emerging from the Narrows, we had a distant view of the Richardson chain of hills, which skirts the western branch of the Mackenzie, and a little before noon reached Point Separation. Mr. Rae observed for the latitude; but the sun being obscured by passing clouds, the observation would have been doubtful, had it not corresponded exactly with the position assigned to the spot in Sir John Franklin's map, of 67° 49′ north latitude. The variation of the compass by the sun's meridional bearing was south 55° east, being five degrees more than in 1826.

In compliance with my instructions, a case of pemican was buried at this place. We dug the pit at the distance of ten feet from the best-grown tree on the point, and placed in it, along with the pemican, a bottle containing a memorandum of the objects of the Expedition, and such information respecting the Company's post as I judged would be useful to the boat party of the "Plover," should they reach this river. The lower branches of the tree were lopped off, a part of its trunk denuded of bark, and a broad arrow painted thereon with red paint. A stake was also erected on the beach, and a paper attached thereto, directing attention to the tree. We considered it likely that the stake might be taken down by Indians or Eskimos; but as the latter people lop trees in the same manner, we did not think the circumstance of one being so cut here would induce people of either nation to dig in the vicinity. To conceal as effectually as we could that the earth had been moved, the soil was placed on a tarpaulin, and all that was not required to fill the hole was carried to a distance. The place then being smoothed down, a fire of drift timber was made over it, that the burnt wood might indicate the exact spot to the "Plover's" party, who were furnished with a memorandum mentioning that these precautions would be used. Along with the pemican, a letter for the purser of the "Herald," which his friends had committed to my care, was placed in the pit.*

In performing these duties at this place, I could not but recall to mind the evening of July 3d, 1826, passed on the very same spot in company with Sir John Franklin, Sir George Back, and Lieutenant Kendall. We were then full of joyous anticipation of the discoveries that lay in our several paths, and our crews were elated with the hope of making their fortunes by the parliamentary reward promised to those who should navigate the Arctic Seas up to certain meridians. When we pushed off from the beach on the morning of the 4th to follow our separate routes, we cheered each other with hearty good-will and no misgivings. Sir John's voyage fell some miles short of the parliamentary distance, and he made no claim. My party accomplished the whole space between the assigned meridians; but the authorities decided that the reward was not meant for *boats*, but for ships. Neither men nor officers made their fortunes; and, what I more regretted, my friend and companion, Lieutenant Kendall, remained in that rank till the day of his death, notwithstanding his subsequent important scientific services. On the present occasion, I endeavored to stimulate our crews to an active look-out, by promising ten pounds to the first man who should announce the Discovery ships.

Most of the islands constituting the delta of the Mackenzie are alluvial, and many of the smaller ones are merely a ring of white spruce trees and willows on a sand or mud bank, inclosing ponds or marshes filled with drift timber. Some of the larger ones have a drier and firmer soil, but are low and even, except near the sea, where a few conical hummocks rise abruptly above the general level to the height of eighty or ninety feet. Sir John Franklin saw these hummocks on Ellice Island; and, as they occur also near the western boundary of the delta, I shall have occasion to notice them again. The Richardson Mountains, which skirt the western channel of the river, appear like a continuous ridge

* Commander Pullen, with two boats from the "Plover," in Sept. 1849, visited this dépôt, and found it safe. The lopped tree had previously, in autumn, 1848, been examined by a party of the Hudson's Bay Company's servants going to Peel River, but they did not discover the pit, having no key to enable them to find it.

when viewed from Point Separation; but it is more probable that they are the termination of a succession of spurs from the main chain of the Rocky Mountains, which come obliquely to the coast of the estuary. The general altitude of the ridges does not apparently exceed one thousand feet; but some peaks, as Mount Goodenough, Mount Gifford, and Mount Fitton, are perhaps considerably higher.

In ascending from the west bank of the delta, the brow of the first range is attained at the distance of about forty miles from the river, of which the first four miles are over a low, marshy, alluvial plain, covered with willows. Two or three almost precipitous ascents and descents across other mountains bring the traveler to a small stream, called the Rat River, which flows to the westward. This is said to issue from a lake, which also gives origin to a still smaller stream, bearing likewise the appellation of the Rat, and taking an opposite course to join the western branch of the Mackenzie. The western Rat River is an affluent of a considerable stream, named the Porcupine, which, running to the west-southwest for two hundred and thirty miles, enters the Yukon, a river emulating the Mackenzie in size, and flowing parallel to it, but on the western side of the Rocky Mountains. Of the country watered by this great river, and its inhabitants, I shall take occasion to speak hereafter.

The eastern channel of the delta of the Mackenzie is also flanked by a ridge, named the Reindeer Hills, which I consider to be a prolongation of the spur that the Mackenzie crosses at the "Narrows." They are not so rugged or peaked in their outline as the Richardson chain, or the spurs which the Mackenzie passes through higher up; and their general height does not appear to exceed seven or eight hundred feet.

Having finished the operations at the *cache*, we resumed the voyage, and, retracing our way for a few miles, entered the eastern channel of the delta, and pursued it until seven in the evening, when we encamped, about twenty-two miles below Point Separation. The banks of the river here, and the numerous islands are well wooded. The balsam poplars rise to the height of twenty feet, and the white spruces to forty or fifty. Numbers of sand-martens burrow in the banks. These birds winter in Florida. Mr. Audubon informs us, that in Louisiana they begin to breed in March, and rear two or three broods in a season. In the Middle States their breeding-time commences a month later and in Newfoundland and Labrador it rarely takes place before the beginning of June. Near the mouth of the Mackenzie, the banks are scarcely thawed enough to admit of excavation by the feeble instruments of this bird before the end of June; and in the beginning of September, the frosts prostrating the insects on which the martens feed, they and their young broods must wing their way southward. I was unable to procure a specimen of this marten, though it breeds in multitudes along the whole course of the Mackenzie, and am therefore unable to decide whether it is the *Hirundo riparia* or *Hirundo serripennis* of Audubon; but from its nearly even tail, I rather incline to think it may be the latter; and if so, it may not be the same species which breeds in the Southern States. The sand-marten was first seen by us on the 28th of May, as we were descending the River Winipeg, near the 50th parallel, and we know, from our observations in 1826, that it reaches the delta of the Mackenzie by the beginning of July; affording thus an index to the progress of spring in different latitudes. On the Winipeg it was accompanied by the purple swift (*Progne purpurea*), whose northern limit we did not ascertain.

We resumed our voyage at three in the morning on the 1st of August, and when we landed to cook breakfast, saw some recent footmarks of Eskimos. As these people are employed at this time of the year, in hunting the reindeer on the hills which we were skirting, we were in constant expectation of seeing some of their parties. The Reindeer Hills, as viewed from the eastern channel, seem to be an even-backed range; but when examined with the telescope, they are seen to consist of many small, oblong, rocky eminences, apparently of limestone, and are sparingly wooded. In the course of the morning we crossed the mouths of three pretty large affluents, coming in from the hills, and also two cross canals, dividing M'Gillivray Island into three sections.

About thirty-five miles from Point Separation, or in latitude

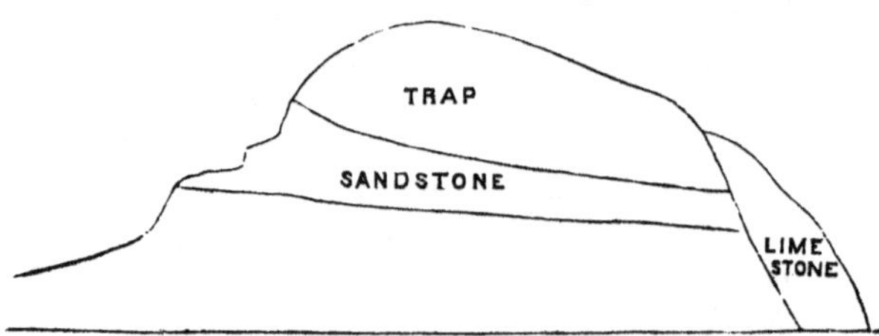

68° 10′ N., the channel washes the foot of a low dome-shaped bluff, in which the intrusion of a mass of trap, which now forms the top of the hill, has tilted up a bed of limestone, and separated it from one of sandstone.

In the afternoon we passed another considerable affluent from the hills in lat. 68° 18′ N.; some hours later, another one of less size; and very soon afterward crossed a channel which bounds Harrison Island on the south. This island, like M'Gillivray's, is divided into several portions by minor creeks. The boats were under sail all the afternoon, and must have been observed, about 5 o'clock, by the hunting-parties of the Eskimos, for at that time we noticed a line of six or eight signal smokes, raised in succession along the hills, and speedily extinguished again. As the Eskimos use fire-wood very sparingly for cooking, and, like the Indians generally, burn only dry wood which emits but little vapor, we knew that the smokes we saw were intended to spread the intelligence of the arrival of strangers in the country, and therefore that we might expect to find a considerable body assembled on some part of the river to meet us. In the evening we landed to cook supper, and re-embarked to continue under sail all night, with a very light breeze; our progress was, however, slow, owing to the uncertain eddies and currents, produced by the junction of the several cross-channels. At midnight we passed the creek which bounds Harrison Island on the north, in 68° 37′ N. Here several gently swelling elevations interpose between the river and the main ridge of the Reindeer Hills. The valleys and borders of the rivers are well wooded, but the summits of the eminences present only scattered spruce firs, with stunted tops and widely spreading depressed lower branches. The canoe-birch (*Betula papyracea*) is frequent, and the trees we measured were about five inches in diameter. The *Populus balsamifera* and *Alnus viridis* grow to the height of twenty feet, and the *Salix speciosa* to upward of twelve. The *Ribes rubrum*, *Rubus chamæmorus*, and *Vaccinium vitis idæa*, bore at this time ripe fruit. The *Rosa blanda*, *Kalmia glauca*, *Nardosmia palmata*, and *Lupinus perennis*, were also observed flourishing in this high latitude, together with several other plants which extend to the sea-coast. Among the birds, we saw the great tern (*Sterna cayana*), the *Coryle alcyon*, and *Scolecophagus ferrugineus*, the latter in flocks.

August 2d.—For five or six hours this morning we ran past the ends of successive ridges separated by narrow valleys. The diagram gives the outlines of one of these spurs seen on the southern flank. It is about three hundred feet high, and its acclivities are furrowed deeply, producing conical eminences which are impressed with minor furrows. The vegetation is scanty; a few small white spruces struggle up the sides; and the soil, where it

SAND HILLS. LAT. 68° 50′ N

is exposed to view, is a fine white sand. Large boulders lie on the sides of the hills, and, judging from the structure of the only point on which time permitted me to land, the whole appears to be similar to the sand deposit with its capping of boulder gravel which covers the shale on Athabasca and Mackenzie Rivers. On the point in question, the white sandy soil was ascertained to come from the disintegration of a sandstone, which has just coherence enough when *in situ* to form a perpendicular bank, but crumbles on being handled. It consists of quartz of various colors, with grains of Lydian-stone loosely aggregated, and having the interstices filled with a powdery matter, like the deposits of some calcareous springs. Similar sandstones occur at the "Narrows." Above it, there is a bed of gravel, also formed of variously colored grains of quartz, mixed with chert from limestone. Most of the quartz is opaque, and veined or banded, but some of it is translucent. Some bits are bluish, others black, and many pebbles are colored of various shades of mountain green. The latter are collected by the Eskimos and worn by them as labrets. The gravel covers the whole slope of the point, which is so steep as to require to be ascended on all fours. In one part a torrent had made a section of a bêd of fine brown sand, twenty feet deep. On this bank I gathered the *Bupleurum ranunculoides*, which grows in Beering's Straits, but had not been found so far westward as the Mackenzie before; also the *Seseli divaricatum*, which had not been previously collected to the north of the Saskatchewan.

In latitude 68° 55′ N. the trees disappeared so suddenly, that I could not but attribute their cessation to the influence of the sea-air. Beyond this line a few stunted spruces only, were seen struggling for existence, and some scrubby canoe-birches, clinging to the bases of the hills. Further on, the Reindeer Hills lowered rapidly, and we soon afterward came to Sacred Island, which with the islets beyond it, is evidently a continuation of the sandy deposit noticed above. Had time permitted, I should have gone past Sacred Island, northward, to deposit some pemican on Whale Island, but at so advanced a period of the summer, I was unwilling to incur the loss of a day which that route was certain to occasion, and perhaps even of two days.

We did not land on Sacred Island, but observed in passing that it still continued to be a burying-place of the Eskimos; two graves covered by the sledges of the deceased, and not of many years' construction, being visible from the boats. This is the most northerly locality in which the common red currant grows on this continent, as far as I have been able to ascertain. Five miles beyond the island, we landed on the main shore, to obtain a meridional observation, by which the latitude was ascertained to be 69° 4′ 14″ N., and the sun's bearing at noon, south 51° east. About three miles further on, we had a distant view of an eminence lying to the eastward, which resembled an artificial barrow, having a conical form, with very steep sides and a truncated summit. This summit, in some points of view, presented three small points, in others, only two, divided from one another by an acute notch. In the afternoon I landed on Richard's Island, which rises about one hundred and fifty or two hundred feet above the water, has an undulated grassy surface, and is bor-

CONICAL HILL BEYOND POINT ENCOUNTER

dered by clayey or sandy cliffs and shelving beaches. The main shore has a similar character. The channel varies in depth from two to six fathoms, but is full of sand-banks, on which the boats frequently grounded.

At ten in the evening we encamped on Point Encounter, in latitude 69° 16′ N., and set a watch at the boats, and also on the top of the bank, which is here nearly two hundred feet high. The tide ebbed at the encampment, from seven in the evening till half an hour after midnight. The ensign was planted on the summit of the cliff all the evening, and was no doubt seen by the Eskimos, who were in our neighborhood, and most probably reconnoitred our encampment, but we saw nothing of them.

The readers of the narrative of Sir John Franklin's Second Overland Journey will recollect that off this point the Eskimos made a fruitless attempt to drag the boats of the eastern detachment on shore, for the purpose of plundering them.

CHAPTER VIII.

Enter the Estuary of the Mackenzie.—Interview with the Eskimos.—Remarks on that People.—Winter-houses near Point Warren.—Copland Hutchison Bay.—Flat Coast with Hummocks.—Level boggy Land.—Mirage.—A Party of Eskimos visit us.—Point Atkinson.—Kashim.—Old Woman.—Old Man.—Young Men.—Cape Brown.—Eskimos.—Russell Inlet.—Cape Dalhousie.—Sabine Xema.—Liverpool Bay.—Nicholson Island.—Frozen Cliffs of Cape Maitland—Rock Ptarmigan.—Eskimo Tents.—Harrowby Bay.—Baillie's Islands.—River of the Toothless Fish or Beghula Tessè.—Eskimo of Cape Bathurst.—Their Summer and its Occupations.—Shale Formation of the Sea-coast.

August 3d, 1848.—Having given some verbal instructions to the crews of the boats, respecting their conduct in the presence of the Eskimos, we embarked at four in the morning, and, crossing a shallow bar at the east end of a sandbank, stood through the estuary between Richard's Island and the main, with a moderate easterly breeze, which carried us gradually away from the main shore. About an hour after starting, we perceived about two hundred Eskimos coming off in their kaiyaks, carrying one man each, and three umaiks filled with women and old men, eight or ten in each. The kaiyaks are so easily overbalanced, that the sitter requires to steady it before he can use his bow or throw his spear with advantage, unless when three lie alongside each other, and lay their paddles across, by which the central man is left at liberty to use both hands. By taking the precaution, therefore, of not allowing the Eskimos to hamper us, by clinging to the boats, and continuing to make some way through the water with the oars, we were pretty sure that they could not take us altogether by surprise; and I felt confident that as long as they saw that we were on our guard, and prepared to resist any aggression, none would be attempted. I had, moreover, especially directed Duncan Clark, who was cockswain of the third boat, in which there was no other officer, to keep close to mine, which he could easily do as his was the swifter of the two; but the novelty of the scene caused him to neglect this command, fortunately with no serious bad consequence, though a conflict might have been the result of his inattention to orders.

Mr. Rae and I carried on a barter with the men in the kaiyaks, paying them very liberally for any thing they had to offer in exchange, such as arrows, bows, knives of copper or of bone, &c., and thereby furnishing them with much iron-work, in the shape of knives, files, hatchets, awls, needles, &c. The articles we received were of no value to us; but a gift is generally considered by the American nations as an acknowledgment of inferiority, and it is better to exact something in exchange for any article that you may wish to bestow. The men were very persevering in their attempts to hold on by the boats, and we were obliged to strike them severely on the hands to make them desist. Previous experience had taught me the absolute necessity of firmness in repressing this practice, and I was pleased as well as surprised to see the patience with which they generally endured this treatment —a few only of the bolder spirits showing a momentary anger, but all acquiescing at length in the rule we had laid down. The freshness of the breeze which blew during our intercourse rendered it easier to deal with them, as they dropped behind directly they had ceased to ply their paddles; but they had no difficulty in out-stripping our boats whenever they exerted themselves; and I have little doubt but that they are able to propel their light kaiyaks at the rate of seven miles an hour.

While we were thus engaged, we heard the report of two muskets from the third boat, which had dropped two or three hundred yards astern. It appeared that the kaiyaks had not been so rigidly kept off by Clark, but had been allowed to hamper the oars so as to retard the boat's way. Some of the Eskimos, paddling close up to the stern, had tried to drag the cockswain overboard by the skirts of his jacket; and performed other various aggressive acts, evidently with the view of ascertaining to what lengths they might proceed with impunity. The umiaks, which had been kept aloof from the foremost boats, made a push for the third one, and one of them running across her bows, the men and women it contained instantly began to plunder the boat, and to struggle with the crew, who, being only six in number, would have been soon overpowered. Immediately on hearing the reports of the muskets, which were fired in the air merely as signals, I wore my boat round amid the shouts of the Eskimos who were hovering near us, and who thought that I was about to comply with their urgent requests that we should land and encamp in their neighborhood. Hauling up under the stern of Clark's boat, I declared that I would immediately fire upon the assailants if they did not desist, and my crew at the same time presenting their muskets, the attempt was at once quelled. I found, on subsequent inquiry, that nothing had been carried away but a small box on which the cockswain sat, which contained his shaving apparatus, and some other trifling articles belonging to himself, together with the boat's ensign. Though vexed at the loss of the latter, I could not but be glad that he was the only one of the crew who suffered, and I looked upon the theft of his property as poetical justice for his want of firmness. The umiaks, not being suffered again to approach our boats, soon pulled toward the shore, but the majority of the kaiyaks continued to keep company with us, the men conversing and begging as if nothing unpleasant had occurred. At length, as we had been drawing away from the shore all the morning, we totally lost sight of the land; and the kaiyaks assembling together, the men held a short consultation, and then paddled toward their encampment, being guided in their course by a dense column of smoke, which their families on shore had raised. Four or five of them continued to follow us for a short time, after the great body had gone away, evincing their boldness, even when much inferior in numbers, but they, also, went off on receiving some presents, which we could then make to them without fear of misconstruction.

Our inquiries were directed chiefly to obtaining information of the Discovery Ships, but the Eskimos, one and all, denied having ever seen any white people, or heard of any vessels having been on their coast. None acknowledged having been present at the various interviews of their countrymen with white people in 1826, and perhaps the circumstances attending those meetings might have deterred them from confessing that they were relatives of the parties that assailed Sir John Franklin's boats at that time; and as most of the men were stout young fellows, and few beyond the prime of life, only two or three of the old men in the umiaks could have been actually engaged in the struggle which then took place. One fellow alone, in answer to my inquiries after white men, said, "A party of men are living on that island," pointing, as he spoke, to Richard's Island. As I had actually landed there on the preceding day, I was aware of the falsehood he was uttering; and his object was clearly to induce us to put about and go on shore, which he and others had been soliciting us to do from the commencement of our conversation. I, therefore, desired Albert to inform him, that I had been there, and knew that he was lying. He received this retort with a smile, and without the slightest discomposure, but did not repeat his assertion. Neither the Eskimos, nor the Dog-rib or Hare Indians, feel the least shame in being detected in falsehood, and invariably practice it, if they think that they can thereby gain any of their petty ends. Even in their familiar intercourse with each other, the Indians seldom tell the truth in the first instance; and if they succeed in exciting admiration or astonishment, their invention runs on without check. From the manner of the speaker, rather than by his words, is his truth or falsehood inferred; and often a very long interrogation is necessary to elicit the real fact. The comfort, and not unfrequently the lives, of parties of the timid Slave or Hare Indians are sacrificed by this miserable propensity. Thus, a young fellow often originates a story of his having discovered traces of an enemy for which there is no real foundation. This tale, though not credited at first, makes some impression on the fears of the others, and soon receives confirmation from their excited imaginations. The story increases in importance, a panic seizes the whole party, they fly with precipitation from their hunting grounds, and if they are distant from a trading post, or large body of their nation, many of the number often perish in their flight by famine.

The Eskimos are essentially a littoral people, and inhabit nearly five thousand miles of sea-board, from the Straits of Belleisle to the Peninsula of Alaska; not taking into the measurement the various indentations of the coast-line, nor including West and East Greenland, in which latter locality they make their nearest approach to the western coasts of the Old World. Throughout the great linear range here indicated, there is no material change in their language, nor any variation beyond what would be esteemed in England a mere provincialism. Albert, who was born on the East Main, or western shore of James's Bay, had no great difficulty in understanding and making himself understood by the Eskimos of the estuary of the Mackenzie, though by the nearest coast-line the distance between the two localities is at least two thousand five hundred miles. Traces of their encampments have been discovered as far north in the New World as Europeans have hitherto penetrated ; and their capability of inhabiting these hyperborean regions is essentially owing to their consuming blubber for food and fuel, and their invention of the use of ice and snow as building materials. Though they employ drift-timber when it is available, they can do without it, and can supply its place in the formation of their weapons, sledges, and boat-frames, wholly by the teeth and bones of whales, morses, and other sea animals. The habit of associating in numbers for the chase of the whale has sown among them the elements of civilization ; and such of them as have been taken into the Company's service at the fur posts fall readily into the ways of their white associates, and are more industrious, handy, and intelligent than the Indians. The few interpreters of the nation that I have been acquainted with (four in all) were strictly honest, and adhered rigidly to the truth ; and I have every reason to believe that within their own community the rights of property are held in great respect, even the hunting grounds of families being kept sacred. Yet their covetousness of the property of strangers and their dexterity in thieving are remarkable, and they seem to have most of the vices as well as the virtues of the Norwegian Vikings. Their personal bravery is conspicuous, and they are the only native nation on the North American continent who oppose their enemies face to face in open fight. Instead of flying, like the Northern Indians, on the sight of a stranger, they did not scruple in parties of two or three to come off to our boats and enter into barter, and never on any occasion showed the least disposition to yield any thing belonging to them through fear.

As the narratives of the recent Arctic voyages contain descriptions of the manners, customs, and features of these people, and the sketches of Captains Beechey, Lyons, Sir George Back, Lieutenant Kendall, and others, give correct delineations of the personal appearance and costume, I shall not say more of them in this place.

As soon as the last of the kaiyaks disappeared beyond our horizon, we struck the boats' masts, and, pulling obliquely in toward the shore, landed to cook our supper at a place where there were three winter Eskimo habitations, and which is situated about seven or eight miles to the eastward of Point Warren. These buildings are generally placed on points where the water is deep enough for a boat to come to the beach, such a locality being probably selected by the natives to enable them to tow a whale or seal more closely to the place where it is to be cut up. The knowledge of this fact induced us generally to look for the buildings when we wished to land. The houses are constructed of drift-timber strongly built together and covered with earth to the thickness of from one to two feet. Light and air are admitted by a low door at one end; and even this entrance is closed by a slab of snow in the winter time, when their lamps supply them with heat as well as light. Ten or twelve people may seat themselves in the area of one of these houses, though not comfortably; and in the winter the imperfect admission of fresh air and the effluvia arising from the greasy bodies of a whole family must render them most disagreeable as well as unwholesome abodes. I have been told that when the family alone are present, the several members of it sit partly or even wholly naked.

As soon as supper was prepared, we withdrew to the boats to eat it, having anchored them under a sand-bank about a bow-shot from the shore. We had scarcely assumed this position when a party of Eskimos were perceived coming round a point a little to the westward, crouching under the bank, and evidently hoping to surprise us at the fire, of which they had seen the smoke. As soon as they knew they were perceived they brandished their bows and knives, made gestures of defiance, and threw their bodies into various most extraordinary attitudes, as they are accustomed to do when they meet with a strange party, of whose intentions they are uncertain.

As we were all, both men and officers, exhausted by the constant clamor, watchfulness, and exertion sustained throughout the morning, and I wished the men to have some repose, we determined on having no communication with this warlike group, since we could hope for no additional information from people residing so near the large body we had so recently parted from. On coming to the winter-houses, they showed themselves, at first cautiously, then openly, and ceased to gesticulate. Soon afterward one began to wade out to the sand-bank, and another to launch a kaiyak which he had brought. Thereupon, to intimate that we declined to receive them, Mr. Rae fired a ball so as to strike the water a few feet on one side of him. The bold fellow leaped into the air, cut several capers of defiance, and then retreated with the others into one of the houses, where they thought themselves safe, and from whence they continued to watch us. After a time one of them waded out to the sand-bank, planted a stick on it, and pulling off his reindeer jacket, hung it up, intimating that he wished to have its value, and then retreated to the shore. We, however, declined bartering, and at 10 P. M. weighed anchor, and, standing out to sea, worked to windward all night against a stiff breeze.

August 4th.—We gained only a few miles by the night's operations, having to contend with a tumbling sea, which drove us to leeward; and at 3 in the morning, on the wind moderating sufficiently to allow us to use the oars, we struck the masts and pulled for three hours across Copland Hutchison Inlet, when we landed on its eastern side to prepare breakfast. The shore in this quarter is for the most part low, but, at intervals of seven or eight miles or more, some of the conical eminences already mentioned occur. They have not the ridged and escarped aspect of sand hills, but, on account of their isolation, look more like artificial barrows, though unquestionably they can not be works of art. I am inclined to think that they are remnants of the sand formation which covers the shale so extensively along the banks of the Mackenzie, and that they have received their conical form from the washing of high tides during the occasional inundations of the low lands by the sea.

Copland Hutchison Inlet is about ten miles across, and its mouth is obstructed by sand-banks. A river seems to flow into it, as we could trace sandy cliffs for some distance inland, like the banks of a stream. Though many small ponds existed at the place where we landed, they were mostly brackish, and we had to search for some time before we obtained fresh water. On, at length, discovering some, we filled our breakers, to avoid a similar detention at our next meal. Drift-wood was also scarce. From these causes we were unusually long at breakfast; and soon after embarking we landed again to observe the meridional altitude of the sun; by which the latitude was ascertained to be 69° 44′ N., and the variation of the needle 58° E.

As we advanced to the eastward, in the afternoon, the coast became still flatter, so that the beach was under the horizon when we were in no more than half a fathom of water. The hummocks above-mentioned, which came into view in succession, looming like conical islands, as we ran along with a light westerly breeze, were the only land in sight. In blowing weather, the only resource for boats on this coast is to keep a good offing, as the surf then breaks high on the shelving flats. During the afternoon the sky was lurid, as if loaded with fog, and, though the horizon was tolerably clear, objects were very much altered by mirage. Altogether I have never, either in this climate or in any other, seen a more disagreeable atmosphere, and we all predicted a coming storm. At night, the wind heading us, so that we could not fetch the easternmost hummock in sight, we ran under a sand-bank, and anchored in a foot and a half of water about a quarter of a mile from the shore. On the tide ebbing, the boats were left dry.

The perfectly flat land here is covered, to the depth of four or five feet, with a moorish or peaty soil, which is much cracked, and in many places treacherously soft and boggy. Small lakes and ponds intersect it in all directions, mostly filled with brackish water, but some of them containing water fresh enough for cooking, though by no means good. The irregular ponds and marshy places make the course to any particular point exceedingly devious, and I had a long walk to reach one of the eminences, though its direct distance from the beach was little more than a mile. This hill rose from the boggy ground, in a conical form, to the height of about one hundred feet, its base having a diameter about equal to its height. A ditch, fifteen or twenty feet wide, which surrounded it, was passable only at two points; and, on ascending the hill, I found that hollow, like the crater of a volcano; in which, about fifteen feet below its brim, stood an apparently deep lake of very clear and sweet water. The interior beach of this curious pond was formed of fine clean gravel, and the hill itself apparently consisted of sand and gravel with a coating of earth. From its summit, I saw many similar heights to the eastward, and southeastward, that is, in an inland direction, and they seemed to rise, like islands, out of a great inlet or bay, for the low land connecting them, if such existed, was not visible. I looked long on the scene, but could not satisfy myself whether what I beheld was actually water, or merely the mirage of a low fog simulating an inlet of the sea. I was inclined to consider that the latter was the real state of the case, since, during the whole afternoon, the hills we had passed, as well as the one on which I then stood, seemed equally to rise out of a hazy sea, when we were at the distance of only four or five miles from them. Could I have convinced myself that the isolated peaks I saw were really islands, I should have thought that I was on the borders of the extensive Eskimo Lake, laid down as problematical in Sir John Franklin's chart, but which I now believe does not exist. The Indians, upon whose report it was indicated, meant, most probably, by the expression they used, the sea itself, or perhaps the inlet of Liverpool Bay, which lies further to the westward. Round the lower part of the hill, about eight or ten feet above the then level of the sea, lay a ring of large drift-timber, showing that, in certain states of the wind, the sea rises so as to inundate all the low lands. Nothing that we observed in the whole course of our voyage led us to think that the spring tides rise more than three feet on any part of this coast, and their rise was more generally only about twenty inches.

Reindeer frequent these flat lands at this season to feed on bents and grasses, and the ponds are full of geese and ducks. Many red-throated divers (*Colymbus septentrionalis*) also resort thither, and utter a most mournful cry when any one invades their retirement; the Lapland finch (*Plectrophanes lapponicus*) and horned lark (*Phileremos cornutus*) make a breeding station of this coast, and we observed the young birds running in numbers over the ground.

While Mr. Rae was out hunting, the crews were cooking supper on shore, and I remained with two boat-keepers at the boats, I received a visit from two middle-aged Eskimos and their wives; the latter being fat, jolly-looking dames, and considerably younger than their husbands. Albert was absent, so that I could not profit by the interview to ask as many questions as I desired to do, but I made out that if we kept in shore of the sand-banks, we should have water enough for our boats when the tide rose, and would find a passage out to sea at a point on which their winter houses stood—or, as they invariably termed such stations in their own language, Iglulik. They had not seen any white men on their coast, but with a ready flattery expressed their affection for them. Having been very liberal to my visitors, not only in purchasing all the small articles of traffic they had to offer, but also by making them some useful presents of files, hatchets, and knives, with a considerable quantity of beads, I became tired of their company and of the constant vigilance required to prevent them from pilfering, and therefore requested that they would go away, since I would give them nothing more. Their smiling countenances and deferential manner prevented me from using any threat, but I had to repeat my request several times, and, at length, from the urgency with which I spoke, one of the women seemed to think that I was afraid of their using violence, and opening the hood of her jacket, she showed me that there was a little naked infant in it. She then explained that when they went to war they left the children behind, and that she and the other woman had brought their young ones as a pledge of their pacific intentions. I prevailed upon them at length to go, by assuring them that I would part with nothing more at that time, but would give them something on a future occasion, if they came again to us.

Mr. Rae went in pursuit of reindeer in the afternoon, but the talking of the men who were wandering about scared away the animals. He was more successful in the night, and, by taking Halkett's portable boat on shore with him, killed nineteen brent geese, and some ducks. The water-fowl at this time were moulting in the small lakes, and became an easy prey.

August 5th.—A strong gale of wind raised a high surf on the shoals under which we lay, so that we could not launch the boats over them seaward. We, therefore, made sail when the tide rose, and ran on the inside of them until we became involved among the flats, and the boats were left quite dry. During low water, we marked off the best channels with poles. The tide began to flow at 11½ A.M., and by 2½ P.M., there was water enough for us to proceed, by launching the boats over some bars which intervened between the place where we were detained, and deeper water. Following a narrow channel till 5 in the afternoon, we secured the boats in a snug harbor under a tongue of sand at Point Atkinson. A strong northwest gale had raised a high sea, and we could do no otherwise than remain here at anchor until the wind and sea subsided.

Point Atkinson is a flat, low piece of ground, with a range of sand hills, forty or fifty feet high, thrown up along its northern side by the winds and waves. When we visited it in 1825, its extreme point was a small island separated from the main by a ditch, but this was now choked up and formed a marshy pond, the water of which being brackish, and fetid as well as greasy, from the quantity of whale oil with which the ground was saturated, was totally unfit for use. The oil had acted as a manure on the soil, and produced a luxuriant crop of grass from one to two feet high (*Elymus mollis, Calamagrostis stricta, Spartina cynosuroides*, and some shorter *Carices*). A small village of seven or eight huts stood on the point, among which is the *Kashim*, or house of assembly, described and figured in the narrative of my former voyage along this coast. I am not aware that a house of this description, appropriated as a council chamber and eating-place for the males, has been described as existing among the more eastern Eskimos. They possess, however, the appellation for it in their language. It is of more importance among the western tribes, as I shall have occasion to mention in a subsequent chapter.

The sea has carried away much of the sandy bank on which the *Kashim* stood, and the spray now washes its walls, so that it will likely be overthrown in a few seasons. It is still supported by whale-skulls built round its outside wall. After seeing the boats secured, Mr. Rae and I walked through the village by a well-beaten path which meandered among the long grass; and on looking into one of the smaller houses, found an old crone sitting warming herself over a few embers. She seemed surprised when she saw our faces, but exhibited no signs of fear, and soon began to talk very volubly. Some rows of very inviting herring-salmon being hung to dry on poles by her tent, she gave me five or six at my request, and, while Mr. Rae had gone to the boats to procure some articles to repay her and for presents, she told me that she must go to a party who were encamped on the shore about a mile off. I endeavored to persuade her to remain till she got the articles I had sent for, but she was bent on going, and, having made up a small bundle of some of her property,. walked off with a quickness and elasticity of step which her first appearance was far from warranting. The deep furrows of her countenance, and its weather-beaten aspect, gave her the appearance of advanced age; and while she sat she seemed to be a woman of ordinary stature, a little bowed down; but when she stood up she looked dwarfish, her height not exceeding four feet. Two tents, toward which she bent her steps, were visible from the boats, but we thought it better to let the inmates recover from their surprise at our arrival, and to seek an interview in their own way.

After a time we saw a man coming toward us. He made his approach very slowly, and by a devious path, keeping a muddy channel between us and him. On Mr. Rae going alone and unarmed to meet him, he waited his approach with evident trepidation, and, when they met, began to express by signs and words that he was old, infirm, and nearly blind, which was by no means the truth. Mr. Rae invited him to come to our tents, which were by this time pitched, and, after much persuasion, induced him to come on, but not until he had, on the request of the old man, blown in each of his ears, tapped his breast, and touched his eyes, as a charm, either to remove his maladies, or more probably to avert any evil influence which the white men might possess. After the performance of these ceremonies he came to us; and as his confidence increased he gradually laid aside the appearance of infirmity, and began to bustle about and pry into every thing, until at length he became troublesome. Though repeatedly spoken to, and told that we would not suffer him to handle any thing belonging to us, he was scarcely restrained, and required constant watching to prevent him from stealing. He took up his abode in one of the huts, and, after we had retired to rest, made an attempt to raise one of the boat's anchors, and continued to prowl about until the sentinel on duty checked him by showing his musket. In the middle of the night he entered one of the tents, with a long knife in his hand, but retreated on being spoken to. Perhaps he had no intention of committing any violence, as he habitually carried this knife, which was of Russian manufacture, either in his hand or up his sleeve. Through Albert I learnt that the two natives whom we had seen on the preceding day were sons of this man, and that they, with their families and some young people who were hunting with them, would come to the village in a few days to engage in the chase of whales, when they would be joined by one or two parties then in the pursuit of game on the east side of a river which falls into M'Kinley Bay. This small community does not wander far from their winter station on Point Atkinson. The hunters pursue reindeer and water-fowl on the neighboring flats in summer, chase the whale during one month or six weeks of autumn, live with their families in the village during the dark winter months, and in spring travel seaward on the ice to kill seals, at which time they dwell in snow houses. The old man and several of his elderly companions who subsequently came to the village declared that they had no remembrance of my former visit to this coast, and said that now was the first time that they had seen white men. They do not go as far as the mouth of the Mackenzie, and dread their turbulent countrymen in that quarter.

As none of the ponds or ditches which intersect Point Atkinson in all directions contain drinkable water, we were on first landing at some loss. The old woman, however, had told us that her people procured water from the seaward side of the sand-hills, and, by following a path which led in that direction, we discovered three wells, carefully built round with drift-timber, below high-water mark; which, when we first saw them, were completely sanded up. On clearing them out water contaminated with fetid whale oil flowed in abundantly, but this being repeatedly drawn off, until the surrounding sand was washed from its impurities, we at length obtained tolerable water for making tea. These wells are evidently supplied from rain falling on the sand-hills, and kept up to the level at which we found it by the pressure of the sea.

August 6th.—The old woman whom we first saw, another still older, and an aged blind man, came to the village this day; and in the afternoon three fine young men brought some ducks, which we purchased from them. They were eager to sell water-fowl for buttons, beads, or any trifle we chose to offer, and our crews eventually obtained a considerable number in that way; but they were very unwilling to dispose of the fish which hung on poles in the village. After letting us have a few, they refused to part with any more, even for a good price, assigning as a reason, that they belonged to a man who was absent. They either prize that kind of food very highly, or are scrupulous about using the property of an absent countryman.

A heavy gale continued all day, and raised a very high surf on the beach. As the weather was extremely cold, we required a considerable quantity of fire-wood, and converted two of the Eskimo scaffolds to that use; but I informed the owners that we would pay for it before we went, to which arrangement they gave a ready assent. At high-water, about a quarter before seven o'clock, A.M., the sea rose so much that the shoal off the point was covered, and the surf began to reach our boats, so that we had to shift them further in. The wind blowing on the shore had increased the rise of water considerably beyond an ordinary tide. It was low-water again a few minutes before one. In the afternoon the gale began to moderate, and in the evening three men came to the village, two of them being of the party who had visited the boats three days before. They told us that their women were coming in the umiaks on the following day.

We had now a pretty numerous body around us, and all were perfectly orderly except the man that we first saw, who seemed to be the chief, or perhaps the *shaman* or conjuror, of the community. His features were forbidding, but the younger men had intelligent countenances, modest and cheerful manners, and were neither forward nor troublesome. They were of moderate stature.

August 7th.—The wind continuing to abate throughout the day, and the sea to subside, we prepared to resume our voyage as soon as the evening tide had flowed sufficiently to admit of the boats being launched over the sand-flats. In the afternoon we saw the women and children approaching slowly in the umiaks. They stopped about a mile off, and, notwithstanding the signals and shouts of their husbands, hung back. The men called out *umiët kai-it*, "Boats, come here," with a peculiar elevated intonation, which could be heard at a great distance, similar to that practiced by the inhabitants of the Swiss Alps. We did not await their arrival, but, having found water enough about half a mile to the westward of the point, pushed out to sea; and partly under sail, partly by rowing, coasted during the night a very low shore, varied by a few higher islands, which the very flat sands surrounding them prevented us from approaching.

At eight in the morning of the 8th August we landed on one of a cluster of clayey and sandy islands, near Cape Brown; and while we were preparing breakfast were visited by four Eskimos, who told us that they had just killed several reindeer on a neighboring island. On desiring to have some of the venison in exchange for a knife they went away in their kaiyaks, and, before we had finished breakfast, returned again, bringing with them only the toughest and most inferior parts of the animals. I paid them what was promised, and made them some presents besides, but reprimanded them for their want of hospitality, and told them that I meant to have given them an ax had they been more liberal. These people, like the other parties we had previously communicated with, declared that no large ships nor boats had been seen on their coasts, and that we were the first white men they had ever beheld. I could not discover that any remembrance of my visit to their shores, twenty-three years previously, existed among any of the parties I saw on the present voyage, though I never failed to question them closely on the subject.

After breakfast we crossed Russell Inlet, and as we passed Cape Brown several Eskimos put off from the shore, and three of them overtook us though we were going with a stiff breeze at the utmost speed of our boats, or about six geographical miles an hour. To those that came up we made presents, and put the usual questions.

In the evening we anchored the boats under the westernmost of two islands lying immediately off Cape Dalhousie, and, having landed, pitched our tents on the beach. The islands here, and the Cape itself, consist of loam or sand, and present steep cliffs toward the sea forty or fifty feet high. The surface is level, except where ravines, occasioned by the melting of the snow in the beginning of summer, intersect it; and all the islands are so surrounded by sandy flats that a boat can not come near the beach. On the present occasion we anchored three or four hundred yards from the shore.

The island on which we encamped is a breeding-place of the *Xema sabinii*, the handsomest of all the gulls. Many of the parents were flying about accompanied by their spotted young, also on the wing. This is the most westerly ascertained breeding station of the species, which has been found at Spitzbergen, Greenland, and Melville Peninsula. Mr. Rae shot some fine male specimens, whose plumage and dimensions agreed exactly with the description in the *Fauna Boreali-Americana*. The eggs are deposited in hollows of the short and scanty mossy turf which clothes the ground.

August 9th.—Through the carelessness of the night-watch the boats were suffered to ground on the ebbing of the tide, and it was not till eight o'clock that the water had flowed sufficiently to permit of our embarking. We then stood across Liverpool Bay with a very light breeze, and about two o'clock had sight of the eastern coast near Nicholson Island, the western shore being visible at that time, but soon afterward sinking below the horizon. At half-past nine, P.M., we reached the eastern shore, and encamped under the frozen cliffs of Cape Maitland. This cape is an island,

and on the former voyage I passed through the channel which divides it from the main. Its surface is nearly level, the soil is loam or clay, and the cliffs which bound it are about eighty feet high, and, being worn at the base by the action of the waves, overhang the narrow beach by eight or ten feet. Landslips are frequent, and occasion a frozen surface to be constantly exposed to view. The island does not differ from the neighboring lands in its subsoil being frozen: since permanent ground ice is found every where at eighteen or twenty inches beneath the surface. All access from below would have been cut off by the overhanging cliff, were it not for deep gullies which here and there afford a steep path to the top. Vegetation was very scanty, and throughout this voyage I observed that flowering plants were more scarce, and the herbage generally thinner, than in 1826. The bulk of the cliff is composed of a black clay or loam, which is disposed in undulated beds, and, in some places, the section exhibits a spherical mass eight or ten yards in diameter, with concentric layers, like the coats of an onion. A few pebbles occur in the loam, and the beach is formed of sand and small pebbles washed from the cliffs, and consisting mostly of trap mixed with quartz, and a little white sandstone and limestone.

In 1826 I observed many slabs of red sandstone in the channel behind the island, but there were none at our present encampment. I caused a pit to be dug at the top of the cliff, and found that the thaw had penetrated sixteen inches. A thermometer laid on the bottom of the pit indicated 33° F., the temperature in the shade being at the time 42° F. High water occurred at 1 A.M., and the ebb flowed to the northward along the island.

The distance we ran from our encampment at Cape Dalhousie to Point Maitland, measured by Massey's log, was thirty-five geographical miles, or thirty-one direct (excluding the angle we made), which agrees exactly with Lieutenant Kendall's map of 1826. Mr. Rae killed some rock ptarmigan in the night. The old males were moulting at this time.

August 10th.—Having breakfasted before embarking, we left our sleeping place at 6 A.M. Soon afterward we saw two Eskimo tents on the extreme point of the island; but as the inmates did not show themselves, and we did not wish to be delayed, we proceeded onward without disturbing them. and crossed the mouth of Harrowby Bay. At 4 P.M. Baillie's Islands came in sight, and we held on our course between them and the main. Some Eskimos coming off here, we learned from them that a large river falls into the bottom of Liverpool Bay; and we had previously received the same information from the party which we saw on the 8th. Eskimos inhabit the banks of this river, but the families residing at Cape Bathurst do not go so far. The river can be no other than the "Begh'ula tessè" of the Hare Indians, who frequent Fort Good Hope. These Indians say that it is a large river, abounding in the fish from which it is named ("toothless fish," *Salmo Mackenzii*); that it rises near Smith's Bay of Great Bear Lake, and is eight or ten days' journey to the eastward of Fort Good Hope overland (one hundred miles or thereabout). They also said, that some years ago they fell in with a party of Eskimos who were hunting on its banks, and, a quarrel ensuing, several of the latter were killed.

We could not find a convenient landing-place either on Baillie's Islands or on the main shore, owing to the flatness of the coasts, and were compelled to anchor the boats nearly a mile from the beach. The men waded ashore to collect driftwood and cook supper, after which we all embarked to sleep in the boats. We had scarcely completed our arrangements for passing the night, when we became aware of a fleet of kaiyaks with three umiaks coming down upon us in a crescentic line, looming formidably in the faint twilight. As I did not wish the men's rest to be interrupted by visitors, a ball was fired across their path to arrest their progress, on which they assembled in a group evidently in consultation. Albert now hailed them by my direction, and said that we were going to sleep, and that the sentinels would fire on any one who came near the boats by night, but that we should be glad to trade with them in the morning. When they fully understood our wishes, after a little further parley, they retired and did not afterward trouble us.

At 2 o'clock A.M., on the 11th, we weighed anchor, and nine Eskimos, with their kaiyaks, and three umiaks, containing the women, having come off, they led us by a sufficiently deep channel to the westward of a dry bank or a sandy island, over a bar on which we found from four to five fathoms of water. On crossing this we passed suddenly from muddy water into a green sea, in which we had no bottom with the hand-lead. In 1826 we sought a way out to the eastward of the island, and, the bar there being muddy and shallow, found no little difficulty in forcing the boats through. The northern channel would form a good ship harbor.

From these people we learnt that during their summer of two moons they see no ice whatever, that they were now assembling to hunt whales, and would go out to sea to-morrow for that purpose. The black whale, their present object, they call *ai-ë-werk*, and the white whale, which also frequents this coast, *keilaloo-ak*. In some summers they kill two black whales, very rarely three, and sometimes they are altogether unsuccessful. In the course of conversation, we were told that the several families have hunting grounds near their winter houses, on which the others do not trespass; and the proprietors of several points of land in sight were named to me. They knew but little of the country beyond their own vicinity; and one of them having told me that Cape Bathurst was an island, I affirmed that it was not, on which, with an air of surprise, he exclaimed, "Are not all lands islands?" None of them could remember my former visit, though I had communicated with a part of their countrymen only a few miles from their present residence. We told them that we were looking for ships and men of our nation whom we expected to meet; and they said they would be glad of the visits of white men, and would treat them hospitably. In exchange for some fish, seal and whale-skin leather, and a few other things, we supplied them well with knives, files, hatchets, and beads. Part of the number who wished to come to the boats the preceding evening had, on our declining the interview, gone to their winter houses on the western shores of Baillie's Islands; and those who accompanied us from our anchorage in the morning landed on the extremity of Cape Bathurst, where their winter houses stood.

It was part of my instructions to bury some pemican at this cape, and to erect a signal-post; but the presence of the natives hindered me from doing so. As soon, however, as we had gone far enough to be, as we supposed, beyond their view, we put on shore, and having dug a hole on the top of the cliff, deposited therein a case of pemican, with a memorandum explaining the objects of the expedition. Every precaution was used to replace the turf, so as not to betray that it had been moved; some drift-timber was piled upon it and set on fire, and a pole, painted red and white, planted at the distance of ten feet. As all the drift-timber at this place had been gathered into a heap by the Eskimos, and the pole was part of it, we hung up some articles of value to them by way of payment, in the hope that it would cause them to respect the signal-post. In the mean time our crews were preparing breakfast, and we had just finished this meal, when we saw some Eskimos from the cape running toward us. They had evidently been watching us, and came in the expectation of receiving some additional articles; nor were they disappointed. The soil here was thawed to the depth of fourteen inches. The deposit was made about five miles from the extreme point of Cape Bathurst.

Many black whales and two white ones were seen this morning. The eider ducks had now assembled in immense flocks, and with the brent geese were migrating to the westward. Both these water-fowl follow the coast-line in their migrations on the Pacific as well as the Atlantic sides of the continent. The eiders are only accidental visitors in the interior, and the brents are not seen inland to the eastward of Peel's River; but Mr. Murray informs me that in their northerly flight they follow the valley of the Yukon, thus cutting across the projecting angle of Russian America.

The surface of the country in the vicinity of Cape Bathurst is level or gently undulated, and the sea cliffs are in many places nearly precipitous, and about one hundred and fifty feet high. The strata, where exposed, are sand and clay, and I believe that this promontory, from its northern point to the bottom of Franklin Bay, is the termination of the sandy and loamy deposit and bituminous shale which throughout the whole length of the Mackenzie rests on the sandstone and limestone beds so frequently noticed in the preceding pages, and fragments of which may be traced among the alluvial islands in the estuary of the Mackenzie, and in Liverpool Bay. A line drawn from *Clowt sang eesa*, or Scented Grass Hill, of Great Bear Lake, to the north-northwest, would form a tangent to the eastern coast-line of Cape Bathurst, and most probably mark the limit of the formation on that side. If so, the river Beghula, which enters Liverpool Bay, will flow through a country similar to that forming the banks of the Mackenzie, and being consequently well wooded, will abound in animals.

As we proceeded to the southeast from Cape Bathurst along the shore, the crest of the high bank rose to about two hundred and fifty feet, and beds of bituminous shale, similar to those on the Mackenzie, are exposed in many places. At Point Trail, in latitude 70° 19′ N., the bituminous shale was observed to be on fire in 1826, and the bank had crumbled down from the destruction of the beds. Selenite, alum in powder, and the wax-colored variety of that salt named "Rock butter," with sulphur, were among the products of the decomposition of the shale which I then collected; and the clays which had been exposed to the heat were baked and vitrified, so that the spot resembled an old brick-field. The sand covering the shale here is coherent enough to be a friable sandstone; and many concretions of clay iron-stone exist in the shale beds, exactly similar to those which are imbedded in the shale of Scented Grass Hill. Wilmot Horton River flows out by a narrow gorge from a flat valley, and the high banks, rising in ridges above the valley, flank it some way inland, as we had noticed them doing on the tributary streams that join the Mackenzie.

Near Point Fitton the cliff is two hundred feet high, and contains layers of rock-butter two inches thick, with many crystals of selenite adhering to the surface of the slates. The cliff is capped by a marly gravel, two yards thick, containing pebbles of granite, quartz, Lydian stone, and compact limestone. To the southward of Wilmot Horton River, portions of the ruined bank continue to emit smoke.

Cape Bathurst has been recently invested with more interest since it is the point of the main shore from whence Commander Pullen was directed by the Admiralty to take his departure in the summer of 1850, in his adventurous attempt to reach Melville Island. By the last accounts from Mackenzie's River, we learn that this enterprising officer received his instructions by express, on the 25th of June, being then in Slave River, on his way to York Factory. He immediately turned back, having been supplied with 4500 lbs. of jerked venison and pemican by Mr. Rae, which he embarked in one of the Plover's boats, and in a barge of the Hudson's Bay Company, being the only available craft. The barge is well adapted for river navigation, but from its flatness unfitted for a sea-voyage, though it may be in some respects improved by the addition of a false keel, which Commander Pullen would probably give it before he descended to the sea. Its weight will render it much less manageable among ice than a lighter boat. No intelligence of this party has reached England since the above date, but we may expect to hear of his proceedings in May or June, 1851, before this volume has passed through the press.

* This anticipation has been realized, as has been mentioned in p. 133. Commander Pullen found the sea covered with unbroken ice all the way from the Mackenzie to Cape Bathurst, a small channel only existing in shore, through which he advanced to the vicinity of the cape. Failing in finding a passage out to sea, to the north of Baillie's Islands, he remained within them, until the advance of winter compelled him to return to the Mackenzie.

CHAPTER IX.

Voyage continued along the Coast.—Franklin Bay.—Melville Hills.—Point Stivens.—Sellwood Bay.—Cape Parry.—Cocked-hat Point.—Cache of Pemican.—Ice Packs.—Archway.—Burrow's Islands.—Darnley Bay.—Clapperton Island.—Cape Lyon.—Point Pearce.—Point Keats.—Point Deas Thomson.—Silurian Strata.—Roscoe River.—Point De Witt Clinton.—Furrowed Cliffs.—Melville Range.—Point Tinney.—Buchanan River.—Drift Ice.—Croker's River.—Point Clifton.—Inman's River.—Point Wise.—Hoppner River.—Wollaston Land.—Cape Young.—Stapylton Bay.—Cape Hope.—Cape Bexley.—Ice Floes.—Point Cockburn.—A Storm.—Chantry Island.—Salmon.—Lambert Island.—Leave a Boat.—Cape Krusenstern.—Detained by Ice.—Basil Hall's Bay.—Cape Hearne.—Peculiar Severity of the Season.—Conjectures respecting the Discovery Ships.—Resources of a Party Inclosed by Ice among the Arctic Islands.—General Reflections.

August 11*th*, 1848.—We sailed along the coast all day with a light breeze, and in the afternoon eleven Eskimos came off from the shore and sold us some deer's meat. A woman of the party ran for two miles along the beach in the hope of receiving a present, and, when quite exhausted with her exertions, stripped off her boots to barter with us. One of the men in the kaiyaks brought them off, but, as they were too small for any of our crew, we returned them with a present of more than their value. These men gave us no additional information, but expressed pleasure when told that they might expect to meet other parties of white men. A scull of eleven white whales were seen in the evening.

We continued under sail all night, and at three in the afternoon of the 12th, landed in a very shallow bay, to the southward of Point Stivens, to cook a meal which served for both breakfast and dinner. Mr. Rae went in pursuit of some reindeer which were seen from the boats, but, owing to the extreme flatness of the land, which afforded no cover, he was unable to approach them.

The high banks of Cape Bathurst are continued to the bottom of Franklin Bay, where they recede a little from the coast, and are lost in an even-backed ridge, apparently not exceeding four or five hundred feet in height. These hills are named the Melville Range, and cross the neck of the peninsula of Cape Parry, appearing again behind Darnley Bay. The peninsula is so flat near its isthmus, and so much intersected by water, that I am still in doubt whether it may not be actually a collection of islands. But if this is the case, the channels which separate the islands are intricate and shallow. To the south of Point Stivens the soil was wholly mud, apparently alluvial; to the northward beds of limestone crop out. In the evening we encamped on Point Stivens, which is a long, narrow, gravel beach, composed mostly of pieces of limestone, some of which contain corals. Sea-weed is very scarce throughout the whole of the Arctic coast, but we saw on the beach here some rejected masses of decayed *Laminaria,* probably *saccharina;* also a stunted white spruce, lying on the beach, still retaining its bark and leaves. Mr. Rae shot a fine trumpeter swan, on which we supped. The only water we could find here for cooking was swampy, and full of very active insects shaped like tadpoles, which were just visible to the naked eye (*Apus; Lepidurus*, probably *Lynceus*).

The crimson and lake tints of the sky, when the sun set this evening, were most splendid, and such as I have never seen surpassed in any climate.

On August the 13th, we embarked at 3 A. M., and at half-past ten landed in Sellwood Bay on some horizontal beds of limestone, which are the first rocks *in situ* that become visible, in tracing this peninsula from the south. No organic remains were detected in the stone. Many very large slabs, moved but a short way from their parent beds, were piled upon each other within reach of a high surf, and among them lay great boulders of greenstone-porphyry and hornblende rock. To the north of the bay, there are high cliffs of limestone, and also a detached perforated rock, which employed Lieutenant Kendall's pencil on my former voyage. Many white-winged, silvery gulls were breeding on the various shelves of its cliffs, and their still unfeathered young were running about, alarmed by the clamor of the parent birds.

In the evening we anchored in a snug boat-harbor, within the westernmost of the two points which terminate Cape Parry. The part of the Cape which will be first visible on approaching from sea, is a hill about five hundred feet high, which far overtops all the neighboring eminences. From it a comparatively low peninsular point stretched west-northwest about half a mile, being connected to the main by a gravel bank, and terminated on its sea-face by a limestone cliff, which in some points of view resembles a cocked-hat. An indented bay, about three miles across, separates this point from another more to the east, which extends fully as far north. Booth's Islands, five in number, form a range nine miles long, whose extremities bear from the hill northwest and southwest respectively. The channels between the islands vary in width from one to three miles. On the east side of the hill, cliffs of limestone, washed by the waves, have been scooped into caves and arches, which, without much aid from the imagination, recalled many fine architectural forms. A boulder of chert, lying on the shore, measured five feet in length, by four in breadth, and exhibited some curious veining.

In approaching our anchorage we shot two seals, and one of them, being thrown on the strand, attracted, late in the evening, the attention of a gray fox, which was prowling about the beach. As soon as the animal saw the carcass, it halted, and, after a momentary survey, leaped lightly behind a large log of drift-timber, from whence it peeped out at it from time to time. While I was watching the fox's reconnoitring tactics with some interest, and speculating upon the time it would devote to the survey before it ventured to approach the carrion, a noise made by the sentinel, who was seated by the fire, scared the wary object of my study, and it fled swiftly up the hill.

August 14th.—This morning we deposited a letter, with a case of pemican, on the verge of the cliff of Cocked-hat Point, covered it with fragments of limestone, and erected in front of it a pile of stones marked with red paint. The beds of this cliff are horizontal, and they are not only interleaved with regular layers of chert, but also contain nodules of the same material. On rounding the eastern point of Cape Parry, we saw packs of drift-ice for the first time since we commenced our sea voyage. Eight or ten miles to the southward of the Cape we passed an isolated rock, perforated by an archway, standing in front of some bold limestone cliffs; soon after noon we were opposite to a point distinguished by a remarkable rounded hummock; and in the evening we encamped on one of Burrow's Islands. This island is composed of cherty limestone, which in decaying acquires a honeycombed surface as hard as a flint or file. Rain fell in the night for the first time for many days.

A fair wind having sprung up in the night, we embarked an hour before midnight, and on the morning of the 15th stood down the bay, with light breezes and hazy weather; landing at 10 A.M. on Clapperton Island to prepare breakfast. The coast in this quarter is similar in character to that on the opposite side of the peninsula in the same parallel, being low and not easily approachable on account of the extensive sandy banks which lie off it. Clapperton Island, itself, is gravelly. From it, we saw land round the bottom of the bay, with some intervals, apparently inlets.

On re-embarking, we steered directly for Cape Lyon, distant about ten miles; but a low thick fog coming on, we got involved in a stream of drift-ice, on which the course was altered a little, so as to fetch within the cape. We made our way through the ice without damage to the boats, and in the afternoon found ourselves about four miles from the pitch of the cape, close in-shore, under some very high mural precipices of a bluish-gray, slaty rock, on which a thick mass of columnar basalt is imposed. The cliffs are about two hundred feet high, and stand out in succession, forming the salient angles of several shallow indentations of the coast-line. A talus of unmelted snow-drift lay under most of them, which being undermined by the action of the waves, would be detached on the first heavy fall of rain, and become icebergs. Toward the bottom of Darnley Bay, the coast-line declines greatly in altitude, but the heights of the Melville Range are dimly seen in the distance. We anchored for the night in the northwest angle of the cape, between two projecting points of basalt. The land here rises between three and four hundred feet above the sea.

On the 16th we continued our voyage to the eastward, and, on landing to prepare breakfast at Point Pearce, ascertained that high-water took place there at eight o'clock. The trap ranges run in this quarter in a southwest and northeast direction, and produce small bays; their precipitous faces are turned toward the west-southwest. Judging from the view we had of them from the boats in passing, the cliffs were mostly composed of greenstone-slate. At Point Pearce, the shore is formed of flesh-colored limestone, whose beds crop out in successive cliffs like stairs, and attain a height of two hundred and fifty feet a short way from the beach. At a small point lying between the one just named, and Point Keats, there are magnificent columns of basalt, with a pillar of the same material rising out of the water immediately in front of them. The bay to the south of these columns is lined with cliffs of flesh-colored limestone ; thin layers of sandstone crop out further to the eastward, and covering them there is an overflow of dark leek-green basalt or greenstone.

A strong head-wind having sprung up and occasioning much fatigue to the rowers, while our progress was small, we put in at Point Keats, and encamped. Mr. Rae and Albert went to hunt reindeer, and I took a short walk inland. I soon came to extensive beds of flesh-colored sandstone, forming the bounding walls of a deep narrow ravine, through which flowed a small shallow river varying in width from ten to thirty yards, but whose channel bore evidence of a considerable body of water passing through it in the spring floods. The ridges of sandstone seem to have a direction from west-northwest to east-southeast, or nearly parallel to the general coast-line here, and are much fissured, the principal fissures being in the line of the strike. The cliffs for the most part face westerly, and the west wall of the ravine is lower than the eastern one. The sandstone is fine-grained, hard, and durable. Some beds are very white, others flesh-colored, and interleaved with them are beds of chert or quartz rocks. The dome-shaped summits of the Melville Range are visible in the distance.

The fragments of sandstone cover many miles of surface, and the limestone, which splits off in thin layers by the action of the frost, becomes by the action of the weather as rough as a file, and soon wears out a pair of shoes. There are some beautiful excavations in the cliffs that are exposed to the waves, and a fine gothic shrine, with a canopy and mouldings supported on slender pillars, attracted our special notice. An isolated column, which stands before it, had been selected as a breeding-place by two ivory gulls, who were very clamorous when any one approached their nest. The young had ash-gray backs, and were nearly fledged. A very clear sunset enabled me to obtain an extensive view to seaward, and I am convinced that no land much above the level of the water lies within forty miles to the northward of this point.

A thick, wet fog, accompanied by a strong head-wind, detained us at our encampment till after breakfast on the 17th. During our enforced stay, Mr. Rae killed a roe reindeer in excellent condition, and we procured also some waveys (*Anser hyperboreus*). These geese and the northern divers (*Colymbus borealis*) were at this time migrating to the southeast, or in the opposite direction with respect to the line of coast that we had observed the eider ducks and brent geese proceeding a few days previously near Cape Bathurst.

About a mile and a half to the westward of Point Deas Thomson, the projecting point of a deep cove is perforated, forming a natural bridge, and not far from it another projection exhibits a less striking opening.

TORSO ROCK.

A detached column of limestone, inclosing masses and layers of chert or quartz rock, is also cut through, forming a pointed arch. The whole coast is composed of limestone, forming high cliffs at intervals. The quartz rock beds acquire occasionally a pistachio-green color, as if from the presence of epidote. A similar stone occurs at Pigeon River on the north shore of Lake Superior; and the limestones and sandstones of the latter district, with their associated trap rocks, as at Thunder Mountain, correspond in most respects with those between Cape Parry and the Coppermine River; and consequently, if we can rely on lithological characters, they may be considered as the oldest members of the silurian series, or as the rocks on which that series is deposited, to which epoch the Lake Superior formation has been assigned. If we had been able to trace up the limestone spurs of the Rocky

Mountains which traverse the Mackenzie, they would most probably have been found running up to and connected with the limestone of this coast.

At six in the afternoon the wind veering round to the southwest, we embarked again, and continued under sail till three in the morning of the 18th, when the wind failing us we dropped anchor in the mouth of Roscoe River. We resumed our voyage at 8 A.M., after preparing an early breakfast.

A little to the westward of Point de Witt Clinton, a range of basalt and limestone cliffs extends for a mile along the beach, under which there lay a talus of unmelted drift-snow. In 1826, though we passed along this coast nearly a month earlier in the season, we observed much fewer of this kind of memorial of the preceding winter, and that year, the flowering plants were more plentiful, and the vegetation generally more luxuriant. The deterioration of the climate after rounding Cape Parry became daily more and more evident to us, though we had decreased our latitude above a degree since leaving Cape Bathurst. The presence of a large body of warmer water, carried into the Arctic Sea by the Mackenzie, may have some influence in ameliorating the temperature within the limits of that wide estuary; but I am inclined to think that Wollaston and Banks's Lands, by detaining much drift-ice in the channels which separate them from one another and from the main shore, have a still more powerful effect in lowering the summer temperature.

At Point de Witt Clinton the cliffs are formed of flesh-colored beds of limestone interleaved with bluish-gray beds, and containing fibrous and compact gypsum in veins. These cliffs are forty or fifty feet high, and are covered to a considerable depth with diluvial loam, containing fragments of sandstone, limestone, and trap rocks, some of them rolled, others angular. The surface of the loam is undulated; and, about a quarter of a mile from the beach, cliffs of basalt protrude, at the height of two hundred and fifty feet above the water. A short way to the westward also of the point, cliffs of basalt rise from the beach. This stone breaks up here into cubical blocks, many of which are piled up at the foot of the cliff. These fragments and the basaltic shelves at the base of the cliff, are sculptured by fine acute furrows, and polished by the action of ice and gravel, the scratches being generally perpendicular to the line of coast, but occasionally crossing each other. The Melville Range is about five miles distant from this part of the shore, and presents many mural precipices and ravines on its acclivity. The highest points did not appear to rise more than seven or eight hundred feet above the sea. An undulated grassy country intervenes between the range and the shore.

In the evening we encamped on a point situated in latitude 69° 30′ N., to the westward of Point Tinney. The sea-bank shelves down from the general level of the country here one hundred and fifty or two hundred feet, and, being cut by ravines, shows conical eminences when seen from a boat. The diluvium is at least forty feet thick.

August 19*th*.—This morning we crossed the mouth of Buchanan River, which is a very small stream in this month, but the channel which it fills in time of flood is one hundred yards wide. I have mentioned, in the preceding narrative, that when we visited Fort Good Hope, no rain had fallen there this season, and a few short showers only occurred after we came to the coast. The banks of this stream gave further evidence of the dryness of the summer, in the clayey soil being cracked every where into round flat cakes, on which the foot-prints of geese, which had walked over them while yet muddy from the melting snow, were sharply impressed. Were such a surface to be covered by drift-sand, the foot-prints might be preserved as in the ancient sandstones. In the present instance the winter frosts set in without any heavy rains having fallen to obliterate the traces, which would consequently remain hard until the following spring.

Mr. Rae brought in two fine reindeer, and several seals also were killed; but none of the men relishing the dark flesh of the seals, while they had abundance of excellent venison, I gave directions that no more should be shot. A meridional altitude was obtained in lat. 69° 19½′ N.

To-day we passed through much drift-ice by very devious channels, and not without risk of the boats being crushed, but fortunately without damage. Croker's River issues from a triangular, level valley, three or four miles wide at the beach, and extending about five miles backward. Over this the stream spreads when flooded, but when low, filters out by narrow channels, barred across by sand-banks. The valley is bounded on the east and west by elevated banks of sand, diluvial loam and boulders, which meet at the Melville Range. The valley at this time presented a singular scene of desolation; for though the summer was now far advanced, its flat bottom was entirely covered with large floes of ice, which had been probably driven over the sand-bar from the sea by northerly winds.

A brood of long-tailed ducks (*Harelda glacialis*) were seen swimming in one of the streams, with the mother bird in the van. Her wariness did not prevent us from laying her flock under contribution for our evening meal.

August 20th.—This day also our voyage was performed among crowded floes of ice, and was consequently slow. When we landed to prepare breakfast, Mr. Rae killed a fine buck reindeer. In this quarter, a skillful hunter, like Mr. Rae, could supply the whole party with venison without any loss of time. A meridional observation was obtained in lat. 69° 9′ N. between Point Clifton and Inman's River, and about two miles from the latter; the variation of the compass, by the sun's bearing at noon, being 61½° E., and Point Clifton bearing north 26° west, distant a mile and a quarter.

A little to the eastward of Point Clifton, there are cliffs of limestone, from whence to Inman's River the beach is alluvial and shingly. The river flows between high gravel banks and alluvial cliffs; and to the eastward of it, the limestone rises in successive terraces to the height of four hundred feet above the sea. The eminences which this formation produces are long and round-backed, and it abounds in narrow deep ravines or fissures.

At Point Wise, the cliffs are composed of crumbling earthy limestone, containing chert in layers and nodules. From this point, at sunset, Wollaston Land was distinctly seen at the distance apparently of thirty miles.

On the morning of the 21st, we passed two ranges of high limestone cliffs, at the second of which, lying to the eastward of Hoppner River, we put ashore to prepare breakfast. The ice under this cliff was loaded with many tons of gravel. Wollaston Land, as seen from hence, appeared to have its summits and ravines covered with snow, but the channel being filled with ice, the ice-blink rendered the true form and condition of distant land very uncertain.

One of our boats having been injured by the ice and rendered leaky, we put ashore early at Cape Young to repair the damage, which was effected in the course of the evening.

August 22d.—On embarking this morning in rather thick weather we struck across Stapylton Bay, for three hours, and then, getting sight of Cape Hope, bearing east-northeast, hauled up for it. The sky was dark and lowering, with occasional thick haze and heavy showers, and a water-spout, seen in-shore, gave intimation of an approaching storm. Ice floes lying close off Cape Hope caused us no little trouble, the passages among them being very intricate, and the perpendicular walls of the masses being too high to allow of our landing, or seeing over them. In the afternoon we passed Cape Bexley, running before a stiff breeze, and at 5 P.M., a storm coming suddenly on, we were compelled to reduce our canvas to the goose-wing of the main-sail, under which we scudded for an hour, and then entering among large masses of ice, about two miles from Point Cockburn, found shelter under some pieces that had grounded. The shore was too flat to admit of our bringing the boats near enough to encamp; the ice-cold sea water chilled the men as they waded to and fro; there was no drift-timber on the beach; and we passed a cold and cheerless night in the boats, the wind being too strong to admit of our raising any kind of shelter. I afterward learned that this storm began at Fort Simpson at 6 A.M. on the 23d, or, making allowance for the difference of longitude, about thirteen hours and a half later. It commenced on the Mackenzie by the wind changing from north-east to northwest, and the sky did not clear up till nine in the morning of the 24th. At the same date an earthquake occurred in the West India Islands, which did much damage.

During the night much ice drifted past, and in the morning of the 23d the sea as far as our view extended was one dense close pack. By 10 A.M. the wind had moderated considerably, and the rising tide having floated some of the stranded pieces of ice, we were enabled to advance slowly along the shore, by moving them aside. In this way every small indentation of the coast-line required to be rounded, and as these were numerous the direct distance made good was small. We encamped, on the tide falling again, at 2 P.M., on a gravel point lying about ten miles to the westward of Chantry Island. Snow, which fell in the night, did not wholly melt this day, and the distant rising grounds were white. The weather continued very cold; drift-wood proved to be exceedingly scanty; and in the night we had high winds and much sleet. The coast-line is more deeply indented in this quarter than the chart* indicates, as chains of low sandy islands which lie across the entrances of the bay hid them from our view on the former voyage. The country is flat and strewed with fragments of limestone.

No lanes of open water could be discerned on the 24th from any of the eminences near the coast. By handing the boats over the flats, where the water was too shallow for heavy ice, we were enabled to round six small bays. In the course of this labor many salmon were seen, a few were killed by the men with their poles, and some were found on the ice, having been left by seals, which were scared away by our approach. The shape of this salmon is much like that of the common sea-trout of England, but its scales are rather smaller. Its flesh is red, and its flavor excellent. A medium-sized fish measured 29 inches in length, and 16 in girth. We encamped at five in the evening a little to the eastward of Chantry Island, having traveled about twenty-four miles round the bays, but gained only eight in direct distance. At this place, beds of compact white limestone crop out on the beach, and the surface of the country is thickly strewed with boulders of bright-red sandstone, some of them very large. Many boulders of basalt and other trap rocks also occur.

August 25th.—A strong west-northwest wind blowing during the night, cleared away much of the ice that pressed immediately on the beach, though all remained white to seaward. We were enabled thereby to run for three hours before the wind, but then came to a bay, through which there was no passage, large floes resting on the rocks of the beach, and no lanes existing outside. A meridional observation gave the latitude of this place 68° 36′ N. Lambert Island lies some miles distant in the offing. The surrounding country, as far as my examination extended, consists of limestone, but many sandstone boulders of various colors, lying on the surface, point out that stone as existing *in situ* in some locality not far distant.

August 26th.—A frosty night covered the sea and ponds with young ice, and glued all the floes immovably together, so that the rise of the tide was no longer serviceable to us. We carried the cargo and launched the boats across a point of land for half a mile in the morning, and spent the rest of the day in the various operations of cutting through tongues of ice, dragging the boats over the

* In Franklin's second Overland Journey.

floes, where they were smooth enough, moving large stones that lay in the way and resorting to every expedient we could devise to gain a little advance. Two more portages were made in the afternoon over rugged paths, and we traveled in all about five miles in a day of very severe labor. A heavy snow-storm converted the surface of the pools of sea water into a thick paste, the water being already cooled down to the freezing point.

By a repetition of the same operations, which occupied us during the previous day, we advanced on the 27th about three miles and a half. After Mr. Rae had attentively examined the sea from a high cliff without perceiving any slackness in the ice, or motion during the flood or ebb tide, I determined on lightening the labors of the men, by leaving one boat and her cargo on a rocky point, which bears north 28° west from Cape Krusenstern, distant twelve miles. Our encampment in the evening was on a flat terrace of slaty limestone under a high cliff of the same rock. The limestone reposes on beds of chert or quartz rock in thin layers, which in some places are detached in large slates. Here we deposited, on a flat shelf of the rock, several cases of pemican, an arm-chest, and some other things that encumbered the boats, and rendered them less fit for launching over the ice.

During the night, a fresh wind from the east-southeast brought much snow, which added to the pasty condition of the surface of the water, and produced a layer of semi-fluid matter that completely deadened a boat's way under oars.

Three hours were consumed on the morning of the 28th in bringing the boats about a hundred yards, the cold weather almost paralyzing the men's powers of exertion.

When the tide flowed in the afternoon, a portage of a thousand yards was made, and the boats being afterward dragged across some smooth floes of ice, we gained a pool of open water, and pulled to the bottom of Pasley's Cove, where we encamped. Mount Barrow is a conspicuous object from the bottom of the cove, as it rises abruptly from the flat limestone strata.

August 29th.—During the night and this morning the same keen frosty east winds continued, with snow showers. The tide fell so much that the boats were left aground in the morning, and we were unable to proceed until 7 A. M. In three hours we came within about a mile and a half of the pitch of Cape Krusenstern, when further progress being barred by ice heaped against the cliffs, we put ashore and drew up the boats on the beach. In the flat limestone beds, of which the country here is formed, I observed a curious variety of structure which I saw no where else, and which I can not satisfactorily account for. It was a diminutive ridge like the roof a house, formed in this manner $\wedge$, of the upper slaty layer of the limestone, its height or the breadth of its sides being about a foot only, yet its length was half a mile. It seemed to be connected with a fissure. Though the layers of the limestone are most extensively detached by the freezing of the moisture, which insinuates itself between them, such a process could not produce any thing so regular as this small anticlinal ridge. If beds already fissured were, in subsiding, to be pressed more closely together, the edges of the fissure might perhaps assume such a form. Elsewhere, in the same formation, straight furrows, as if drawn by a plow, are common, and evidently proceed from the small fragments which cover the ground, filling a crack, nearly to the brim. Judging from the whole surface here being covered with thin pieces of limestone to the exclusion of soil, I should infer that the frost splits off the layers and breaks them up more effectively than any agent to which rocks are exposed in warmer climates, and that the scantiness of the soil is owing to the shortness of the season of growth of the lichens and other plants, which have the power of decomposing the surface of the stones and so producing a little mould. The frost breaks up the stone before the lichens have time to establish themselves.

The limestone which forms the cliffs of Cape Krusenstern, and the other cliffs on the coast between it and Cape Kendall, contains many thin slaty beds of chert or quartz rock, either bluish-white, or colored reddish-brown by oxide of iron.

We remained all the 30th in an encampment, watching the ice outside, or making excursions across the cape to examine the sea in various directions. Some small lanes of water were visible, and the ice was moved to and fro by the flood and ebb, but no channel was discovered by which we could hope to make any progress toward the Coppermine River. The wind continued in the east-northeast quarter, and the weather was very chilling. We employed the men in erecting a column of stone near the tents. It was on this cape that Mr. Rae spent a month of the following summer in anxiously watching for an opening in the ice, by which he might cross to Douglass Island and Wollaston

Land The true position of Douglass Island is ten miles from Cape Krusenstern.

At 4 P.M. on the 30th, a sudden movement of the ice having opened a narrow channel, we hastened to launch and load the boats; and, pushing them through, succeeded in rounding the cape. We then ran under sail with a favorable breeze till 11 P.M., when the night being dark we got involved among drift-ice, and not being able to reach the shore dropped anchor off Point Lockyer, and went to sleep in the boats.

We resumed the voyage at 4 A.M. on the 31st, and, getting inside of fields of ice which covered the sea as far as our view extended, we ran along the coast until we came to an island in Basil Hall's Bay, on which the sea ice rested, barring our further progress on its outside. On the former voyage, this island was thought to be part of the main shore; but on ascending to its summit, which rises about three hundred feet above the sea, we discovered its insular nature, and perceived that the ice within it was not only smoother, but lay less compactly. We therefore took that direction, and found that the inlet runs about five miles behind the island into a narrow valley, bounded by hills between three and four hundred feet high.

In the afternoon we reached Cape Hearne, and ascended its high grounds to look to seaward, from whence we beheld the same impacted floes of ice to which we had of late been accustomed. The cliffs of this cape are composed of a shingly or slaty limestone, and the beach presents much greenish slate-clay, which breaks down like wacké, and becomes brown on the surface, but its relations to the limestone in respect to position could not be made out. The extreme point of the cape is low and sandy; and the country lying immediately to the southward of the limestone ridge, that constitutes the high grounds of the promontory, is flat, grassy, and marshy, forming a fine feeding ground for reindeer, of which we saw several herds. A considerable stream winds through the plain, and enters the sea about two miles to the southward of the cape. Its mouth, which is barred by a sand-bank, is marked by two cliffs of sand, and it pours out water enough to render the sea clay-colored for two or three miles, and fresh enough to be drinkable. We encamped three miles beyond it, on a point formed of slate-clay, of which the beach, after we had passed Cape Hearne, seemed every where to consist. Here we found a decayed sledge, that was put together with copper nails marked with the broad arrow, which must have been extracted from the boats I abandoned on the Coppermine River in 1826.

Since rounding Cape Parry, we had seen very few traces of Eskimos, and had not met a single individual of that nation; but we had now entered a better frequented district, in which traces of the natives abounded. There was a hard frost in the night, with a sharp east-southeast wind blowing from the ice.

The coast being flat, and the water within the ice very shallow, the officers and most of the men walked along the shore, on the morning of the 1st of September, leaving two of the crew in each boat to pole them along. The country is level and swampy, and is crossed by long channels like ditches, on whose banks shale and slate-clay are occasionally exposed. It would seem that on this eastern flank of the limestone formation there is also a shale deposit, but not so extensive a one as that seen on its western side.

In the course of our walk we passed an Eskimo stage, on which, among deer-skins and other effects, we observed the skin of a white bear. We had previously found a skull of this animal on the beach, so that there is no doubt of its frequenting this coast.

After breakfast we made very slow progress, having to cut a way through new ice. It did not exceed an inch in thickness, but, being formed on a foundation of snow, did not crack readily, while, at the same time, it was hard enough to cut the planks of the boats through, rendering them scarcely sea-worthy, though we had strengthened them on the water line with sheets of tin beat out from the pemican cases. In dragging them over the floes they were much shattered. At noon, finding that we could not advance further, in the present condition of the ice, without pulling the boats to pieces and running the risk of losing all our stores and provisions, we encamped about eight miles from Cape Kendall, which bore southwest.

On viewing the sea from the high grounds behind our encampment, and ascertaining that no lanes of open water were visible in any direction, I determined, after consulting with Mr. Rae, to leave the boats at this place, and commence the overland march in the course of two days, if no amelioration of the weather or alteration in the state of the sea occurred during the interval. f the weather should improve, it was our intention to remain some days longer, to watch its effects on the ice. The higher grounds at this time were covered with snow, but the lower lands were mostly bare.

The unavoidable conclusion of our sea voyage while still at some distance from the Coppermine River was contemplated by me, and I believe by every individual of the party, with great regret. I had hoped, that by conveying the boats and stores up the Coppermine River beyond the range of the Eskimos we could deposit them in a place of safety to be available for a voyage to Wollaston Land next summer. But abandoned as they must now be on the coast, we could not expect that they would escape the searches of the hunting parties who would follow up our foot-marks, and who were certain to break up the boats to obtain their copper fastenings. The unusual tardiness of the spring, and our unexpected delay on Methy Portage for want of horses, caused our arrival on the Arctic coast to be considerably later than I had in secret anticipated, though it differed little from the date I had thought it prudent to mention when asked to fix a probable time. Even a few days, so unimportant in a year's voyage elswhere, are of vital consequence in a boat navigation to the eastward of Cape Parry, where six weeks of summer is all that can be reckoned upon. Short, however, as the summer proved to be, neither that nor our tardy commencement of the sea voyage would have prevented me from coasting the south shore of Wollaston Land, and examining it carefully, could I have reached it, for the distance to be performed would have been but little increased by doing so. The sole hinderance to my crossing Dolphin and Union Straits was the impracticable condition of the close packed drift ice. In wider seas, where fields and large floes exist, these offer a pretty safe retreat for a boat party in times of pressure, and progress may be made by dragging light boats like ours over them : but the ice that obstructed our way was composed of hummocky pieces, of irregular shape, and consequently ready to revolve if carelessly loaded or trod upon. At certain times of the tide, moreover, they were hustled to and fro with much force.

As only small packs of ice and few in number were seen off the Coppermine by Sir John Franklin in 1820, by myself in 1826, and by Dease and Simpson in 1836 and 1837, being four several summers, the sight of the sea entirely covered so late in August was wholly unexpected, and I attributed so untoward an event to the northwest winds having driven the ice down from the north in the first instance, and to the easterly gales, which afterward set in, pressing it into that bight of Coronation Gulf; but Mr. Rae's experience in the summer of 1849 shows that in unfavorable seasons, the boat navigation is closed for the entire summer, and we learned from a party of Eskimos whom we met in Back's Inlet, as I shall have occasion to mention hereafter, that the pressure of the ice on the coast this summer was relieved only for a very short time.

The state of the straits produced the melancholy conviction that a party, even though provided with boats, might be detained on Wollaston Land, and unable to cross to the main; but yet at that time my apprehensions for the safety of the missing ships were less excited than they have been since. For then their absence had not been extended much beyond the time that their provisions were calculated to last: and, being ignorant of Sir James C. Ross having been arrested in Barrow Straits, I hoped that the accumulation of ice which annoyed us might be the result of a clearance of the northern channels, and that the two ship expeditions might have happily met at the very time that we were no longer able to keep the sea. It is now known that the season was equally unfavorable throughout the Arctic seas north of America.

The idea of a cycle of good and bad seasons has often been mooted by meteorologists, and has frequently recurred to my thoughts when endeavoring to find a reason for the ease with which at some periods of Arctic discovery navigators were able to penetrate early in the summer into sounds which subsequent adventurers could not approach, and to connect such facts with the fate of the Discovery ships. But neither the periods assigned, nor the facts adduced to prove them by different writers, have been presented in such a shape as to carry conviction with them, until very recently. Mr. Glaisher, in a paper published in the Philosophical Transactions for 1850, has shown, from eighty years' observations in London and at Greenwich, that groups of warm years alternate with groups of cold ones, in such a way as to render it most probable that the mean annual temperatures rise and fall in a series of elliptical curves, which correspond to periods of about fourteen years; though local or casual disturbing forces cause the means of particular years to rise above the curve or fall below it.

The same laws doubtless operate in North America, producing

a similar gradual increase and subsequent decrease of mean heat, in a series of years, though the summits of the curves are not likely to be coincident with, and are very probably opposed to those of Europe; since the atmospherical currents from the south, which for a period raise the annual temperature of England, must be counterbalanced by currents from the north on other meridians. The annual heat has been diminishing in London ever since 1844, according to Mr. Glaisher's diagram, and will reach its minimum in 1851.

It can be stated only as a conjecture, though by no means an improbable one, that Sir John Franklin entered Lancaster Sound at the close of a group of warm years when the ice was in the most favorable condition of diminution, and that since then the annual heat has attained its minimum, probably in 1847 or 1848, and may now be increasing again. At all events, it is conceivable that, having pushed on boldly in one of the last of the favorable years of the cycle, the ice, produced in the unfavorable ones which followed, has shut him in, and been found insurmountable; but there remains the hope that if this be the period of the mean heat in that quarter, the zealous and enterprising officers now on his track, will not encounter obstructions equal to those which prevented their skillful and no less enterprising and zealous predecessor in the search, from carrying his ships beyond Cape Leopold.

With respect to the maintenance of a party detained on the islands north of Coronation Gulf, reindeer and musk-oxen may be procured by skillful hunters; but unless the chase were duly organized, and only the most expert marksmen and good deer stalkers suffered to go out, there would be a danger of the animals migrating from feeding grounds on which they were much disturbed. With nets a large quantity of salmon and other fish might be captured in Dolphin and Union Straits, and doubtless also in the various channels separating the islands; with percussion guns we had no difficulty in killing seals, and we might, had we chosen, have slain hundreds, though, as they dive at the flash, the chance of shooting them with a ship's musket having an ordinary lock, would be greatly diminished. Swans, snow geese, brent geese, eiders, king ducks, cacawees, and several other waterfowl, breed in immense numbers on the islands: and the old ones when moulting, and the young before they are fledged, fall an easy prey to a swift runner, and still more surely to a party hemming them in and cutting off their retreat.

To people acquainted with the Eskimo methods of building ice and snow houses, shelter may be raised on the bleakest coast, except in the autumn months ; but, unless blubber were used as fuel, there would be a difficulty in maintaining fire for cooking by any one who has not the genius for turning every thing to account which Mr. Rae evinced, when he boldly adventured on wintering on a coast bearing the ominous appellation of Repulse Bay, with no other fuel than the *Andromeda tetragona*—an interesting and beautiful herb in the eye of a botanist, but giving no promise to an ordinary observer that it could supply warmth to a large party during a long Arctic winter. To apply it, or any of the other polar plants, to such a purpose, a large quantity must be stored up near the winter station before the snow falls.

I have thought it right to throw these few observations together in this place, that a reader unacquainted with the natural resources of the country, may judge of the probability of the whole or part of the crews of the Erebus and Terror maintaining themselves there, supposing the ships to have been wrecked. Of course, as long as the vessels remained, they would afford shelter and fuel ; but the other contingencies would come into consideration, if parties went off in various directions in quest of food. One great purpose of the expedition which I conducted along the coast, was to afford relief to such detached parties, or to the entire crews, had they directed their way to the continent, and our researches proved at least that none of the party, having gained that coast, were dragging out a miserable existence among the Eskimos, without the means of repairing to the fur posts. In the following summer of 1849, Mr. Rae ascertained that the Eskimo inhabitants of Wollaston Land had seen neither the ships nor white men. The knowledge of these facts had an influence with the Admiralty in concentrating the future search in the vicinity of Melville Island ; Captain Collinson and Commander Pullen being directed to approach its coasts from the westward, while Captain Austen, and the squadron of hardy navigators in his wake, were to trace the Discovery ships from the eastward. A more ample and noble effort to rescue a lost party was never made by any nation, and it has been humanely seconded from the United States of America. May God bless their endeavors !

CHAPTER X.

Preparing for the March.—Sleep in Back's Inlet.—Eskimo Village.—Eskimos Ferry the Party across Rae River.—Basaltic Cliffs.—Cross Richardson's River.—March along the Banks of the Coppermine.—Geese.—First Clump of Trees.—Musk-oxen.—Copper Ores and Native Copper.—Kendall River.—Make a Raft —Fog.—Pass a Night on a Naked Rock without Fuel.—Fine Clump of Spruce Firs.—Dismal Lakes.—Indians.—Dease River.—Fort Confidence.—Send off Dispatches and Letters.

On the 1st and 2d of September, we had northerly and north-east winds, with a low temperature, sleet, snow, and occasional fogs. We were all employed in preparing the packages for the march, consisting of thirteen days' provision of pemican, cooking utensils, bedding, snow-shoes, astronomical instruments, books, ammunition, fowling-pieces, portable boat, nets, lines, and a parcel of dried plants. These were distributed by lot, each load being calculated to weigh about sixty or seventy pounds. Mr. Rae voluntarily resolved to transport a package nearly equal to the men's in weight; but, distrusting my own powers of march, I made no attempt at carrying such a load, as I had done on a former voyage, and restricted myself to a fowling-piece, ammunition, a few books, and other things which I thrust into my pockets. Six pieces of pemican were buried under a limestone cliff, together with a boat's magazine full of powder. The tents were left standing near the boats, and a few cooking utensils and hatchets deposited in them for the use of the Eskimos.

After an early breakfast on the morning of Sunday, the 3d of September, we read prayers, and then set out at six o'clock. At first we pursued a straight course for the bottom of Back's Inlet, distant about twelve geographical miles; but finding that we were led over the shoulder of a range of hills on which the snow was deep, we held more to the eastward, through an uneven swampy country, where we saw many deer feeding; but made no attempt to pursue them.

The men, with a few exceptions, walked badly, particularly the two senior seamen, and after we had gone a few miles, were glad to lighten their loads by leaving their carbines behind. At half-past three we reached the inlet, about seven miles from the pitch of Cape Kendall, and halted for the night under a cliff of basalt two hundred feet high. The inlet and the sea in the offing were full of ice, and the weather continued cold ; but some scraps of drift-wood, chiefly willows, being found on the beach, we managed to cook supper ; and, selecting the best sleeping places we could find among the blocks of basalt, passed a pretty comfortable night.

We started a quarter before six on the morning of the 4th, to walk round the inlet ; and Frazer having sprained his knee on the preceding day, we were constrained to lighten his load by leaving a large hatchet, and distributing a portion of his pemican among the others. Our course along the inlet was south 74° west for four miles and a half, when we perceived ten Eskimo tents on the opposite shore. Mr. Rae and Albert went ahead of the men, who were straggling very slowly along ; and on coming opposite to the tents, and shouting, three Eskimos crossed the inlet in their kaiyaks, and cordially consented to ferry the whole party over. This small tribe have no "umiaks ;" and, as the kaiyaks carry only one person, some contrivance was requisite to render them available as ferry-boats. Our friends had already learned how to effect this from their intercourse with Mr. Simpson and his party in 1838, viz., by placing two poles across a pair of kaiyaks, and lashing them firmly together. In this way a single paddler could take over a sitter and his bundle. Four kaiyaks, being all they could muster, were brought into requisition, by which, with the addition of Lieutenant Halkett's portable boat, three men with their loads could be ferried over at each trip. At the place where we crossed, the inlet had contracted to the breadth of four hundred yards ; and is there, in fact, a river, since its water is fresh. The whole party was landed on the southern shore by eleven o'clock. On the river I bestowed the name of my active, zealous, and intelligent companion Mr. Rae, as a testimony of my high sense of his merits and exertions, which had been called forth to the uttermost in our late endeavors to push on through the ice. It was mainly through his skill and perseverance that we had been enabled to travel as far as we did by sea, and thus shorten the land journey ; which, with an increased distance, and, conseqently, proportionably augmented loads, would have been a very arduous undertaking indeed to some of our party. We considered ourselves as very fortunate in obtaining the assistance of a friendly party of Eskimos at this place, on learning from them that the river kept its width, and was not fordable for a long way up the country. Mr. Rae, in the succeeding spring, ascended it for twenty miles, and ascertained that it flowed directly from the west, and was about the size of the Dease, or about one hundred and twenty yards wide. Its bed is limestone; and a range of basaltic cliffs, varying from fifty to two hundred feet in height, skirts its northern bank. These cliffs are a continuation of the magnificent precipices, which, commencing at Cape Kendall, rise at intervals of three or four miles on the north shore of Back's Inlet, their faces being to the southward, and their line of direction or strike nearly due east and west. At Cape Kendall the basalt is obscurely columnar, and rests on a bed of compact felspar, containing minute grains of a green mineral. At a cascade in Rae River, ten miles above its mouth, walls from eight to twenty feet high, of bluish-gray quartz rocks in thin layers, hem in the stream. Salmon and other fish ascend a shelving shoot of the cascade. At this place Mr. Rae discovered, among the limestone and quartz rock, layers of asparagus-stone, or apatite (phosphate of lime), thin beds of soap-stone, and some nephrite, or jade—a group of minerals which belongs to primitive formations; and from the similarity of the various rocks associated in this quarter to those occurring at Pigeon River, and other parts on the north shore of Lake Superior, I am inclined to consider that the two deposits belong to the same geological era, both being more ancient than the silurian series. Neither Mr. Rae nor I discovered any organic remains in the limestone.

Among the Eskimos here encamped, we recognized two mentioned by Mr. Simpson, one having a wen on his forehead, and the other being a very old man who walked on crutches. The kind treatment and presents they received from Messrs. Dease and Simpson had impressed them with a favorable opinion of the dispositions of white men, and doubtless was the cause of their readiness to come to our assistance, and to put themselves and their families so completely in our power. Our men bought seal-skin boots from them, which proved very useful; and we paid the man with the wen, who was the leader, for his services in ferrying us over, with two hatchets, which were of great value to him. I had cautioned every one against offering this harmless, good-natured people any offense; and I must give our men the credit of having strictly adhered to the orders they received. I believe I was the only one who entered any of their huts; and I did so for the purpose of presenting some needles and other articles to the women, and obtaining a glimpse of their ménage. In one tent six or seven women were seated in a circle sewing. They were nearly naked, very dirty, hung their heads down, and seemed to be much afraid. As the females we met on the coast, who showed neither fear nor shamefacedness, were generally clean, I believe that the apprehensions of these poor women had caused them to rub ashes or mud on their faces and persons. They received my presents, but seemed to be relieved when I took my leave. Before we quitted the encampment, several younger men joined from the northern shore of the inlet; and we learnt that we had interrupted their day's occupation in killing reindeer. The more active among them go at this season to the meadows which we had crossed on the previous day, and gradually drive the animals to the inlet, hemming them in, and compelling them, with the aid of their dogs, to take the water. As soon as this takes place, the rest of the party, who are lying in wait in their kaiyaks, paddle toward the herd, and spear as many of them as they can. A considerable quantity of deer's meat was hanging to dry on stages; and we purchased a little of it for our evening meal.

These people told us, as I have mentioned already, that the ice had parted from the shore only a very short time this season, which, they added, was almost unprecedented within their recollection. Their migrations extend only to the lower part of the Coppermine River on one side, and a short way along the coast on the other. They communicate occasionally with the Eskimos of Wollaston Land, but none of them had been so far to the westward as the sources of Rae's River. The want of umiaks was a sufficient indication of the shortness of their migrations seaward.

Our friend with the wen accompanied us three or four miles on our journey, to show us a ford across Richardson's River; but the number of questions he put to Albert respecting the boats, showed that his thoughts were directed to the treasures he expected to find in them; and at length he turned back, after pointing out the direction in which we ought to go. Albert had been told not to mention the place where the boats were left; but the Eskimos could without the slightest difficulty trace up the foot-marks of so large a party as ours; and I believe that by the evening, or early next day, most of the party were assembled in our deserted tents.

We arrived on the banks of Richardson's River about three o'clock, but failed in finding a ford; and, the walking being bad, some of the men lagged far behind, which induced us to encamp early. Richardson's River, as well as Rae's, is flanked by lofty precipices of basalt, which, coming successively into view, produce striking vistas in a bleak and otherwise uninteresting country. From the summit of one of these eminences near our encampment I obtained a wide view of the land, and saw a line of cliffs running along the Rae from Cape Kendall; another rank marks out the course of the Richardson, from Point Mackenzie* up to the junction of its two branches, where the cliffs also fork off at an acute angle, a series of them skirting the valley of each branch. A range of cliffs, but of a less imposing character, forms the western boundary of the valley of the Coppermine, separating it from that of the Richardson. All these rows of precipices face toward the south-southeast, or east-southeast, and radiate between west and south-southwest from a point in Coronation Gulf, at which they would meet if prolonged. The western boundary of the granite formation appears in the islands of that gulf, associated with many trap rocks; in the form of lofty hills at Cape Barrow; again at the bend of the Coppermine, on the south side of Kendall's River; on the northeast and eastern arms of Great Bear Lake; on Point Lake; in country round Fort Enterprise; and from thence to Fort Providence and across Great Slave Lake to the mouth of Slave River, and so onward to Athabasca.

Richardson's River was discovered in 1822 by some hunters of Sir John Franklin's party, and, on their report, it received its present name from that officer; but its outlet was erroneously supposed to be only four or five miles to the west of the Coppermine. In 1826, I ascertained that its supposed mouth was only a shallow bay; and, in 1838–9, Mr. Simpson examined the river, and proved that it falls into Back's Inlet; on which occasion he confirmed the appellation which Sir John Franklin had given it.

* At this point the basalt is superimposed on a dark bluish-gray crystalline limestone.

Its junction with the inlet was ascertained by Mr. Simpson to be in lat. 67° 53′ 57″ N., long. 115° 56′ W.

Commencing the day's journey at six in the morning of the 5th, we crossed, about an hour afterward, a small tributary of the Richardson; and at nine, having then walked about four miles and a half from our sleeping place without discovering a ford, we determined on crossing in Lieutenant Halkett's boat, though, as it could carry no more than two men at a time, the operation was likely to be tedious. Some tall willows (seven or eight feet high), growing on the muddy banks of the river, afforded us the means of making a fire and preparing breakfast. In the mean time, all the net lines, spare lines, and carrying slings were united to form a hawser, wherewith we might draw the boat backward and forward. Mr. Rae and Albert crossed first; and, owing to the man to whom the paddles had been assigned as part of his load having left them behind, they had no other instruments for propelling the boat than two tin dinner-plates. They succeeded, however, in crossing, though their hands were much chilled by the ice-cold water; and subsequently the whole party were drawn across. The width of the stream, by measuring half the line, was ascertained to be one hundred and forty yards.

At one o'clock all had crossed; and, the bundles being again duly arranged, we resumed the march, and in a short time gained the summit of the ridge dividing the valley of the Richardson from that of the Coppermine. The latter was clothed with snow, the climate being seemingly more severe, though the distance between the streams is so small. The plain which lay at our feet, between the ridge and the Coppermine, is so much intersected by small lakes, that we chose the driest line of march, rather than the most direct, to avoid the necessity of fording the lakes, or losing ground by rounding them. At three o'clock we reached the banks of the river, three or four miles above Bloody Fall; and, having found a sufficiency of wood, made a good fire, which of late had been a very rare luxury. Many deer, Hutchin's and snow geese were seen; and Mr. Rae having killed nine or ten of the latter, we enjoyed an excellent supper.

The country within the influence of the sea-breezes which come from the icy surface of Coronation Gulf, has the barren aspect and poor climate of the *tundras* of the Siberian arctic region. The moister tracts, where the soil is clayey, retain so much ice-cold water in the short summer, that a sparing vegetation exists only in the hassocks, which bear, among the Chepewyan tribes, the name of "women's heads," and render the footing of pedestrians insecure and dangerous. "You may kick them," say the ungallant Indians, "but they cause you to stumble and never go out of the way." In the drier, sandy, and gravelly spots, which are more common among the primitive rocks, the ground is covered with the lichens on which the musk oxen and reindeer feed. Of these the *cornicularia* and *cetraria* are the most important; and they are most prized by the animals when the melting snow in spring renders them soft and tender. As the season advances, the grasses and bents which flourish in sheltered valleys furnish the chief food of the herbivorous animals; and, when the snows fall, the reindeer retreat southward to the woody districts, into which they penetrate deeper in severe weather, and in the milder intervals return to the barren grounds to scrape the hay from beneath the snow. The suddenness of the winter in these high latitudes serves the important purpose of arresting the juices of the grasses and freezing them, so that until late in spring they retain their seeds and nutritive qualities without withering. It has the same effect on the berry-bearing plants. The crow-berry (*Empetrum nigrum*), bleaberry (*Vaccinium uliginosum*), and cranberry (*Vaccinium vitis idea*), which grow in profusion among the lichens of the arctic wastes, not only furnish fruits for the bears and geese in autumn, but retain them in perfection until the ground begins to dry up under the influence of the hot summer suns, and the new flowers are expanding. In the month of September the snow-geese (*Anser hyperboreus*), and Hutchin's geese (*Anser Hutchinsii*), feed much on the crow-berries, which render them fat and well-flavored. The first-named geese breed in Wollaston Land, to which they cross in the beginning of June. We had noticed, while on the coast of Dolphin and Union Straits, the earliest bands traveling southward again in the middle of August, so that their stay in their native place falls short of three months. The Hutchin's geese and brent geese breed on the coasts of the Arctic Sea, and the laughing geese (*Anas albifrons*) resort to the country north of the Yukon, beyond the Arctic circle. The Canada geese, or "bustards" of the Canadians (*les outardes*), breed throughout the woody districts, but do not reach the vicinity of the Arctic Sea, except on the banks of some of the large rivers. The most northern localities in which we observed them were the channels between the alluvial islands which form the delta of the Mackenzie.

On the 6th we had clear weather with a hard frost, and gladly welcomed the face of the sun, which had been a stranger to us for more than a fortnight. The swamps being frozen over so as to support a man's weight, the party generally walked more briskly than usual; but three of the seamen and two of the sappers and miners were so lame, that we were obliged to make long and frequent halts to allow them to close in, and were unable to accomplish two geographical miles in the hour. To spare their strength, we encamped at the early hour of 2 P.M., having marched about ten miles and a half. Deer, geese, and ptarmigan, were seen in abundance during the day. In the evening the weather became cold, with rain, snow, and hail.

On the 7th our morning's march was performed in a snow storm, with a chilly northerly wind. About four miles from last night's sleeping-place, we came to a chain of narrow lakes, lying parallel to the river, and emptying themselves into it by a small stream which issues from their northern extremity. They are three miles in length, and lie about a mile from the river. We afterward forded two rapid torrents full of large greenstone boulders. One of them flows through a narrow chasm in friable dark-red sandstone, and the other is bounded by cliffs of red quartz rock, or perhaps of trap, but I could not approach them near enough for examination. The discomfort of the march was greatly augmented by the freezing of our clothes, wet in crossing the streams, and we gladly encamped, at two o'clock, on coming to a clump of stunted white spruce trees, where we arranged a comfortable bivouac, by placing small branches between the frozen ground and our blankets. In the existence of many scattered stumps of decayed spruce fir trees, and the total absence of young plants, one might be led to infer, that of late years the climate had deteriorated, and that the country was no longer capable of supporting trees so near the sea-coast as it had formerly done. Many plants of different species of *Pyrola* grow on the sea-shore; and as these are most abundant in forest lands, it is possible that they may be the memorials of ancient woods. The largest tree in the clump in which we bivouacked had a circumference of thirty-seven inches at the height of four feet from the ground.

Its annual layers were very numerous and fine, and indicated centuries of growth, but I was unable to reckon them. This place lies in lat. 67° 22′ N.

The evening proving fine, Mr. Rae and Albert went out to hunt, and both had the pleasure of seeing the musk-ox, for the first time in their lives. The *uming-mak* is known by name and reputation to all the Eskimo tribes; but as it does not exist in Greenland, or Labrador, nor in the chain of islands extending north from that peninsula along the west side of Davis Straits, Albert, who was a native of East Main, now for the first time approached its haunts. Mr. Rae, with the feelings of an ardent sportsman, had longed to encounter so redoubtable an animal; and the following is an account of the meeting:

On perceiving a herd of cows, under the presidency of an old bull, grazing quietly at the distance of a few miles from our bivouac, he and Albert crept toward them from to leeward; but the plain containing neither rock nor tree behind which they could shelter themselves, they were perceived by the bull before they could get within gun-shot. The shaggy patriarch advanced before the cows, which threw themselves into a circular group, and, lowering his shot-proof forehead so as to cover his body, came slowly forward, stamping and pawing the ground with his fore-feet, bellowing, and showing an evident disposition for fight, while he tainted the atmosphere with the strong musky odor of his body. Neither of the sportsmen were inclined to irritate their bold and formidable opponent by firing, as long as he offered no vital part to their aim; but, having screwed the bayonets to their fowling, pieces, they advanced warily, relying on each other for support. The cows, in the mean time, beat a retreat, and the bull soon afterward turned; on which Mr. Rae fired, and hit him in the hind quarters. He instantly faced about, roared, struck the ground forcibly with his fore-feet, and seemed to be hesitating whether to charge or not. Our sportsmen drew themselves up for the expected shock, and were by no means sorry when he again wheeled round, and was, in a few seconds, seen climbing a steep and snow-clad mountain side, in the rear of his musky kine.

These animals inhabit the hilly, barren grounds, between the Welcome and the Copper Mountains, from the sixty-third or sixty-fourth parallels to the Arctic Sea, and northward to Parry's Islands, or as far as European research has yet extended. They travel from place to place in search of pasture, but do not penetrate deep into the wooded districts, and are able to procure food in winter on the steep sides of hills which are laid bare by the winds, and up which they climb with an agility which their massive aspect would lead one ignorant of their habits to suppose them to be totally incapable of. In size they are nearly equal to the smallest Highland or Orkney *kyloes;* but they are more compactly made, and the shaggy hair of their flanks almost touches the ground. In structure they differ from the domestic ox, in the shortness and strength of the bones of the neck, and length of the dorsal processes which support the ponderous head. The swelling bases of the horns spread over the foreheads of both sexes, but are most largely developed in the old males. The musk-ox has also the peculiarity in the bovine tribe of the want of a tail; the caudal vertebræ, only six in number, being very flat, and nearly as short, in reference to the pelvis, as in the human species; the extreme one ending evenly with the tuberosities of the ischium. A tail is not needed by this animal, as in its elevated summer haunts musquitoes and other winged pests are comparatively few, while its close, woolly, and shaggy hair furnishes its body with sufficient protection from their assaults. The fore-pasterns are provided on their outsides with a slender accessory bone, of about half their length. The fossil Irish elk and musk-deer have also rudimentary toes, but of a different form. Though I have not been able to ascertain that the range of the species was ever greater than it is known to be at present, I have read somewhere of a skull having been found in Greenland. One in tolerable preservation, but defective in the nose, was procured by Captain Beechey, from that very curious deposit of bones in the frozen cliffs of Eschscholtz Bay of Beering's Straits. That skull is now preserved in the British Museum, and a perfect skeleton of the recent animal exists in the museum at Haslar Hospital.

Sept. 8th.—A meridional observation was obtained to-day in lat. 67° 17′ N. We crossed a projection of the Copper Mountains, to cut off a considerable bend of the river; and, at four in the evening, reached its banks again, and encamped. While among the hills we had to walk in snow shoes, with much fatigue; but in the afternoon a thaw took place in the low grounds, under the influence of a warm sun; and we were annoyed by sand-flies in the evening. I noticed that the upper branches of the scrubby spruce firs, among which we encamped, were confined to their southeast and southern aspects. The lower branches, as usual in such exposed situations, lay close to the ground, and spread widely, considering the small height of the tree.

The effect of the last two or three days' march proved to me that I had over-calculated my strength, in loading and clothing myself too heavily. I therefore transferred my gun and part of my clothing to Dore, an active young seaman, who was always at the head of the line, and whose load, as well as that of the others, had been reduced by the consumption of pemican. Some of the worst walkers had already been eased of every thing but their blankets, spare clothing, and a few pounds of pemican, but they still lagged in the rear.

In this neighborhood, in 1826, we found a vein containing malachite and other ores of copper, with some of the native metal scattered in detached pieces. The Indians procure the metal on both sides of the Coppermine, in a district which requires several days to traverse. A rolled piece of chromate of iron was picked up on the banks of the river by Mr. Rae. This mineral, so valuable on account of the beautiful pigments which are manufactured from it, is found, according to Jameson, in primitive porphyry, and in beds between clay-porphyry and wacké, and more abundantly in America than on the Old Continent.

The 9th proved to be another fine day. Commencing our march a little before six, we halted at noon for an hour and a half, and encamped at five. A meridional observation gave the lat. 67° 14′ 32″ N. In the afternoon we passed the boat left by Dease and Simpson in 1839, which required too much repair to render it water-tight, or we should have availed ourselves of it for the remainder of the river course we had to follow.

Starting at the usual hour on the 10th, we struck the Kendall about a mile and a half from its junction with the Coppermine, after a march of five hours and a half. Mr. Rae went down to its mouth to look for a note which we expected to find, as I had directed James Hope, with two or three Indians, to meet me there; or, if he arrived earlier than us, to leave a memorandum and descend the river as far as Bloody Fall. This arrangement, which was made in anticipation of our bringing the boats up the river, was my chief reason for making the circuit of the Coppermine; for our most direct course, after leaving Back's Inlet, would have been by tracing up Richardson's River, and crossing the mountains more nearly in the parallel of Fort Confidence. As we had discovered no foot-marks of the party on our march, we concluded that they had not arrived; and Mr. Rae confirmed this opinion by his report of the absence of any signal mark at the mouth of the Kendall. From specimens of the rocks obtained by this gentleman, I ascertained that the walls of the gorge by which the stream enters the Coppermine are composed of red quartz rock disposed in thin layers. The mouth of the Kendall is laid down by Mr. Simpson in lat. 67° 7′ N., long, 116° 21′ W.; and a meridional observation gave 67° 06′ 43″ N., as the latitude of the place where we fell upon the stream.

We walked for nearly three miles along its banks to look for a crossing-place; but, finding that it was nowhere fordable, we resolved to construct a raft, as there was a sufficiency of dry timber for the purpose. We therefore encamped, and Mr. Rae superintended the operation of raft-making. The weather being mild we were again troubled with sand-flies.

Sept. 11*th.*—During a fine night we enjoyed the light of a full moon; but toward the morning the wind veered to the north west, and a moist, chilling fog enveloped us. Our raft could support three at a time, and enabled us all to cross by seven o'clock. A fresh arrangement of the loads was made here; and, to lighten them as much as possible, I deposited my packet of dried plants and some books in a tree, intending to send for them in the winter. After breaking up the raft to recover the lines by which we had fastened it, we piled the logs up on the bank to attract the attention of Hope's party, should we happen to miss them.

Our course was shaped directly across the country for Dease's River; and as we ascended the high grounds the fog became more dense, so that by noon we could not see beyond two or three yards. We steered by the compass, Mr. Rae leading, and the rest following in Indian file. I kept rather in the rear to pick up stragglers; but, though we walked at a much brisker pace than usual, there was little loitering. The danger of losing the party made the worst walkers press forward. On the hills the snow covered the ground thickly; and it is impossible to imagine any thing having a more dreary aspect than the lakes which frequently barred our way. We did not see them until we came suddenly to the brink of the rocks which bounded them, and the contrast of the dark surface of their waters with the unbroken snow of their borders, combined with the loss of all definite outline in the fog, caused them to resemble hideous pits sinking to an unknown depth. The country over which we traveled is composed chiefly of granite; and after walking till half-past five without perceiving a single tree, or the slightest shelter, we came to a convex rock, from which the snow had been swept by the wind. On this we resolved to spread our blankets, as it was just big enough to accommodate the party. There being no fuel of any kind on the spot, we went supperless to bed. Some of the party had no rest, and we heard them groaning bitterly; but others, among whom were Mr. Rae and I, slept well. We learned afterward that a clump of wood grew within a mile and a half of our bivouac; but even had we been apprised of its existence, we could scarcely have found it in the fog. Several showers of snow occurred in the day, and some fell in the night.

Had it not been for the fog, we should have met James Hope and two Indians this day, for they were not many miles distant in the morning; but, notwithstanding their acquaintance with the country, they went astray in the thick weather, and did not reach the place where we crossed the Kendall till the second day afterward. Perceiving then by the remains of the raft that we had crossed, they traced our foot-marks, and, following with their utmost speed, reached our bivouac on the rock two days after we left it.

Commencing the day's march at half-past four in the morning of the 12th, we came to a tributary of the Kendall at eight. In fording this, the water came up to our waists, and we were all more or less benumbed; but a few trees on the bank furnished us with the means of making good fires; and by the time that we had finished breakfast we were comfortably dry. A meridional observation gave us lat. 67° 09′ N.

At two we came to another branch of the Kendall, which runs through a ravine of red and spotted sandstone, under whose shelter there grew a remarkably fine grove of white spruces. The best-grown tree measured sixty-three inches in circumference, and did not taper perceptibly for twenty feet from its root. Its total height was from forty to fifty feet. Other trees of equal girth tapered more, and one decayed trunk, which lay on the ground, looked to be considerably thicker. We encamped in this snug place, and Mr. Rae and Albert, employing the evening in the chase, killed a reindeer and some snow geese.

Mr. Rae endeavored in the winter to measure the height of the creek on which we encamped this night, and of other remarkable places on the route between Great Bear Lake and the Coppermine River, by the aneroid barometer; but that instrument during the journeys underwent such a change, that no reliance could be placed on its indications, when they were compared with those of the barometer at Fort Confidence. The same inconvenience, however, did not materially affect observations made on it at short intervals of time; and in this way the brow of the hill to the south of the creek was ascertained to be six hundred and seventy feet above the stream.

Onward from the level of this brow the country is a gently undulated plain, which is bounded on the south at the distance of a few miles by an even range of hills two or three hundred feet high, and far to the north by the Coppermine Hills, which Lieutenant Kendall and I crossed in 1826, as mentioned in the narrative of Sir John Franklin's second overland journey. A range of lakes, named by Mr. Simpson the Dismal Lakes, lies between these hills and our line of route. They are skirted by broken belts of wood, but the rest of the country is quite naked, the few dwarf trees that exist on the plain being concealed in the depressions of the water-courses of the small rivulets.

The comfortable supper of venison, a sound night's rest in an encampment where nothing was wanting, and the lighter loads, had such an effect on the spirits of the party, that we mounted the hill above the ravine on the morning of the 13th with unusual alacrity, and kept together in close single file. Traveling in this way, our line, as it undulated over the gentle swellings of the plain, was seen from afar, and we were discovered very soon after emerging from the ravine by a party of Indians, encamped on the side of a hill about six miles distant. Happily for these people they knew we were now on the march, and expected to see us at this time; for had it not been so they would have fled instantly with their wonted timidity, and most probably have left every thing they possessed behind them. As it was, we were not many minutes in sight before they signaled their position by raising a column of smoke. This was replied to by us as soon as we could strike a light and gather a few handfuls of moss; and our answer was immediately acknowledged by them with a fresh column. They were encamped nearly at right angles to our line of route; but I thought it better to join them for the purpose of obtaining intelligence, and we accordingly struck off in that direction.

We reached their tents a little before noon, time enough for us to make a meridional observation, by which we ascertained that the latitude was 67° 11′ 30″ N., and the sun's bearing at noon S. 50° E. These Indians informed us that James Hope and his companions had been with them five days previously, and that he had then been two days absent from the fort.

The site of their encampment was selected for the commanding view it possessed of the neighboring country, so that they could mark the movements of the herds of reindeer and musk-oxen that at this season were numerous. Their chase was successful, and their condition and that of their dogs showed that they were reveling in abundance. No doubt this party might now have laid up a sufficiency of venison to feed them, with due economy, all the winter; but such is not the habit of the nation. When the pressure of want ceases to be felt their exertions flag, and they consider it useless to store up provision which, according to their custom, is at the mercy of every idle and hungry person of the tribe.

They gladly sold us some meat for ammunition, and would readily have parted with their whole stock on hand, but I had no desire to load my party again. We agreed, however, with one of the young men to accompany us to the fort, that he might lead us by the best paths, and waited for an hour until he had prepared a heavy load of half-dried meat, to carry with him as an article of trade. In the afternoon our way lay over hills of spotted gray sandstone, sandy shale beds, and toward the evening over knolls of gravel. The day's journey was seventeen geographical miles.

Our march on the 14th was made in a southwest by west direction, and was short, for our guide complained of being fatigued by his load. We relieved him of a part, by distributing about forty pounds of it among the men for their supper.

The country we crossed in the course of the day is composed of sandstone, with gravel banks, and undulates, but is not mountainous. Thin groves of trees occur here and there, especially on the borders of rivulets, and many dwarfish and ancient dead stumps remain on the sides of the eminences. The soil is cracked, hummocky, and swampy, and affords uneasy footing to pedestrians. I found much comfort by walking immediately behind the Indian, that I might avail myself of his quick eye, and tread exactly in his footsteps.

We set out early on the 15th, that we might reach the fort betimes. We lost, however, a considerable time, while the guide went in pursuit of several bands of deer that crossed the path. His skill in hunting was indifferent, and he had no success. The morning was snowy. Before noon we forded a branch of the Dease, and at two o'clock came to the banks of that river at the first rapid. Here we found a barge moored for our use, and, embarking the whole party in it, reached the house at 4 P.M. We were happy to find Mr. Bell and his people well, and the buildings much further advanced than we had expected. All the houses erected by Dease and Simpson had been burnt down, except part of the men's dwelling. Mr. Bell reached the site on the 17th of August, and immediately set to work. Since that time he had built an ample storehouse, two houses for the men, and a dwelling-house for the officers, consisting of a hall, three sleeping apartments, and store-closet. This building was roofed in when we arrived, but the flooring and ceiling of the rooms were not yet laid, though planks had been sawn for that purpose; the kitchen was still to be built, and tables, chairs, and other articles of furniture, to be made. In the log houses, which are commonly erected in this country, the chimneys are massive affairs of tempered clay and boulder stones, and require to be leisurely constructed. The Canadians, who are all practiced in the use of the ax, soon set up the wood-work; and Bruce, the guide, who superintended the operations, and indeed did two men's work himself, advanced them rapidly.

Mr. Bell and Mr. Rae quartered themselves with Bruce in the store room, and I took possession of my sleeping room, which was put temporarily in order. I could there enjoy the luxury of a fire while I was preparing dispatches for the Admiralty and writing my domestic letters, though the walls not being as yet clayed, the snow drifted in between every log. The 16th of September was employed in writing, and on the 17th, being Sunday, we assembled in the hall, where I read divine service and returned thanks to the Almighty for our safety. The fishermen who were stationed about five miles from the house came in on this day, so that the whole party were met together. The Canadians, though Roman Catholics, were present on the occasion; and most of them regularly attended our Sunday services in the winter. In addition to the party from the coast, Mr. Bell had with him here fourteen men, with three women and four children; so that we had in all forty-two souls to provide for, exclusive of Indians coming casually on our store.

On Monday the 18th of September, the packet of letters was placed in charge of François Chartier and Louis La Ronde, who were directed to carry it on without delay to Isle à la Crosse, where the wife of the latter resided. Henry Smith, Joseph Plante, and Henry Wilson, Canadians, accompanied them for the purpose of wintering at the fishery on Big Island, Great Slave Lake; and with them I sent the following men of the English party, whose services could be well dispensed with at our winter quarters: Stairs, Sully, and Clarke, *seamen;* Frazer, Dall, Dodd, Sulter, Hobbs, Ralph, Geddes, Webb, Weddell, and Bugbee, *sappers and miners.* Being thus relieved from the maintenance of eighteen people, the resources of the post were considered equal to feeding the remainder, and I looked forward to the winter without anxiety.

Mr. Bell had placed two fishermen, by my desire, at the west end of Great Bear Lake, near its outlet, to be ready to feed my party, had I found it necessary to return up the Mackenzie. I judged it prudent to continue these men there, not only as their fishing hut would be a convenient station for parties traveling to and fro, between Fort Confidence and the posts on the Mackenzie, but also that they might give aid, should our fisheries near the fort fail.

CHAPTER XI.

ON THE ESKIMOS OR INUIT.

The four Aboriginal Nations seen by the Expedition.—Eskimos.—Origin of the Name.—National Name *Inu-it*.—Great extent of their Country.—Personal Appearance.—Occupations.—Provident of the Future.—Villages.—Seal Hunt.—Snow-houses.—Wanderings not extensive.—Respect for Territorial Rights.—Dexterous Thieves.—Courage.—Traffic.—Compared to the Phœnicians.—Skrellings.—Western Tribes pierce the Lips and Nose.—Female Toilet.—Mimics.—Mode of defying their Enemies.—Dress.—Boats.—Kaiyaks.—Umiaks.—Dogs.—Religion.—Shamanism.—Susceptibility of Cultivation—Origin.—Language.—Western Tribes of the Eskimo Stock.—Tchugatchih.—Kuskutchewak.—A Kashim or Council House.—Feasts.—Quarrels.—Wars.—Customs.—Mammoth's Tusks.—National Names.—Namollos or Sedentary Tchuchke—Reindeer Tchukche.—Their Herds.—Commerce.—Shamanism.—Of the Mongolian Stock.

To keep the interruptions of the narrative within reasonable limits, I have hitherto avoided saying much of the native tribes that occupy the countries through which the Expedition traveled, and shall here supply that deficiency by giving some details of the manners and customs of the four nations whose boundaries we crossed in succession.

Reversing the order of our journey, the first of the native nations that presents itself in descending from the north, is that of the *Eskimos*, as Europeans term them. This appellation is probably of Canadian origin, and the word, which in French orthography is written *Esquimaux*, was, probably, originally *Ceux qui miaux* (*miaulent*), and was expressive of the shouts of *Tey-mō*, proceeding from the fleets of kaiyaks, that surround a trading-vessel in the Straits of Hudson, or coasts of Labrador. The sailors of the Hudson's Bay Company's ships, and the Orkney men in the employment of the Company, still call them *Sŭckĕmòs* or *Seymòs*. Some writers, however, have thought the word to be a corruption of the Abenaki term *Eskimantik*, signifying "eaters of raw flesh," which is certainly a habit peculiar to the Eskimos. But be the origin of the name what it may, it certainly does not belong to the language of the nation, who invariably call themselves *Inu-it* (pronounced *Ee-noo-eet*), or "the people," from *i-nuk* "a man," though families or tribes have, in addition, local designations.

The Eskimos offer an interesting study to the ethnologist, on account of the very great linear extent of their country—of their being the only uncivilized people who inhabit both the old and new continents—and of their seclusion to the north of all other American nations, with whom they have a very limited intercourse; so that their language and customs are preserved more than any other from innovations.

They are truly a littoral people, neither wandering inland, nor crossing wide seas; yet the extent of coast-line which they exclusively possess is surprising. Commencing at the Straits of Bellisle, they occupy the entire coast of the Peninsula of Labrador, down to East Main in Hudson's Bay; also, both sides of Greenland, as far north as they have been examined; and they also inhabit the islands which lie between that land and the continent, and bound Baffin's Bay and Davis's Straits on the west. On the main shore of America, they extend from Churchill, through the Welcome, to Fury and Hecla Straits; thence along the north shore to Beering's Straits, which they pass, and follow the western coast, by Cook's Sound and Tchugatz Bay, nearly to Mount St. Elias; members of the nation have also possessed themselves of the Andreanowsky Islands, Unalashka, and Kadiak. They even cross the Straits of Beering, a part of the nation dwelling on the Asiatic coast, between the Anadyr and Tchukotsky Noss, where they are known to the Russians by the names of *Namollos* or *Sedentary Tchukche.* Outside of Beering's Straits, on the North Pacific, their language and customs have undergone considerable changes, as we shall have occasion to notice; but elsewhere there is no substantial variation in either; the modes of life being uniform throughout, and the differences of speech among the several tribes not exceeding in amount the provincialisms of English counties.

The Greenlanders have been known to Europeans longer than any of the other North American nations, and full accounts of their manners and customs have been given to the world long ago. All the recent voyages in search of a northwest passage, also, contain characteristic portraits and descriptions of the Eskimos that reside on the west side of Davis's Straits and Melville peninsula. I shall not, therefore, attempt a systematic account of the nation, but shall confine myself chiefly to what fell under my personal notice in the central parts of the northern coast-line, where the Eskimos, from their position, have little or no intercourse with other nations, and have borrowed nothing whatever, either from the Europeans or '*Tinnè*, the conterminous Indian people.

The faces of the Central Eskimos are, in general, broadly egg-shaped, with considerable prominence of the rounded cheeks; but few or no angular projections even in the old people. The greatest breadth of the face is just below the eyes; the forehead is generally narrow and tapers upward; and the chin conical, but not acute; most commonly the nose is broad and depressed, but it is not always so formed. Both forehead and chin in general recede, so as to give a more curved profile than is usually to be observed in any variety of the Caucasian race, or among the male Chepewyans or Crees, though some of the female 'Tinnè have countenances approaching to the egg-shape. As contrasted with the other native-American races, their eyes are remarkable, being narrow and more or less oblique. Their complexions approach more nearly to white than those of the neighboring nations, and do not merit the designation of "red," though from exposure to weather they become dark after manhood. As the men grow old, they have more hair on the face than Red Indians, who take some pains to eradicate it, but I observed none with thick bushy beards or whiskers like those of an European who suffers them to grow. An inspection of the portraits in "Franklin's Second Overland Journey," and in "Back's Great Fish River," will show that in elderly individuals both the upper lip and chin have a tolerable show of hair, though none have the flowing beard which was productive of so much benefit to Richard Chancellor and his countrymen.

Dr. Pickering says of the Mongolian, with which, in common with other ethnologists, he classes the Eskimos and the major part of the other American nations, that both sexes have a feminine aspect; that the stature of the men and women is nearly the same; and that the face of the male is pre-eminently beardless. These peculiarities are but faintly developed among the Central Eskimos, and the females are uniformly conspicuously shorter than the males. Most of the men are rather under the medium English size; the defect in height being, perhaps, attributable to a disproportioned shortness of the lower extremities, though this opinion was not tested by measurements. They are broad-shouldered, and have muscular arms; so that, when sitting in their kaiyaks, they seem to be bigger men than they do when standing erect. Some individuals, however, would be considered to be both tall and stout even among Europeans, and they certainly are not the stunted race which popular opinion supposes them to be. The comparative shortness of the females is common to them and the neighboring 'Tinnè (Hare Indians and Dogribs), whose women are of small stature.

In both sexes of Eskimos the hands and feet are small and well-formed, being less than those of Europeans of similar height. The boots which we purchased on the coast were seldom large enough in the feet for our people, none of whom were tall men.

The Central Eskimos, when young, have countenances expressive of cheerfulness, good nature, and confidence; and the females, being by no means inclined to repress their mirth, are wont to display a set of white teeth that an European belle might covet. The elderly people have features more furrowed than those we see in civilized life, as we might expect when the passions are not habitually repressed; and in some of the old men the lines of the countenance denote distrust and hatred. These ill-favored individuals were, happily, not numerous, and several of the patriarchs we communicated with had a truly benevolent aspect. The weather-beaten faces of some of the old women, gleaming with covetousness, excited by seeing in our possession wealth beyond the previous creations of their imagination, lead one to believe that the poet who sang, "Old age is dark and unlovely," had drawn his picture from a people equally hard and unsoftened by the cultivation of intellect; and I feel no surprise that Frobisher's people should have suspected the unfortunate elderly woman who fell into their hands of being a witch, while they let the young one go free.

Year after year sees these people occupied in a uniform circle of pursuits. When the rivers open in spring they resort to rapids and falls, to spear the various kinds of fish that ascend the streams at that period to spawn. At the same date, or a little earlier in more southern localities, they hunt the reindeer, which drop their young on the coasts and islands while the snow is only partially melted. Vast multitudes of swans, geese, and ducks, resorting to the same quarters to breed, aid in supplying the Eskimos with food during their short but busy summer of two months. In the beginning of September the reindeer assemble in large bands and commence their march southward; and then the Eskimos reap a rich harvest by waylaying them at established passes on the rivers or narrow places of a lake. On parts of the coast frequented by whales, the month of August is devoted to the exciting pursuit of these animals, a successful chase insuring a comfortable winter to a whole community. Throughout the summer, the families associated by twos and threes live in tents of skins, and generally enjoy abundance of food, while they carefully lay up what they can not consume for after use. In this respect they are more provident than the Hare Indians, or Dog-ribs, who seldom trouble themselves with storing up provisions. This difference of the habits of the two nations, which greatly influences their general characters, has perhaps originated in the different circumstances in which they are placed. The Eskimos, wintering on the coast, are in darkness at mid-winter: the reindeer and musk-oxen have then retreated into the 'Tinnè lands, and fish can not at that season be procured in their waters; life, therefore, can only be maintained in an Eskimo winter by stores provided in summer.

In the country of the 'Tinnè, on the contrary, the winter fishery is productive, and animals are by no means scarce at that season, but they require to be followed in their movements by the hunter and his family, often to a great distance. In such a case, any surplus of food that has been procured must be placed *en cache*, as the term is, where it is exposed to the depredations of *wolverenes*, or the still more irresistible attacks of their hungry fellow-countrymen, who are wont to track up a successful hunter in order to profit by his labors. The 'Tinnè, therefore, have practically decided that it is better for them to live profusely while they have venison, and then to go in search of more. Were they to be content with the product of their fisheries, they might build villages, and live easily and well, so productive are the boundless waters of the north; but they like variety of diet, and prefer the chase, with the hazard of occasional starvation which follows in its train.

The villages of the Eskimos are, therefore, a feature in their domestic economy in which they differ wholly from their neighbors. The houses are framed strongly of drift timber, are covered thickly with earth, and are used only in winter. They have no windows, and are entered by a low side door, or, when they stand in situations where the drift-snow lies deep, by a trap-door in the roof. The floor is laid with timber, and they have no fire-places; but a stone placed in the centre serves for a support to the lamp, by which the little cooking that is required is performed. For the site of a village, a bold point of the coast is generally chosen where the water is deep enough to float a whale; and to the eastward of Cape Parry, where we saw no whales, we met with no villages, although solitary winter-houses occur here and there on that coast. The association of a number of families is necessary for the successful pursuit of the whale. When the villagers of the estuary of the Mackenzie, or of Cape Bathurst, are fortunate enough to kill one or more of these marine beasts, they revel in greasy abundance during the dark months, and the ponds and the soil around are saturated with the oil that escapes.

In March the seals have their young, and soon afterward they become the principal objects of chase to the Eskimos, who greatly esteem their dark and unsightly flesh, reckoning it as choice food. The seal, being a warm-blooded animal, respiring air, requires a breathing-hole in the ice, which it has the power of keeping open in the severest frosts, by constant gnawing. It is a watchful creature, with acute senses of sight and hearing; but it is no match for the Eskimo hunter, who has carefully studied all its habits from his infancy. As the days lengthen, the villages are emptied of their inhabitants, who move seaward on the ice to the seal hunt. Then comes into use a marvelous system of architecture, unknown among the rest of the American nations. The fine, pure snow has by that time acquired, under the action of strong winds and hard frosts, sufficient coherence to form an admirable light building material, with which the Eskimo master-mason erects most comfortable dome-shaped houses. A circle is first traced on the smooth surface of the snow, and the slabs for raising the walls are cut from within, so as to clear a space down to the ice, which is to form the floor of the dwelling, and whose evenness was previously ascertained by probing. The slabs requisite to complete the dome, after the interior of the circle is exhausted, are cut from some neighboring spot. Each slab is neatly fitted to its place by running a flenching-knife along the joint, when it instantly freezes to the wall, the cold atmosphere forming a most excellent cement. Crevices are plugged up, and seams accurately closed by throwing a few shovelfuls of loose snow over the fabric. Two men generally work together in raising a house, and the one who is stationed within cuts a low door, and creeps out when his task is over. The walls being only three or four inches thick, are sufficiently translucent to admit a very agreeable light, which serves for ordinary domestic purposes; but if more be required a window is cut, and the aperture fitted with a piece of transparent ice. The proper thickness of the walls is of some importance. A few inches excludes the wind, yet keeps down the temperature so as to prevent dripping from the interior. The furniture, such as seats, tables, and sleeping-places, is also formed of snow, and a covering of folded reindeer skin, or seal skin, renders them comfortable to the inmates. By means of ante-chambers and porches in form of long, low galleries, with their openings turned to leeward, warmth is insured in the interior; and social intercourse is promoted by building the houses contiguously, and cutting doors of communication between them, or by erecting covered passages. Storehouses, kitchens, and other accessory buildings, may be constructed in the same manner, and a degree of convenience gained which would be attempted in vain with a less plastic material. These houses are durable, the wind has little effect on them, and they resist the thaw until the sun acquires very considerable power.

The success of the seal-hunt depends much on the state of the ice, and should it fail, great misery results; the spring being, in fact, the time of the year in which the Central Eskimos incur the greatest risk of famine. When the thaw lays the ground in the valleys bare, reindeer and wild-fowl return to the sea-coast, and plenty follows in their train.

It will be evident, from the account of the yearly round of the lives of these people, that their movements are restricted to narrow limits, as compared with the 'Tinnè, who pursue the chase over tracts of country hundreds of miles in diameter, as necessity, fear, or caprice, drives them. A strict right to hunting grounds does not seem to be maintained by the several members of the widely spread 'Tinnè nation, so as to hinder several tribes from resorting to the same districts in pursuit of deer, and meeting each other in amity, unless an actual feud exists. Thus our presence at Fort Confidence was sufficient to determine various bands of Hare Indians, Dog-ribs, and Martin-lake Indians to resort to the northeastern arm of Great Bear Lake; and but for a deadly feud with the Dog-ribs, which twenty years ago greatly reduced the numbers of our old friends, the Copper Indians, we should have had their company also. The Eskimos, on the contrary, have a strong respect for their territorial rights, and maintain them with firmness. We learned at Cape Bathurst, that each head of a small community had a right to the point of land on which his winter house or cluster of houses stood, and to the hunting grounds in its vicinity. We had also evidence, at various places on the coast, of the unwillingness of these people to appropriate the goods of their absent neighbors, even when we, not knowing the proper owner, tempted them by the offer of a price much beyond the value of the article in their eyes. The answer on such occasions was, "That belongs to a man who is not here." We also saw on the coast stages on which provisions, furs, lamps, and other articles were placed, while the owners had gone inland; and hoards of blubber, secured from animals by stone walls, but without any attempt at concealment. "Tiglikpok" (he is a thief) is a term of reproach among themselves; but they steal without scruple from strangers, and with a dexterity which training and long practice alone can give. Nor did they appear ashamed when detected, or blush at our reproofs. I believe that on this point their code is Spartan, and that to steal boldly and adroitly from a stranger is an act of heroism.

In personal courage, the Eskimos are superior to the Chepewyans, Crees, or any other Indian nation with whom I am acquainted. The Hare Indians and Dog-ribs dread them, and even when much superior in numbers, would fly on their approach. Nor do the fire-arms which the bolder Kutchin have lately acquired enable that people to lord it over the Eskimos, or encroach on their grounds.

The populous and turbulent bands which inhabit the estuary of the Mackenzie carry on a traffic with the Western Eskimos from the neighborhood of Point Barrow and Beering's Straits, whom they meet midway on the coast; and though often at feud with the Kutchin have occasionally commercial relations also with them. But they who dwell to the eastward of Cape Bathurst communicate with none of their own nation except the families in their immediate vicinity, and speak of the distant Eskimos as of a bad people. The reputation of the *Kablunaht* or *Kablunèt* (white men) is superior among them to that of the remote tribes of their own nation. With the *Allani-a-wok*, as they term the inland Indians, they have no intercourse whatever.

The Central Eskimos have had no traffic with Europeans, except with those employed on the recent voyages of discovery, until the last year (1849), when a family from the coast to the west of the Mackenzie, having gone inland with a party of Kutchin, were visited at their tents by a trader sent out from La Pierre's house, which is an outpost to the Hudson's Bay Company established on the Western Rat River.

Articles of Russian manufacture, procured by barter coastwise, were traced by us in an easterly direction no further than Point Atkinson. Previous to the recent establishment of the Russian Fur Company's post in the vicinity of Beering's Straits, the objects exchanged at Barter Island, on the 144th meridian, were brought on the Asiatic side from the fair of Ostrownoie near the Kolyma, by the Tchukche, who passed them in the first instance to the Eskimos of Beering's Straits, by whom they were bartered at the island in question, for furs brought thither by the Eskimos of the estuary of the Mackenzie. In like manner various wares of English make found their way, through the Kutchin and Mackenzie River Eskimos, coastwise to the Russian establishments on the Pacific.

From the predilection for commercial pursuits shown by the Eskimos, Von Bäer compares them to the Phœnicians, and, referring back to very early times, finds traces of their voyages along the eastern coasts of America, as far south as the present state of Massachusetts. There the Scandinavian discoverers of Vinland (Rhode Island) had many skirmishes with the Skrellings (*Skrällingern*), whose identity with the Eskimos Von Bäer considers as established by the recorded descriptions of their personal appearance and dress, and the appellation given to them being the same as that applied to the Greenlanders.

From Beering's Straits, eastward as far as the Mackenzie, the males pierce the lower lip near each angle of the mouth, and fill the apertures with labrets resembling buttons, formed often of blue or green quartz and sometimes of ivory. Many of them also transfix the septum of the nose with a dentalium shell or ivory needle. These ornaments have perhaps been adopted from the Kutchin and Pacific coast tribes south of Mount St. Elias, since they have not extended to the Eskimos of Cape Bathurst or more eastern members of the nation. Most of the women are tattooed on the chin, but they have not adopted the unsightly gash and extension of the under lip on which the Kolushan ladies pride themselves.

Unlike the Hare Indian and Dog-rib women, who neglect their personal appearance, the Eskimo females turn up and plait their hair tastefully, ornament their dresses, and evidently consider their toilet as an important concern : hence we may judge that more deference is paid to them by the men. Egede informs us, that unmarried Greenland women are modest, both in words and deeds, but that greater laxity exists among the wives, with the connivance of their husbands, who are not jealous. I fear that so much, scanty as the praise is, can not be justly said in favor of the fair sex on the northern coast. The gestures and signs made by young and old when they came off in the *umiaks* were most indelicate, and more than once a wife was proffered by her husband without circumlocution in the presence of his companions and of the woman herself. I understood, indeed, from Augustus, our interpreter in 1826, that such an offer was considered by the nation as an act of generous hospitality ; and similar customs are said to exist among the inhabitants of Tartary.

Almost all savage people are excellent mimics, and the Eskimos are not defective in this accomplishment. They imitated our speech and gestures with success and much drollery ; and the men excel the other native Americans in the art of grimacing. When they wish to defy strangers who intrude into their country, they use the most extraordinary gestures and contortions of the body and limbs, making at the same time hideous faces. This was evidently practiced systematically to terrify invaders ; for such as resorted to it on their first interview with us, the moment that they were made to understand our friendly intentions, instantaneously relaxed their features into a broad, good-natured grin, and came alongside our boats without further hesitation.

The dress of the two sexes is much alike, the outer shirt or jacket having a pointed skirt before and behind, those of the females being merely a little longer. The Kutchins also wear these pointed skirts, but they have not been adopted by the Hare Indians or any of the Chepewyan tribes, who in common with the more southern Indians cut their shirts or frocks evenly round at the top of the thigh. I suspect that the long skirts of the Kutchin or Eskimos have given origin to the fabulous account of men with tails, thought by the Kolushes of the Pacific coast to inhabit the interior in the direction of Mackenzie's River.

The Eskimo boots are also peculiar to the nation, being made of seal-skin so closely sewed as to be water-tight, and coming up to the hips like those used by fishermen in our own land. The Chepewyans and Crees manufacture no leather that resists water; the deer-skin dressed by them like shammy absorbs water like a sponge, and hardens and spoils in drying. Neither have these Indians boots, but merely shoes or mocassins, with soft tops that wrap round the ankle, and are unconnected with the leggins or trowsers.

The Eskimos show much skill in the preparation of whale, seal, and deer-skins, using the first for thongs and lines employed in the capture of sea-beasts, also as harness for dog-sledges, soles for boots, and other purposes where strength and durability are required.

Their skin *kai-yaks* and *u-mi-aks* are also peculiar to the nation, and can be formed only by a people who dress hides so as to be water-proof. The kaiyaks are impelled by a double-bladed paddle, used with or without a central rest, and the umiaks with oars; neither of which are employed by the inland Indians, except where they have been adopted from Europeans. The use of a light waterproof outer dress, formed of the intestines of the whale, and secured to a ledge round the aperture of the kaiyak so as perfectly to exclude the water in a stormy sea, is also an Eskimo invention; and the address which is acquired in the management of the light, swift, but unstable kaiyak, contributes to the education of a race of fearless seamen.

The dogs of the Eskimos along their whole line of coast are superior in strength to those of the neighboring nations, and are used in sledges and also in the chase of reindeer and musk-oxen.

With respect to the religion of the Eskimos I could obtain personally no satisfactory answer to my inquiries; but it is certain that belief in witchcraft and the agency of evil spirits prevails throughout the nation, except in Greenland and Labrador, where demon worship has been combated by Christianity. Connected with this belief is the Shamanism, or influence which certain individuals claim to possess over the evil spirits. Sorcery has been reduced to a system on the shores of Beering's Sea; and that it is not unknown even on the Labrador coast, the following words, collected from an Okkak dictionary, will show. *Ange-kok*, "a shaman;" *Elihètak*, "one killed by sorcery;" *I-yèrok*, "the devil's servant or messenger;" *Nang-iner-minik*, "an appearance produced by a sorcerer;" *Torngak*, "a devil or evil spirit;" *Torngiwòk*, "he performs the office of a sorcerer."

As to intelligence and susceptibility of civilization, I consider the Eskimos as ranking above the neighboring Indian nations, though my personal experience on this head, being confined to the interpreters employed on the several expeditions to which I have been attached, is perhaps too limited to found much upon. These individuals, however, showed a docility, industry, steadiness of purpose, a ready adoption of European customs, and an amiability which I did not observe among the Northern Indians or Crees in the course of several years' study of their characters.

The success of the Moravian Missionaries, in introducing Christianity and the arts of reading and writing among the population of the Labrador coast, is a strong inducement to attempt an extension of the same system of instruction to the well-fed multitudes that frequent the estuary of the Mackenzie.

The origin of the Eskimos has been much discussed, as being the pivot on which the inquiry into the original peopling of America has been made to turn. The question has been fairly and ably stated by Dr. Latham, in his recent work "On the Varieties of Man," to which I must refer the reader; and I shall merely remark that the Eskimos differ more in physical aspect from their nearest neighbors, than the red races do from one another. Their lineaments have a decided resemblance to the Tartar or Chinese countenance. On the other hand, their language is admitted by philologists to be similar to the other North American tongues in its grammatical structure; so that, as Dr. Latham has forcibly stated, the dissociation of the Eskimos from the neighboring nations, on account of their physical dissimilarity, is met by an argument for their mutual affinity, deduced from philological coincidences.

The comprehensiveness of the Eskimo language and its artificial structure are curious, when we take into our consideration the isolated position of the people, and the few objects that come under their observation. In 1825, I devoted the whole winter to the formation of an Eskimo vocabulary and grammar, with the aid of our very intelligent interpreter Augustus, who was a native of the shores of Sir Thomas Roe's Welcome, and having resided at Churchill, had acquired the power of expressing his meaning in very tolerable English. The book containing the results of his labors and mine was unfortunately stolen from me in the following summer by the Eskimos of the estuary of the Mackenzie; but through the kindness of the Reverend Peter Latrobe, the philanthropic secretary of the Moravian Mission, I was provided for use on the present expedition with an excellent grammar, and a pretty full dictionary, formed by some of the industrious missionaries of the Labrador coast. By carefully perusing these volumes, together with Captain Washington's extensive vocabulary, published under sanction of the Admiralty in 1850, I feel justified in maintaining the assertion I have already made, that the Eskimo language does not materially vary throughout a line of coast longer than that which any other aboriginal people possesses. Many seeming discrepancies I have been able to trace to the genius of the language, by which the same object receives a distinct appellation for every different aspect and condition which it assumes; and the formers of the vocabularies have seldom given the precise translation such a language requires. Thus *a-niö* signifies "the snow;" *ap-ut*, "snow," a general name for snow on the ground, whence *ap-uti-tut*, "as white as snow;" *kan-ek*, "snow falling;" *aki-lokak*, "new fallen white snow;" *auma-ÿali-wok*, "a great fall of damp snow;" *siko*, "ice;" *tu-wak*, "solid ice;" *nilak*, "light ice;" *ka-cho-ak*, "drift-ice;" *sir-mek*, "thin ice." We have already remarked, that the Eskimos of Labrador and Beering's Straits retain the name of the musk-ox, though the Central Eskimos alone come into contact with the animal, (page 193.)*

The inhabitants of the northwestern coasts from Tchugatsky Bay (or as it is named in the English charts, Prince William's Sound), northward, including the peninsula of Alaska and the

* The following are some of the local designations of tribes of the central Eskimos. The *A-hak-nan-helet* reside near Repulse Bay; the *Ut-ku-sik-kaling-mè-ut*, or "Stone-kettle Eskimos," live further to the westward; the *Kang-or-mè-ut*, or "White Goose Eskimos," dwell to the eastward of Cape Alexander; those who frequent the mouth of the Coppermine River call themselves *Na-gè-uk-tor-mè-ut*, or Deer-Horn Eskimos;" and the numerous tribe that resorts to the eastern outlet of the Mackenzie call themselves *Kittè-gà-re-ut* or "inhabitants of land near the mountains."

islands in Beering's Sea and Straits, are considered by Baron Wrangell, Bäer, and others acquainted with them, to be of the Eskimo stock.* Captain Beechey believes that the Western Eskimos who meet the Mackenzie River tribes at Barter Island have their western boundary at Cape Barrow; there they have commercial intercourse with the tribes described by Wrangell and Bäer, who, in their turn, barter with the Asiatic Tchukche and with the Russians settled on the American coast, and their neighbors the Kolushans. The tribes crowded together on the shores of Beering's Sea within a comparatively small extent of coast-line exhibit a greater variety, both in personal appearance and dialect, than that which exists between the Western Eskimos and their distant countrymen in Labrador; and ethnologists have found some difficulty in classifying them properly. The appellations they have assumed, or which have been bestowed upon them correctly and incorrectly, have increased the confusion. They are, however, like the other Eskimos, a littoral people, who, in their skin kaiyaks, pursue all kinds of sea-beasts—seals, sea-lions, walruses, polar-bears, sea-otters, and whales—clothing themselves in their spoils and in bird-skins, and making much less use of the leather of the reindeer skin than their southern and eastern neighbors of a different stock. The *Tchugatchih* of King William's Sound are the most southern of several tribes, and state that, in consequence of some domestic quarrels, they emigrated in recent times from the Island of Kadyak,† and they claim, as their hereditary possessions, the coast lying between Bristol Bay and Beering's Straits. They believe that their nation originally sprung from a dog, in which respect they agree with the Chepewyan tribes, and differ from the Kolushes.

The Tchugatchih are of middle stature, slender, but strong; with skins often brown, but in some individuals whiter than those of Europeans, and with black hair. The men are handsomer than the women. They pierce the under lip and septum of the nose, filling the apertures with corals, shells, bones, and

* "The inhabitants of the Aleutian Islands (*i. e.* Beering's and Copper Islands), of the Rat Islands, Andreanowsky Islands, and Prebülowüni Islands, of Unalaska and Kadiak, are all Eskimo; a fact which numerous vocabularies give us full means of ascertaining. In respect of the difference of speech between particular islands, there is external evidence that it is considerable."—*Dr. Latham*, *Varieties of Man*, *&c.*

† *Kikhtak* of the English maps.

stones. Their manners were originally similar to those of the *Kuskutchewak* and other communities living more to the north; but in later times they have carried off the women of the more southern tribes, and from their intermarriages with the captives, combined with their long intercourse with the Russians, their opinions, customs, and features have undergone a change, so that they have now a greater resemblance to the inland Indians than to the northern Eskimos.

Bäer's work, which is my chief authority with regard to the inhabitants of Russian America, contains some interesting details of the habits of the Kuskutchewak alluded to above, from which I shall make a few extracts for the pnrpose of comparison with the better-known manners and customs of the eastern Eskimos. The *Kuskutchewak* inhabit the banks of a river which falls into the sea on the 60th parallel, between the island of Nuniwak and Cape Newenham. They are neither a nomadic nor hunter folk, but dwell in winter in stationary villages built on the river, and in summer disperse themselves inland to collect provisions. They have a strong attachment to their ancestral abodes. Their winter dwellings are partly sunk in the earth, as on the Eskimo coasts, but nowhere else to the eastward of the Rocky Mountains. On the west coast this mode of building extends as far south as Unalaschka; and in Cook's Third Voyage there is a representation of a winter house at that place far superior in size, accommodation, and furniture, to any that we saw on the northern shores.

In each village of the Kuskutchewak there is a public building, named the *Kashim*, in which councils are held and festivals kept, and which must be large enough to contain all the grown men of the village. It has raised platforms round the walls, and a place in the centre for the fire, with an aperture in the roof for the admission of light.*

I have mentioned such a building as existing at Point Atkinson (page 155), but that was of inferior size, being indeed suited to a smaller community. In the language of the Labrador Eskimos *Kashiminwik*, or *Kashimin-wikhak*, signifies a place where men assemble in council;" and *Kaschim-i-ut*, "an assemblage of men for council;" from which we derive additional evidence of the national identity of the two people.

* In Franklin's Second Overland Journals there is a plan of the Point Atkinson kashim, which answers to the above description.

The kashim is the sleeping apartment for all the adult able-bodied males of the village, who retire to it at sunset; while the old men, women, children, and the shaman sleep in the ordinary dwellings. Early in the morning the shaman goes to the kashim with his drum, and performs some religious ceremony, varied as his fancy prompts, for the shamanism of the tribes of the Eskimo stock is said not to be guided in its ceremonials by any fixed practice.* The only women who are allowed to enter the kashim to eat with the men are those who have been initiated in a certain formal manner.

Feasts are held in the kashim, and particularly a great festival or harvest-home which recurs annually at the close of the autumn hunting season. Then the produce of the chase of each hunter is proclaimed before the assembly in detail, down to the small birds or mice killed by the children, and the generosity of the contributors to the feast is lauded. Many are thereby excited to give profusely, and to pinch themselves and their families for the whole of the ensuing winter. Minor feasts are held on various occasions, and the hospitality of the Kuskutchewak and neighboring tribes is said to be very great, not only at festival occasions, but at all other times.

On the murder of a relation, retaliation is decided upon at a council held in the kashim, and is generally blood for blood. In their wars they do not slay old people or children, and instead of killing women, they lead them into slavery. On the north coast, in 1826, we observed that the old men and women were more forward in provoking a fray, in anticipation of plunder, than the young men, and perhaps they reckoned upon personal immunity in the contest. Disputes between parties of long standing are settled by dual combat in a ring of the people. Augustus, our interpreter, told me that the Eskimo of the Welcome decided their quarrels by alternate blows of the fist, each in turn presenting his head to his opponent; and Cunningham says, that the

* Augustus informed us that in his tribe, which occupies the coast of Hudson's Bay between Churchill and Knap's Bay, there were sixteen men and three women who were acquainted with the mysteries of shamanism. The women exhibited their skill on their own sex only. When the shaman was sent for to cure a sick person, he shut himself into a tent with his patient, and, without tasting food, sung over him for days together. The shamans also swallowed knives, fired bullets into their bodies, and practiced various other deceptions to show their powers.

natives of New South Wales have a similar practice, but use the waddie instead of the unarmed hand, their thick skulls being able to resist blows with that formidable weapon. Both people consider it cowardly to evade a stroke. In these primitive methods of settling their points of honor we may perceive the germ of the mediæval combats in lists, and of the more absurd modern duels, which the light of Christianity has not yet abolished.

When a Kuskutchewak hunter returns from the chase, he steps from his kaiyak or dog-sledge, and goes straight to the kashim, while his wife dries and secures the kaiyak, or unharnesses the dogs, and lays up the produce of his hunt. She cooks for him, and makes and mends his clothing. The husbands visit their wives, like the Spartans, by stealing out of the kashim at night, when the others have gone to rest. Every hunter preserves some remembrance of each reindeer that he kills. He either scratches a mark on his bow, or draws out a tooth of the beast, and adds it to a girdle which he wears as an ornament.

The mode of treating infants is one of the national customs of a people that changes most slowly. It does not appear that any branch of the Eskimo nation flatten the heads or repress the growth of the feet of their children, like the Tchinuks and Kutchin. The Central Eskimo women carry their nurslings in the hoods of their shirts, and the figures in Cook's Third Voyage show that custom to be practiced as low down as Unalaschka.

The Kuskutchewak are passionately fond of the vapor bath, and often use it three or four times a day, occasionally in the kashim, more frequently in small inclosures, which can be formed in every hut, and in which the steam is raised by throwing water on hot stones. If a father happens to be on bad terms with his grown-up son, he invites his most intimate friend into the bath, discloses his grievance to him, desires him to inform his son why his father is displeased, and what he ought to do to appease him. A secret which no one will tell elsewhere is revealed in the bath. This is also, I believe, a Turkish custom.

The Kuskutchewak indicate the times of day or night with great accuracy, and they can even distinguish some stars and planets.*

* Viz., *Tuntonok* (Reindeer), the Great Bear; *Mi-seuschit* (the Rising), Orion; *Ka-wegat* (the Fox-earth), the Pleiades; *As-guk*, Aldebaran; *Uleuchtugal-ya* (Fox and Hare Killer), Venus; *Ag-yach-laik* (Abundance of Wild Beasts), Sirius.

Before concluding my extracts from Bäer relating to this tribe, I may remark that mammoth-teeth are numerous in crevices of the sandy banks of the river Kuskokwim. The natives have a tradition that the great animals to which the tusks belonged came in old times from the East, but that they were destroyed by a shaman of the river Kwichpak. Some of them, however, say that the herd was merely driven into the earth, and that it comes up in one night of the year. Elsewhere I have alluded to the singularity of no tusks nor fossil bones having been hitherto discovered in Rupert's Land, though they abound on the coast of Beering's Sea.

The various small tribes or communities nearly related to the Kuskutchewak, enumerated by Baron Wrangell, are inserted in a note at the foot of the page.* Their name for people or men is *Tā-tchut*, which corresponds in signification with the Eskimo *Inu-it;* and among the inhabitants of the Aleutian Archipelago, the word is modified into *Tā-gut* and *Yagut*. The similarity of this term to the national appellation of the Lena Yakuts of Turkish stock is worthy of notice, though it may probably be no more than a mere accidental coincidence.† The syllable *ta* in the language of the *Kutchin*, who are the inland neighbors of the Kuskutchewak, signifies water, and *Ta-kutchi* denotes "the water or ocean people." *To* (or *Ta* in composition) means "water" in the 'Tinnè or Chepewyan tongue also.

The Sedentary Tchukche, who inhabit the shores of the Gulf of Anadyr, and assume, according to Sauer, the national appellation of *Namollo*, are a tribe of Eskimos. They seem once to have possessed the coast of Asia as far westward ‡ as the 160th

* Agolegmeuten, Kiyataigmeuten or Kiyaten, Mayimeuten, Agulmeuten, Paschtoligmeuten, Tatchigmeuten, Malimeuten, Anlygmeuten, Tschnagmeuten, Kuwichpack-meuten, are the designations of the communities most closely allied to the Kuskutchewak by neighborhood and identity of manners. The Tchugatschen and Kadyaken, the Inkaleuchlenäten and the Inkaliten reside at greater distances to the southward, and have some diversities of customs.

† The Turkish words *Yacubi* and *Yakutski* signify the "Sons of Jacob," and, on applying to several well-informed Turks, they could recollect no other words so similar in sound to *Yakut*. A Christian community named *Yakubi* or *Yakupi* reside at the present time in Jerusalem.

‡ Commencing at the east coasts of Labrador and Greenland, the Namollos become the western members of the family, as Dr. Latham has noticed; and even in respect to Europe they are *less eastern* than their American brethren. Is *Namollo* from *nuna-mullo*, "distant land?"

parallel, traces of Eskimo dwellings having been found up to the mouth of the Kolyma. The more powerful Reindeer Tchukche oppress and restrain them within narrow limits, and are therefore considered as the invaders of the Namollo territory.

With the mention of the Asiatic detachment of the nation, and without entering into the question of whether it ought to be considered as the remains of the ancient trunk, or merely a decaying branch, we close our remarks on the Eskimos.

REINDEER TCHUKCHE.

Before proceeding to give an account of the *Kutchin*, the second of the native nations whose lands were traversed by the Expedition, I shall introduce a brief notice of the Asiatic Reindeer Tchukche,* who designate themselves by the appellation of *Tchekto*, "people." Mr. Matiuschkin describes them as being a remarkably strong and powerful race, resembling the Americans in their physiognomy. They once owned the whole country from Beering's Straits to some distance westward of the Kolyma, having dispossessed, according to tradition, a once numerous people, named *Omoki*, who are now extinct, and also the *Namollos* of the coast.† The advance of the Cossacks and Russians has driven them back beyond the Kolyma into the northeastern corner of Asia; ‡ but they resist the invaders with firmness, and maintain a greater degree of independence than any other native Siberian tribe.

Neither the Eskimos nor any other North American nation have domesticated any animal except the dog; but the Asiatic *Tchekto* are a truly nomadic people, and have tamed the reindeer, of which they have numerous herds. The rearing of the deer, which constitutes their wealth, requires the command of a woody country and also of barren grounds or *tundras*. Commodore Moore, during the winter that the "Plover" passed in Emma's Harbor, not far from Cape Tchoukotsky, purchased reindeer from

* The notices of this people are taken from Matiuschkin's description of them in "Wrangell's Expedition to the Polar Sea." In the orthography of the name I have followed the English translation of Wrangell's book. The French translator writes *Tçhouktchas*, and Bäer *Tchuktschen*. In Cook's Third Voyage it is written *Tschutski*, and by Dr. Latham *Tshuktshi*.

† *Omoki* has an Eskimo sound: thus *oma*, "he," *okköa* or *omoköa*, "they." And as the Central Eskimos soften *k* and *g* into *l* and *m*, a little etymological coaxing might produce a word like *Namollo*.

‡ For the use of the relative terms east and west see note, p. 219.

the inhabitants of a village near his anchorage, to the great benefit of his crew, and at the low rate of twelve carcasses for a ship's musket. The Tchukches are skillful traders. Those who frequent the fair of Ostrownoie bring thither furs and walrus-teeth, and receive in return tobacco, iron articles, hardware, and beads. They are accompanied by their women and children, and bring with them their arms, skin tents, and household goods, all conveyed on sledges drawn by reindeer. The journey occupies six months, for they have to make circuitous routes in search of pasture; and they also visit by the way Anadyrsk and Kamenoie, where inferior markets are held. After remaining eight or ten days at Ostrownoie, they commence their return; so that their life is actually passed on the road, allowing barely time for the necessary preparations and for their visits to Beering's Straits. These are made in summer in baidars, or skin-boats, and in winter over the ice on sledges, with which they carry Russian wares, to the Gwosden Islands in the Straits. There they are met by people from Cape Prince of Wales, with furs and walrus-teeth, collected from the dwellers in Kotzebue Sound, and from the inhabitants of the coast still further north. The Tchukche trade with St. Lawrence Island also, and with Ukiwok, a rock of not more than three miles in circumference, but rising seven hundred and fifty-six feet above the sea. It is destitute of vegetation, and yet a body of two hundred people, from the American coast, have formed a settlement on it, at the height of one hundred and fifty feet above the water, for the purposes of trade. They inhabit caves of the rocks, and procure clothing, tobacco, and other necessaries by the sale of walrus-teeth. Sledge Island, equally small, is also inhabited by skillful traders, who are employed by the Tchukche as factors, to exchange the articles intrusted to them for furs collected on the banks of the Kwichpak, Kuskokwim, and neighboring rivers of America. By this channel, previous to the formation of the Russian Fur Company, wares brought from Asia were distributed over seventeen hundred miles of American coast.

The Reindeer Tchukche practice shamanism; and though they occasionally beat their shamans to compel them to bring about some event which they desire, this treatment may be considered as evincing a belief in the powers of the sorcerer, and in times of general fear and calamity the shamans have put forth their pretensions with great effect. In 1814, an epidemic having carried off many of the Tchukche assembled at the fair of Ostrownoie, the shamans held a consultation, and decided that Kotchen, the most respected of their chiefs, must be sacrificed to appease the wrath of the spirits. Neither presents nor severe treatment could prevail on the shamans to alter their decision, and Kotchen, like another Curtius, devoted himself to the infernal gods. The love of his people, however, was such, that none could be prevailed upon to execute the sentence, until his own son, incited by the exhortations of his father, and terrified by his threatened curse, plunged a knife into his heart, and gave the body to the shamans.*

The tents of the Tchukche, called *namet*, have a fire in the centre with an opening for the smoke to escape, and inclose several apartments named *pologs*, or square closets of skins, stretched over laths, and so low that the inmates must remain in a crouching position. The polog is heated by a lamp, and its temperature is so high, and the air so close, as to be scarcely endurable by a person unaccustomed to breathe so impure an atmosphere.

Dr. Latham considers the Tchukche as the northern branch of the *Koriaks*, the southern branch being named *Koræki*, which is said to be an indigenous appellation†; and he reckons the Koriaks as a division of the Peninsular Mangolidæ. It is probable that on further investigation the Reindeer Tchukche will be found to be the connecting link between the Asiatic and American Mongolidæ. In their attachment to commercial pursuits, fondness for beads, and in their bold independent character, they have a resemblance to the *Kutchin*, described in the following chapter. The similarity of the appellation Tchukche, derived from *Tchekto*, "people," to *Tchutski* or *Ta-kutski*, "water-people," tribes of the Kutchin, is, however, in Dr. Latham's opinion, merely an indirect glossarial affinity. The great variety of dialects which prevail in the Aleutian Archipelago and neighborhood of Beering's Straits is most probably the result of the active commerce there carried on, having brought several nations into contact with each other.

* Wrangell's Polar Sea, translated by Mrs. Sabine, p. 119.

† *Kora*, "a reindeer."

CHAPTER XII.

ON THE KUTCHIN OR LOUCHEUX.

Designations.—Personal Appearance.—Tattoo.—Employ Pigments.—Dress.—Ornaments.—Beads.—Used as a Medium of Exchange.—Shells.—Winter Dress.—Arms.—Wives.—Treatment of Infants.—Compress their Feet.—Lively Dispositions.—Religious Belief.—Shamanism.—Anecdotes.—Treachery.– Contests with the Eskimos.—Occupations.—Traffic.—Beads and Shells.—Tents.—Vapor Baths.—Deer Pounds.—Oratory.—Talkativeness.—Dances.—Manbot or Blood-money.—Ceremonies on meeting other People.—Population of the Valley of the Yukon.—Same People with certain Coast Tribes.—Kolusches.—Kenaiyers.—Ugalents.—Atnaër.—Koltshanen.—Persons and Dress.—Deer-Pounds.—Passion for Glass Beads.—Kolushes descended from a Raven.—Courtship.—Wives.—Revenge.—Murder.—Burn the Dead.—Mourning.—Do not name the Deceased.—Custom connected therewith.—Winter Habitations.—Journeys of the Kenaiyers inland.—Porcupine Quills.—Slavery.

From Churchill River in Hudson's Bay round northward to the estuary of the Mackenzie, the only nation that the Eskimos come in contact with is that of the 'Tinnè or Chepewyans, and even with them they have no friendly intercourse, nor do they meet except at the trading post of Churchill, and within its influence. To the west of the Mackenzie, however, another people interpose between them and the Tinnè, and spread westward until they come into the neighborhood of the coast tribes of Beering's Sea, which have been already noticed. On Peel's River they name themselves *Kutchin*, the final *n* being nasal and faintly pronounced. It is dropped altogether further to the westward, on the banks of the Yukon. They are the *Loucheux* of Sir Alexander Mackenzie, and the *Di-go-thi-tdinnè* of the neighboring Hare Indians.

Of this people I have but little personal acquaintance, having had only brief interviews with the families that frequent the banks of the Mackenzie for a hundred and fifty miles or so above its delta. My information respecting them is derived from my friend Mr. Bell, who has traded with them for many years, and is the first European who penetrated into their country from the eastward; and from Mr. Murray, who is now, and has been for some seasons, resident among them. It is to this gentleman's very able letters, which I have had the advantage of perusing, through the kindness of chief factor Murdoch M'Pherson, that I am indebted for descriptions of the tribes dwelling on the Yukon.

The few members of the nation that I saw on the Mackenzie had much resemblance in features to their neighbors the Hare Indians, but carried themselves in a more manly manner. Being, however, merely outliers of the Kutchin, they were a less favorable example of that people than the dwellers on the Yukon that came under Mr. Murray's observation. He states that the males are of the average height of Europeans, and well-formed, with regular features, high foreheads, and lighter complexions than those of the other Red Indians.* The women resemble the men, and Mr. Murray speaks of the wife of one of the chiefs as being so handsome that, setting aside her Indian garb and tattooed face, she would have been considered a fine woman in any country. All the females have their chins tattooed, and when they paint their faces they use a black pigment. The men employ both red and black paints on all occasions of ceremony, every one applying them according to his fancy ; and that they may always have them ready, each has a small bag containing red clay and black lead suspended to his neck. Most commonly the eyes are encircled with black ; a stripe of the same hue is drawn down the middle of the nose ; and a blotch is made on the upper part of each cheek. The forehead is crossed by many narrow red stripes, and the chin is streaked alternately with red and black. The Chepewyans and Crees paint their faces in a similar manner.

The outer shirt of the Kutchin is formed of the skins of fawn reindeer, dressed with the hair on, after the manner of the Hare Indians, Dog-ribs, and other Chepewyan tribes, but in its form it

* Mr. Isbister, speaking of the Kutchin who frequent Peel's River Fort, says, " They are an athletic and fine-looking race, considerably above the average stature, most of them being upward of six feet in height, and remarkably well proportioned. They have black hair, fine sparkling eyes, moderately high cheek bones, regular and well set teeth, and a fair complexion. Their countenances are handsome and pleasing, and capable of great expression. They perforate the septum of the nose, in which they insert two shells joined together, and tipped with a colored bead at each end."—*Rep. of Brit. Ass. for* 1847, p. 122.

resembles the shirts of the Eskimos, being furnished with peaked skirts, though of smaller size. The men wear these peaks before and behind; the women have larger back skirts but none in front. A broad band of beads is worn across the shoulders and breast of the shirt, and the hinder part of the dress is fringed with fancy beads and small leathern tassels, wound round with dyed porcupine quills, and strung with the silvery fruit of the oleaster.* The inferior garment of both sexes is a pair of deer-skin pantaloons, the shoes being of the same piece, or sewed to them. A stripe of beads, two inches broad, strung in alternate red and white squares, runs from the ankle to the hip along the seam of the trowsers, and bands of beads encircle the ankles. The poorer sort wear only a fringe of beads, and sometimes only porcupine quills. The wealthy load themselves with beads, strung in every kind of pattern, on the breast and shoulders; and sometimes immense rolls of this valuable article are used as necklaces. Headbands are formed of small, various-colored beads, mixed with dentalium shells, and the same kind of shells are worn in the nose and ears. The hair is tied behind in a cue, bound round at the root with a fillet of shells and beads, and loose at the end. This cue is daubed by the tribes on the Yukon with grease and the down of geese and ducks, until, by repetitions of the process continued from infancy, it swells to an enormous thickness; sometimes so that it nearly equals the neck in diameter, and the weight of the accumulated load of hair, dirt, and ornaments, causes the wearer to stoop forward habitually. The tail feathers of the eagle and fishing-hawk are stuck into the hair on the back of the head, and are removed only when the owner retires to sleep, or when he wishes to wave them to and fro in a dance. Mr. Murray, when he went among these people, found that they attached nearly as much honor to the possession of their cues as the Chinese do to their pig-tails, but he in a short time acquired sufficient influence to persuade a young but powerful chief to rid himself of the cumbrous and uncleanly appendage; his example was followed by the rest of his band, and will, it is to be hoped, spread through the nation. The mittens, which the men always carry with them, are also adorned with shells, and some of these expensive appendages are even attached to their guns. The

* *Elæagnus argentea.*

women wear fewer shells and beads, both of which have a high value in the nation, especially the shells.*

In winter, shirts of hare-skin are worn, and the deer-skin pantaloons have the fur next the skin. On their journeys, travelers carry with them their dress clothes, which they put on every evening after encamping, and when they come to the trading posts. None of the neighboring nations pay so much attention to personal cleanliness and appearance.

The arms of the men are a bow and arrow, a knife, a dagger, and a spear, with a quiver hanging on the left side, and suspended by an embellished belt, which passes over the right shoulder. Fancy handles and fluted blades are more valued than the good temper of a knife; and this people complain of the trouble of shapening a hard steel weapon. Not so the Central Eskimos, who try one knife against another, and will purchase a well-tempered blade at a high price. Guns have been lately introduced among the Kutchin, and are in great demand. All the men carry powder and ball, whether they own a gun or not, and for it obtain a share of the game killed by the possessors of fire-arms. The same custom exists among the Dog-ribs.

The husbands are very jealous of their wives, but in general treat them kindly, contrasting in these respects with their neighbors. The Chepewyans treat their wives indifferently, and are jealous: the Eskimos treat them well, and are not jealous. The principal men of the Kutchin possess two or three wives each, and Mr. Murray knew one old leader who had five. Poor men, whose abilities as hunters are small, and who have been unable to accumulate beads, remain bachelors;† but a good wrestler, even though poor, can always obtain a wife. In winter the women do all the drudgery, such as collecting the fire-wood, assisting the dogs in hauling the sledge, bringing in snow to melt for water, and in fact perform all the domestic duties except cooking, which is the man's office; and the wives do not eat till the husband is satisfied. In summer the women labor little, ex-

* The shells, being several species of *Dentalium* and *Arenicola*, are collected in the Archipelago lying between the Oregon and Cape Fairweather, and pass by trade from tribe to tribe. The large-ribbed Dentalium is most prized.

† Scilicet uxorem cum dote, fidemque, et amicos,
Et genus, et formam, regina pecunia donat,
Ac benè nummatum decorat Suadela Venusque.

cept in drying meat or fish for its preservation. The men alone paddle, while the women sit as passengers; and husbands will even carry their wives to the shore in their arms, that they may not wet their feet. The Eskimo women row their own umiaks, and the Chepewyan women assist the men in paddling their canoes. On the whole, the social condition of the Kutchin women is far superior to that of the Chepewyan women, but scarcely equal to that of the Eskimo dames.

The Kutchin women do not carry their infants in their hoods or boots after the Eskimo fashion, nor do they stuff them into a bag with moss, as the Chepewyan and Crees do, but they place them in a seat of birch bark, with back and sides like those of an arm-chair, and a pommel in front, resembling the peak of a Spanish saddle. This hangs at the woman's back, suspended by a strap which passes over her shoulders, and the infant is seated in it, with its back to hers, and its legs, well cased in warm boots, hanging down on each side of the pommel. The child's feet are bandaged to prevent their growing, small feet being thought handsome; and the consequence is, that short, unshapely feet are characteristic of the people. A practice so closely resembling the Chinese one, though not confined, as with them, to females, may interest ethnologists.

The Kutchin live more comfortably than the Hare Indians or Dog-ribs. They are a lively, cheerful people. Dancing and singing, in which they excel other Indians, are their favorite amusements, and they practice leaping, wrestling, and other athletic exercises. All these are called into play when different bands meet on friendly terms. They are inveterate talkers. Every new-comer, as he arrives at a trading post, halts at the door of the house and makes a speech, in which he tells where he has been, what he has done, how hard he has labored to obtain furs, and urges the propriety of his being well paid for his exertions, relating also the news he obtained from other tribes, and any thing that has checkered his life or crossed his thoughts since his former visit. Established etiquette forbids any one to interrupt him until he has concluded.

Of their religious notions no full account has yet been obtained, but they speak of good and evil spirits, and belief in shamanism is common to them, the Eskimos, and the Chepewyans. The evil spirit whose malevolence they dread is propitiated through their shamans, who profess to have the sole power of communicating with the unseen world, and foreseeing deaths and foretelling events. Such powers clothe the shamans with authority and awfulness. Should any one have a quarrel with the members of another tribe, his death is attributed to sorcery, or, as the interpreters render it, to "evil medicine." A strong party is forthwith mustered, to seek the band which the shamans have designated, and to demand blood-money for their relative, or to avenge his death should compensation be denied. The amount claimed varies with the rank of the deceased, and the estimation in which he was held, from twenty "skins" of beads to thrice that quantity. Mr. Murray mentions a bloody instance of this superstition which occurred in 1847. A woman of the *Kutcha-kutchi* tribe dying suddenly, her death was at first attributed to the presence of white people on their lands, but the matter being debated, this opinion was overruled, and the blame was attached to a band named *Teytse-kutchi*, residing further down the river, some of whom had a dispute with the husband of the deceased. Upward of thirty warriors started on the blood-quest, and five of the unsuspecting Teytse-kutchi happening to approach a sleeping-place of the war-party were waylaid. Four of them were dispatched silently on their landing, and the fifth, who was a little behind the others, not seeing his companions when he came up, suspected that evil had befallen them, and, landing on a sand-bank, interrogated the war-party across the stream. While his attention was engaged by the conversation that ensued, two of his foes carried a canoe through the willows to the other side of a point higher up the stream, and, having embarked, drifted leisurely down the river, as if they belonged to another party. On approaching the sand-bank, they called to the Teytse-kutchi man, that they were going further down, and would be glad of his company. He waited till they came up, and as he was stepping into his canoe, one of the Kutcha-kutchi tripped him up, and the other stabbed him to the heart as he lay. Having accomplished these murderous feats, the war-party resumed their voyage, but meeting afterward only with numerous bodies of the Teytse-kutchi, they concealed their evil intentions, and returned to their own lands.

Mr. Thomas Simpson, in his "Narrative of Discoveries in the Polar Sea," relates an instance of the Peel River Kutchin demanding blood-money from the Eskimos, and receiving it for several years, for one of their countrymen, who they asserted had died of wounds received in a contest between the two nations. The Eskimos having at length discovered that the man for whose death they had been paying was still living, reviled the Kutchin for their falsehood and extortion, and then took their revenge by killing three of the party who had come to demand the compensation for the following year.

Mr. Murray reports that the Kutchin are a treacherous people, that they never attack their enemies in open fight, and only when they consider themselves to be unquestionably superior, either by numbers or in position. They boast of their successes, but seldom tell of their reverses, which are nevertheless frequent, as their wars are chiefly with bands of their own nation, who are as wary and treacherous as themselves. By these feuds one half of the population of the banks of the Yukon has been cut off within the last twenty years. Little value is set by this turbulent people on human life, and the constant dread of ambuscade deters them from traveling except in large parties. They have not as yet imbrued their hands in European blood. Messrs. Dease, Bell, and others of the Company's officers, who have resided at Fort Good Hope on Peel River, have used their influence, and distributed large presents among the tribes, for the purpose of establishing peace, but with only temporary success. The pretensions and arts of the shamans are fertile sources of mischief.

The Peel River Kutchin, in speaking to Mr. Bell of their contests with the Eskimos, always charged the latter with treachery, but it is more likely that they were themselves the aggressors. One of their encounters with that people deserves to be mentioned here, because of its resemblance in some particulars to the meeting of Joab and Abner recorded in the Second Book of Samuel. A party of each of the two nations having met, the young men rose up to dance, as if the meeting had been entirely amicable; but the Eskimos having, as they are accustomed to do, concealed their long flenching knives in the sleeves of their deer-skin shirts, drew them, in one of the evolutions of the dance, and thrust them into their opponents. A general conflict ensued, in which the Kutchin were the victors, owing to their guns—that is, according to their report of the affair; but had the Eskimos been the tellers of the story, the circumstances might have been related differently.

Another incident, which occurred on the banks of the Yukon in 1845, gives a further insight into the suspicious and timorous lives of these people. One night four strangers, from the lower part of the river, arrived at the tent of an old man who was sick, and who had with him only two sons, one of them a mere boy. The new-comers entered in a friendly manner, and, when, the hour of repose, came lay down ; but the sons perceiving that their guests did not sleep, and suspecting from their conduct that they meditated evil, feigned a desire of visiting their moose-deer snares. They intimated their purpose aloud to their father, and went out, taking with them their bows and arrows. Instead, however, of continuing their way into the wood, they stole back quietly to the tent, and listening on the outside, discovered as they fancied, from the conversation of the strangers, that their father's life was in danger. Knowing the exact position of the inmates, they thereupon shot their arrows through the skin covering and killed two of the strange Indians ; the other two, in endeavoring to make their escape by the door, shared the fate of their companions. This is spoken of in the tribe as an exceedingly brave action. The golden age of innocence and security is not to be sought for among a savage people ignorant of the precepts of divine truth.

The Yukon Kutchin pass the summer in drying white-fish (*Coregonus*) for winter use. For the purpose of taking these fish, they construct weirs by planting stakes across the smaller rivers and narrow parts of lakes, leaving openings in which they place wicker baskets to intercept the fish. This practice is common in Oregon and New Caledonia,* but does not exist to the eastward of the Rocky Mountains. On the other hand, the inhabitants of the Yukon are unacquainted with nets, so largely employed by the Chepewyans and Crees. The Kutchin take the moose-deer in snares, and, toward the spring, most of the nation resort to the mountains to hunt reindeer and lay in a stock of dried venison.

Beads are the riches of the Kutchin, and also the medium of exchange throughout the country lying between the Mackenzie and the west coast, other articles being valued by the number of strings of beads they can procure. No such near approach to money has been invented by the nations residing to the eastward

* Cook observed fishing weirs at Nootka Sound.—*Third Voy.*, vol. ii. p. 281.

of the Rocky Mountains, though their intercourse with the fur-traders has given them a standard of value in the beaver's skin. Their accounts at the posts being reckoned by the number of "beavers" they owe, and the Company's tariff fixes the value of a "beaver." To be accounted a chief among the Kutchin, a man must possess beads to the amount of 200 beavers. The standard bead, and the one of most value, is a large one of white enamel which is manufactured in Italy only, and can with difficulty be procured from thence in sufficient quantity. Fancy beads, *i.e.*, blue and red ones of various sizes, and the common small white ones, are, however, in request, for ornamenting their dresses.

Dentalium and *Arenicola* shells are transmitted from the west coast in traffic, and are greatly valued. None of the Chepewyan tribes wear nose-ornaments, neither have the latter people the same passion for beads. A supply of them is indeed sent to all the trading posts frequented by the 'Tinnè, but they are mostly purchased by the wives of Canadian voyagers or half-breeds residing in the establishments, and if desired by the natives for the same purposes they are given to them as presents, or exchanged for articles of small value, and never, I believe, for furs. The Kutchin, on the contrary, will not part with their furs unless they receive most of the price in beads or shells, as they have not yet learnt to value English cloth and blankets above their skin dresses. Ammunition, by the instructions of the Hudson's Bay Company, is given in exchange for provisions, or, when the natives are in want, gratuitously, that they may be able to support themselves.

Each family possesses a deer-skin tent or lodge—the skins used in winter being prepared without removing the hair, that the cold air may be more effectually excluded. In summer, when the family is traveling in quest of game, the tent is rarely erected. A winter encampment is made usually in a grove of spruce firs. The ground being cleared of snow, the lodge-skins are extended over flexible willow poles, which take a semicircular form, and are transported with them from place to place. The hemispherical shape of their lodges is not altogether unknown among the Chepewyans and Crees, being that generally adopted for vapor baths, which are framed of willow poles stuck into the ground at each end; but the lodges used by these nations for dwelling places are cones, formed by stiff poles meeting at the top. The lodges of the Kutchin resemble the Eskimo snow huts in shape, and also the *yourts* of the Asiatic Anadyrski. When the Kutchin winter-lodge is raised, snow is packed on the outside to half its height, and it is lined equally high within with the young spray of the spruce fir, that the bodies of the inmates may not rest against the cold wall. The doorway is filled up by a double fold of skin, and the apartment has the closeness and warmth, but not the elegance of a snow house. Mr. Murray remarks, that though only a very small fire is usually kept in the centre of the lodge, the warmth is as great as in a log house. The provisions are stored on the outside, under a covering of fir branches and snow, and further protected from the depredations of the dogs by the sledges being placed on the top.

Mr. Bell informed me that, on the open hilly downs frequented by reindeer, the Kutchin have formed pounds, toward which the animals are conducted by two rows of stakes or trunks of trees extending for miles. These rows converge, and as the space between them narrows, they are converted into a regular fence by the addition of strong horizontal bars. The extremity of the avenue is closed by stakes set firmly in the ground, with their sharp points sloping toward the entrance, so that when the deer are urged vehemently forward they may impale themselves thereon. The hunters, spreading over the country, drive the deer within the jaws of the pound ; and the women and children, ensconced behind the fence, wound all that they can with arrows and spears. These structures are erected with great labor, as the timber has to be brought into the open country from a considerable distance. Some of the pounds visited by Mr. Bell appeared to him, from the condition of the wood, to be more than a century old. They are hereditary possessions of the families by whom they were constructed.

Mr. Murray's letters describe the meetings of several tribes which he witnessed. On one occasion two parties who had been at war with each other, and had not yet arranged their differences, met at his encampment. Very long harangues were made by different members of the two bodies before they landed. After this, one party, stepping from their canoes, formed a circle and pranced round, yelling and shouting furiously. The other party landed a little way off, and ranging themselves in Indian file, the chief in front and the women and children in the rear, danced forward slowly, until they came up to the others; when the whole, joining in one circle, capered for half an hour, uttering the most horrible cries, the two chiefs meanwhile keeping in the centre.

The formal dance is always in a circle, but the gestures and the songs which accompany them vary. After a ball kept up vigorously by man, woman, and child, for hours, the parties retire to their tents, and raise the song at intervals till the morning. At the festivals held on the meeting of friendly tribes, leaping and wrestling are practiced.

Several other illustrations of the superstitions and manners of this people might be quoted from Mr. Murray's letters, but, as they relate only to peculiarities already mentioned, I shall restrict my extracts to two other anecdotes. A young chief was at Mr. Murray's encampment when a party of another tribe, named the *Vanta-kootchi*, arrived, one of whom had married the chief's sister, and was reported to have killed her. The young chief demanded an expiatory offering in beads for his sister's death, which was refused; and an altercation ensuing, something was said which insulted him. He immediately drew his knife and walked boldly up to the others, who would have cut him in pieces but for the immediate intervention of the white men. A few words of explanation from a hunter in Mr. Murray's employment calmed the storm. The woman had not been killed, but was drowned in crossing a river, through the upsetting of a canoe. A present of a large Eskimo spear, valued at ten skins, was made to the brother, and peace was restored.

A body of the *Han-kutchi*, residing at the sources of the Yukon, came to visit Mr. Murray early in August. Rumors of their hostile intentions preceded them. The sudden death of their chief in summer had been attributed by them to the shamanism of the *Kutcha-kutchi*, and also to the presence of white people in the country. The Canadian voyagers looked on, therefore, with apprehension, which was not quieted by the first movements of their visitors. Twenty canoes first appeared, gliding stealthily down the river to a point above the encampment, on which the party landed and assembled in silence. Mr. Murray walked up to them, and expressed his pleasure at the meeting; but, rushing past him, the whole body ran in full career to the lower end of the encampment and back again to their landing-place, shouting and whooping in a peculiar manner. Then they formed a half-circle, and danced with great energy for a few minutes, beating time to their songs with their feet. Their dresses, tinkling with beads and brass trinkets, and their long clouted hair shaking in the wind, gave them a wild and savage aspect. They were armed with pikes of their own manufacture, also with similar weapons, mounted with sheet-iron, procured from the Russians. On the dance ceasing, Mr. Murray presented the brother of the deceased chief with twelve inches of tobacco to smoke over the grave, which produced a favorable impression, and called forth the remark that *now* he could consider the white people as his friends. These Indians afterward became troublesome, asked for goods on credit, and on being denied threw out some significant innuendos, saying, among other things, that the Russians had used them so at first, but had become more civil since they had cut off one of their outposts. At the great dance in the evening, the deceased chief's brother did not join the circle, but retired to a corner, and made piteous lamentations.

Mr. Murray, from information collected from the natives, estimates the population of the banks of the Yukon at about one thousand men and boys able to hunt. They are distributed as follows. Between the upper branches of the river and the coast of the Pacific, on or near the 62d parallel, reside the *Artez-kutchi*, or "tough and hard people," numbering 100 men. The *Tchŭ-kutchi*, "people of the water," of about the same numbers, inhabit the banks of Deep River, a western affluent of the Yukon, opposite Comptroller's Bay. The *Tathzey-kutchi*, "people of the ramparts," known to the traders and Canadian voyagers by the name of "*Gens du Fou*," number about 230 men, who are divided into four bands, the uppermost of which is called *Trătzè-kutchi*, "people of the fork of the river."* The *Tathzey-kutchi* inhabit a wide country, which extends from the sources of the Porcupine and Peel to those of the river of the Mountain-Men. They visit the Russians on the coast of the Pacific, and trade with the intervening tribes. The Indians of the lowlands at the

* One of '*Dtchè-tà-ùt 'tinnè*, or Mountain Indians, who inhabit the conterminous mountainous country that lies between the *Noh'hannè* and the *Becàtess* ("Gull" or "Gravel River" of the voyagers), called the *Gens du Fou*, *Ey-unè 'tinnè;* and a Mountain Indian hunter, employed by Mr. Bell, called them *Tratzà-ut 'tinnè*, which is evidently another mode of pronouncing the designation they give themselves, substituting for *Kutchi* the synonymous appellation '*Tinne*.

influx of the Porcupine River, named *Kutcha-kutchi*, number 90 men. Further down the Yukon are the *Zèkā-thaka* or *Zi-unka-kutchi*, "people on this side," or "middle people," numbering only 20 men. West of these reside the "people of the bluffs," *Tanna-kutchi*, 100 strong. And further down, at the influx of Russian River, are the *Teytsè-kutchi*, "people of the shade," or "shelter," of about 100 men. Nearer the mouth of the Yukon are two bands, usually called *Tlagga-silla*, or "little dogs," but not by themselves. These trade their furs with the *Kuwichpack-meuten*, mentioned in page 219. The banks of the Porcupine and country on the north of it belong to the *Vanta-kutchi*, "people of the lakes," having 80 men; and to another band, named *Neyetsè-kutchi*, "people of the open country," who have 40 men.

Though the short Kutchin vocabulary formed by Mr. Murray, which will be printed in the Appendix, contains some Eskimo words, the language, as far as one can judge from so brief an example, is more nearly related to the 'Tinnè. How far the Kutchin are to be considered as actually a distinct nation from the 'Tinnè or Chepewyans, must be decided hereafter when their two languages are better known.

At my suggestion Mr. Murray made, by the aid of his Athabascan interpreter, the following collection of words, having a similar sound and signification in the Kutchin and Dog-rib languages.

English.	Kutchin.	Dog-rib.
One	Tech-lagga	Inch-lagga or Ingeh-lagga.
Two	Nāk-heiy	Nak-hè.
A hare	Kė	Kā.
Fire	Kon or Khon	Khu or Kun.
Wood	Tutshun	Tutshin.
A blanket	Tsattè	Tsat-hè.
Reindeer (male)	Bat-zey-tcho	Bet-sich-tcho.
A dog	Chli-en or Chlyn	Chli or Thling.
My country	Sun-nun	Sa-nun-nā.
Moss	Ni-en	Ni.
A goose	Chrè	Chrā.
A snow goose	Ku-kè	Ko-ka.
Thunder	Nach-thun	Nach-thun.
Willow	Khy-i	Khy-i.
Snow shoes	Ai-i or Ay-i	Ah.
A poplar tree	T'ho	Tőë or T'höe.
An Indian cap	Tsè	Tsa.
White	Tā-kynè	Tek-kynè.

English.	Kutchin.	Dog-rib.
Man	'Tingi	'Dinnè, or 'Tennèh, or Dunneh
A quill or feather	Teh or Tay	Tah.
A sinew	Tchè	Tlè.
You	Nuhn or Nunn	Nunney.

In this vocabulary the *ch*, except when immediately preceded by *t*, is pronounced as in the Scottish "*loch*," or Irish "*och*;" *u* is sounded as *oo* in good, except before double consonants; and *i* as *ee* in "see," or *e* in "me."

Mr. Murray remarks that, though the above words and a few names of trading goods are similar in sound, the languages of the two nations are very different. More resemblances, he thinks, might be traced through the Mountain Indian speech (*Naha 'tdinnè* or *Dtchè-ta-ut 'tinnè*) than directly between the Kutchin and Dog-rib tongues. The *Han-Kutchi*, of the sources of the Yukon, speak a dialect of the *Kutcha-Kutchi* language, yet they understand and are readily understood by the Indians of Frances Lake and the banks of the Pelly. Now these converse freely with the *Naha-* or *Dtchè-ta-ut 'tinnè*, and other Rocky Mountain tribes, whose language resembles the Dog-rib tongue, and who are, in fact, acknowledged members of the Chepewyan nation. Again, the Frances Lake Indians understand the *Netsilley*, or Wild Nation, who trade at Fort Halkett, on the River of the Mountains; these again are understood by the *Sikānis;* and the *Sikānis* by the Beaver Indians, whose dialect varies little from that of the Athabascans, the longest-known member of the 'Tinnè nation.

From the great resemblance in manners, customs, and person of the '*Tnaina** or *Kenaiyer* of Cook's Inlet, the *Ugalakmutsi*

* *Tnai* signifies "men," and is used when the Atnäer speak of themselves. Their Eskimo neighbors the *Kadyakers* call them *Kenai-yut*, and the Russians have adopted this latter appellation. *Koltshanen* means "strangers" in the Atnäer dialect, and *Gultzanen* "guests" in the language of the Kenaiyer. These people, coming from the interior, about the sources of the Copper River and the water-shed between it and the valley of the Yukon, have commercial relations with the Kutchin who dwell on Deep River, an affluent of the Yukon. Their accounts of the pointed skirts of the Kutchin shirts have given rise to the fable already alluded to of men with tails dwelling beyond the mountains, current among the Kenaiyer. And their reports of the cannibalism of the tribes of the interior have a similar foundation. Lynn's Canal is not mentioned as the ascertained southern limit of the tribes akin to the Kutchin, but because down to that inlet the language of the Yukon Kutchin seems to be readily comprehended. The traveling merchants who go from thence

or *Ugalents* of King William's Sound, the *Atnüer* of the mouth of the Copper River, the *Koltshanen* or *Galtzanen* of the sources of that stream, and the other Kolusch tribes as far as Tchilkat or Lynn's Canal, on the 54th parallel and 135th meridian, to those of the Kutchin, I am inclined to consider them as all of the same stock. Captain Cook, from whom we have the earliest accounts of these people, remarks their dissimilarity in person and language to the *Wakash* nation who inhabit Vancouver's Island.

The *Kenaiyer* and *Ugalents* are described by Captain Cook* and Baron Wrangell as a moderate-sized race, occasionally tall, with women of equal height, and, when young, handsome; particulars in which they agree with the Kutchin, whose women have finer persons than the 'Tinnè females. Both these authors, however, remark that there are many deformed individuals among

to the banks of the Pelly and sources of the Yukon meet there with the *Tratze-kutchi* or *Gens du Fou*, by whom they are understood—as has been mentioned in page 105 of the narrative. Dr. Scouler and others maintain, and probably with justice, that the Kolusch language is spoken as low as Queen Charlotte's Island and the Observatory Inlet.

* "The natives who came to visit us in the Sound were generally not above the common height, though many were under it. They were square, or strong-chested; and the most disproportioned part of their body seemed to be their heads, which were very large; with thick short necks; and large, broad, or spreading faces, which upon the whole were flat. Their eyes, though not small, scarcely bore a proportion to the size of their faces; and their noses had full round points hooked or turned up at the tip. Their teeth were broad, white, equal in size, and evenly set. Their hair was black, thick, straight, and strong; and their beards in general thin or wanting; but the hairs about the lips of those who have them, were stiff or bristly, and frequently of a brown color. And several of the elderly men had even large and thick but straight beards." "Their countenance indicates a considerable share of vivacity, good-nature, and frankness. Some of the women have agreeable faces."—*Cook*, *Third Voyage*, vol. ii. p. 366.

Wrangell says, "The Kenaiyer are in general of middle size, slightly built, and betray a true American descent in their features and color of the skin. Many among them are of gigantic stature, and I have never seen so many deformed persons among any people in the colonies."

These quotations will give the impressions the coast tribes made on two accurate observers at distant periods. With respect to the shortness and thickness of the neck, I may mention that the skeletons of the races which flatten their foreheads artificially have very short necks, the bodies of the vertebræ being unusually thin. This is doubtless the result of the process resorted to in infancy; and the same effect may be produced by masses of dirt and hair adding weight to the head.

the coast tribes, contrary to what occurs among the inland Indians. Both sexes of the Kenaiyer and of the allied tribes powder their hair with the down of birds, and smear their faces with black, blue, and red pigments. The blue pigment was noticed by Cook, and Mr. Bell informed me that some of the Peel River Kutchin possess it, but he did not ascertain where they procured it, and I have not been able to obtain a specimen of it, so as to ascertain its nature. The men wear dentalium shells in the nose, and also ear-pendants, and the unsightly labrets. They are a cheerful people, sing during labor, and when it is over recreate themselves with dancing. Cook notices the thick short necks of the people of Prince William's Sound. The muscles of the necks of the Kutchin are called strongly into action for the support of their weighty ties of hair, and, in consequence thereof, increase in size; and we might have been inclined to attribute the disproportion noticed by Cook to a similar cause, had he not mentioned that the men wear their hair cropped round the neck. Some of the women, however, clubbed it behind; and this looks as if, at that period, the tribe had found out the inconvenience of the unwieldy and uncleanly cues, and that a few only of the women retained them.

Among the Atnäers and neighboring tribes who hunt reindeer, pounds, formed of hedges converging thus >, are in use. Weirs and wicker-baskets for taking fish are also constructed by the coast tribes. The shirts of the Ugalents reach to the knee or lower, and are cut evenly round without peaks. They are made of skins, with the fur turned outward; but the coast people wear also a waterproof outer dress, made of whale-gut or other thin membraneous substance, which they have most probably adopted from their Eskimo neighbors. The dwellers of the interior do not use it, and indeed have not the material. The Koltshanen construct birch-bark canoes; but on the coast skin boats or baidars, like the Eskimo kaiyaks and umiaks are employed. Another opening has, however, been added to the kaiyak, which is therefore longer, and carries two sitters. It would seem that the constant commercial intercourse between the coast tribes had led them to adopt whatever they thought worthy of imitation from one another.

The passion for glass beads extends from the coast to the Mackenzie. The Atnäers bury their accumulations of this treasure in the earth, and leave the hoard to their children. In Cook's time a large light-blue bead was in the greatest request.

All the coast tribes now under consideration claim descent from the Raven; and it would strengthen the opinion we entertain of their affinity with the Kutchin, if the same belief existed among the latter. This, however, has not been ascertained. The west coast Eskimo and the Chepewyan tribes agree in tracing the origin of their respective nations to the Dog. The Raven, say the Northern Kolushes, Prometheus-like, stole the elements one after another, out of which he made the world. The Kenaiyer tradition is, that the Raven also made two women, one of whom is the mother of six races, and the other of five.* It was the custom that the men of one stock should choose their wives from another, and the offspring belonged to the race of the mother. This custom has fallen into disuse, and marriages in the same tribe occur; but the old people say that mortality among the Kenaiyer has arisen from the neglect of the ancient usage. A man's nearest heirs in this tribe are his sister's children, little going to his sons, because they received in their father's lifetime food and clothing.

Courtship is a simple affair with these people. Early in the morning the lover makes his appearance at the abode of the father of the object of his choice, and, without a word of explanation, begins to heat the bath-room, to bring in water, and to prepare food. Then he is asked who he is, and why he performs these offices. In reply he expresses his wish to have the daughter for a wife, and if his suit be not rejected, he remains as a servant in the house a whole year. At the end of that time he receives a reward for his services from the father, and takes home his bride. No marriage ceremony takes place. Rich men have three or four wives. The wife, though the most industrious worker in the family, is not the slave of her husband. She may

* The stock descended from one female ancestor are, 1. the *Kachgiya*, from *Gekaihze*, "the raven;" 2. *Tlachtana*, so called from their being weavers of grass-mats; 3. *Montochtana*, who take their name from a corner in the back part of their huts; 4. *Tschichgi*, named from a color; 5. *Nuchschi*, "descendants from heaven;" 6. *Kali*, "fishermen."

The races sprung from the other women are, 1. *Tultschina*, "bathers in cold water;" 2. *Katluchtna*, "lovers of glass beads;" 3. *Schischlachtana*, "deceivers like the raven, who is the primary instructor of man;" 4. *Nutschichgi*, and 5. *Zaltana*, named from a mountain on the borders of Lake Skiläch toward the sources of the river Katnu.

return, if dissatisfied with the treatment she receives, to her father's house, and then she takes with her the dowry the husband received at the conclusion of his year of service. The wife retains as her own property whatever she gains by her labors, and it often happens that the husband makes purchases from her. If there be several wives in a family, each has her own household stuff, which may not be meddled with by the other wives, or by any member of the household.

Banquets, accompanied by dances, songs, and distribution of presents, take place on various occasions. A man, on recovering from sickness, will give a feast for the benefit of those who have shown him most sympathy during his illness. One who spends freely on these occasions is looked up to by his fellow-countrymen, and his advice sought. This is the origin of *Toyonhood*, or chieftainship. Though the power of the petty chief does not depend on descent, it frequently passes to his heir; but submission to him is conditional, and any one may attach himself to another leader.

If a man be murdered or injured by one of his own clan, the nearest relative revenges it without seeking aid; but if the injury be perpetrated by one of another clan, the allied families are called together to consult on the defense of their honor. The feud that ensues is sometimes bloody, but seldom of long duration, and any prisoners that are taken are set free for a ransom, or retained as slaves. Before the Russians came, all the tribes were at war with the Kadyakers and others of the Eskimo nation, who on that account received from the Kenaiyer the denomination of *Ultsehna*, or *Ultsehaga*, "slaves."

One who dies is mourned by his whole clan. The mourners assemble in the dwelling of the nearest kinsman, sit round the fire, and howl. The master of the house, dressed in his best garments, leads the lamentation, having his face blackened, an eagle's feather in his nose, and a cap of eagle feathers on his head. Ringing a bell which he holds in each hand, he raises the voice of mourning, making at the same time violent contortions of the body, and stamping continually on the ground with his feet. He recounts in his song the famous actions of the deceased; stanzas are improvised by the other mourners, and sung to the accompaniment of drums. A general cry of grief is raised at the end of every verse, and during its continuance the chief mourner pauses from his exertions, and lets his head sink on his breast.

The clothing and rest of the property of the deceased are divided among mourning relatives. The body is then burnt, and the bones are collected and interred by friends who are not of kin to the deceased. At the end of the year the nearest relative celebrates a festival to the memory of the departed. From that time the dead man's name must never be pronounced in the presence of that relative, and *he* even changes the name by which the deceased had been accustomed to address him. If a relation transgresses this law, he is reproved ; but if it be a more distant friend he is challenged by the kindred, and must buy himself off. Poor men will sometimes endeavor to entrap a rich relative into a breach of this custom, to obtain the redemption-money. The Kenaiyer suppose that after death a man leads, in the interior of the earth, where a sort of twilight reigns, a similar life to his former one, but that he sleeps when those on the surface are awake, and wakes when they sleep.

One other way, among several, in which a poor Kenaiyer endeavors to improve his condition is, to invite his rich friends of another family to a festival. Melted snow only is set before them, and the relatives, watching for any sneering expressions respecting this Barmecide feast, report them to the host. He then rushes out with angry gestures, and challenges his mockers, wounding himself at the same time with an arrow, to signify that he prefers death to degradation. This scene was expected, and the mockers having their presents prepared, declare their readiness to make reparation.

The winter huts of the Northern Kolushes are high, large, and roomy, built of wood, with the hearth in the middle, and the sides divided into as many compartments as there are families living under the roof; the number varying from two to six. Two or three bath rooms are constructed at the end, and in them much of the winter is spent. These vapor baths resemble the den of a bear, and open into the hut by a small aperture through which a man can creep with difficulty. They are covered on the outside with earth, and heated within with hot stones.

Baron Wrangell, from whom most of the preceding details have been translated, is of opinion that the Kenaiyer came originally from the interior, bringing with them their bark canoes, and borrowing the skin kaiyaks from the Eskimo Tchugatschih, who are more expert and bolder navigators of the sea. The following notice of the journeys of the Kenaiyer, given by the author above named, will show that they must come occasionally into contact with the dwellers of the Yukon Valley, as they approach its watershed. The people on the south side of Cook's Inlet hunt mountain sheep in the neighboring hills. Those that dwell on the north side travel much further. Striking off to the northeast, after seven days of rapid march, or ten of ordinary traveling, in which they can accomplish about one hundred and thirty or one hundred and forty miles, they arrive at the foot of a high mountain chain, where, on the banks of Lake Knitiben, the women, children, and less skillful hunters are left. The others cross the mountains in a southerly direction, and in seven days more arrive at Lake Chtuben, on an elevated plateau not far from the source of the Suschitna, and fourteen days' march from the northern arm of Cook's Inlet.

There they kill reindeer, which winter in that district in numbers. The Atnäer and Galzanen of Nutatlgat, on the Copper River come, after ten days of rapid traveling, to the same lake. From thence the Kenaiyer go six days' march further to a small lake, where they are met for the purposes of trade by the more distant Koltshanen, who supply them with articles of English manufacture. Porcupine quills, colored by the Atnäer with moose berries, used for embroidering seal-skin shirts, are valued articles of commerce; and in their traffic all parties are wary and skillful.

The fatigues of this excursion render the Kenaiyer hunters lean and exhausted; but they kill beavers as they return home, and continue to do so up to the beginning of winter, when they hold the annual festival, announce the produce of the chase, and give themselves up to recreation and pleasure.

Slavery, to a certain extent, prevails among the coast tribes, and the slaves, who are originally prisoners taken in battle and not redeemed, are transferred from one tribe to another by barter. Captain Cook mentions, that on his first intercourse with them none would come into his ship, until a seaman had gone into one of their boats, when an Indian was sent up into the ship. This was considered by the English officers as an exchange of hostages, but perhaps the Indians reckoned it to be a barter of slaves, as they would not part with the seaman until muskets were presented at them; and when he returned on board they took their own man and departed. The slaves are named *Kalgen*, and are occasionally sacrificed to the manes of deceased chieftains by the Koltshanen and Koluschen, but not by the Atnäer. Previous to the arrival of Europeans, the Atnäer were workers in copper, and supplied the neighboring tribes with weapons. They have since acquired the art of fabricating Russian iron into various articles. They are a milder tribe than the other Kolushes, and are generally on good terms with the rest.

CHAPTER XIII.

OF THE 'TINNÈ OR CHEPEWYANS.

Geographical Position.—National Name.—Tribes.—Hare Indians and Dog-ribs.—Personal Appearance.—Women.—Dress.—Dispositions.—Wars.—Socialism.—Improvidence.—Suffering.—Affection for their Children.—Hospitality feeble.—Falsehood.—Honesty.—Religious Belief.—Volatility.—Marriages.—Wrestling for a Wife.—Dogs.—Moose-hunting.—Public Opinion the only Rule of Conduct.—Chiefs.—Introduction of Christianity.—Horses.—Houses.—Dawnings of Civilization.—Members of the 'Tinnè People west of the Rocky Mountains.—Southern Athabascans.

'TINNÈ or *'Dtinnè*, Athabascans, or Chepewyans. Under these national appellations I have to speak of a people whose southern border is the Churchill River, or the *Missinipi*, as it is termed by the Eythinyuwuk, to whom it is also a boundary line. Every where, in the country lying east of the Mackenzie, the *'Tinnè* lands are conterminous with the Eskimo coast, and, to the westward of the Rocky Mountains, with the Kutchin grounds, though the precise geographical limits of the two nations in that direction have not yet been correctly ascertained. The *'Tinnè*, however, extend across the continent, since the *Tă-kuli* and almost the entire population of New Caledonia have been referred by ethnologists to their nation.

The name by which the 'Tinnè designate themselves has, as is usual with the native Americans, the signification of "people," or "the people," and its proper application, when ascertained with care, would seem, at first sight, to be a good test for fixing the nationality of some tribes whose position in the ethnological scale is still uncertain. But as our acquaintance with the various American languages extends, and the way in which the pronunciation of the same word in the mouths of different tribes is gradually modified becomes known, doubts arise as to the value of such a test, or, rather, the opinion of the intimate connection between the various tongues is strengthened, though it may be difficult to trace their links in vocabularies compiled by Europeans. Thus, though no two languages can be apparently more dissimilar than the harsh, guttural, unpronounceable, and unwritable 'Tinnè speech, and the flowing, harmonious, and easily acquired tongue of the Eythinyuwuk, yet the '*Thinyu* (man) of the latter may be resolved into the '*Tinnè*, '*Tinye*, or '*Dunnè*, of the former, and the *Ting-i* of the Kutchin, without much philological artifice.*

Various tribes have been distinguished by peculiar names, but there is little variety in their general appearance, and few discrepancies in their dress, customs, or moral character. The Hare Indians (*Kă-cho-'dtinnè*) inhabit the banks of the Mackenzie, from Slave Lake downward, and the Dog-ribs (*Thling-è-ha-'dtinnè*) the inland country on the east, from Martin Lake to the Coppermine. There is no perceptible difference in the aspect of these two tribes. They meet in the same hunting-grounds at the north end of Great Bear Lake, intermarry, and their speech scarcely differs even in accent. The Hare Indians, frequenting a thickly wooded district in which the American hare abounds, feed much on that animal, and clothe themselves with its skins, while the Dog-ribs depend more upon the reindeer for a supply of winter dresses, but in all essential respects they are the same people. To the eastward of the Dog-ribs are the Red-knives, named by their southern neighbors the *Tantsa-ut-'dtinnè* (Birch-rind people). They inhabit a stripe of country running northward from Great Slave Lake, and in breadth from the Great Fish River to the Coppermine. They were also formerly in the habit of resorting to the north end of Great Bear Lake, to kill musk-oxen and reindeer; but many of their influential men being cut off by treachery in a feud with the Dog-ribs, they have lately kept more toward the east end of Great Slave Lake. These three tribes roam northward to the Eskimo boundary line, but mutual fears cause the two people to leave an ample neutral ground, on which neither party are willing to venture.

Other members of the 'Tinnè nation inhabit the country at the mouth of the Missinipi, and carry their furs to Fort Churchill, where they meet the Eskimos that come from the north-

* Mr. Isbister says the Chepewyan tongue is "harsh and guttural, difficult of enunciation, and unpleasant to the ear." "As a language it is exceedingly meagre and imperfect."—*Rep. Brit. Ass.* for 1847. Mr. M'Pherson pointed out to me, as a curious coincidence, the similarity in sound of the Gaelic word for people, with the '*Dunnè* of the Dog-rib Indians.

ward, and, through the influence of the traders, carry on an amicable intercourse with them, so that the Tinnè families occasionally accompany the Eskimos to their hunting-grounds. A wide tract of barren lands intervenes between the Churchill 'Tinnè and the Red-knives, and the tribes on the Slave and Elk Rivers which resort to Fort Chepewyan. These "barren-grounds" are very thinly peopled, and rather by isolated families who resort thither for a year or two to hunt the reindeer than by parties associated in such numbers as to deserve the name of a tribe. Part of these wandering, solitary people resort at intervals of two or three years to Churchill for supplies, and part to Fort Chepewyan, where, from the direction in which they came, they are named *Sa-i-sa-'dtinnè* (Eastern or Rising Sun folks). The Athabasca 'Tinnè, named also Chepewyans, frequent the Elk and Slave Rivers, and the country westward to Hay River, which falls into Great Slave Lake. There is some difference between their dialect and that of the tribes on the Mackenzie, but not so much as to occasion any difficulty to an interpreter, versed in either tongue. The name *Chepewyan* has no relation to the word *Ojibbeway* or *Chippeway*, which designates an Eythinyuwuk people frequenting the coasts of Lake Superior, but has rather, I believe, its origin in the contempt felt by the warlike Crees for the less manly Tinnè, whom they oppressed by their inroads, before commerce introduced peace between them. *Chi-pai-uk-'tim* (you dead dog) is a most opprobrious epithet. The appellation of "slave" given to the Dog-ribs by the same people, whose war-parties penetrated even to the banks of the Mackenzie, has a similar origin; and it has been stated in a preceding page, that the Kolushes also called the Eskimo Kadyakers with whom they warred "slaves." To the south of the Athabascans, a number of 'Tinnè frequent the upper part of the Missinipi, where they mingle with the Crees, and in common with them trade with the posts on Lac la Ronge and Isle à la Crosse. (See p. 63). The *Sarsis* or *Circees*, who live near the Rocky Mountains, between the sources of the Athabasca and Saskatchewan Rivers, are said to be likewise of the Tinnè stock.

Between the Peace River and the west branch of the Mackenzie are the Beaver Indians, who take their name from an affluent of the latter. Their dialect is reported to be softer than that of the other 'Tinnè, having probably been modified by their intercourse with the Crees of the prairies. Other tribes on the mountain branch of the Mackenzie differ somewhat, either in language or manners, from the eastern part of the nation, and have peculiar designations. The *Noh'hannè* inhabit the angle between that branch and the great bend of the trunk of the river, and are neighbors of the Beaver Indians. Higher up are the *'Dtcha-ta-ut-'tinnè*, "Mountain Indians" or "Strong-bows," who keep tò the ranges of the Rocky Mountains, and the *Tsilla-ta-ut-'tinnè*, or "Brushwood-people."

Between the trunk of the Mackenzie, on the 65th parallel, and the Rocky Mountain ranges, dwell a tribe named *Dahā-'dtinnè* by the Dog-rib Indians, and *Noh'hai-è* by the Kutchin. They descend the Gravel River to come to Fort Norman, and are ill understood by the Dog-rib interpreters there. In p. 112 I have mentioned, on the authority of Mr. M'Kenzie, that the *Dāha-'dtinnès* name themselves in their own tongue *Cheta-ut-tdinnè*, which indicates their identity with the Strong-bows, both being mountaineers. Further down the Mackenzie, near the 65th parallel, another small tribe also descends from the mountains to visit Fort Good Hope, and is named *Amba-ta-ut-'tinnè*, or "Sheep-people," because they hunt the *Ovis montana* on the mountain-tops. These people speak a dialect of the *'Tinnè*, which is well understood by the Hare Indians.

This enumeration of the various 'Tinnè tribes dwelling on the east side of the Rocky Mountains, all of whom believe that they are sprung from a dog, will give some idea of the geographical extent of the nation. It is not my intention to speak of them severally, as my personal acquaintance is too partial to enable me to state correctly in what respects, they differ from each other. The Athabascans or Chepewyans proper have been so long known, and so often mentioned by writers on the fur countries, and Hearne has given so many details of the habits of the, 'Tinnè of Churchill, and of the tribes he encountered in his journey over the barren grounds, that I could add little of importance; I shall, therefore, restrict my remarks to the Dog-ribs and Hare Indians, who resorted to Fort Franklin and Fort Confidence during my residence on Great Bear Lake.

These people possess more regular features than the Eskimos, with, at the same time, a greater variety among individuals, many of whom have good profiles. Taken as a whole, they exhibit all the characteristics which we observe in the red races dwelling further south; but their inattention to personal appearance, want of cleanliness, and their abject behavior, give them a very inferior aspect, particularly when in company of white people. For they possess the whine and air of accomplished beggars, and their solicitations are constant as long as they have any hope of gain. The women are inferior to the men in height, features, and care of their dress; for, dirty as the men generally are, they do paint their faces and wear ornaments on festive occasions, while few of the women take so much trouble. Most of the latter, however, are tatooed on the chin, or at the angles of the mouth.

The clothing of the men in summer is reindeer leather, dressed like shammy, and is beautifully white and soft when newly made. A shirt of this material, cut evenly below, reaches to the middle; the ends of a piece of cloth secured to a waistband, hang down before and behind; hose or Indian stockings descend from the top of the thigh to the ankle; and a pair of mocassins or shoes of the same soft leather, with tops which fold round the ankle, complete the costume. When the hunter is equipped for the chase, he wears, in addition, a stripe of white hare-skin, or of the belly part of a deer-skin, in a bandeau round the head, with his lank, black elf-locks streaming from beneath; a shot pouch, suspended by an embroidered belt, which crosses the shoulder; a fire-bag or tobacco pouch tucked into the girdle; a pair of mittens; and a long fowling-piece, in its coat, thrown carelessly across the arm or balanced on the back of the neck. The several articles here enumerated are ornamented at the seams and hems with leathern thongs wound round with porcupine quills, or are more or less embroidered with bead-work, according to the industry of the wife or wives. One of the young men even of the slovenly Dog-rib tribe, when newly equipped from top to toe, and tripping jauntily over the mossy ground with an elastic step, displays his slim and not ungraceful figure to advantage. But this fine dress, once donned, is neither laid aside nor cleaned while it lasts, and soon acquires a dingy look, and an odor which can be perceived from some distance. In the camp a smoky, greasy blanket of English manufacture is worn over the shoulders by day, and forms, with the clothes, the bedding by night.

In winter the skins of fawn reindeer, retaining the hair, are substituted for the shammy leather, and a large robe of the same material is thrown over the shoulders, and hangs down to the feet, in place of the blanket. As the preparation of so much leather and dressed fur keeps the women busy, they are glad to use English cloth, of blue, red, or green colors, or Canadian capots of white or blue cloth, which they acquire at the trading posts in exchange for venison or furs. But, with regard to the winter dress especially, the substitution of the produce of the English loom for their native leather is a loss both of comfort and of appearance.

The women's dress resembles the men's, except that the shirt is somewhat longer, and, for the most part, is accompanied by a petticoat which reaches nearly to the knee.

The form of the dress here described is common to the whole 'Tinnè nation, and also to the Crees and Dakotas, though the material varies with the district; moose deer, red deer, or bison leather, being used in the south and west, where those animals abound; and the Hare Indians make their shirts of the skin of the hare. This, being too tender to be used in the ordinary way, is torn into narrow strips, which are then twisted slightly, and plaited or worked into the required shape. I have noticed no process among the northern Indians that approaches so nearly to weaving as the manufacture of these white hare-skin shirts.* Such is the closeness and fineness of the fur, that they are exceedingly warm, notwithstanding the looseness of their texture. Though the dress of the southern Indians is after the same pattern with that of the 'Tinnè, the Kutchin, both in the interior and on the coast, form, as has been already mentioned, the hose and shoes of the same piece; thus imitating the Eskimo boot, though with a different material.

The Dog-rib men and women leave their hair without other dressing than simply wiping their greasy hands on the matted locks, when they have been rubbing their bodies with marrow, which they occasionally do.

The Hare Indian and Dog-rib women are certainly at the bottom of the scale of humanity in North America. Not that they are treated with cruelty, for the 'Tinnè are not a cruel people, but that they are looked upon as inferior beings, and in this be-

* The Kenaiyer of Cook's Inlet are said to weave the wool of the mountain goat (*Capra americana*) into a stuff used for clothing.

lief they themselves acquiesce. In early infancy the boy discovers that he may show any amount of arrogance toward his sisters, who, as soon as they can walk, are harnessed to a sledge, and inured betimes to the labors which are their inevitable lot through life; while the future hunter struts in his tiny snow-shoes after the men, and apes their contempt of the women. The women drag the sledges alone or aided by dogs, clear the ground for the tent, cut poles to extend the lodge or tent-skins upon, collect firewood, bring water, make all the dresses and shoes, clean the fish, and smoke or jerk the venison for its preservation. They also cook both for themselves and their husbands, the 'Tinnè not holding the opinion of the Kutchin that a man ought not to eat meat prepared by a woman. Neither are the 'Tinnè women altogether precluded from eating with the men; though in times of scarcity the man would expect to be first fed, as it is a maxim with them that the woman who cooks can be well sustained by licking her fingers. The women are not, however, generally discontented with their lot, and better days are certainly dawning upon them, as the opinions of the traders are beginning to tell visibly on the whole nation. Notwithstanding their servile condition they are not without influence over the stronger sex; and they seldom permit provisions or other articles to be disposed of without expressing their thoughts on the matter with much earnestness and volubility.

Few traces of the stoicism popularly attributed to the red races exist among the Dog-ribs: they shrink from pain, show little daring, express their fears without disguise on all occasions, imaginary or real, shed tears readily, and live in constant dread of enemies, bodied and disembodied. Yet all, young and old, enjoy a joke heartily. They are not a morose people, but, on the contrary, when young and in a situation of security, they are remarkably lively and cheerful. The infirmities of age, which press heavily on the savage, render them querulous. They are fond of dancing, but their dance, which is performed in a circle, is without the least pretensions to grace, and is carried on laboriously with the knees and body half bent and a heavy stamping, having the effect of causing the dancers to appear as if they were desirous of sinking into the ground. It is accompanied by a song resembling a chorus of groans, or pretty nearly the deep sigh of a pavier as he brings his rammer down upon the pavement. They are great mimics, and readily ape the peculiarities of any white man; and many of the young men have caught the tunes of the Canadian voyagers, and hum them correctly.

They are an unwarlike people, and averse to shedding blood; yet, as they do not meet their foes in open warfare, or man to man, their very timidity impels them to treachery or a violation of the laws of hospitality, when, by long-continued oppression and the loss of relatives, they have been driven to retaliate upon the few individuals or families of the domineering tribe who were living in confidence among them. This remark applies directly to their feud with the Red-knives, who for many years resorted to the hunting-grounds of the Dog-ribs, tyrannized over them, and carried away their women. This was long borne; but, at length, some lives having been lost in the contests which occasionally ensued, the Dog-ribs, watching their opportunity, cut off several leading Red-knives and their families, who, not dreading any thing at the time, were scattered among the Dog-rib encampments. The details of these reprisals give a curious insight into the character of the people. Some of the victims, deprived of the means of resistance, and aware of their intended fate, traveled for a whole day with the hostile party; but the latter required to have their passions roused by altercation before they acquired sufficient boldness to perpetrate the deed, and were finally incited to its commission by the sufferers demanding to be killed at once if their death was intended, for they would go no further. When the husbands and grown men were killed, the Dog-ribs argued that pity impelled them to slaughter also the wives and children, who would be unhappy and perish for want, having lost their means of support. To a people who could no longer support the tyranny of their bolder neighbors, nor combine so as to repel aggression by force, treachery seemed to be the only mode of obtaining redress; and, in fact, the extent to which they carried their reprisals, effectually broke the spirit of the Red-knives, and drove them to a distance.

The Dog-ribs are practical socialists; and, as much of the misery they occasionally experience may be traced to this cause, the study of the working of such a system may be instructive in a community like this, whose members owe their condition in the social scale solely to their personal qualities, and not to inheritance, favor, or the other accidents which complicate the results in civilized life. Custom has established among them a practice universally acted upon—that all may avail themselves of the produce of a hunter's energy and skill; and they do not even leave to him the distribution of his own game. When it is known in the camp that deer have been killed, the old men and women of each family sally forth with their sledges, and, tracing up the hunter's footsteps to the carcases of the animals he has slain, proceed to divide them among themselves, leaving to the proper owner the ribs, which is all that he can claim to himself of right. He has also the tongue, which he takes care to cut out on killing the deer. It is not in the power of these people to restrain their appetites when they have abundance; and the consequence is, that when the chase is successful, all the community feast and grow fat, however little many of the men—and there are not a few idle ones—may have contributed to the common good. The hunter's wife dries the rib-pieces, after cutting out the bone, in the smoke, or over a fire, to carry to a fort for the purposes of trade; but, unless there is a superabundance, little provision is made by the party for a time of scarcity, which is sure to arrive before long; since the deer, when much hunted, move to some other district. Taught by their frequent sufferings on such occasions, the more active hunters frequently withdraw themselves and their families from the knowledge of the drones of the community, leaving them at some fishing station, where, with proper industry, they may subsist comfortably. A fish diet is not, however, agreeable to the palates of these people for any length of time; and, as soon as rumors of a hunter's success reach them—which they do generally much exaggerated by the way—a longing for the flesh-pots is instantly excited, especially among the old, and a general movement to the hunting-ground ensues. If, on their march, the craving multitude discover a hoard of meat stored up by any of the hunting parties, it is devoured on the spot; but they are not always so fortunate. Before they reach the scene of anticipated abundance, the deer may have gone off, followed by the hunters, with uncertain hopes of overtaking them, and nothing remains for the hungry throng, including the old and the lame, but to retrace their steps, with the prospect of many of them perishing by the way, should their stock of food have been quite exhausted. Such occurrences are by no means rare; they came several times under our immediate notice during our winter residence at Fort Confidence, and similar facts are recorded by Mr. Simpson of the same tribe. This gentleman expresses his opinion that the charge made against this nation, of abandoning their infirm aged people and children, had its origin in the *sauve qui peut* cry raised during a forced retreat from some one of these most injudicious excursions; and I am inclined fully to agree with him; for I witnessed several unquestionable instances of tenderness and affection shown by children to their parents, and of compliance with their whims, much to their own personal inconvenience. The grief they show on the loss of a parent, is often great and of long continuance, and it is the custom, both for men and women, to lament the death of relations for years, by nightly wailings.

Hospitality is not a virtue which is conspicuous among the Dog-ribs, who differ in this respect from the Eythinyuwuk, in whose encampments a stranger meets a welcome and a proffer of food. It is not customary, however, for the Dog-rib to receive the traveler who enters his tent with the same show of kindness. If he is hungry, and meat hangs up, he may help himself without eliciting a remark, for the 'Tinnè hold it to be mean to say much about a piece of meat; or he may exert his patience until some cookery goes on, and then join in the meal; and should there be venison at hand, he will not have long to wait, for every now and then some one is prompted to hang a kettle on the fire, or to place a joint or steak to roast before it.

Another habit which darkens the shade in the character of these Indians is that of lying, which they carry to such an extent, even among themselves, that they can scarcely be said to esteem truth a virtue. If a young man has been successful in his morning's hunt in a time of famine, he does not rush into his family circle with joy beaming on his countenance, to tell that there is food, but, assuming an aspect of sadness, squats himself in silence beside the fire. The women with doubt and anxiety examine his shoes and dress for spots of blood, that may betoken the death of an animal, but discovering none, put the question, "Did you see no deer?" "Not one, the deer are all gone, not a single footstep was to be seen." When the colloquy has continued for a time, and hope seems to be extinct, he then draws out from beneath his shirt two or three tongues, as the case may be, and says with an air of the utmost indifference, "You may go for the meat." It is not, however, merely at such times, and to enhance the pleasure by previous disappointment, that truth is violated, but on almost every occasion; and the skill of an Old Bailey practitioner would find exercise in eliciting facts from the mass of contradictions with which they overload them. A story which was at first a pure invention, or perhaps, a perversion of some simple occurrence, becomes so changed by the additions it receives in its transmission from individual to individual, that it deceives the originators, and if it bears on the safety of the community, may spread consternation among them, and occasion a hasty flight.

It is pleasant, instead of dwelling longer on this defect, to turn to another feature—their strict honesty; the practice of the 'Tinnè with regard to the property of white people differing remarkably from their northern neighbors, the Eskimos, and their southern ones, the Crees, though the temptations to which they are exposed are equally great. No precautions for the safety of our property at Fort Confidence were required. The natives carefully avoided touching the magnetic instruments, thermometers, and other things placed outside the house, and could be trusted in any of the rooms without our finding a single article displaced. Our dining-hall was open to all comers; and though the smallness of our separate apartments caused us to exclude hangers-on, new comers were permitted to satisfy their curiosity respecting our occupations, and they always squatted themselves down at the door, and looked on in silence, wondering, as we were told, at our constant writing. From M. La Flèche, the intelligent missionary at Isle à la Crosse, I received a similar character of the southern part of the nation, who, if they find any article left by the voyagers on the portages, are sure to bring it in to be claimed at the forts.

Of the peculiarities of their religious belief I could gain no certain information. The interpreters to whom I applied for assistance disliked the task, and invariably replied, "As for these savages, they know nothing; they are ignorant people." The majority of the nation recognize a "Great Spirit," at least by name, but some doubt his existence, assigning, as a reason for their atheism, their miserable condition; or they say, "If there be such a being, he dwells on the lands of the white people, where so many useful and valuable articles are produced." With respect to evil spirits, their name in the Dog-rib country is legion. The 'Tinnè recognize them in the Bear, Wolf, and Wolverene, in the woods, waters, and desert places; often hear them howling in the winds, or moaning by the graves of the dead. Their dread of these disembodied beings, of whom they spoke to us under the general name of "enemies," is such, that few of the hunters will sleep out alone. They never make any offerings to the Great Spirit, or pay him an act of adoration; but they deprecate the wrath of an evil being by prayer, and the sacrifice of some article, generally of little value, perhaps simply by scattering a handful of deer-hair or a few feathers.

The dead are not burnt, after the manner of the Kolushes, but are buried. In lamenting for deceased relatives the mourners sometimes gash their bodies or limbs with knives, but more rarely now than in old times. It was formerly the custom, on a death occurring, for the family to abandon every article they possessed, and betake themselves in a perfectly destitute condition to the nearest body of their own people, or to the trading post. The advice of the traders is gradually breaking down this practice.

Shamanism does not seem to exert the important influence upon the 'Tinnè that it does among the Asiatic Tchukche, the Kutchin, or the Eskimos. There are men in the nation, with the reputation of sorcerers, who profess to have power over spirits; but they have but little personal influence, and are generally of small repute, to which, perhaps, the contempt of the white people for their arts contributes. A belief, however, in the power of the Eskimos and of strange Indians to hurt them by incantations, or "bad medicine," prevails. White people are said to be exempt from such dangers, their "medicine" being the most powerful. The "conjurers" are occasionally employed to cure the sick, and I suppose on such occasions receive some reward; but I heard of no instance of their being beat and coerced to influence the spirits favorably, in the manner that the Asiatic Tchukche are reported in Baron Wrangell's work to deal with their shamans.

Among the Crees the conjurers perform a much more prominent part than with the 'Tinnè, and their practices come frequently under the observation of residents on the lands of that people; but I never saw one exhibit among the Hare Indians, Dog-ribs, or Red-knives, in the course of four or five years passed among them, though I have many times seen some of the old men throw trifling articles into the water, to procure a fair wind, or secure a safe passage across a lake or down a rapid.

From a people so liable to be actuated by fears of imaginary evils no steady line of action can be expected, and the Dog-ribs are in reality as volatile as children. When accompanied by a white man, they will perform a long journey carefully to a distant post; but we found, by experience, that however high the reward they expected to receive on reaching their destination, they could not be depended upon to carry letters. A slight difficulty, the prospect of a banquet on venison, or a sudden impulse to visit some friend, were sufficient to turn them aside for an indefinite length of time.

In general, the 'Tinnè have only one wife, the numbers of the sexes being equal, or the males rather predominating. The women are married very young, but the man must have shown some skill in hunting before he obtains a helpmate readily. The consent of the parents is usually gained by the suitor, and is seldom withheld from a man whose activity promises the old folks some addition to their comforts or consequence. The woman's wishes have, perhaps, some weight with her parents, but I could not ascertain that any show of courtship * was made, or that her disinclination was allowed to interfere with the man's determination to take her, if the parents did not oppose. No ceremony attends the union. Hearne says, that it is the established etiquette among the Eastern 'Tinnè for the woman to affect unwillingness to change her condition, and for the man to rush into her father's tent, and drag her off by the hair of the head. We witnessed no scene of this kind among the Dog-ribs, but more than once saw a stronger man assert his right to take the wife of a weaker countryman. Any one may challenge another to wrestle, and, if he overcomes, may carry off his wife as the prize. The younger children generally follow the fortunes of the mother, but the father may retain them if he chooses. In such contests, it is suspected that the wife sometimes prompts the aggressor; but I have been told—for I never actually witnessed one of these wrestling matches—

* The term "dear," or "beloved," is said to be unknown in the language; and Captain Lefroy, who tried to ascertain if it was so, says, "I endeavored to put this intelligibly to Nannette, by supposing such an expression as *ma chère femme; ma chère fille*. When at length she understood it, her reply was (with great emphasis), '*I' disent jamais ça; i' disent ma femme; ma fille.*'"

that she looks on with composure and impartiality, and does not insult her late master with a display of pride on being the object of such a struggle, the *causa teterrima belli.* The bereaved husband meets his loss with the resignation which custom prescribes in such a case, and seeks his revenge by taking the wife of another man weaker than himself. From a passage in one of Mr. Murray's letters, I infer that this practice extends to the Kutchin, but it is unknown among the Cree tribes, and does not exist among the Eskimos. The 'Tinnè are said to be jealous of their wives; but rather, I believe, lest they should be enticed away, than from any nice sense of honor. The laxity of morals, however, with respect to female chastity, which prevails in the Eskimo tribes is not conspicuous in the 'Tinnè, and is, perhaps, contrary to the national character, though some corruption may have crept in through their acquaintance with white people.

Before the introduction of articles of European manufacture, the 'Tinnè caught fish with hooks of bone, or speared them with weapons pointed with bone or copper. Some of their fish harpoons were constructed very artistically. They also used, and still continue to use, nets made of lines of twisted willow bark, or thin stripes of deer-hide cut very evenly. Nets are unknown among the northern tribes west of the Mackenzie, and some of the parties of the Eskimos that we saw declared their ignorance of their use. On the banks of the Mackenzie and other rivers frequented by moose-deer, these animals are hunted in spring by a small breed of dogs, which run lightly over the crusted snow, and hold the animal at bay until the Indian comes up in his snow-shoes. At other times of the year, the success of the Hare Indian's and Dog-ribs in killing the moose is small, as they have not the skill of approaching so wary an animal which the Athabascans and Crees possess. Reindeer are captured in pounds and by nooses, but are in the present day more generally killed with the fowling-piece, which is also the weapon used against the musk-ox. The pounds are formed on the verge of the woods, and are made with much less labor than those of the Kutchin; yet, as they need the exertions of all the community for their construction, the indolence of the major part causes them to be rarely made. The black bear is snared or shot, but few of the Dog-ribs will venture to attack the "brown barren-ground bear," whose fierceness, or, as they say, "potent medicine," appalls them. It is killed by them, however, without risk when it is detected hybernating under the snow in spring.

Order is maintained in the tribe solely by public opinion. It is no one's duty to repress immorality or a breach of the laws of society which custom has established among them, but each opposes violence as he best may by his own arm or the assistance of his relations. A man's conduct must be bad indeed, and threaten the general peace, before he would be expelled from the society; no amount of idleness, nor selfishness, entails such a punishment. Superior powers of mind, combined with skill in hunting, raise a few into chiefs, under whose guidance a greater or smaller number of families place themselves; and a chief is great or little, according to the length of his tail. His clients and he are bound together only by mutual advantage, and may and do separate as inclination prompts. The chief does not assume the power of punishing crimes, but regulates the movements of his band, chooses the hunting-ground, collects provisions for the purchase of ammunition, becomes the medium of communication with the traders, and extends his sway by a liberal distribution of tobacco and ammunition among his dependents. At present, the rank of a chief is not fully established among his own people until it is recognized at the fort to which he resorts. The Company send in annually a number of red coats, ornamented with lace, for presents to the chiefs, which are worn as badges of office on great occasions. The power of a chief varies with his personal character. Some have acquired an almost absolute rule, by attaching to themselves in the first instance an active band of robust young men, and using them to keep in order any refractory person by claiming his wife, after the custom of the tribe. It is in vain in such cases that the poor husband, dreading to be deprived of his most valuable property, retires to a remote hunting-ground; for he is sure to receive a message, from some passing Indian, expressive of the chief's intentions; and he generally comes to the conclusion that submission is the best policy. He is certain to fall in with the chief and his band sooner or later, either as he goes to the fort for supplies of ammunition or elsewhere. A free expenditure by the chief of the presents he receives from the traders, and even of the produce of his furs, is a main bulwark of his authority, in addition to the skill which he must possess in the management of the various tempers with which he has to do.

The sounds of the 'Tinnè language can scarcely be expressed by the English alphabet, and several of them are absolutely unpronounceable by an Englishman. In my attempts to form a vocabulary I had great difficulty in distinguishing several words from one another which had dissimilar sounds to the native ear, and were widely different in their signification. A Dog-rib or Athabascan appears, to one unaccustomed to hear the language, to be stuttering. Some of the sounds must have a strong resemblance to the Hottentot cluck, and palatal and guttural syllables abound in the language. Vocabularies of this tongue can not be greatly depended upon, as no two people will agree on the orthography.

With respect to the future prospects of the 'Tinnè, the nation in general may be said to be more docile and confiding, and more directly under the influence of the traders, or of missionary exertions, than their southern neighbors, the Crees. As yet Roman Catholic missionaries alone have entered the 'Tinnè country, and they have already a large number of nominal converts. For some years Canadian priests from the Red River colony went annually to Methy Portage, where many of the Athabascans and Churchill River 'Tinnè congregate at the usual season of transporting the outgoing furs and incoming supplies. On these occasions, numbers of the Indians were baptized, a considerable inducement to submit to the rite being the present of a piece of tobacco, or perhaps some vague notion of the protection thereby afforded against evil influences. There was no time to instruct them in the truths of the Christian religion, and this could be but very imperfectly done through the medium of interpreters. In 1846, however, the Roman Catholic mission under Monsieur La Flèche was established, as has been mentioned in a preceding chapter. This gentleman and his associate, Monsieur Taschè, members, I believe, of the Society of Jesus, applied themselves to the study of the 'Tinnè language, and were soon enabled to teach many of their converts to read and write.

By sympathizing with their people in all their distresses, taking a strong interest in every thing that concerns them, by acting as their physicians when sick, and advisers on all occasions, the priests of the mission have gained their entire confidence. It is not likely that Protestant missionaries, coming later into the field, will succeed in introducing their more spiritual but less imposing form of worship among a people whose first teachers have been so successful.

When the fur traders first penetrated to the Elk River, the Athabascans had only a small breed of dogs useful for the chase, but unfitted for draught; and the women did the laborious work of dragging the sledges. Now the cultivation of a stouter race of dogs has in some respects ameliorated the lot of the females, and within a few years the acquisition of horses by many of the natives on that river, has introduced a still greater improvement. Houses are beginning to be built, and the more provident and staid of the people have fixed homes to retire to. With the means of securing their property and provisions, new ideas respecting them spring up, and a revolution in the opinions of the nation is evidently in progress. Recently, also, it has been the policy of the Hudson's Bay Company to employ many of the young natives, during the summer, in navigating their boats to the dépôts, and back again to the outposts. By these trips prejudices are broken down; the youth acquire information and habits of labor and steady industry, and, being well paid, the clothing they purchase gives them respectability in the eyes of their countrymen. A generation has passed away since the whole Indian country was demoralized by the opposition of trading companies, and the present race of Chepewyans are ignorant of the use of spirituous liquors.

Of the nations belonging to the 'Tinnè stock who inhabit the country west of the Rocky Mountains, the *Tă-kuli* or Carriers occupy the greater part of New Caledonia. They subsist chiefly on fish, and their name denotes people employed on the waters. They burn their dead; the widow becomes the servant of the relations, is harshly treated, and is compelled to carry about with her for several years the ashes of the deceased. When the time of her trial ends, a feast is made by the kindred, and she is at liberty to marry again. A custom somewhat similar prevails among the Chippeways. The *Tsitka-ni*, who dwell between the Stikeen and Simpson's Rivers, to the north of the Carriers, are said also to be of the same stock. They bury their dead and are hunters.

In addition to these tribes, a detached portion of the 'Tinnè people is mentioned by Dr. Latham, under the name of Southern Athabascans. They occupy the sea-coast from the north bank of the Oregon, southward to the River Umqua, in 43½° of lat. For an account of these, I must refer the reader to the works of the author just named, and to the Transactions of the American Ethnological Society from which he quotes.

Dr. Latham may, also, be consulted for an account of four or five isolated languages, spoken by tribes that interpose between the North and South Athabascans to the west of the Rocky Mountains, and for notices of the inhabitants of the Archipelago skirting that coast. The Kolush language ends, he thinks, at the north end of King George's Archipelago.

The *Chenooks*, one of these isolated people, are noted for their habit of flattening the foreheads of their infants artificially, a custom which crosses the continent southward to the coast of Florida, and was practiced, though not exactly in the same way, by the extinct Peruvian races of Lake Titicaca.*

* Among some good examples of flattened skulls from the west coast of America, in the Museum at Haslar, there is the remarkable one of Comcomly, the hero of Washington Irving's Astoria.

CHAPTER XIV.

EYTHINYUWUK, OR CREES AND CHIPPEWAYS.

National Names.—Division.—Tribes.—Territory.—Wars with the Mengwè.—Conventional Character not true.—Persons.—Gait.—Crimes.—Wabunsi.—Wigwams.—Religious Belief.—Vapor Baths.—Everlasting Fire.—Its Rites.—Used in Sickness.—Its Priests.—Its Origin.—Chief Sun.—Policy.—Calumet.—Maize.—Food.—Reindeer.—Bison.—White-Fish.—Earth-Works.—Pottery.—Language.—Half-breeds.—Colony of Red River, or Osnaboya.—Spirituous Liquors.

THE people who designate themselves *Eythinyuwuk* or *Ininyuwë-u*, occupy the country lying between the Rocky Mountains and Hudson's Bay, and reaching from the 'Tinnè boundary down to the plains of the Saskatchewan and valley of the St. Lawrence; their hunting-grounds on the plains interlocking with those of the Dakotas or Sioux. They are identified as a nation with the Algonkins and Lenni-lenape or Delawares, who once owned the whole country east of the Mississippi as far south as Carolina, but who, blighted by the precocious expansion of the Anglo-Saxon colonists, have dwindled down to a few remnants of mixed blood. The generic term Algic, taken from the root of the word Algonkin, has been employed by the philologists of the United States to comprehend all the tribes who speak dialects of the Algonkin tongue, and whose southern limits are stated by Schoolcraft to be conterminous with the *Catawbas*, *Creeks*, *Cherokees*, *Chactas*, and *Chickasas*. The tract which they occupied in the year 1600, includes the whole area of the United States east of the Mississippi, north of these nations, excepting the grounds of several tribes of the Iroquois race, north and south.* In 1603, when the French settled in Canada, the Algonkins, according to Colden, were "the most warlike and polite nation in all North America."

The national name of this people is derived, according to the

* The *Abenakis*, *Etchemins*, and some kindred tribes located to the south of the Gulf of St. Lawrence, and the *Hochungarras*, or *Winnebagoes*, and *Wyandots*, to the northwest, belonged to the Iroquois stock.

custom of the Americans, from the word "man," which is in different dialects *Ethinyu*, *Ethin-u*, *Inin-yu*, or *Ininè*.* According to Schoolcraft, they do not call themselves *Unìschauba*,† or "aborigines," but, on the contrary, have a tradition current among the southern members of the nation, that the country they now hold was previously possessed by the *Alligèwi*, of whom the name only remains in the appellation of the Alleghany Mountains.

Before the European invasion, the Dakota, Huron, Oneida, Mohawk, and Iroquois association, or *Mengwè*, ‡ generally known as the "Five Nations," had penetrated into the *Eythinyuwuk* territory by way of the Missouri and St. Lawrence. The contests by which the Mengwè established themselves in a district, surrounded on all sides by their enemies, must have been severe; and they are not even now ended, but are carried on in the country between the Saskatchewan and Missouri, notwithstanding the persevering efforts of the Hudson's Bay Company, and the officers of the American outposts, to suppress them. Deadly feuds exist between the Blackfeet Eythinyuwuk, and the Mandans, Minetares,§ and other Dakota tribes which frequent the bison plains; and on the Red River of Lake Winipeg fatal conflicts took place in the last year between the Chippeways and the Sioux, who are Dakotas.

Dr. Latham states that the *Shyennes*, who dwell on the head waters of the Yellow Stone and Platte Rivers, are of the Algonkin race, though insulated by other people from the rest of their nation; and that there is in like manner a southern detachment of Iroquois (*Tuscaroras*, &c.), between whom and their countrymen the Delawares interpose; but as I mean to restrict myself to the St. Lawrence Valley, and the country lying north of it, I

* Dr. Latham traces affinities between the terms for "people," in several languages. The similarity of the terms *Inuk* or *Inuit*, and '*Tinnè*, or '*Tinyè*, to some of the above is obvious. Mr. Howes makes *Ethĭn'u* = εθν-ος. From *Ethinu* comes *ethinìseu*, "manly;" "wise," indicating the opinion the Crees have of their own nation.

† From *unisha*, "common" or "general," *ininè*, "a man," and *aub*, a generic particle denoting "light," "virility," or "life."—*Schoolcraft*.

‡ This confederacy assumed the appellation of *Mengwè* from the ancient Iroquois title, *Ongwè-honwe*, which signifies, according to Colden, "men surpassing all others."

§ Called also *Absoroka*. These and the Mandans are the so-called Welsh Indians, said to be descended from Madoc and his followers. The same origin has been attributed to the southern *Tuscarora*, of the Iroquois stock.

must refer the reader who wishes for a general classification of the native American races, to the "Natural History of the Varieties of Man," by the learned author just named.

The various tribes of Eythinyuwuk assume local designations from the rivers or other remarkable features of the districts they inhabit, and they have also names of more general import. Thus the northern ones, who border on the 'Tinnè, call themselves *Nathèwy-withinyu*, *Nehethè-wuk*, or *Nithè-wuk*,* "Exact or complete men." These are the Crees of the fur-traders; and Mr. Howse, though he does not publish the grounds of his opinion, considers them to be the stem of the Algonkin race. On the south of these, in the country extending from Lake Winipeg to the south side of the basin of Lake Superior, dwell the *Odchipewa* (Chippeways or Ojibbeways, called also *Sauteurs* † or *Sotoos*). A third great division of the nation name themselves *Lenni-lenape* (Delawares), which denotes "Uncommon men." ‡

* From *Ni*, "exactly."

† Spelt by some Canadian writers *Saulteaux*. A populous Chippeway tribe frequent the Saut Ste. Marie to feed on the *Adikumayg* or *Attikamaig* (White-fish, *Coregonus sapidissimus*, Agassiz), whence the name of "Cascade people" (*sauteurs*).

‡ The following list, drawn up in 1770 by Mr. Hutchins of the Hudson's Bay Service, gives the names of the Eythinyuwuk tribes then trading with the Hudson's Bay Company:

KEISCATCHEWAN NATION.

Names of Tribes or Places.	Districts they inhabit.
Muska-siskow	Saskatchewan prairies.
Athăpèskow	
Omiska-sipi	Beaver River and Lake.
Pegogĕ-mè-u nipi	Muddy Lake, Moose Lake.
Misi-nipi	Churchill or English River.
Wuskèsew-sipi	Red-deer River.
Po-i-thinnè-kaw-sipi	Nelson or North River.
Pemmichi-ke-mè-u	Cross Lake (Nelson River).
Maskègowuk	Swampy or low grounds near Hudson's Bay.
Ne mè-u sipi	Sturgeon River.
Chuki-tanu sipi	Hill River.
Penesay-wichewan sipi	Hay's River.
Washè-u-sipi	Severn River.
Wewanito-wuk	
Kà-stitchewanuk	Albany River. Here Hudson had his first interview with the natives, among whom traditions of the circumstances attending it were current in 1770.

NAKA-WE-WUK, OR NORTHERN UTAWAWA.

This people inhabit the country lying between Christianux Lake (Lake

The Iroquois name the Algonkin race *Adirondak*, and Mr. Schoolcraft thinks that this appellation, being still retained for the Highlands at the source of the Hudson, countenances the

Winipeg) and James's Bay, approaching within one hundred miles of the latter. They speak the Odchipewa tongue.

Names of Tribes or Places.	Districts they inhabit.
Namèkusi-sipi	Trout River.
Wà-pusi-sipi	Hare River.
Christianux	Lake Winipeg.
Weniska-sipi	Badger River.
Odchipewè-sipi	River Winipeg.
Mistĕhè-saka-hegen	Great Lake Winipeg.
Mith-kwa-ga mè-u-sipi	Red or Bloody River.
Shama-tawa	Henly House River.

UPE-SHI-POW.

This people resort to the eastern coast of Hudson's Bay, between Rupert's and Whale Rivers. Their language differs in some words both from the Keiskatchewan and Nakawawa. (They border on the Eskimos of the Labrador peninsula.)

Muswà-sipi	Moose River.
Winne-peskowuk	East Main.

La Hontan enumerates the tribes speaking the Algonkin language in 1700 as follows:

In Acadia (Nova Scotia): *Abenakis;** *Mickemac; Canibas; Mahingaus; Openangos; Soccokis* (*Sokokies*, living eastward of Boston, New England—Colden); *Etechemins.*† These seven tribes are brave warriors, more expert and less cruel than the Iroquois. Their language differs little from the *Algonkin.*

On the St. Lawrence, from the Sea up to Montreal: *Papinachois; Montagnois; Gaspesiens; Abenakis* of Sciller; *Algonkins.*

On Lake Huron: *Outaouas; Nockes; Missisagues; Attekamek; Outchipoues* (*Odchipewa*), called *Sauteurs*, brave warriors.

On the borders of Lake llinois (Lake Michigan): Some *Ilinois* of Chegakou; *Oumamis*, brave warriors; *Maskoutens; Kikapous*, brave warriors; *Outagamis*, brave warriors; *Malomimis; Pouteouatamis; Ojatinons*, brave warriors; *Sakis.*

On the borders of Lake Frontenac (Lake Ontario): *Tsonontouans; Goyoguans; Onontagues;* all of whom speak a language differing from the Algonkin; *Onnoyoutes* and *Agnies.*

On the Outaouas (Uttawa): *Tabitibi; Monzoni; Machakandibi; Nopemin* of Achirini; *Nepisirini; Temiskamnik* (Lake Temiscamaing). These six tribes speak Algonkin, and are all cowards.

* This tribe are of the Iroquois stock, according to Schoolcraft, who says that *Abenaki* is a derivative from *Wabanung*, "the east," and *ahkė*, "earth," and signifies "eastlanders." The *Abenakis* were called Tarrenteens by the early English colonists, and formerly inhabited part of the present States of Maine and New Hampshire. They were divided into several sub-tribes, of whom the best known are the *Penobscots*, *Norridgewocks*, and *Ameriscoggins*. About the year 1754, all but the Penobscots withdrew into Canada. The fullest vocabulary of the Abenaki language is furnished by the manuscripts of Father Rale, and has been published by the American Academy of Arts and Sciences. The language is peculiar, from the frequent use of the rolling sound of *r*, or a burr.

† The *Etchemins* are of the Iroquois race, according to Schoolcraft.

traditions current among the western Algonkins, that their ancestors came from the eastern coast. It is probable that the Mengwè, on the contrary, advanced from the west, if we may judge by the way in which their tribes are distributed. The aggressive movements, however, of the two nations, would throw little light on the primary peopling of the continent, even were they ascertained, since there are traces of their respective districts having been previously occupied by a people of small stature but superior in the arts, who have left memorials of their existence, in numerous and extensive earth-works, and mounds of ancient date, wherein copper bosses overlaid with silver have been found. The shafts or galleries lately discovered at the copper mines on the south side of Lake Superior, containing immense quantities of stone chisels, betoken a people more advanced than the Canadians were on the first arrival of the French; and are said to be now followed by the American miners, as guides to the most valuable deposits of native copper.

The wars of the *Eythinyuwuk* and the *Mengwè* with each other, or with Europeans, have been recorded by many pens, and have supplied incidents for numerous works of fiction, in which the writers have ascribed a loftiness of soul and other noble qualities to these people, of which it would be in vain to seek traces in the present day; and we may without much skepticism assert, that they never really possessed them. Actions prompted only by the caprice of a barbarous people, have been considered as the results of refined sentiment; and savage cunning, seen through a false medium, has been elevated to the promptings of far-seeing policy. The revolting cruelty with which they tortured prisoners of war, and the stoicism with which, when vanquished, they endured such treatment in their turn, are more certain traits of character. A few men, remarkable for their powers of mind, have certainly appeared among the Eythinyuwuk nations, and

On the north of the Mississippi, and in the country bordering on Lake Superior and Hudson's Bay: *Sonkaskitons; Ouadbatons; Atintons; Clistinos*, brave and skillful warriors.

The *Assimpouals*, (*Assinipoytuk*, or Stone Indians) and Eskimaux are struck out from La Hontan's list, as they belong to other nations. The chart appended to his book gives the positions he assigns to these several tribes; but Schoolcraft, whose authority is of the greatest weight, says that the list contains many errors. The chart, of course, has all the imperfections which attached to the geography of the great lakes and more northern country then, and for more than a century afterward.

from them the abstract idea of a North American Indian has been formed by Europeans.

Among this people there are to be found finer examples of the human figure, handsomer countenances, and a more manly and independent carriage, than among the Eskimos and 'Tinnè; and West's exclamation on seeing the Apollo Belvidere, that he was a young Mohawk warrior, may be adduced as evidence of the natural grace which a ranger of the woods, unfettered by artificial restraints, may possess. In fact, the attitudes of the Eythinyuwuk are occasionally, and especially when actuated by strong passion, striking, and sometimes elegant; yet the habitual gait of the Red Man is not a graceful one. The toes are turned in; the step, though elastic, has an appearance of insecurity, and is by no means majestic, nor even pleasing, to one unaccustomed to see the centre of gravity thrown so much forward. Even though the palm of personal appearance be given to the Eythinyuwuk, in moral conduct I hold them to be decidedly inferior to the Eskimo and 'Tinnè. They are less honest, and though, perhaps, not so much given to falsehood as the 'Tinnè, are more turbulent and more prompt to invade the rights of their countrymen, as well as of neighboring nations. Their wars are carried on by ambuscade and treachery, seldom in open field: they spare neither infants nor women in their forays; and instances of personal bravery, such as the Eskimos often exhibit, are rare indeed among them. The worst of the vices of which St. Paul accuses the heathen world are said to exist among the Crees of the plains, and gambling is practiced to excess by the whole nation.* One game in which the odd or even number of pebbles, and the hand in which they are held, are to be guessed, is constantly resorted to whenever two or three meet together, and it is accompanied by singing and gestures, indicating some kind of divination. They will pass a whole day so occupied, and will stake all that they hold most valuable on the result.

As the Narrative of Sir John Franklin's First Overland Expe-

* A society named *Wabuno* is said to have been formed among the Chippeways, for the practice of certain nocturnal orgies called *Wabunsi*—an appellation signifying "not yet light," from *wauben*, "daylight," and the negative suffix.—See Schoolcraft. The appellation of *Wahunsenacawh*, by which Powhattan's subjects were accustomed to address him, had probably reference to his being chief of a society of this kind. (See Virginia by Strachey).

dition in 1819–21, contains all the particulars of the manners and religious belief of the Crees, or *Nethewuk*, that I had then collected, I shall not here repeat them; but shall merely allude in a brief way to such of their habits and usages as have not been noticed among the 'Tinnè.

The ordinary wigwams,* skin tents, or "lodges," as they are called, of the two people are exactly alike as to form, being extended on poles set up in a conical manner; but, as a general rule, the tents of the Crees are more commodious, and more carefully and frequently supplied with a fresh lining of the spray of the balsam fir. This people also occasionally erect a larger dwelling of lattice-work, covered with birch bark, in which forty men or more can assemble for feasting, debating, or performing some of their religious ceremonies. These erections are chiefly made on the skirts of the bison plains, or in localities where a large number of the nation are accustomed to assemble together. The entire nation of the Eythinyuwuk cultivate oratory more than their northern neighbors, who express themselves much more simply, and at the same time with much less readiness.

Neither among the Eskimos nor 'Tinnè did I observe any image or visible object of worship; but most of the Crees carry with them one or more small wooden figures rudely carved, some of which they state to be representatives of a malicious, or, at least capricious being named *Kepuchikan*,† to whom they make offerings. They acknowledge other spirits or *Manito-wuk*, and demons or vampires called *Witako;* but I could not ascertain that prayer was ever made to the *Kitche-manito*, the "Great Spirit" or "Master of Life." The vapor-bath, which is comparatively seldom used by the 'Tinnè, is in frequent request with the Crees, and is more or less connected with religious observances. It is the great medium by which the shamans or conjurors cure the sick. The operator in this case shuts himself up with his patient in the sweating-house, where he shampoes him, singing all the

* *Wiggè* or *Wigwap* signifies a dwelling. Most of the Indian words and names of places adopted into the languages of the United States and Canada are of Algic origin. The Mandans and other Dakotas of the Missouri build more substantial huts, with dome-shaped roofs, covered with earth, to which, as to look-out places, the men resort.

† Or *Gepuchikan*. The propensities as well as designation of this being resemble those of "Puck." Dr. Johnson derives the latter word and Pug, from the Icelandic and Gothic word *Puke*, signifying "a hobgoblin."

time a kind of hymn. As long as the shaman can hold out, so long must the sick man endure the intensely hot atmosphere, and then, if the invalid be able to move, they both plunge into the river.

One custom of the Chippeways which fell into desuetude after the arrival of the French on the Great Lakes, is still preserved by tradition in the tribe. It was an institution for preserving an eternal fire named *Kagagish'koda.** Mr. Schoolcraft gives the following account of the rites and duties connected with it, which I make no apology for quoting, as so singular a custom, related on such good authority, deserves to be mentioned, when the peculiar habits of the race are spoken of:

"The Chippewa tribe had its council-house, and the seat of eternal fire, on the south side of Lake Superior, west of Keeweenau Point. Here lived the principal chief, called the *Mutchèkewis*, who exercised more authority, and assumed more state, than would be compatible with the present feelings of the Indians. The designation was official, and not personal, and the office was hereditary in the direct male line. He was supported by voluntary contributions, his *mushinawa*, or provider, making known from time to time his necessities by public proclamation. Whatever was required on these occasions, whether food or clothing, was immediately furnished. He appears to have been the chief priest, and could neither engage in war nor hunting."

"In the village where he resided, and near his cabin, the eternal fire was kept burning. The altar was a rude kind of oven, over which no building was erected. Four guardians were selected by the Mutchèkewis, to take care of the fire. Two of these were men, and two women. They were all married; but the wives of the men employed on this service, were required to cook and do the necessary domestic work, while the husbands of the women destined to the sacred duty were always engaged in hunting, and in providing whatever else was wanted. The four persons devoted to the altar were thus left without any secular cares to divert their attention from the holy trust committed to them. A perpetual succession was kept up in the priesthood, by a prerogative of the Mutchèkewis, and the principal head woman; the former selecting a husband, and the latter a wife for the survivor,

* From *Ka-gi-gi*, "everlasting," and *iskōda*, "fire." The corresponding words in Cree are *kă-ki-ki* and *iscu-teyu*.

whenever one of these eight persons died. The chain was thus always unbroken, and the traditionary rites transmitted unimpaired. Death was the penalty for any neglect of duty, and it was inflicted without delay and without mercy."

"The council fires were lighted at the great fire, and carried wherever the council was held. After the transaction of the business, a portion of it was carefully returned, and the remainder extinguished. Whenever a person became dangerously ill, if near enough, he was taken to the house of the Mutchèkewis, where his fire was extinguished, and a brand was brought from the altar and a fire kindled, at which a feast was prepared. A great dance was then held, and the viands consumed. And it is added that the patient seldom failed to recover.

"Once in eight years, the whole Chippewa tribe assembled at their principal village, about the season of the swelling of the buds. Early in the morning the great pipe was lighted at the sacred fire, and delivered to the Mutchèkewis. He took one smoke, and then handed it to the women, and these to the men, by all of whom it was in like manner smoked. It was then passed to the children. This ceremony consumed the day, and early next morning a feast was held, at which the men and women and children sat in separate groups. This feast was partaken silently, and without singing or dancing. In the evening they departed to their different villages."

"The principal male attendant on the Kaugagiskoda was the *Kauga gizhek*, or 'Everlasting Sun;' and his assistant was named *Kanawaudenk-shkuda*, or the 'Fire Keeper.' The principal female was called *Gaubewekwa*, or the 'Everlasting standing woman;' and her assistant *Kabagaubewekwa*, 'The woman who stands all the time.'"

"The Chippeways assert that they received this custom from the *Shawnees*, who are the most southern of the western Algonkins, their country being in the present State of Kentucky. Traces of its prevalence at a former period among other North American nations exist. The Natchez and most of the Louisiana tribes are represented by Charlevoix as having had a perpetual fire in their temples. Both he and Du Pratz were eye-witnesses of the rite. The hereditary ruler, or 'Chief Sun,' whose title was equivalent to that of Inca or Emperor, exercised a more despotic power than appears to have been permitted in any other nation north of Mexico. This power and this worship were kept up with an oriental display of honor and ceremony long after the French had settled in the valley of the Mississippi, and indeed up to the destruction of the nation by them in 1729. 'The Sun has eaten,' proclaimed an official functionary daily, before the Ruling Sun, after his morning's repast; 'the rest of the princes of the earth may now eat.'"

From this interesting extract we may gather, that the Algic race were much more advanced in the forms of government and association of tribes than the more northern nations, and especially than the 'Tinnè, who had no villages when first known to Europeans. Cultivation of the earth was not carried on to the north of the Chippeway country, since maize does not prosper in America beyond the 52d parallel.

M'Kenney relates that a Chippeway widow must carry a bundle of rags, or a doll, which is called her husband, constantly in her arms, until the relations of the deceased think that she has mourned long enough, when one of them releases her from it. This occurs generally at the expiration of a year, and the widow is then allowed to marry again; but the probation may be extended much longer, if her husband's relations choose.

The use of the *Uspogan*, or Calumet, which forms so important a part of every ceremony among the Eythinyuwuk, was not an original practice of the 'Tinnè, but was introduced to that people by Europeans along with tobacco, whereas this weed must have been grown from the most ancient times by the Chippeways, if the traditions which Mr. Schoolcraft collected during his long residence with that people are to be trusted. Maize is more used on the Missouri than in the proper Chippeway country, its cultivation forming a part of the regular economy of the Dakota tribes; the Chippeways, however, do not admit that they received it from that quarter; but, in a legend related by Mr. Schoolcraft, ascribe its origin to one of their own chiefs, who received it as the prize of a victory he obtained over a spirit. Hence its name of *Mondamin*, or the Spirit's Grain. The Delawares had extensive fields of maize at the time of the discovery of America, and to them the early Virginian colonists were indebted on their first landing for food, which being afterward withheld, produced extreme misery and famine.

From some of the details of Mr. Schoolcraft's account of the rites of Kagagish'koda, we may infer that the national polity and social condition of the Chippeways have greatly deteriorated since their acquaintance with Europeans. The contact with civilized man has induced among them an incontrollable desire for intoxication, unaccompanied by any real benefit. For though missionaries have made a number of nominal converts, the blessings of vital Christianity are confined, as far as I could ascertain, to only a few Chippeway communities on Lake Huron, and to some of the Crees in the Hudson's Bay Company's territory. The well-fed *Sauteurs* of the River Winipeg, who are independent of the traders, repel the missionaries; and the same is the case with the bison-hunters on the prairies.

Throughout the whole eastern wooded and barren country, down to the 42d parallel of latitude, the reindeer was, three centuries ago, the most abundant of the deer kind, and, being the most easily approached, furnished the staple provision for the Eskimos, 'Tinnè, and Eythinyuwuk. On the wide prairies of the Missouri and Saskatchewan, the populous Sioux, Stone-Indians, or Assini-poytŭk, and other Dakota tribes, fed on the countless herds of bison which pasture there. Next to the reindeer in importance in the eastern districts, is the species of *Coregonus*, named "white-fish," to which the Chippeways and Nithè-wuk have given the figurative appellation of "reindeer of the waters," *Adikumaig* or *Atih-hameg*.* On referring to Strachey's account of Virginia, I do not find this word, nor the name of the reindeer, in his vocabulary of the Delaware tongue; the white-fish indeed not being an inhabitant of the southern waters. The Chippeways have a legend, which relates that the white-fish sprung first into existence at the outlet of Lake Superior, being produced from the scattered brains of a woman, whose head, for some very guilty conduct, was doomed to wander through the country, but, coming in its travels to the Falls of St. Mary, was there dashed in pieces. A crane, by virtue of that inherent power so frequently attributed to birds and beasts by the aboriginies of America, instantly transformed the particles of brain into the roe of a white-fish, to the wide-spread benefit of the Indian nations.†

Though the earth-works already alluded to are supposed to

* *Adikumaig*, from *adik*, a "reindeer," and *guma*, a generic word for "water" in composition, and the animate plural *ig*, (Schoolcraft). *Athik* or *atik*, "a reindeer," in Cree.

† Schoolcraft.

have been raised by a people more ancient than the Eythinyuwuk, yet the fact of their northern limits being within the Chippeway lands is worthy of note ; and vestiges of pottery works, apparently of a rude kind, have been found on the south branch of the Saskatchewan within the Nithè-wuk bounds, but not further north,* the substitute for earthenware among the Eskimos being vessels of potstone, and among the Tinnè water-tight baskets, in which the fluid was warmed by hot stones dropped into it.

I have already alluded to the softness and harmony of the Cree language. It differs in construction from the Eskimo tongue, in the personal pronouns being prefixes, not suffixes, and in other particulars ; but both have the polysynthetic character of the other American idioms. The sounds of the English *f* and *v* do not occur in the Cree; *l* and *r* are also wanting in the pure Cree of the plains. Other Algic tribes substitute *y*, *n*, or *l*, for the Cree *th*, and instead of *k*, the inhabitants of East Maine use the sound of *tch*. The Chippeway is distinguished from the Cree by the frequent omission of *s* before *k* and *t*, and the insertion of *m* before *b*, and of *n* before *d* and *g*. The permutations of the Cree and its cognate dialects chiefly affect the linguals ; but the Mohawk and Huron languages have none of the labials, neither *b*, *p*, *f*, *v*, nor *m*. When conversing, the teeth of these people are always visible ; the auxiliary office usually performed by the lips being by them transferred, or superadded, to that of the tongue and throat.† Of the grammar of the 'Tinnè I know little, but the nouns seem to be much more frequently monosyllabic than in the Algonkin dialects. The Appendix contains some portions of a Cree vocabulary, which I formed in 1819–20.

It is from among the Eythinyuwuk that most of the servants of the Fur Companies, who have married native women, have selected their wives ; few of them having chosen Chepewyan females, and no one, I believe, an Eskimo maiden. From these marriages a large half-breed population has arisen, which will ere long work a change in the fur trade, and in the condition of the whole native population. In character, the half-breeds vary ac-

* On the east side of the Rocky Mountains. The Eskimos on the western coast of Russian America manufactured a very rude pottery when first visited by the Russians.

† Mr. Howse, from whose Grammar much of this paragraph has been borrowed.

cording to their paternity; the descendants of the Orkney laborers, in the employ of the Hudson's Bay Company, being generally steady, provident agriculturists of the Protestant faith; while the children of the Roman Catholic Canadian voyagers have much of the levity and thoughtlessness of their fathers, combined with that inability to resist temptation, which is common to the two races from whence they are sprung. Most of the half-breeds have been settled by the Hudson's Bay Company in the colony of Osnaboya, which extends for fifty miles along the banks of the Red River of Lake Winipeg. Of the six thousand souls, to which the mixed population of this settlement is said to amount, three-fifths are stated by Mr. Simpson to be Roman Catholics; while the valuable property is mostly in the hands of the remaining two-fifths, who own sixteen out of eighteen wind and water mills, erected within the precincts of the colony.

The settlement is under the government (it can scarcely be said the control) of a governor, council, and recorder, all nominated by the Hudson's Bay Company. The recorder is the civil and criminal judge, presides at jury trials, and is aided by justices of the peace, and a constabulary in the Company's pay.

In 1849 a bishop was sent from England to oversee the Episcopal church. There are also some ministers of the Wesleyan persuasion; and the Roman Catholic worship is maintained by two bishops, a staff of priests, and a nunnery. The Hudson's Bay Company aid the clergymen of all the persuasions by free passages, rations, and other advantages, besides granting salaries to those employed at their fur posts, whether Protestants or Roman Catholics. There are also various educational establishments in the colony for the settlers and native population; and most of the children, both male and female, of the Company's officers are now instructed in a boarding-school in the colony, of a high character, a few of them only being sent to Great Britain or Canada. Many of the young men so educated have entered the Hudson's Bay Company's service as clerks, and some have attained the rank of chief traders and chief factors; while the young women, in their vocations as wives of the officers and clerks, diffuse a knowledge of Christianity, and a taste for domestic comfort and decorum, to the remotest posts. The present state of society in the fur countries contrasts most favorably with the almost general heathenism which prevailed during the murderous contests between the trading companies by which the country was demoralized when I first traversed it thirty years ago.

The half-breeds, as a class, show great quickness in acquiring a knowledge of letters, as well as skill in the mechanical arts. As joiners, workers in iron, and boat-builders, many of them would rank high among European craftsmen; and, taught by necessity, they have generally the advantage of being able to work at all the several branches of the carpenter's and blacksmith's arts, even to the forging of their tools.

At the Wesleyan Missionary establishment of Rossville, near Norway House, and round the Episcopal church at the Pas on the Saskatchewan, native villages have sprung up, and agriculture to a small extent is practiced. Though the cerealia and leguminous vegetables thrive well at Red River, and horses, cattle, hogs, poultry, and sheep flourish, agriculture is eschewed by the large section of the population, who are descendants of the Canadian voyagers. The pleasures of the precarious chase are preferred by this part of the community to steady industry, and every summer there is accordingly an extensive movement to the plains to dry bison meat for winter use.

As to the effect of the colony on the neighboring natives, Mr. Simpson, who from his residence in the settlement had an opportunity of becoming acquainted with the facts, speaks as follows: "Nothing can overcome the insatiable desire of the Indian tribes for intoxicating liquors; and though they are interdicted from the use of spirits, and the settlers are fined when detected in supplying them with ale, yet, from the great extent of the colony, they too often contrive to gratify that debasing inclination, to which they are ready to sacrifice every thing they possess. They feel no gratitude to their benefactors or spiritual teachers; and while they lose the haughty independence of savage life, they acquire at once all the bad qualities of the white man, but are slow indeed in imitating his industry and virtues." It appears from this testimony that the Chippeways have not the friendly feelings toward their instructors which the 'Tinnè, according to Monsieur La Flèche, manifest; but Mr. Simpson speaks more favorably of the Crees, who are in general better disposed than the Chippeways.

Goods for the use of the colonists are imported both by the Company and by individual store-keepers in the ships that come annually to York Factory; but the distance is too great, and the inland navigation too difficult, to admit of agricultural produce being carried down profitably in return. Hence most of the half-breed settlers, encouraged by some of the colonial merchants and Roman Catholic priests, have made strenuous attempts to share the fur trade with the Hudson's Bay Company, who at present have the monopoly of that traffic; and the Company do not seem to possess a force adequate to prevent their eventually succeeding in their object.

Of late years, a communication has been formed between the colony and the United States by way of the plains and St. Peter's River. This furnishes a channel for the disposal of peltry without detection; and through the relationship existing between the half-breeds of the colony and the various tribes of Indians as far north as Methy Portage, no great difficulty is experienced by them in withdrawing a considerable quantity of the most valuable furs from the Company's trade.

In the winter of 1848 a half-breed was summoned before the Recorder of Osnaboya for a breach of the Company's regulations in this respect, and on the day of trial, five hundred of his class, armed to the teeth, surrounded the court-house. The Recorder was obliged to secrete himself, and the matter was finally compromised by the Company's agent purchasing the furs from the delinquent. Secretly or openly, this contravention of the right of exclusive trade in fur claimed by the Company is sure to proceed, and, emboldened by success, the young half-breeds are not likely to acknowledge any law that is contrary to their own will. They hold that the territorial right derived from their Indian ancestry is theirs, and not the Company's; and their claims have been supported by a philanthropic body in England, and advocated in parliament. Without entering into the question of the chartered rights of the Hudson's Bay Company, or the propriety of maintaining a monopoly of the fur trade, it is my firm conviction, founded on the wide-spread disorder I witnessed in times of competition, that the admission of rival companies or independent traders into these northern districts would accelerate the downfall of the native races. This has been rapid on the confines of the settled parts of the United States and of Canada, and has been stayed only by the extinction of the fur-bearing animals, by which the power of the Indians to purchase spirits has been cramped. Even the benevolence of the English government in making annual presents of clothing and blankets to the Indians of Canada is converted into an injury by a set of unscrupulous petty dealers, who hang about the encampments to purchase these articles as soon as they come into the possession of the Indians, by supplying them with the baneful liquid they so ardently covet. This is punishable by the colonial laws; but when crimes are committed beyond the pale of civilization, conviction is difficult. By the laws of the United States, also, it is penal to supply Indians with spirits; but according to general report this benevolent enactment is extensively violated by their fur traders; and it is greatly to be regretted that competition for the Indian trade in that quarter should induce the Hudson's Bay Company to follow so bad an example, after having abolished the use of spirits with so much advantage in the north, where they have no rivals.

I was informed that in 1848 the natives at the Red River colony of Osnaboya were paid a high money price for their furs by the Company's agent, and that they immediately crossed the boundary-line to purchase rum at the American post with their money; but it would be better to seek for the redress of such an abuse by a representation to the United States government, than resort to retaliatory measures of the same nature.

CHAPTER XV.

OCCURRENCES IN WINTER.

Fort Confidence.—Its Situation.—Silurian Limestone.—Lake Basin.—Trees.—Dwelling-house.—Occupations.—Letters.—Galena Newspaper.—Oregon Spectator.—Extent of the Hudson's Bay Company's Territory.—Fisheries.—Venison.—Wolverenes.—Native Socialism.—Provisions collected at Fort Confidence.—Fêtes.—Winter Fishery.—Eskimo Sleds.—Reindeer.—Wolverene.—Wolves.—Honesty of the Dog-ribs.—Their Indolence.—Provisions not individual Property.—Indians move off.—An Accouchement.—Cœlebs in Search of a Wife.—Might makes Right.—None but the Brave deserve the Fair.—Progress of the Seasons.—Temperature.—Arrival of Summer Birds.—At Fort Confidence.—At Fort Franklin.—On the Yukon.

THE site selected for our winter residence was about three miles from the mouth of Dease River, on a peninsula having an undulating surface, which, at the distance of three or four miles from the lake, attained a height of about three hundred feet. In front, or to the south, and separated from the main by a strait five or six hundreds yards in width, lies Fishery Island, elevated toward its centre two hundred and forty-five feet above the water.*

The peninsula is composed of limestone, which forms low precipices at the edge of the water, as well as in various places of the interior; and the same rock appears in higher cliffs on the borders of the lake, about eight miles to the westward, at Limestone Point. Six or seven miles back, on the banks of Dease River, red sandstone is the prevailing rock. The soil generally is a mixture of gravel and loam; and boulders of granite and trap rocks are scattered over the surface of both hill and valley.

Ten miles to the eastward, a range of primitive rocks rises gradually from the borders of the lake, to the height of, perhaps, six hundred or seven hundred feet, and separates Dease's Bay from the northern arm of M'Tavish's Bay. This rising ground is a continuation of the "intermediate primitive belt" mentioned in page 189, and many other parts of the preceding journal, and which will be described more fully in the Appendix. The nearest

* This altitude was ascertained by Mr. Rae, in the spring of 1848, by the aneroid barometer.

pyrogenous or metamorphic rocks to Fort Confidence that we observed are about four miles off, in a bay on the southeast side of Fishery Island.

The limestone is probably the remains of the silurian strata, which were removed when the basin of the lake was excavated. On the south side of the lake, about ninety miles distant in a direct line from Fort Confidence, stands the Scented Grass Hill, between Smith's and Keith's Bays. It consists of bituminous shale, and is one of the extreme points of that shaly formation, which constitutes so large a part of the banks of the Athabasca and Mackenzie Rivers, and which has been thought to be equivalent to the Marcellus shale of the New York system of rocks.

The summits of the higher eminences are mostly naked, but on the edges of streams and small lakes a thin forest of spruce fir covers the ground. In wet places there is a tolerable growth of willows. Little underwood of any other kind exists. Birch is very scarce; neither the balsam spruce nor banksian pine were observed on the lake, and only a few young aspens. Except where the forest has been destroyed by fire, the spruce firs are from three to four hundred years old, as ascertained from their annual rings. One of the best-grown trees that I saw, measured fifty-seven inches in circumference, at the height of four feet from the ground. The tallest of them are between forty and fifty feet high. The observations of Mr. Simpson in 1837–8 place Fort Confidence in 66° 54′ of north latitude, and 118° 49′ of west longitude, which corresponds pretty closely with the position I assigned to the mouth of Dease River on the chart constructed in 1825. The mean of Mr. Rae's observations for latitude gave about a quarter of a mile more northing than Mr. Simpson's.

Our winter dwelling, though dignified, according to custom, by the title of "the fort," had no defensive works whatever, not even the stockade which usually surrounds a trading post. It was a simple log-house, built of trunks of trees laid over one another, and morticed into the upright posts of the corners, doorways, and windows. The roof had considerable slope: it was formed of slender trees laid closely side by side, resting at the top on a ridge-pole, and covered with loam to the depth of six or eight inches. A man, standing on the outside, could touch the eaves with his hand. Well-tempered loam or clay was beat into the spaces left in the walls by the roundness of the logs, both on the outside and inside, and as this cracked in drying, it was repeatedly coated over, for the space of two months, with a thin mixture of clay and water, until the walls became nearly impervious to the air. The rooms were floored and ceiled with deal. Massive structures of boulder stones and loam formed the chimney-stacks, and the capacious fire-places required three or four armfuls of fire-wood, cut into billets three feet long, to fill them.

The building was forty feet long by fourteen wide, having a dining-hall in the centre, measuring sixteen by fourteen, and the remaining space divided into a store-room and three sleeping apartments. A kitchen was added to the back of the house, and a small porch to the front. Mr. Rae's room and mine had glazed windows, glass for the purpose having been brought up from York Factory. The other windows were closed with deerskin parchment, which admitted a subdued light. Two houses for the men stood on the east, and a storehouse on the west, the whole forming three sides of a square which opened to the south. The tallest and straightest tree that could be discovered within a circuit of three miles was brought in, and, being properly dressed, was planted in the square for a flag-post; and near it a small observatory was built, for holding magnetic instruments.

Of the buildings which Dease and Simpson erected, Mr. Bell, on his arrival in the middle of August, found only part of the men's house and a stack of chimneys standing; the others having, through the carelessness of the Indians, been destroyed by fire. Our predecessors had cut down most of the timber within a mile of the house, and what we needed had consequently to be brought in from a wider circle. A part of Mr. Bell's people were constantly engaged with the fisheries, but the others had worked so diligently, that the buildings were all covered in on our arrival, and the flooring, ceiling, and partitions were shortly afterward completed. Two of the sappers and miners, Mackay and Brodie, carpenters by trade, were employed to make tables and chairs: and Bruce, the guide, acted as general architect, and was able and willing to execute any kind of joiner's work that was needed. Two men were constantly employed as sawyers; four as cutters of fire-wood, each of them having an allotted task of providing a cord of wood daily; others were occupied in drawing it home on sledges; and four men were continually engaged in fishing. On the Sunday no labor was performed, the fishing party came in, and all were dressed in their best clothes. Prayers were said in the hall, and a sermon read to all that understood English ; and some of the Canadians, though they were Roman Catholics, usually attended. James and Thomas Hope, who were Cree Indians, having been educated at Norway House as Protestants, and taught to read and write, were regular attendants; and James Hope's eldest son, a boy about seven years of age, who had already begun to read the Scriptures, frequently recognized passages in the lessons that he had previously read.

During the winter Mr. Rae and I recorded the temperatures hourly, sixteen or seventeen times a day ; also the height of the mercury in Delcre's barometer ; the degrees of the aneroid barometer, the declinometer, and dipping-needle. Once in the month a term day, extending to thirty-six hours, was kept, in which the fluctuations of the magnets were noted every two and a half minutes, and various series of observations were made for ascertaining the magnetic intensity with the magnetometer, the vibration apparatus, and Lloyd's dipping-needle. Mr. Rae ascertained frequently the time and rates of the chronometers by observations of the fixed stars ; and a register of the winds and weather and appearances of the aurora was constantly kept.*

From this sketch of our occupations, it will be seen that our time was filled up, and that we had no leisure for ennui in the long winter. In fact, we enjoyed as much comfort as we could reasonably expect, and had our postal arrangements succeeded as well as the others, we should have had little more to desire. Our schemes for sending and receiving letters were, however, failures, and productive of much subsequent disappointment.

The packet of the Admiralty dispatches and private letters sent off on the 18th of September, 1848, on the third morning after our arrival from the coast, was placed in the charge of François Chartier and Louis le Ronde, with directions for them to proceed with all speed to Isle à la Crosse, at which place Chartier's wife was residing. I wrote to Mr. M'Pherson, requesting him to forward the party without delay ; and Mr. Rae, who put up the packet, inclosed, I believe, a circular, soliciting the gentlemen at the several posts to send the packet on, as quickly as possible. Mr. Rae himself was of opinion that he

* The magnetic observations are now in process of reduction at Woolwich, and will soon be published under the superintendence of Lieutenant Colonel Sabine, along with an abstract of the meteorological observations.

inclosed such a document, though he does not perfectly recollect that he did so. But whether the circular was inclosed or not in the first instance, or afterward left out, the circumstance of a packet being sent express for fifteen hundred miles ought to have insured its being forwarded from the further posts. No delay occurred at Fort Simpson, Mr. M'Pherson sending the party on as soon as their provisions could be prepared. Chartier and his companion reached Fort Chepewyan by open water, and were dispatched to Isle à la Crosse as soon as the ice was strong enough for traveling over. At Isle à la Crosse the letters were put *en route* again after a fortnight's detention, and at Carlton House they were kept two months. This last delay was unaccountable. When they did reach Red River they were sent on; but instead of reaching England in April or May, as we had a right to expect, and when a knowledge of our proceedings was much desired by the Admiralty previous to the sailing of the "North Star," they did not arrive till the middle of July, and our families were nearly twelve months without intelligence from us. We were also unfortunate with our subsequent letters, which were not, however, sent by special express, but were left to the chance of the ordinary conveyance through Rupert's Land.

On the 31st of October, two men and an Indian guide were sent with a second packet of letters to Fort Simpson, hoping that they would be in time for an express which leaves that post annually for the south on the 1st of December. The Indian lost himself, or rather, I believe, went willfully astray, for the purpose of falling in with some hunters that he expected to find. In this he failed; and the party, after suffering some privations, were saved from starvation by killing a deer. They did not reach Fort Simpson till some time after the winter express had left; and as the letters were not of public importance they remained there until the spring, when they were forwarded along with some others that we subsequently sent to Fort Chepewyan, that they might go down with the first boats. On my way out in the summer, finding part of these letters at one of the posts, I took them on with me; the others reached England by the same mail packet that I crossed the Atlantic in, and were delivered on the day after my arrival at home.

The only letter-bag from England that we received during our stay at Fort Confidence came in on the 12th of April, 1849, and brought us home news up to the 22d of June, 1848, ten months old. This came by the usual canoe route, and was brought up from Canada with the Red River mail; but at the same time we received a single newspaper, which gave us some English intelligence as late as the 15th of September. The history of this newspaper is that of the triumph of the electric telegraph. While the English mail packet was steaming up the sound of New York, on the 30th of September, a summary of European news having been carried on shore by an express steam-vessel, was in the act of being transmitted by telegraph to the banks of the Mississippi. Within a few hours, it was published there in the Galena "Advertiser," of which it filled one entire folio. This paper, being carried over the plains to Red River, by a party which set out on the day following its publication, was sent to Great Bear Lake, and gave us the first intimation of a rebellion in Ireland. The other newspapers that we received at the same time were of very old date, but every paragraph of them, as well as of our letters, was read again and again with a keenness that can be understood only by those who have undergone similar privations of intelligence. We heard of an old resident in Rupert's Land, who was philosophic enough to extend this pleasure over the whole year, by laying up his annual file of newspapers, and taking one down daily for perusal according to its date, so that he had just mastered the news of the preceding year when a new file arrived. Our impatience was too great to permit us to follow an example so systematic.

By the return of our packet men from Fort Simpson in January, we received the Oregon "Spectator," dated Oregon City, February 10th, 1848, with the motto "Westward the star of Empire takes its way." It was a creditable production for so young a State, remarkable for the extreme dearth of "news," but a strenuous advocate of temperance and morality, and curious for the insight which it gave of the first movements of a community destined at no distant period to play a conspicuous part among the nations of the world. The State is already involved in an Indian war, which will not cease until the Red Men are hunted from their native soil. The cause of hostility was one of those unavoidable accidents which the vicinity of white people entails on the Indian race. A large body of emigrants brought small-pox and measles with them, which, spreading among the populous and warlike Kaiyuses or Black-feet, cut off many of the tribe. By the Indian moral code, the death of their brethren was to be revenged by the slaughter of people from whom the injury came, and as it was sufficient if the victims were of the same nation with the offenders, the Kaiyuses fell upon the nearest and most defenseless. A missionary and his family, to the number of twelve persons, were cut off, and their property and some women and children seized. Through the interference of the Hudson's Bay Company's Governor at Fort Vancouver, the captives were redeemed, but five hundred of the inhabitants of Oregon marched to chastise the Indians. The paper says, "the thunders of war have commenced; let them be continued until American property and American life shall be secure upon *American soil.*" We afterward learnt that the demonstration had little effect upon the Indians, who, being well provided with horses, shunned the encounter, or returned to harass the Oregon army at their pleasure. Sooner or later, however, the Kaiyuses will feel the strong arm of the white man, and be compelled to cede their native lands to the emigrant hordes that are pressing westward.

As the crow flies, the distance between Fort Vancouver, on the Oregon, and Fort Confidence exceeds 1350 geographical miles, and the space between the Company's posts on the Labrador coast, or on Lake Huron, and their advanced station on the Porcupine, measures about 2500 miles. Throughout this vast extent of territory, a regular communication is kept up between the Governor and the numerous scattered posts, and supplies are forwarded to all the districts annually, with a regularity which can not be interrupted without hazarding the lives of both traders and natives. Besides the establishment of fisheries for our winter support, Mr. Bell employed several of the most active Dog-ribs in the capacity of fort hunters, furnishing them with clothing, guns, and ammunition, to be repaid in venison. He also gave large credits of ammunition and other articles of trade to several leaders of small bands, for the same object. In the end of September and in October, which is the best hunting season, we heard of great success. Two hundred carcasses of reindeer were reported as having been put *en cache* for us, which we were to send for as soon as the snow was sufficiently deep to permit the dog-sledges to run. A few animals killed near the fort were brought in, and our prospects looked flourishing. In the mean time the Indians sent a sick man and a very aged woman to be nourished by us through the winter, and a large body of old men, elderly widows, and children settled down near us at a fishing station behind Fishery Island. From this body visitors came to us almost daily, begging for a meal or two of dried meat to vary their diet, or bringing in a trout or two for sale. The fish were always purchased, and then the seller invariably asked for a bit of meat, as he could not walk back without eating. The simple cunning by which these poor folks endeavored to accomplish their ends, and to move Mr. Bell to be liberal, was amusing, and generally in the end successful, for his habitual good-nature was not long proof against their varied entreaties.

In the first two weeks of October, the ice driving about compelled the fishermen to take the nets out of the water, and during that time the Indian party subsisted mostly on rations from us, being supplied with both meat and fish. As soon as the straits separating Fishery Island from the main froze completely over, which occurred on the 16th of the same month, the old men were well supplied with trout-hooks to set under the ice, and they caught, I believe, fish enough for their wants, but they concealed their success that they might continue to draw aid from our store. By dint of much talking, occasionally withholding supplies, and threatening to do so entirely, Mr. Bell at length persuaded most of the party to move toward Cape M'Donald, where fish were reported to be more plentiful. Our own fisheries, however, as well as the Indian ones, declined as the winter wore on, and in February scarcely sufficed to furnish a meal daily to the fishermen themselves.

The snow having by the middle of October smoothed the inequalities of the surface, and covered the stones and stumps, Mr. Bell sent out parties to bring in the venison that had been stored up for us; but instead thereof we received a very beggarly account of empty *caches*. The wolverenes had destroyed some; our Indian friends at the fishery had eaten up a greater quantity, having, unknown to us, made several excursions for the purpose; and we did not take into the store-house a tithe of what had been reported to us. The hunters by whom the *caches* had been made came in for fresh supplies of ammunition, and, on being remonstrated with, merely said, what could they do if hungry Indians came their way? they must eat. This socialist practice presses heavily on the industrious hunter, and encourages the lazy individuals in their idleness; but its continuance in force after so long an intercourse with white men is a proof of a fund of good-nature at the bottom of the national character. It is of itself sufficient evidence against the imputation that the Chepewyan tribes habitually desert the old and infirm. We saw on several occasions children attending their sick or aged parents with tenderness and solicitude. Instances of desertion, which have undoubtedly occurred, are to be ascribed to the pressure of famine, which has urged the able-bodied to hurry on in quest of relief, disregarding those who were unable to keep up with the line of march.

Our intercourse with the Indians continued throughout the winter in the way that has been stated. The more industrious among them resorted to good hunting stations, generally in parties of two or three families together, and also in two more numerous bands, under the direction of two chiefs. Most of them resorted to the confines of M'Tavish's Bay, where the animals are plentiful in winter. From them we received occasional supplies of venison, and two or four of our men were employed for a considerable part of the winter in bringing it in with the dog-sledges. From two steady old men, who had been furnished with nets, we purchased some hundreds of fine trout, together with a quantity of white-fish and freshwater herrings. Some of the Martin Lake Dog-ribs also, though not fitted out by us with ammunition, found it convenient to bring their meat to Fort Confidence, instead of going to Fort Simpson. In this way we obtained more than we required for our present and future wants, including the eleemosynary demands of the Indians, which were, however, kept within bounds by Mr. Bell's careful management.

The following table, extracted from Mr. Bell's journal, gives a summary of all the provisions received into our store-house up to the middle of April, 1849 :

Months.	No. of Fish.	Fresh Venison.	Half dry Venison.	Pounded Meat.	Reindeer Fat.	Reindeer Tongues.
	No.	*lbs.*	*lbs.*	*lbs.*	*lbs.*	*No.*
1848 September	420	500	1500	200	150	120
October	2370	130	2100		170	105
November	1163	4330	570		18	65
December	560	1830	140			10
1849 January	279	2005				
February	223	1850	560			40
March	176	3165	1680	105	10	125
April			2670	55	5	160
Totals	5191	13,810	9220	360	353	625

In addition to the above, Mr. Bell brought up, in autumn,

1200 lbs. of dried meat from Fort Simpson, 6 cwt. of barley-meal, and three kegs of rough barley, several 90 lb. bags of flour, some bags of potatoes, with tea and sugar, together with a full supply of pemican for Mr. Rae's summer expedition, and for the provisioning of the men returning to England.

So well provided, we had no dread of want at any time, and passed the winter in abundance Our men had each a daily ration of 8 lbs. of venison on five days in the week, and on the other two from 10 lbs. to 15 lbs. of fish. The women also received rations, and the children smaller allowances. Barley and potatoes were issued in addition as long as they lasted, and flour occasionally. All the men preferred barley-meal to wheaten flour, as it answered better for thickening the soup, and they thought that it was a more substantial article of diet. The rough barley was beaten in a wooden trough until the husks separated, and then boiled whole along with venison, in which way it made a nourishing soup, that was much relished by all the party. Few of the Europeans consumed the whole of their provisions, and the Indians were generally in attendance at their meals to receive the surplus.

Several feasts varied the monotony of our winter life: one was given as a house-warming when the buildings were finished; another, as is customary at all the posts, on the first day of the new year; and two others when the winter was further advanced. On these occasions, the fishermen and wood-cutters were called in, and the whole establishment, man, woman, and child, supped at long tables placed temporarily in the hall. Preparations for the feasts were made by a great baking of bread, pies, and tarts* for two days previously; and tea was served liberally as long as any of the party felt an inclination to drink. The tables were then cleared away, and the dance was kept up with vigor to a late hour, or rather to an early one, for the party did not separate till the morning was advanced. Mr. Bell and Bruce were the musicians. The latter, with that aptness which the half-breeds show to learn any thing that comes under their observation, had made his own fiddle, and taught himself to play upon it.

* We had large supplies of cranberries, bleaberries, and the fruit of the amelanchier, the produce of the country; which, with a few pounds of Zante currants, served for tarts and pies all the winter.

A short description of the modes of fishery by which most of the fur posts in Rupert's Land are supported may not be inappropriate in this place. The nets, formed like those used in the herring fishery, measure, before mounting, one hundred and twenty yards in length, but are gathered in to eighty yards by the introduction of the backing-line along the upper edge. The depth of the net varies with that of the waters in which it is to be employed, from two to four yards. For the capture of white-fish, of the ordinary size of three or four pounds, the mesh is five and a half inches long, and where these fish are very large it is increased to six. For taking the Bear Lake herring, and the small *coregoni* of other localities, the meshes vary from two inches to two and three-quarters. In open waters the nets are shot, as in the herring fishery; the upper margin being buoyed with cedar or fir floats, and the lower one depressed by stones. The fish hang themselves in the meshes, being unable, from the form of the gill-plates, to withdraw their heads after having once passed them through. Trout of 15 lbs. weight may be taken in the white-fish nets, and also *inconnu* (*Salmo mackenzii*) weighing 25 lbs.; but the meshes will not admit the heads of the larger trouts (*namaycush*), which weigh from 30 lbs. to 50 lbs. These are caught with cod-hooks.

In winter the nets are set under the ice. The first step is to make a series of holes, about fifteen feet apart. A pole is then introduced, and conducted along the surface of the water from hole to hole, carrying with it a line, which serves to haul in a string of nets, properly buoyed and loaded, but seldom exceeding five in number. The rope is then detached, and each end of the net is fastened to a piece of wood, laid across its respective hole, or to a stake driven into the ice. On visiting the nets next day only the extreme holes are opened, the rope is attached anew at one end, and is veered away as the nets are withdrawn by the opposite hole. The fish that have been caught being removed, the nets are drawn back to their places by the line. A line of nets reaches about 400 yards, and the fisherman generally endeavors to carry it entirely across a strait or pass in the lake which fish are known to frequent.

Every second or third day, fishermen who are careful, take their nets out of the water to dry and repair them. If this be not attended to, the threads swell and rot and few fish enter the meshes; the floats also become water-logged if not often dried. In severe weather, the fisherman erects a canvas or skin screen to windward, to shelter him while he overhauls his nets. Eskimo snow barricades are much more effective, but pride will not permit the Orkney or Canadian fishermen to turn the useful expedients of the Eskimos to account.

In the winter, Albert built a snow wall very neatly round the water-hole by which the fort was supplied, to keep off the snow-drifts, cut steps through the ice down to the water, and then fitted to the aperture a light snow-lid, that could be easily removed. By this contrivance the water-hole required little clearing for a week, and the convenience was great; but after the first admiration of his ingenuity subsided, the cover was thrown aside, and the hole allowed to fill up with snow-drift. The consequence of this neglect was, that the first man or woman who went for water in the morning had half an hour's hard work to procure it, and then it was necessary to remove all the impurities left by the dogs which had resorted there to drink on the preceding day.

Albert found more ready imitators in another practice which he taught the men. He was appointed to attend to the officers' fires, and immediately set about preparing his wood-sledge according to his own fashion. He first coated the runners with earth or clay tempered with water, coat after coat freezing as rapidly as it was applied. Hot water was used in this operation, otherwise it would have frozen too quickly for him to give it the convex form and smoothness that were necessary. He next washed the runners with water, polishing the ice with his naked hand as it formed. Canadians and Europeans looked on carelessly, merely saying to one another, "What can the savage be about?" but none of them having the most distant idea that they would follow his example next day. The four sledges employed by the woodmen were of equal size, and each was drawn by two men. The drawers of wood went out together, were equally loaded, and, to their extreme surprise, Albert and his companion outstripped them all on the journey home; their emulation was excited, they labored hard the whole day, and at night confessed that they were fairly beaten. Then they tried Albert's sledge, and found it run so easily, that forthwith they requested him to prepare their sledges in the same way; and during the winter the sledgemen invariably dressed their runners in the Eskimo mode.

The venison that we obtained was the flesh of the small or barren ground reindeer, which drops its young on the coasts or islands of the Arctic Sea. This kind does not penetrate far into the forest even in severe seasons, but prefers keeping in the isolated clumps or thin woods that grow on the skirts of the barren grounds, making excursions into the latter in fine weather. A full-grown, well-fed buck seldom weighs more than 150 lbs. after the intestines are removed. The bucks of the larger kind, which were mentioned in a preceding chapter as frequenting the spurs of the Rocky Mountains near the Arctic Circle, weigh from 200 lbs. to 300 lbs., also without the intestines. Whether these be the same with the woodland reindeer, which inhabit the southern districts of Rupert's Land and the adjoining parts of Canada, and of the United States, has not been determined, no comparisons having been instituted. The small barren ground deer are generally in excellent condition in the proper season, and yield the very finest venison, hence they can scarcely be supposed to have been dwarfed through defective pasture ; and it is probable that a rigid comparison of examples of the several kinds would elicit specific differences. The reindeer that visit Hudson's Bay travel southward toward James's Bay in spring. In the year 1833, vast numbers of them were killed by the Cree Indians at a noted pass three or four days march above York Factory. They were on their return northward, and were crossing Hayes River in incredible multitudes. The Indians, excited by the view of so many animals thronging into the river, committed the most unwarrantable slaughter ; man, woman, and child rushed into the water and stabbed the poor deer wantonly, letting most of the carcasses float down the stream or putrefy on the beach, for they could use only a small number of those they slew. From that date the deer did not use the pass until last year, when a few resumed their old route, and were suffered to go unmolested, the Indians not being prepared for their coming. Mr. Rae made two successful excursions in search of deer in the winter, and the Hopes likewise went a short way into the barren grounds on the same errand, but though the latter also killed animals, their *caches* being carelessly made, were invaded by the wolverenes, and our store reaped no advantage from their efforts. The wolverene is extremely wary, and shows extraordinary sagacity and perseverance in accomplishing its ends. The Indians believe that it is inspired with a spirit of mischief, and endowed with preternatural powers. Though more destructive to their hoards of provision than the wolf, or even the bear, and able to penetrate fences that resist their powerful efforts, it is only about thirty inches long, a foot high at the shoulder, and one foot six inches at the rump, but it is very compactly made.* With teeth that do not seem to be peculiarly fitted for cutting wood, it will sever a log equal to a man's thigh in thickness by constant gnawing. In selecting the spot it intends to breach it shows as much skill as the beaver, generally contriving to cut a log near one end, so that it may fall down into some void space, and thus open an entrance into the hoard. The animal works so hard in carrying on this operation that it causes its mouth to bleed, as the ends of the logs and the snow often testify. Once admitted into the hoard, it has to gnaw the pieces of meat asunder, as they are generally frozen together, and then it proceeds to drag them out one by one, and to bury them in the snow, each in a separate place. As it travels backward and forward over the meat, it smears it with a peculiarly fetid glandular secretion, after which no other animal will touch it. In this way one of these beasts will spoil a large *cache* in an hour or two, and wholly empty it in a few nights. The pieces which are carried off are so carefully concealed in the snow, and

* Its Dog-rib name is *Nòh-gaiye*, pronounced from the depths of the throat with a strong aspirate on the *h*. The exact dimensions of a wolverene, which was surprised in one of our *caches*, and killed in the month of March, were as follows:

	Inches.
Length from root of tail, measured along the back and between the ears to the point of the nose	30·5
Length of stump of tail	6·5
" of long hairs at top of ditto	5·5
Total length of animal, including the entire tail	42·5
Height at fore shoulder	12·0
" at rump	12·6
Breadth of fore paw, the toes moderately spread	2·5
" of hind paw	2·7

The legs are remarkably muscular, the fore ones in particular, when skinned, having a strong resemblance to a finely proportioned, muscular human arm, rather than to the limb of a quadruped. The paws are covered underneath with long matted fur in winter, so that the small callosities that do exist are discovered with difficulty. The anal glands, one on each side, of the size of an olive, are filled with a very fetid secretion of a yellowish color, and the consistence of cream. A young female produces two young; an older one four; and it brings forth later in the summer than other ferine animals inhabiting the same districts.

the wolverene makes so many tracks in the neighborhood, that it is difficult to trace out the deposits, and they are seldom found.

Where there are trees, the meat *caches* are generally made with logs let into each other at the corners by notches, as in building a log-house. This, as we have seen, can be invaded by the wolverene. Mr. Rae, however, made a safe cellar by cutting a hole in the ice, covering it thickly with snow, and then pouring water over all, until the frost had rendered the whole a solid mass.

Wolves also follow the hunter, and lurk in his neighborhood, to share in the produce of his gun. Their strength enables them to break occasionally into a *cache*, but they have neither the skill nor the tenacity of purpose of the wolverene, and the damage they commit is generally on the carcasses of deer recently slain, while the hunter has gone for a sledge to bring them in. On his return in an hour or two he often finds only the well picked bones. These wolves, though of large size, are a timid race, and seldom or never exhibit the ferocity and ravenous boldness of their Pyrenean brethren. When reduced by famine they are very abject and unresisting. Mr. Bell once, while residing on Mackenzie's River, caught a full-grown but famished wolf in a marten-trap tied to a small log, which it had not strength to carry away. He went to the fort for a line to lead it home, and the children who accompanied him back assisted in bringing it in, by pushing it on from behind. It made no resistance, and suffered itself to be tied quietly to the stockades of the fort. The experiment of taming it was not, however, made, and after the curiosity of the people was satisfied, it was killed. At another time, a wolf, driven by hunger, was prowling about Fort Edmonton, when, being scared by some of the people who were passing, it took shelter in the kitchen. The cook, an old Canadian, who was busily engaged in frying pancakes, was frightened by the aspect of his visitor, and oversetting the frying-pan in the fire, and leaping into bed, he hid himself beneath the blankets. The poor wolf, astonished at the novelty of the scene, and amazed by the blaze of the flaming grease, and the screams that issued from the bed, retreated into the square of the fort, and was there killed by the people who had rushed from their several houses on the alarm being raised.

One of Mr. Rae's hunting and exploring excursions was made in the month of December, in the coldest period of the winter;

and he informed me that at that time the vapor which rose from the reindeer completely hid the individuals of a herd; so that, unless he could approach a detached one, or get into the midst of them, he could not take an aim.

Notwithstanding the indolence of a large part of the Dog-rib and Hare Indians who resorted to Fort Confidence, and their total disregard of truth, they had the merit of being strictly honest with respect to property, and also of being quiet and unwilling to offend. No precautions were taken to guard knives and other articles used in the house to which they had easy access, but they meddled with nothing, and we missed none of our effects. The thermometers, of which at least a dozen were constantly hanging up outside, were never touched, and none of the natives ever intruded into the magnetic observatory, after a general intimation that they were not to do so. When parties of them came in with venison, they slept in our dining-hall, and their friends from the fishery joined them to hear the news, and to talk for a great part of the night, yet, though the place was crowded, they gave us very little trouble. Some of the new-comers would frequently enter the sleeping apartments, and crouching down against the wall, remain in perfect quietness for an hour together, gazing at the books and other things exposed to view, and watching Mr. Rae and myself writing.

In December, January, and February, the Indians pressed heavily on our store, as the fishery was at a low ebb, and they had either consumed, or we had brought in, all the venison that had been put *en cache* for us. The more active hunters, with their families, had followed the game to a greater distance, and several detachments of stout but idle young men had joined the fishery encampment. From these able-bodied fellows we steadily withheld rations, though they were more than once furnished with provisions and ammunition to go out on a hunting excursion. The efforts to get them to do any thing were, however, ineffectual. They generally returned when their food was exhausted, saying that the animals were all gone. They were afraid, I believe, to venture far in the proper direction, lest they should meet enemies of whom they are in constant dread, and especially of their spiritual foes. Their complaints of the want of game were, however, proved to be unfounded, by the excursions of Mr. Rae and the two Hopes.

In January, intelligence came that the hunters had stored up a number of deer at a considerable distance from the fort, but the news reached Cape Macdonald as soon as Fort Confidence, and a party from the fishery set off in quest of it. They were, nevertheless, too late, some of their wandering countrymen having discovered and consumed the store before their arrival, and they were compelled to return, in a famishing condition, to the fishery. Our men had also a fruitless journey to the empty *cache*. The lies that were told on these occasions were innumerable, and every one was ready to clear himself, and inculpate some other party. We generally, however, succeeded after a time in finding out the real delinquents, who consequently were coolly received, and had short rations when they came to the fort.

As the months of February and March rolled away, and the days lengthened, cheering reports from the hunting parties came in, and some of the more active of the fishery party went to join them. Mr. Bell, on two several occasions, fitted out the whole of the residents at Cape Macdonald with provisions enough to take them to the hunting stations; but their hearts failing them, they consumed what was given to them, and came with their daily petitions as before. During this time they were taking trout, and denying that they did so, though they occasionally sold us a few. At length, toward the end of March, two of our fort hunters, who had left their wives at the fishery, coming to fetch them, the desire of the whole party to eat venison became uncontrollable, and they came *en masse* to the fort, with their sledges and all their movables, to receive another fit-out. They remained encamped near the house on this occasion for about three weeks before they took their departure, subsisting chiefly on the produce of their trout-lines. During this time some of the young men made two excursions to plunder the *caches* of the hunters employed by us, but were foiled in the attempt, and came back fasting; for Mr. Bell, taught by former losses, had been very prompt in sending men for the venison, and had secured it before the marauders reached the spot.

At this time some considerable supplies were brought in by the Martin Lake Indians, and we had several opportunities of observing the way in which these people act toward each in regard to provisions. The venison intended for us was neatly packed on the sledges, but each Indian generally carried a kind of knapsack on his shoulders, containing some choice pieces of meat to be consumed on the journey back to his tent. The sharp eyes of the hungry party at the encampment, discovered the approach of strangers while yet at a great distance, and the rumor of an arrival spreading with rapidity, men, women, and children crowded into the square of the fort. The first act of the new-comers was to run the loaded sledges at once into the store-house, which was opened to receive them ; but as they arrived in succession, the women from the camp generally pressed in, and throwing their arms round a young hunter, with much kindness of manner would say to him, "It is long since we have seen you, my relation; how have you fared since we met? You are a generous man!" and so on. While his attention was thus engaged, and before he could free himself from the unwashed sirens, whose unwonted softness of speech led him to suspect either ridicule or plunder, one of the females, having cut the strings of his knapsack, would carry it off, amidst the laughter of the crowd. The young fellow, thus despoiled of his provisions, however much he might be vexed in secret, was obliged to join openly in the mirth: and the expression of face of some of the youths thus preyed upon, as they endeavored to force a smile in their distress, was irresistibly comic. The loss fell ultimately on the store, as Mr. Bell had to furnish the party with food for their return, though in that case he did not certainly select titbits. When at length the great move was made, and the fishery party, exceeding forty in number, went off, in the end of March, news came that the wife of one of the hunters, a very small woman, had been taken with the pains of labor. The three females belonging to the fort went to her aid, and found the new-born infant, the mother's first child, wrapped in a deer's skin, and stuck into a hole made in the snow. It was brought into the house, and dressed by the fire. This event delayed the mother's departure about two hours. She then set out dragging a sledge, and having her first-born suspended between her shoulders, in a bag or Indian cradle. She was not suffered, however, to profit by the well-beaten path pursued by the rest of the party, who had gone before her, but had to make a new track parallel to it through the loose snow, always a laborious task, even to a stout man. Want of success in hunting, or some other calamity, was sure to befall an Indian who should incautiously tread in her footsteps. This was the custom; no slight or unkindness was shown to her; her husband was, I believe, really fond of her; and her sledge was a light one, being loaded only with things belonging to herself.

Soon after what may be considered a great event in the simple annals of Fort Confidence—the departure of so numerous a body of hangers-on—we had a visit from a Dog-rib, who had been residing on the other side of the lake, at the plentiful fishery opposite Fort Franklin. Our visitor, a stout, able-bodied fellow, came empty-handed, and introduced himself as a very serviceable man, who had been of great use to our two fishermen stationed at Fort Franklin. We readily understood that the truth concealed under this self-praise was, that he had been living by their labors, perhaps bringing in a little wood for them occasionally, but really sharing in the produce of their nets. On questioning him more closely as to the object of his coming, he at length said that he came to look for a wife. Had he no wife? Yes, he had had one, but an Indian had taken her from him. Had he an aversion to his wife that he had parted from her without a struggle? No; his wife was a very good wife; she suited him very well. Then why did he not fight for her? "You see," said he, "it was a big Indian that took her! I am a little fellow, what could I do?" This hero met with little countenance from us, and after being fed for two days, an intimation was made to him that he would receive no more rations; on which he followed the others, and, as we afterward learnt, soon procured another helpmate.

The singular national custom of the women being the property of the strongest, had been acted upon a short time previously at the fishery. A blind man, who was more assiduous in setting trout lines, and more successful as a fisherman than most of the others, was deprived of his wife on the general movement of the party, by an old fellow who wanted her to drag his sledge. Mr. Bell questioned the ancient Paris about the truth of the report when he came to the fort, and instead of denying the fact, or seeming ashamed, he gloried in the deed as a manly action. A child, the offspring of her former connection, followed the fortunes of the mother.

After Mr. Bell and I quitted the fort in May, the reindeer migrating toward the sea-coast left the Indians again in straitened circumstances. The active hunters moved toward the open country in quest of musk-oxen, and the more helpless threw themselves on Mr. Rae's compassion at the fort. He distributed among them the remains of the dried meat brought from Fort Simpson, and a quantity of musk-bull meat, which was too strong for his own men, but which the natives relished greatly. Thus furnished they betook themselves reluctantly to various fisheries. Such is the life which these poor creatures lead. Occasional feasting and rejoicing, with intervening periods of want, sometimes of absolute famine. With proper management, the natural resources of the country would support a population ten times as great; but as long as all the drones of the community claim a right to appropriate to their own wants the produce of the exertions of an industrious hunter or fisherman, no certain provision for the future will be made. The first step in advance will be the formation of fishing villages, and the culture of barley and potatoes; and, under the guidance of intelligent missionaries, this might be effected without much difficulty; while, at the same time, the truths of Christianity might be brought to bear on the heathenism and moral defects of the 'Tinnè nation.

The preceding details may lead the reader to imagine that the Dog-ribs or Hare Indians are an unhappy race; but such is not the case. They are timid, and assume the attitude and solicitations of beggars in their intercourse with white people; but among themselves they are lively, volatile, and full of fun and mirth, which even an empty stomach can not suppress.

With regard to the progress of the seasons, the "Indian summer," as it is called, brought us three weeks of fine weather after our arrival in September. The centre of Bear Lake usually remains open till late in December, but by the middle of October the bays and straits are frozen across. As the structure of ice has of late years attracted the attention of speculative geologists, principally in connection with the movements of glaciers, I am induced to mention here a few facts which intruded themselves on my observation during my residences in the fur countries.

The first step in the freezing of rivers in this rigorous climate, after the water has been cooled down to 32° by a succession of cold weather, is the formation of somewhat circular plates of ice, six or eight inches in diameter. These drift for a time with the current, until they have become numerous enough to cover the surface of the water, when they are arrested in a narrow part of the river, or by any slight obstacle, and speedily adhere to each other, after which the interstices between the circles fill rapidly with crystals that bind all firmly together. The sheet of ice thus produced is at first nearly opaque; but when, in the course of a day or two, it has acquired the thickness of a few inches, it becomes transparent, and remains so until a fall of snow has obscured the surface. In unsheltered lakes the wind drifts the snow to the beach, and would perhaps keep the ice clean for great part of the winter, were it not that in certain hygrometric conditions of the atmosphere small starry tufts of most beautiful tabular and latticed crystals are deposited at short intervals on the ice, and freeze firmly to it. In a dry atmosphere these crystals evaporate again, but should a fall take place of the fine, dust-like snow, which is the most common kind in the high latitudes, they serve to detain it until it consolidates, so as to resist the wind. It is rare, however, for the snow to lie more than a foot deep on any of the large lakes, unless where it has drifted under the lee of piled-up slabs of ice, or of rocks, islands, or other shelter.

During winter the ice receives an increase of thickness from beneath, and at the same time evaporates above; the latter process going on with a rapidity that would scarcely be credible to one ignorant of the extreme dryness of the air in an arctic winter. The ice acquires a thickness of from four to eight feet, according to the severity of the season, the depth of the lake, and other modifying circumstances; and I desire here to advert especially to the fact, that although it is constructed of successive horizontal additions beneath, when it decays in spring it consists of vertical prisms, penetrating its whole thickness, and standing side by side like the columns of a basaltic cliff; which, in their mode of formation, have, I imagine, a close analogy. Dr. Slagintweit informed me that neither the ice nor the basalt forms exact prisms, the angles never having the precise measurements of true crystals. In this condition the ice may be strong enough to support a considerable weight; and I have traveled over it with a large party on several occasions when the prisms on which the foot rested were depressed at every step, and a pointed stick could be driven through the whole thickness into the water beneath, with as much ease as into a bank of snow. The ice then, in fact, presents the physical characters of a semi-fluid mass, as pointed out by Professor Forbes, its parts being movable on each other, not only vertically, but as in the case of traveling glaciers, capable of gliding past one another horizontally.

In spring, when the action of the sun-light is very powerful, an incipient thaw takes place at mid-day on the surface of the snow, which, on freezing again, acquires a glassy crust. As the season advances, but while the temperature of the air is still, even at noon, far below the freezing point, the crust in clear weather becomes penetrated in the direction in which it is struck by the sun's rays at mid-day by innumerable canals, and finally crumbles into a granular mass like the *firn* of the high Swiss glaciers, that crackles under the feet as soon as the sun sinks toward the horizon. This firn is not universal; it is more common within the Arctic circle, and in situations where there seems to have been originally a certain looseness in the texture of the snow, and where its surface is so much inclined that the sun's rays do not fall on it obliquely about noon. I did not notice it in any quantity on the level surface of a lake.

The rapid evaporation of both snow and ice in the winter and spring, long before the action of the sun has produced the slightest thaw or appearance of moisture, is made evident to residents in the high latitudes by many facts of daily occurrence; and I may mention that the drying of linen furnishes a familiar one. When a shirt, after being washed, is exposed in the open air to a temperature of 40° or 50° below zero, it is instantly rigidly frozen, and may be broken if violently bent. If agitated when in this condition by a strong wind, it makes a rustling noise like theatrical thunder. In an hour or two, however, or nearly as quickly as it would do if exposed to the sun in the moist climate of England, it dries and becomes limber.

Mr. Rae mentioned to me another example of the same fact which bears on the transportation of boulders, and may interest geologists. During his memorable residence on the shores of Repulse Bay, he noticed several large boulders which were partially exposed at low water. When the sea froze they became engaged in the ice, and were lifted with it from the bottom by the flood-tides. The ice gaining at each tide in thickness beneath, and losing above by superficial evaporation, the boulders in process of time came to rest in pits on its surface.

In consequence of the extreme dryness of the atmosphere in winter, most articles of English manufacture made of wood, horn, or ivory, brought to Rupert's Land, are shriveled, bent, and broken. The handles of razors and knives, combs, ivory scales, and various other things kept in the warm rooms, are damaged in this way. The human body also becomes visibly electric from the dryness of the skin. One cold night I rose from my bed, and, having lighted a lantern, was going out to observe the thermometer, with no other clothing than my flannel night-dress, when, on approaching my hand to the iron latch of the door, a distinct spark was elicited. Friction of the skin at almost all times in winter produced the electric odor.

In November the snow was deep enough for sledges to run without receiving much injury. On the 1st of December the sun was just visible for an instant at noon, from an eminence behind the house. This month was a cold one, and the coldest days in the year were the 17th and 18th, when the average temperature for forty-eight hours was 55½° below zero of Fahrenheit. The lowest observation made was at seven in the afternoon of the 17th, at which time the temperature was 58·9° F. The thermometer, one of Adie's making, was hung by the side of a dozen by the same artist, none of which differed a degree from it. When tested by freezing mercury, this one stood at 36°, which is considered to be from 4° to 6° too high; so that, making the smallest of these corrections, the actual minimum temperature we observed was 65° F.* This is one of the greatest colds on record, and pains were taken to ascertain it correctly.†

Even at mid-winter we had three hours and a half of daylight. On the 20th of December I required a candle to write at the window at ten in the morning. On the 29th the sun, after ten days' absence, rose at the fishery where the horizon was open; and on the 8th of January, both limbs of that luminary were seen from a gentle eminence behind the fort, rising above the centre of Fishery Island. For several days previously, however, its place in the heavens at noon had been denoted by rays of light

* Mr. Saunders, commander of the North Star, records 64½° F. as the lowest temperature observed in Wolstenholme Sound in the winter of 1850.

† In two thermometers, made by a London artist, and hung up beside the others, the spirit retreated into the bulb, though the scales were graduated down to 73°. In freezing mercury, these thermometers indicated 55°, being about 15° too low. Mr. Adie's thermometers were constructed under the superintendence of Professor Forbes of Edinburgh. The precautions used to insure their correctness will be detailed when the meteorological observations at Fort Confidence are published.

shooting into the sky above the woods. The lowest temperature in January was 50° F.

On the 1st of February the sun rose to us at 9 o'clock and set at 3, and the days lengthened rapidly. On the 23d I could write in my room without artificial light from 10 A.M. to half-past 2 P.M., making four hours and a half of bright daylight. The moon in the long nights was a most beautiful object; that satellite being constantly above the horizon for nearly a fortnight together in the middle of the lunar month. Venus also shone with a brilliancy which is never witnessed in a sky loaded with vapors, and unless in snowy weather, our nights were always enlivened by the beams of the Aurora.

In February the lowest temperature was 56° F. (or 62° corrected), and in March, 44° F. (48°) was the lowest observed. On the 20th of April, signs of the snow softening on the south side of the house, contiguous to the walls, were perceived. The day by this time had lengthened so far, that I was able to read off the degrees of the thermometer by daylight at nine in the evening. Snow-birds arrived in small flocks, and on the 27th the snow began to melt in sheltered places, exposed to the direct rays of the sun. Ospreys, ger-falcons, eagles, and gulls appeared on the 17th of May; and, from the 19th to the 23d, melted snow stood deep on the surface of the ice. On the prior of these two dates, the first goose was seen. The geese approach the high latitudes as soon as the swamps are uncovered, when they feed on the undeveloped flowering stalks of the cotton-grass and other *Cyperaceæ.* Their arrival is thus an indication of the progress of spring, and frosty weather will sometimes drive them back for a week or so to a milder district. The impulse, however, by which they are urged to their breeding stations is so uncontrollable, that in backward seasons they are driven to the sea coast before the snow is gone, and then, from want of food, they are in a very lean condition. Their incubation, the fledging of the goslings, and the moulting of the parents, has to take place before the end of August, when old and young pass southward to spend the winter in more genial climes. An indelible attachment leads them back to their natal places, and the ensuing summer sees them winging their way northward in cuneal bands, with unerring instinct. Their arrival in a district enlivens white man and Indian: during their passage, plenty reigns in every encampment;

and the dingy, pot-bellied children run about with smiling, greasy faces, brandishing in each hand the leg or wing of a goose.

The Canada geese come in the van, and remain breeding in the woody country; snow geese next arrive, and pass onward to Wollaston's Land; then the laughing geese come and go, holding a northwest course; and at the same time with the latter, the Hutchin's geese speed to the sea coast.

On the 22d, pin-tail ducks were seen; on the 24th, swans; and, on the 30th and 31st, large flocks of snow geese and brown cranes passed northward. On the 1st of June, bees, sandpipers, long-tailed ducks or cacawees, eider and king ducks, and northern divers were seen: the catkins of the earliest willows also burst their envelopes on this day. On the 5th, teal, widgeon, scaup-ducks, shovellers, and jagers arrived; but, on the 8th, the fur of the polar hare was still white.

The progress of spring at Fort Confidence, subsequent to the 7th of May, is recorded from Mr. Rae's notes, as on that date Mr. Bell and I commenced our journey southward. To contrast with the above the dates of the arrivals of the migratory birds at Fort Franklin in the same season are here added, the difference of latitude between the two places being a degree and three quarters.

On the 11th of May, under a hot sun, a pool of water had formed on the ice near the bay of the Deer Pass. We bivouacked on the shore beside it, and had not yet arranged our sleeping-places, when a Canada goose alighted in the pool. It was scarcely allowed to settle before it was shot, and, with a celerity unknown in civilized lands, stripped of its feathers and committed to the cooking-kettle. This was evidently a straggler, and must have seen the small pond in which it alighted from a great distance; for, on our arrival at Fort Franklin on the following day, we learnt that neither our two fishermen employed there, nor an Indian residing near them, had as yet seen any of the spring birds. On the 14th the Indian saw gulls; on the 18th, snow geese and various small birds came, together with the pretty little gull named *Xema bonapartii*, which in large flocks sought for insects in the open water now forming along the shores of the smaller lakes. On the 22d, bands of snow geese passed to the northwest, flying high. They evidently found the country about Fort Franklin still too closely wrapped in its winter garb, and were winging their way to the valley of the Mackenzie, where the season is earlier.

Geese, according to Mr. Bell's information, arrive at Peel River Fort, upward of two degrees farther north, from the 12th to the 15th of May, rarely varying above a day or two, the 15th being the date of their coming in backward seasons. At that time they find the marshy places bare of snow, and can procure the roots of bents and other plants on which they feed. There, as elsewhere, the Canada geese precede the snow geese a day or two. The Hutchin's geese* come later, and pass high overhead toward the north. The Indians believe that a small finch (*Plectrophanes lapponica*) avails itself of the strength of wing of the Hutchin's goose, and nestles among its feathers during its flight. When a goose is shot, they often see the small bird flying from it. Neither Mr. Rae nor I noticed such an occurrence, nor did I obtain a confirmation of it from the personal observation of any of the gentlemen resident in the country, but it is generally affirmed by the Indians.

While we were at Fort Franklin, Mr. Bell, who was employed all day in shooting geese, pointed out to me a fact in the natural history of the snow and laughing geese. Though they migrated in large flocks, each had already selected its mate; and if the female was shot, the male bird instantly separated from the rest, and descended to look after her. In this way, he often fell a victim to his conjugal fidelity; but if he escaped the shots aimed at him, and became shy, he would still continue for hours, and even days, searching the neighborhood for his lost mate. The case was different if the male bird fell first. The female, it is true, also left the flock, but she kept more aloof, and generally, after making a circle or two round the spot where the body of her partner lay, went off with the next flock that came up.

The laughing geese passed Fort Franklin a few days later than the snow geese, but a single individual was often seen some days before the arrival of the main body, associated with a flock of snow geese, and generally acting as leader by assuming a station in the apex of the angle in which they fly. About the same time, the American robin, or migratory thrush, came with the yellow-poll and black and yellow warblers (*Sylvia æstiva et*

* These are commonly called "Eskimo geese" in Rupert's Land.

maculosa). The latter fed on the berries of the Alpine arbutus, as did likewise the golden plovers, whose stomachs also contained the juicy fruit of the *Empetrum nigrum*. The Eskimo curlew at this time fed on large ants. It would appear that these insects descend to the stomachs of the curlews alive, since I found that several, having taken fast hold of the lining of the gullet with their mandibles, remained sticking there, and even after death required some force to detach them. The tree bunting (*Fringilla canadensis*), black-finch (*F. hyemalis*), and white-crowned finch (*F. leucophrys*)* were also early visitors, and soon after their arrival began to construct their nests. The Lapland finch was also seen, but only on its passage to the coast. The *Lestris richardsonii* flew about in pairs, and was observed to have the habit of quartering the ground like the hen-harrier. In the stomach of one which I killed, there were the skin and some of the bones of a mouse, rolled into a ball, like the pellets that are rejected from the stomach of an owl. The purple-throated diver visited Bear Lake River in considerable numbers. This species is easily distinguished from the great northern diver (*C. glacialis*), while flying, by its swollen, bluish-gray neck. Almost all the summer birds arrived before we left that neighborhood; but I have enumerated only the earliest comers, or those which I had not previously seen in so high a latitude, and whose range is, therefore, not correctly given in the *Fauna Boreali Americana*.†

* I have already mentioned the nocturnal song of this bird, which breeds throughout Rupert's Land. In attempting to express its clear, loud notes by syllables, the nearest approach I could make was *cheet-cheet, tareet, cheet, cheet*. The first two syllables are loud and high, the next two short, and the two last lower and softer.

† From Mr. Murray, I have received the following account of the arrivals of the water-fowl in the Valley of the Yukon. "Of the two kinds of swan, only the largest sort (*Cygnus buccinator*) are seen here; they pass on to the northward of the Porcupine River, to breed among the lakes. Bustards (*i. e.* Canada geese) are plentiful, and breed every where, from Council Bluffs on the Missouri to the vicinity of the Polar Sea. On the ramparts of Porcupine River they frequently build high up among the rocks, where one would suppose only hawks and ravens would have their nests. How they take their young down is unknown to me, but they must be carried somehow. Ravens and large gulls are very destructive to young geese. With respect to the breeding-quarters of the laughing geese (*Anser albifrons*), I am able to inform you correctly, having myself seen a few of their nests; and, since the receipt of your letter, made further inquiry among the northern Indians. Their nests are built on the edges of swamps and lakes, throughout most of the country north of the Porcupine, where

Great Bear Lake is navigable for its whole extent for only fifty days in the year. It is frozen over later than shallower pieces of water in its vicinity, but the ice remains longer unbroken, and drift ice continues on its surface till the middle of August. On the 7th of May, the day on which Mr. Bell and I set out to travel to Fort Franklin, the ice was as firm as in mid-winter; and, though the snow softened in fine weather, it still covered the ground deeply.

the ground is marshy. It is only near the most northerly bends of that river that any are seen in the breeding season, and these are male birds. They pass to their breeding-places in the beginning of June, and make their nests among long grass or small bushes, where they are not easily seen. They are shy birds when hatching; and, when any one comes near the nest, manage to escape unperceived, and then show themselves at a distance, and manœuvre like grouse to lead the intruder away from the place. Notwithstanding our ruthless habit of collecting eggs of all kinds to vary our diet, I have often felt for a laughing goose, whose anxiety for the safety of its eggs was frequently the means of revealing to us the situation of its nest. When the bird was swimming some hundreds of yards off, immediately that any person in walking round the lake came near its treasure, the poor bird began to make short, impatient turns in the water, resuming her calm demeanor if the intruder passed the nest without seeing it. As soon as the eggs are taken, the goose rises out of the water and flies close to the head of the captor, uttering a frightened and pitiful cry. These geese are more numerous in the Valley of the Yukon than any other kind; and the numbers that pass northward there are perhaps equal to that of all the other species together. The *Gens du large* (*Neyetse-kutchin*), who visit the north coast regularly to traffic with the Eskimos, say that they have never seen any flying northward over the sea in that quarter. White geese (snow geese, *Chen hyperboreus*) are also passengers here, and there are likewise *black geese*, which I presume you have never seen. A few of them pass down Peel's River, but they are more abundant on the Yukon. They are very handsome birds, considerably smaller than the white geese, and have a dark brown or brownish black color, with a white ring round the neck, the head and bill having the shape of that of the bustard." (This description applies pretty well to the brent goose *Anser bernicla*). The black geese are the least numerous and the latest that arrive here. They fly in large flocks with remarkable velocity, and generally pass on without remaining, as the others do, some days to feed. When they alight, it is always in the water; and, if they wish to land, they swim ashore. They are very fat, and their flesh has an oily and rather disagreeable taste. Bustards, laughing geese, ducks, and large gulls make their appearance here from the 27th to the 29th of April. Snow geese and black geese about the 15th or 16th of May, when the other kinds become plentiful. They have mostly passed by the end of the month, though some, especially the bustards, are seen in June. The white geese and black geese breed only on the shores of the Arctic Sea. They return in September and early in October, flying high, and seldom halting."

CHAPTER XVI.

Mr. Rae's Expedition in the Summer of 1849.—Instructions.—He crosses to the Coppermine.—Descends that River.—Sea covered with Ice.—Surveys Rae River.—Eskimos.—Cape Kendall.—Cape Hearne.—Basil Hall Bay.—Cape Krusenstern.—Douglass Island.—Detention.—Dangerous Situation.—August 23, return.—Author and Mr. Bell leave Fort Confidence.—Cross Great Bear Lake.—Descend Bear Lake River.—David Brodie lost in the Woods.—His Adventures.—Fort Simpson.—Methy Portage.—Receive English Letters.—Norway House.—Part from the Seamen and Sappers and Miners.—Continue the Voyage to Canada.—Boston.—Land at Liverpool.—Summary of the present State of the Search for Sir John Franklin.

Having in the preceding chapter mentioned such occurrences during our winter's residence at Fort Confidence as seemed most worthy of notice, this is the most appropriate place for introducing an account of Mr. Rae's endeavors, in the summer, to reach Wollaston Land, and complete the search in that quarter. Had we succeeded in taking our boats up the Coppermine, beyond the reach of the Eskimos, according to our expectations when the plan of search was formed, the voyage might have been resumed in the summer of 1849 with two or three boats; and, in that case, both Mr. Rae and I would have gone, that we might aid each other among the ice. But, having been compelled to leave our craft in September, without the smallest hope of their being found again in a seaworthy condition, and having only one boat remaining that could be employed on the service, it became necessary to determine which of us should take charge of that vessel and of the small party it could contain. Setting all personal considerations aside, and looking solely to the means of providing for the examination of as large a portion of the Arctic Sea as could be accomplished, I had no hesitation in deciding in favor of Mr. Rae. His ability and zeal were unquestionable; he was in the prime of life, and his personal activity and his skill as a hunter fitted him peculiarly for such an enterprise. The arrangement I made for withdrawing the European party, and employing volunteers from the men engaged in the country, was a considerable pecuniary saving, which I was bound to consider, as far as I could, without cramping the means of search.

It has been already mentioned that Mr. Rae explored the country between Fort Confidence and the Coppermine River in winter, to select the best route for dragging the boat over in spring. In April he conveyed provisions, boat stores, and other necessaries across to the Kendall on dog-sledges, and left two men at Flett's station in charge of them, together with two Indian hunters. Before the end of the month we learnt that they had obtained both musk-ox and reindeer meat, and were drying part of it for summer use. The subsequent proceedings are told in the following official documents, which I quote entire, premising that the names of the men composing his party were—

Boat's Crew.

Neil M'Leod	Orkney man.
James Hope	Cree Indian,
Thomas Hope	"
Halcro Humphrey	Orkney man.
Albert One-eye	Eskimo.
Louis Olivier	Canadian.

Left in charge of Fort Confidence.

Baptiste Paul	Canadian.
Louis Dubrill	"

(*Memorandum*).

"Fort Confidence, Great Bear Lake, May 1, 1849.

"As in the prosecution of the search for traces of the Discovery Ships under command of Sir John Franklin, the continental coast line between the Mackenzie and Coppermine Rivers has been carefully examined, the only part of my instructions not complied with, is the examination of the adjoining shores of Wollaston and Victoria Lands, which the state of the ice in Dolphin and Union Straits rendered inaccessible last autumn. That these two islands are separated from each other by a strait lying between the 111th and 113th meridians, is rendered almost certain by a consideration of the direction of the flood tide, which, on the west side of these parallels, sets to the westward through Dolphin and Union Straits, and to the eastward of them, sets to the eastward toward Cape Alexander; coming, we must conclude, from the northward between the lands in question; for the survey by Messrs. Dease and Simpson has shown that the coast of Victoria Land is continuous up to the 111th parallel; and the latter gentleman records his opinion, that much of the heavy drift ice that encumbers Coronation Gulf descends between these lands from the north.

"The exploration of the shores of this strait is of much importance in the search for the Discovery Ships, for the following reasons: Sir John Franklin having been directed to steer to the southwest after he had passed Barrow's Straits, would be led directly into it, and he would be deterred from attempting a more westerly course by the circumstance of Sir Edward Parry having found that route impracticable for two successive seasons. Should there be several islands between Wollaston and Banks's Lands, and the channels between them be intricate, it is not unlikely that the ships may have been shut up therein by ice. It was the intention of Sir James Ross, in the event of his reaching Winter Harbor last year, to send a party across the ice this spring to pass between Victoria and Wollaston Lands toward Cape Krusenstern and the Coppermine River. To co-operate with that party and to aid it with provisions, or to supply its place, should circumstances have prevented its being sent, it is expedient that a party should go from hence, and as you and a sufficient number of men have volunteered for this service, I hereby, in virtue of the clause of my instructions which authorizes me to detach you and a party of volunteers under your command, appoint you to this duty. You are therefore to descend the Coppermine River; and as soon as the sea opens in July, are to proceed to explore the strait in question, endeavoring to communicate with any parties of Eskimos you may meet with on Wollaston or the neighboring islands. Should you reach the northern coast of Bank's Land, you are there to erect a pile of stones, and deposit a memorandum of your object and proceedings at the distance of 10 feet from its base, marking that side of the pile with a broad arrow in red or white paint. You are also to erect similar piles, and deposit in the same manner on conspicuous headlands, memoranda for the guidance of the party detached by Sir James Ross, when you can do so without materially delaying your progress. Should you discover any signal posts erected by that party, and learn from the memoranda deposited near them that the strait has been sufficiently explored down to that place, you are to proceed no further in that direction; and you are at liberty to use your own judgment in deviating from this route, if, from information given by the Eskimos, or obtained from other sources, you are of opinion that the ships, or part of their crews, may be found in another quarter.

"Having the fullest confidence in your judgment, experience, and prudence, I shall not name a period to your advance, further than by requesting you not to hazard the safety of the party intrusted to your care, by delaying your return too long. The last season furnishes a strong instance of the early date at which the winter occasionally commences in these seas.*

"Having performed this service, or prosecuted it as far as practicable, with a due regard to the safety of your party, you are to return with all speed to Fort Confidence, and, embarking without delay the instruments and stores remaining at that post, to proceed forthwith to Fort Simpson. Such of the stores as are useful to the Company are to be valued and handed over to them, and the instruments are to be forwarded to England, addressed to the Secretary of the Admiralty. The men are to be sent to winter at some fishing station sufficiently to the southward to insure their early arrival at Norway House next spring; and you are to direct them to be furnished with nets, that they may provide for their own sustenance during the winter, with as little expense to the Government as possible.

"Immediately on your return from the coast, you are to communicate an account of your proceedings to the Secretary of the Admiralty, for the

* The 25th of August was the date at which I considered it prudent that Mr. Rae should endeavor to be on the south side of Dolphin and Union Straits, and I expressed this opinion in a private note to him.

information of their Lordships; and you are also to transmit him a chart of any hitherto unexplored coasts or straits you may discover, as soon as you have had leisure to construct it.

"Given under my hand, at Fort Confidence, 1st May, 1849.

(Signed) "JOHN RICHARDSON,

"Commanding Arctic Searching Expedition.

"John Rae, Esq."

Copy of a Dispatch from Mr. Rae to the Secretary of the Admiralty, narrating the Proceedings of the Expedition under his Command to the shores of the Arctic Sea, in the Summer of 1849.

"Fort Confidence, Great Bear Lake, Sept. 1, 1849.

"SIR—I have the honor to acquaint you, for the information of my Lords Commisioners of the Admiralty, that the expedition under my command, which descended the Coppermine this summer to the Arctic Sea, for the purpose of examining the shores of Wollaston and Victoria Lands, in search of Sir John Franklin and party, returned to this place to-day, having been quite unsuccessful in its object, and with the loss of Albert, the Eskimo interpreter, who was drowned at the Bloody Fall, the particulars of which unfortunate accident I shall mention hereafter.

"Having made every requisite arrangement at Fort Confidence for facilitating our progress across land to the Coppermine, I waited impatiently for the disruption of the ice on Dease River, to which our boat was hauled on the 7th of June. Next day we learned that the upper parts of the river were clear of ice; and on the following morning I started in company with four men and two Indians, and a couple of sledges on which our baggage and provisions were stowed. The Dease was still covered with strong and solid ice for two miles up its course, over which we hauled the boat before getting to open water.

"Our ascent of the stream was extremely slow, owing to the many barriers of ice (some of them six or eight miles long) over which we had to launch the boat, and it was the 15th before arrived at the forks of the river, where it was my intention to diverge from the route of Dease and Simpson. They followed the north branch, while we ascended the south-east one. This stream was also much obstructed by ice, and so very shallow, consequent on the coldness of the weather, which prevented the snow on the high grounds from thawing, that the whole party were almost continually up to their knees among water and snow engaged in launching the boat. In ordinary seasons it was evident, by marks along the banks, that there is sufficient water for a boat drawing some inches more than ours throughout its whole length, which is little more than 15 miles, including its various curves. On the 17th we passed over the ice on the lake from which the stream flows. It contains many islands, and its breadth, where we crossed it in a nearly south direction, is 3½ miles.

"Indian report had led me to believe that there was a creek, having sufficient depth of water for the boat, flowing from this lake into the south branch of the Kendall, which we were to descend to the Dismal Lakes; but in this we were disappointed, and consequently had a portage of six geographical miles to make overland nearly due east. I had examined this place in the winter, but the ground being then covered with snow and ice, I could not form a correct opinion on the subject. The west end of the portage is situated in latitude 67° 10′ 48″ N.; longitude by account* 117° 18′ W.; and the variation of the compass 50° 49′ E.

"Crossing the portage occupied us two days; the ice had not yet broken up in many places on the river on its east side, and the water was ten inches lower than when Sir John Richardson and party forded it last autumn; we consequently found some difficulty in descending it. Its general course is northeast by east, and the length from where we entered it, to its influx at the west end of the most easterly of the Dismal Lakes, 17½ miles. On the 21st we arrived at the station on the banks of the Kendall River, to which provisions for the sea voyage had been hauled on dog sleds in April, and found the two men who had been left in charge quite well. We descended the Kendall next day to the Coppermine, which was still covered with ice, so firm and solid that a person might have crossed the river without being more than ankle deep in water.

"During five days that we were detained here, we were occupied repairing the injuries received by the boat, shooting deer to save our pemican, and making observations when the weather would permit. The result of three meridian observations of the sun gave mean latitude 67° 07′ 20″ N., and the mean variation from five sets of azimuths on different days (the extremes being 49° 38′, and 51° 55′) was 50° 37′ 48″ E. On the 28th the dwarf birch was observed to be in leaf, and the leaf-buds of the willows began to develop. In the afternoon of the same day the river was thought sufficiently open to permit us to descend it for some distance among the driving ice; but after proceeding six miles, we found the stream again blocked up. We were so often and so long detained by interruptions of this kind, that it was the 11th of July before we arrived at the Bloody Fall, having been fourteen days in doing the work of one. Notwithstandstanding the inefficiency of our steersman James Hope (one of Dease and Simpson's men), we ran all the rapids, including the Escape, without shipping much water, and with all the cargo in the boat.

"Hitherto deer had been so numerous that we could easily have shot enough for the maintenance of a party double or treble our numbers. Here they had become more scarce and shy, which could be only accounted for by the proximity of the Eskimos, no recent traces of whom could, however, be seen. From the fall to the sea the ice remained fixed until the 13th, when it cleared away, a circumstance that was very soon indicated by the numbers of fish which appeared below the fall. With the aid of Halkett's air-boat, which had been brought from a hill some miles distant, where it had been left last season, a net was set in one of the eddies, and before the men had finished arranging it, seven fine salmon and two white fish were caught, and we afterward obtained a supply for several days' consumption.

"On the 14th we entered the sea, and found a narrow and very shallow channel along the shore of Richardson Bay, until we came to its north side, where the ice lay against the rocks. Here the latitude 67° 51′ 19″ N. was observed; and two azimuths of the sun, the one on the meridian and the other when on the prime vertical, gave variation 57° 04′ and 56° 25′ E. While encamped at this place, we were visited by seven Eskimos, one of whom I at once recognized as the active, intelligent man who had afforded Sir John Richardson's party such efficient assistance last season,

* The rates of the chronometers had become so irregular, or had altered so much, that they were of no use until rated anew.

when crossing the river at the head of Back's Inlet. On inquiry I learnt that they had been well supplied with provisions in the early part of winter and in spring; but that in the interval they had nearly starved, owing to the scarcity of seals, having had to subsist for some time on the skins of the larger species of these animals, which they had reserved for making boots. In the winter they had communicated, either directly or indirectly, with the natives of Wollaston Land, none of whom had ever seen whites, large boats, or ships. They were all made happy by some small presents, and a supply of fish, which they ate raw, and appeared to relish much. They left us near midnight, promising to return next day with some boots and shoes for sale.

"On the 16th, by making a number of portages over the ice, we rounded Point Mackenzie and entered Back's Inlet, which was partially open. Having a fine breeze from the east, we set sail and soon ran to its head, when we entered Rae River (discovered and named by Sir John Richardson last autumn); and on proceeding three miles up it, came to the lodges, six in number, of our Eskimo visitors, who said that they had been so much alarmed at seeing the boat under sail that they were on the point of running away.

"As there was no possibility of our making much progress along shore until the ice wasted a little more, I devoted the two following days to an examination of this river; the Eskimo whom I have already mentioned as our active assistant last year willingly agreeing to accompany us. At the distance of $9\frac{1}{2}$ miles from the river's mouth, there is a perpendicular fall of 10 feet, which extends across the stream, except a few yards on the north side, where the rock slopes so much that, during the spring-floods, salmon and white fish are able to ascend, affording the natives a fine opportunity of spearing them. Here I left the boat and four of the men, while, in company with the other two, and our Eskimo guide, I traced the river $19\frac{1}{2}$ geographical miles further. Its course is nearly due west, and very straight; about the size of the Dease River, and varying in breadth from 80 to 200 yards, with a very strong current, and sufficient depth of water for a boat drawing 14 or 15 inches. It flows over a bed of limestone, and is bounded on the north, at the distance of half a mile or less from its banks, by precipices of basalt from 100 to 200 feet high, superimposed on limestone of the same kind as that which forms the bed of the river.

"At the extreme west point of our journey, we found a party of ten Eskimos with their families, who informed us that the stream maintained the same course and size as far as they had seen it, which was somewhat more than three days' march, or about sixty miles; how much further they knew not, as they had never been to its source. Two of this party returned with us to the boat, where they received presents, which I had some difficulty in getting them to accept, as they said they had nothing to give in exchange. On the evening of the 18th we ran down stream, and landed our guide among his friends, who seemed very glad to see him return safe. They were now much less timid than when we first met them, and we pitched our tent close to theirs, without causing any apparent alarm, although it was afterward observed that two of the men kept watch during the night.

"Early on the 19th we took a friendly leave of these simple and inoffensive people, and pulled down to the mouth of the river, where I staid until noon, when the latitude 67° 55′ 20″ N. was observed. The cloudy state of the weather during the two preceding days had prevented any observations being made. New moon occurring to-day, it was high water a few minutes before 1 P.M.; the rise being 10 inches. In the afternoon, when on our way to Cape Kendall, we experienced a severe thunder-storm from north-northwest, which obliged us to land for shelter.

"Our advance along the coast was so slow, that we did not arrive at the place where the boats were left last autumn until the 24th. We found that they had been much broken up by the Eskimos to obtain the ironwork. The tents, oil-cloths, and part of the sails, still remained uninjured, and were of much value to us, as we were ill-provided with the two first of these articles. The 'cache' of pemican and ammunition was also untouched, having apparently escaped notice from being covered with snow. The latitude 68° 10′ 44″ N., and variation 56° 8′ E., were observed here.

"On the 27th a west-northwest breeze having cleared away the ice for a short distance from the shore, we continued our course toward Cape Hearne, which we reached before noon, and found its extreme point to be in latitude 68° 11′ 17″ N. Basil Hall Bay being filled from side to side with unbroken ice, we encamped here. Next forenoon a light south-southeast breeze opened a crack in the ice, wide enough for the boat to cross to an island in the middle of the bay, on the north side of which we found some open water, which enabled us to get two miles beyond it. At 3 P.M. on the 30th, we arrived at Cape Krusenstern, and when opposite its high cliffs a strong breeze sprung up from north-northeast, which drove the ice so forcibly against the rocks, that we were obliged to unload with all haste, and haul the boat up on a drift-bank of snow to save it from being squeezed. Here for the first time this season we found the ice broken up in the offing, caused evidently by the strong currents of the ebb and flood tides; whereas on looking in the direction from which we had come, all, except immediately along shore, was smooth, white, unbroken, and apparently as firm as in winter.

"We were now at the most convenient though not the nearest point for making the traverse to Wollaston Land, passing close to Douglas Island, and there was no necessity for our proceeding further along the shore, even had we been able to do so, which at present was impossible; the high rocks presenting an insurmountable barrier on the one hand, and the ice by its roughness equally impassable on the other. We pitched our tents on the top of the cliffs, in the ascent of which the before-mentioned snow-bank served as a ladder, and waited for the first favorable change in the ice.

"A few days afterward, Albert (the Eskímo interpreter) and one of the men, when some distance inland looking for deer, overtook five Eskimos, traveling to the interior with loads of salmon, which they had speared in a rivulet that falls into Pasley Cove. From these the interpreter learnt that the sea ice had commenced breaking up only one day before our arrival, and that they had been in company with the natives of Wollaston Land during the winter, none of whom had ever seen Europeans, large ships, or boats.

"During our long and tedious detention here, several gales of wind occurred, principally from the northward, but the space of open water was so small that they produced little effect upon the ice. Our situation was most tantalizing to all the party: occasionally at turn of tide a pool of water, a mile or more in extent, would appear near us, and every thing would be prepared for embarkation at a minute's notice, in expectation of the opening increasing and permitting us to cross to Douglas Island, but our hopes were always disappointed. A number of observations were obtained, which placed our encampment in latitude 68° 24′ 35″ N., the longitude being very nearly the same as that assigned to it by Sir John Richardson and Mr. Kendall. The mean variation of the compass, from eight sets of azimuths, on different days, and at different hours, was 59° 8′ 08″ E., the extremes being 57° 42′ and 61° 25′.

"The ice continued driving to and fro with the tides, without separating sufficiently to allow of the practicability of passing among it until the 19th August, when there was more open water to seaward than we had yet observed, caused by a moderate southerly breeze that had been blowing for the last two days. After waiting some hours for a favorable opportunity of forcing our way through a close-packed stream of ice that was grinding along the rocks as it drove to the northward, we at last pushed off, and after more than once narrowly escaping being squeezed, we reached comparatively open water, where we had room to use our oars. We had pulled more than seven miles and were still three miles from the island (Douglas), when we came to a stream of ice, so close packed and so rough that we could neither pass over nor through it; a thick fog had come on, and the ebb tide was carrying us fast to the southeast. Under these circumstances I thought it advisable to return toward the main shore, on which we landed early on the morning of the 20th, a short distance to the south of the place from whence we had started. A northwest breeze came on some hours after landing, and cleared away the ice a few yards' distance from the beach, of which we took advantage to shift our quarters, which being under some crumbling cliffs, were neither safe nor convenient. We poled along shore for three miles or more toward Point Lockyer (the only direction in which we could go), and then took shelter in a small bay, into which we had scarcely entered when the wind changed to east-northeast, and in a very short time left not a spot of open water visible, either near shore or at a distance.

"The wind continued to blow from east and northeast for the two following days, during which, when the tide was in, we advanced a few miles to the south, principally by launching the boat over the ice. On the evening of the 22d I ascended a hill near the shore, from which a fine view was obtained. As far as I could see with the telescope in the direction of Wollaston Land, nothing but the white ice forced up into heaps was visible, while to the east and southeast there was a large space of open water, between which and the shore a stream of ice, some miles in breadth, was driving with great rapidity toward Cape Hearne and its vicinity.

"As the fine weather had now evidently broken up, and as there was every appearance of an early winter, I thought it would be a useless waste of time to wait any longer in hopes of being able to cross to Wollaston Land; I therefore gave orders for our return toward the Coppermine, at which I did not expect to arrive in less than a week, as the ice wore as unfavorable an aspect as it did last year.

"At an early hour on the 23d the men commenced carrying the baggage to Point Lockyer, still more than a league distant, and afterward hauled the boat to the same place; doing this gave all the party ten hours' hard work; but our fatigues were soon forgotten on finding some open water on the south side of the point, in which we were speedily afloat and sailing before the fast-decreasing breeze. It fell calm in half an hour; but we plied our oars to such good purpose, that we pitched our tents late at night on Cape Hearne. Here I had expected to find the ice close packed on the shore, and quite impassable, but I was agreeably surprised to discover as we advanced, that there was a lane of open water between the beach and the pack, wide enough for us to pass through. The only way that I can account for this is, by supposing that the gale of wind that had been blowing from northeast and east-northeast had shifted to the north and northwest as it approached the land, and carried the ice along with it.

"By working seventeen hours next day we came to the mouth of the river, and on the following morning ascended to the Bloody Fall. Here fish were still very numerous, and while some of the party were cooking, others set a couple of nets to obtain a supply for some days. I may here mention that when on the coast, we obtained as many salmon and herring as we could consume, wherever there was a piece of open water large enough for setting a net to be found.

"Dease and Simpson, after hauling their boats over the rocks opposite the lower and strongest part of the fall, had them towed up the remainder by water; and as our boat was now much weakened by the rough usage it had unavoidably been exposed to, for the purpose of saving time, and the tear and wear that would be caused by having it dragged over the portage, I was desirous of doing the same. I was the more led to do this, as the men were of opinion that it might be effected with safety. All that appeared in any way difficult was easily done, and there was only one short place to be ascended, which was so smooth that a loaded boat might have passed it; here, however, from some unaccountable cause, the steersman was seized with a sudden panic, and called to those towing the boat to slack the line. This was no sooner done sufficiently to allow him to get firm footing, than he leaped on shore, followed by the bowman, and allowed the boat to sheer out into the current, where the line broke, and the boat soon oversetting, was carried into one of the eddies some distance down stream; to which Albert and I ran, and stationed ourselves at two points of rock near which the wreck would pass. It drove to where Albert was, and he hooked it by the keel with an oar until I came to his aid, when I fixed a pole in a broken plank and called my companion to assist me in holding on; he either did not hear me correctly, or thinking that he would be of more use on the bottom of the boat, sprung to it, and before I had time to call him off, or even think of his danger, they were carried by a turn of the current into a small bay, where I believed both were safe; not so, however, for the next minute they were swept out again, and the last I saw of our excellent interpreter was his making a leap toward the the rocks; he missed them and disappeared, nor did he rise again to the surface.

"This melancholy accident grieved me much, as the brave lad was universally liked for his activity, lively and amiable disposition, and extreme good-nature. On James Hope, the steersman, rests all the blame of the loss of the boat; his carelessness in using a small towing-line when there was a much stronger one in the boat, and his cowardice when there was no danger, can admit of no excuse.

"On taking up our nets (which we laid carefully on the rocks for the Eskimos), they were found full of herring-salmon, in fine condition. We commenced our journey, across land, toward Great Bear Lake, on the 26th, each of the men carrying about 90 lbs., and my own bundle being nearly 50 lbs. Three days' easy march brought us to that part of the Coppermine (lat. 67° 12′ N.), from which it was my intention to make a straight course to Fort Confidence. Here we expected to meet with some Indians, but we did not fall in with them until the following evening, when our loads were much lightened, and we arrived here early this afternoon (being our seventh day from the Bloody Fall), accompanied by upward of thirty of the natives, who had joined us at different parts of the route.

"The two men who had passed the summer at Fort Confidence were well; and having all the stores ready packed, I shall leave this to-morrow, after supplying the Indians with ammunition, to enable them to hunt their way to the trading-posts. I have, &c.,

(Signed) "JOHN RAE.

"**Fort Simpson, September 26, 1849.**

"P.S.—I arrived here this afternoon, and intend sending off the expedition men to Big Island, Slave Lake, on the 28th, with an ample supply of nets and twine, so as to enable them to procure sufficient fish for their winter provisions, without being any additional expense to Government.

(Signed) "J. RAE."

Mr. Rae's failure in crossing to Wollaston Land is to be attributed solely to the strait being filled with impracticable ice. I know from his private letters that the mortification he experienced in the result is much more severe than he has thought proper to express in his official dispatch. The presence of ice in Dolphin and Union Straits and in Coronation Gulf for two or more successive seasons, where the experience of former years had led us to expect a comparatively open sea, is suggestive of the manner in which a party may be shut up in these regions, and leads to many melancholy reflections.

Every reader of my narrative of the proceedings of the expedition, will be aware of how much I was indebted to Mr. Rae's activity and intelligence throughout its progress; and this seems to be the appropriate place for me to express formally my sense of obligation to him for his sound advice and co-operation on every emergency. His society cheered the long hours of an Arctic winter's absence from my wife and family, and it was in a great measure owing to his skill and assiduity in observing, that our experiments on magnetism, during our stay at Fort Confidence, were carried on so as to be productive of scientific results.

In consequence of intelligence obtained from newspapers on my journey home, I addressed the following letter to Mr. Rae:

"Lake Winipeg, August 19, 1849.

"My Dear Rae,

"As I learn from the newspapers which I have just read, and shall forward for your perusal, that Sir James C. Ross did not reach Barrow's Strait till after the 28th of August, and that it is probable that he may have been arrested short of his intended wintering quarters at Melville Island or Banks's Land, and could not, consequently, send off his proposed spring party to the Coppermine River this season, I consider it likely that he may determine on sending that party next spring; and if so, by the present arrangements they will, on their arrival at Great Bear Lake, find Fort Confidence deserted.

"I therefore think it important that you should engage either the chief of some band, or two expert hunters, to pass the months of June and July, 1850, on the portage between Bear Lake and the Coppermine River, promising them a handsome reward if they render any assistance to the expected white party, and paying them such moderate sums, in addition to a full supply of ammunition, as may content them for spending the summer on such excellent hunting-grounds.

"You will have no difficulty in engaging either Martin Lake or Bear Lake Indians for this service; and there is abundance of time, after the arrival of the March packet, by which you will receive this letter, for them to reach Fort Confidence long before the snow begins to melt. I will thank you to furnish them with five or six memoranda in waterproof cases, with directions to plant them in conspicuous places at the mouth of the Kendall, Flett's Station, Fort Confidence, and elsewhere.

"These precautions may prove to be unnecessary, as Ross's party will most likely, early in their march, discover some of your landmarks, and learn, by the notes you have left, your intention of quitting Fort Confidence this season, and thereupon turn back to the ship. But, at a small expense, if the Indians carry their instructions out fully, they will save the party, should it come on, from having to make the long journey round Bear Lake without assistance.

"I remain, &c.,

(Signed) "John Richardson.

"P.S.—Mr. M'Pherson recommends Tecon-ne-betah for this service."

Extract of a Letter from Mr. Rae to Sir John Richardson.

"Portage La Loche (Methy P.), July 30, 1850.

"When the winter packet arrived and I received your instructions respecting the establishment of an Indian party on the Coppermine, the Martin Lake Chief Tecon-ne-betah was at Fort Simpson, and I had no difficulty in engaging him to pass the summer at certain stations on the route between the Kendall and Bear Lake. He received three notes which were to be delivered to any Europeans he might meet, in which the strangers were requested to put themselves wholly in the hands of the Indians, who would guide them by the best road, and feed them by the way. In the event of his not seeing any parties, other notes, with a rough chart of the best route, were supplied, well wrapped in oil cloths, which were to be placed on a pole in conspicuous stations.

(Signed) "J. Rae."

Having by the preceding quotations brought the narrative of the search made by the expedition to a close, I now revert to the period at which Mr. Bell and I left Fort Confidence. A party of men preceded us by a week, taking with them the baggage, which they were directed to carry on to Fort Norman, along with some pemican and stores, left at Fort Franklin. This was intended as a precautionary measure to avoid the delay that might be caused by Bear Lake River being late in breaking up.

On the 7th of May we took leave of Mr. Rae, and left the fort. Bruce, Mitchell, Brodie, M'Leod, and Mastegon accompanied us in our journey over the ice, which was completed on the 12th, having occupied five days and a half. On the northern shores of the lake, the snow had lost little of its depth, and we had to clear it away to the thickness of five feet in making our encampment on Cape M'Donald; but on the southern shores we found some exposed sandy spots of ground bare, and pools of water forming at mid-day. Mastegon shot the first goose of the season on the 10th, a straggler that was tempted to cross our pathway by a pool of water, produced under the influence of a powerful sun (p. 302). Cloudy weather followed, and five or six days elapsed before the geese began to arrive in earnest.

The men who preceded us were at the fishing hut, near the site of Fort Franklin, when we arrived there. They had transported three sledge loads of pemican about seventeen miles down Bear Lake River to the usual winter crossing-place, but, finding the stream open, they had put them in *en cache* on the right bank, and returned to wait further orders. According to their report the rivulets were swollen with melting snow, and traveling by land with dog-sledges was at an end for this season.

On the 14th, being Sunday, we assembled to read prayers; and, early on the following morning, Bruce set out for Fort Norman, taking with him Dore, Cousins, Thomas Hope, Mastegon, Plante, and M'Leod. They traveled light, carrying with them merely their blankets and provisions. By the arrangement I had made with Mr. M'Pherson they were to find at Fort Norman a small barge, which Bruce was directed to bring up to us as quickly as he could. The only difficulty we anticipated was at the rapid in Bear Lake River, where lofty walls of ice remain to a late date, covering the tracking-ground. Hope and M'Leod, being part of Mr. Rae's boat's crew, were to return overland immediately with some articles from the store at Fort Norman. They did come back nine days afterward, and passed onward to Fort Confidence; and we learnt in the sequel that the Mackenzie did not break up at Fort Simpson till the 23d of May, being fifteen days later than Mr. M'Pherson had known it to do during twenty years' residence on its banks.

We remained waiting nearly a month for the barge, having with us Mitchell and Brodie, with the two Fort Franklin fishermen, Hector Morrison and Narcisse Tremblè. Our diet consisted of trout, white-fish, Bear-lake herring and geese, the latter being Mr. Bell's contribution to the common stock. Of Fort Franklin the only vestige remaining was the foundation of a chimney stack; and the fishing hut not being large enough to hold us all, we bivouacked under the shelter of a boat's sail, as a substitute for a tent. When the water had run off the surface of the ice on the lake, so that we could transport our effects across it without wetting them, we moved to the banks of Bear Lake River; being glad to quit the vicinity of the hut, which, like all fishing quarters, became extremely disagreeable as soon as the accumulated impurities of the winter were revealed to view by the wasting of the snow. The marshy places or dry sandy banks first became bare, but many wreaths of drifted snow continued unmelted till the end of the month. We stayed at the encampment on Bear Lake River till the 8th of June, much surprised that the boat did not arrive, and forming various conjectures to account for the delay. On that day, Bruce and Mastegon came to us on foot, bringing information of the lower part of the river being still covered with ice, and that they had left the barge a few miles within its mouth. They had been four days on the march, one of which was occupied in examining the ice, which they stated formed a bridge at the rapid many feet in thickness. On receiving this intelligence, I resolved on descending to the mouth of the river on foot, and after going in the barge to Fort Norman, where I could join Mr. M'Pherson, to send her back for the stores. These were accordingly placed in a secure *cache*, and left under the charge of Narcisse, to whom we transferred our surplus stock of fish and geese, being sufficient to maintain him eight or ten days without touching the pemican, of which there was a large reserve. The snow drifts formed excellent ice cellars for preserving fresh provisions, the only precaution necessary being to protect them from the dogs by branches of trees.

On the 9th, Mr. Bell, Bruce, and Mastegon, embarked with me in the fishing coble, and Morrison, Mitchell, and Brodie, were directed to walk along the bank of the river, each of them carrying his own bedding and clothing. I cautioned them against going inland, and promised that we would wait from time to time at particular points for their coming up. Half an hour after setting out, finding the river smoother than we expected, and Bruce being of opinion that we could embark all the party, we put ashore, and in a short time Morrison and Mitchell joined us, but David Brodie, having struck into the woods with the view of making a straighter course, did not arrive in the hour that we remained waiting for him. Supposing then that he had gone past, we resumed our voyage, taking into the coble Morrison and Mitchell.

The rate of our descent of the stream rather exceeded four miles an hour, and at half-past six A.M. we reached the *cache* situated fourteen miles from the lake. Brodie not arriving in the course of the day, I became convinced that he had lost himself in the woods, and therefore sent Morrison and Mitchell back to the lake to acquaint Narcisse with what had happened, and to endeavor to engage an Indian who was residing at the fishery to go in quest of Brodie. In the mean time we fired our fowling-pieces at intervals, and set fire to some trees that the smoke might be seen from a distance.

Mastegon, who knew that part of the country, informed me, that ten or twelve miles back from the river, there was an extensive swamp, from which there flowed one stream that fell into the Mackenzie, and another, named the Black River, which joined Bear Lake River about four miles below the *cache*. The latter stream was so rapid, he said, that Brodie would be unable to cross it on a raft, and it was not fordable ; he would therefore, by following it, be certainly led to Bear Lake River.

Next day our two men came back from the lake, having themselves gone astray for some hours in the attempt to make a short cut across a neck of land. After placing written directions for Brodie in the *cache*, we all embarked again, and in a short time came to the influx of the Black River, which was evidently flooded. Here I left another paper of instructions for Brodie, directing him to the *cache* for provisions, and to remain with Narcisse until the barge came for them. The incident of Brodie's straying gave me much uneasiness, as I feared he would experience some suffering, though I did not apprehend that he would lose his life. He was a man of much personal activity and considerable intelligence, and though his judgment would be probably at fault when he first became conscious that he was lost, I knew that as soon as he was in a condition coolly to consider his position, he would be enabled to shape a course for the river by the sun, and following its bank to return to the lake. And so it eventually happened. When he discovered that he was walking in a wrong direction, he began to mend his pace and to run as is usual in such cases, but took an inland course, and at length came to the borders of the swamp above mentioned. Here the woods being more open he obtained a distant view of the "Hill at the Rapid," which he recognized from having seen it on his former journey to the *cache*, and as he knew that we must pass it in descending the river, he resolved on walking straight for it in the hope of arriving there before us. After this he came to the Black River, and being a fearless swimmer, swam across it, carrying his clothes on his head. The stream, being very tortuous, came again in his way, when he crossed it a second and a third time in the same manner, but on the last occasion, his bundle slipping off floated away, and he regained the bank with difficulty in a state of perfect nudity. After a moment's reflection, he came to the conclusion that without clothes he must perish, and that he might as well be drowned in trying to recover them as to attempt proceeding naked. On which he plunged in again, and fortunately landed this time safely with his habiliments. He now refreshed himself with part of a small piece of dried meat, which, in his anxiety he had hitherto left untouched, and forthwith decided on finding the *cache*, and returning from thence to the lake. On the third day (11th of June), he found my note together with some provisions which we had suspended to a pole for his use, but he had so husbanded his own small supply that he had still a morsel of dried meat remaining. He had no difficulty afterward in joining Narcisse, by keeping sight of the river the whole way.

This adventure is recorded as an example of what happened to all the seamen and sappers and miners of the expedition, each in their turn. Four of them were lost in the winter of 1848–49, for three days, having mistaken their way to the boat encampment in Cedar Lake. The straggling of the others was of less account, but none of them could be taught that they were liable to such accidents, until they learned it by experience. One man who strayed in the winter on Cedar Lake, when found, was contentedly steering for the moon, which, being near the horizon and gleaming red through the forest, was mistaken by him for the fire of the men's bivouac. The snow which covered the ground at the time fortunately enabled the Indian who went in pursuit of him, to trace his steps before he had gone many miles.

About twelve miles below the mouth of Blackwater River, we came to the commencement of the rapid, and hauling the coble on the beach there, proceeded on foot to the "Hill," immediately below "the Rapid," where we encamped. While on our march we perceived that the bridge of ice was giving way, and the river directly fell some feet. Early next morning Bruce went back with the men for the coble, and brought it down by the time that Mr. Bell and I had prepared breakfast for the party. After concluding that meal, we crossed the river in her obliquely, among high walls of ice; and a mile or two below came to the barge which the crew had brought up so far two days previously. A short time sufficed to launch her, embark the tents kindly provided for us by Mr. M'Pherson, and descend to the mouth of Bear Lake River, where we passed the night. Next day we went on to Fort Norman, where I purposed to wait for Mr. M'Pherson, who had gone to Fort Good Hope, to bring up the year's returns of furs from that post, and from the Peel and Porcupine Rivers.

On his arrival on the 14th, I sent back the small barge, to bring down Narcisse, Brodie, and the stores, and embarking with Messrs. Bell and M'Pherson, proceeded to Fort Simpson. The rest of the journey homeward, being by the same route as the outward voyage, need not be mentioned in detail.

On the 25th we left Fort Simpson, having previously been joined by the men who wintered at Great Slave Lake, and also by the small barge, bringing Brodie and Narcisse Tremblè, from Bear Lake. We were detained by drift ice at the west end of Great Slave Lake till the 6th of July, and did not reach Fort

Resolution till the 11th. On the 19th we arrived at Fort Chepewyan, and on the 26th at Methy Portage, which we crossed on the 27th with all our baggage, on horses hired from the Indians. From L'Esperance, who was encamped with his brigade on Methy Lake, I had the pleasure of receiving English letters, brought up from Canada by the governor's light canoe, which leaves La Chine in May. Mr. Bell at the same time received instructions to return to Mackenzie's River, to conduct the Company's affairs there. This was unpleasant tidings to him, since, having spent the greater part of his life in that northern region, he had been soliciting a change, but the mortification was softened by the society of his two daughters who had been sent from Isle à la Crosse to meet him. In taking leave of this gentleman, I must express my obligations to him for his assiduous endeavors to forward the interests of the expedition, and my high sense of his excellent management of the Indians at Fort Confidence, to which we owed a winter of abundance, and the excellent condition in which the store was left in spring. I had enjoyed much pleasure in his society, and parted from him with regret.

The remainder of the voyage down was performed in company with Mr. M'Pherson, who was going down to Canada on furlough with his family. At Norway House, where we arrived on the 13th of August, the men of my party who had been engaged there in 1847 were discharged, and the Europeans were sent down to York Factory to go to England in the Company's ship. During the time these men were under my command not a single act of disobedience occurred. Crews better fitted for heavy portage work, and for the ordinary duties of a winter's residence in the north, might doubtless have been selected in the country, but none that I could have depended upon with so much confidence in adverse circumstances. Dore and Cousins, the two younger seamen, who were extremely serviceable from their activity, intelligence, and willingness to perform any duty that was required, have gone again on the same errand to Beering's Straits.

From Norway House I proceeded to Canada in a brigade of three light canoes, manned by voyagers who were returning thither at the close of their engagements in the country. On the 14th of September we came to Fort William, and on the 25th to Saut Ste. Marie. From thence we went in a steam-vessel to the lower end of Lake Huron, and taking a stage coach there for

Orillia, crossed Lake Simcoe in a steamboat. Then we traveled by coach through Young Street to Toronto, a distance of about forty miles, and there embarked in one of the steam-packets that ply daily between that port and Montreal.

After a few days passed at La Chine with Sir George Simpson in revising the outstanding accounts between the Company and the Expedition, I went to Boston, and embarking in the British mail steam-packet, crossed the Atlantic, and landed at Liverpool on the 6th of November, 1849, after an absence of nineteen months, twelve of them passed in incessant traveling.

Without delay I presented myself at the Admiralty, and, having laid before their Lordships a narrative of my proceedings, had the honor soon afterward to receive a letter announcing their approbation of my conduct.

Here the journal of the transactions of the expedition ends, but a summary of the present condition of the search may not be unacceptable to the many who take an interest in the fate of our absent countrymen.

Sir James C. Ross, with the "Enterprise" and "Investigator," reached the three Islands of Baffin in lat. 74° N., on the 26th of July, 1848, but was not able to cross the "middle ice" till the 20th of August, on which day he attained open water in lat. 75½° N., and long. 68° W. He then steered for Pond's Bay, and examined the coast carefully from thence to Possession Bay, in which he landed on the 26th. There he found a memorandum left by Sir Edward Parry in 1819, but no trace of Sir John Franklin. On the 1st of September, the ships arrived off Cape York, where a conspicuous land-mark was erected. Sir James next examined Maxwell Bay, and some smaller indentations of the north coast of Barrow's Strait, but was prevented by a firm barrier of ice from approaching Cape Riley at the entrance of Wellington Channel. Neither could he get near Cape Rennell, because of compact, heavy ice extending from Wellington Channel to Leopold Island. Not being able to penetrate to the west, the ships were run into Port Leopold on the 11th, and on the following morning the main pack of ice closed in with the land and shut them in for the season. On the 12th of October the ships were hove into their winter quarters. During the winter many white foxes were taken in traps; and copper collars, on which were inscribed notices of the situation of the vessels, and of the depots of provision, having been secured round their necks, they were set at liberty again.

In May, 1849, Sir James Ross and Lieutenant M'Clintock thoroughly explored on foot the west coast of North Somerset down to lat. 72° 38′, N., and long. 95¾° W., where a very narrow isthmus separates Brentford Bay of the Western Sea, from Cresswell Bay of Prince Regent's Inlet. They returned to the ship on the 23d of June.

In the mean time, Lieutenant Robinson examined the western side of Regent's Inlet down to Fury Beach, and several miles beyond it. Lieutenant Brown had crossed the inlet to Port Bowen, and Lieutenant Barnard had traversed Barrow's Strait to the vicinity of Cape Hurd, but was unable to reach Cape Riley on account of the hummocky state of the ice. By these excursions, taken in conjunction with Mr. Rae's expedition in the spring of 1847, the whole of Prince Regent's Inlet and the Gulf of Boothia was examined, with the exception of one hundred and sixty miles between Fury Beach and Lord Mayor's Bay; and as there were no indications of the ships having touched on any part of the coast so narrowly traced, it is certain that they had not attempted to find a passage in that direction. Sir James caused a house to be built at Port Leopold, and covered with housing cloths, in which he left provisions and fuel for twelve months, together with the Investigator's launch and steam-engine. He then proceeded to cut a way out for the ships through the ice, which was not effected until the 28th of August, 1849. On leaving the harbor he crossed over toward Wellington Channel, where he found the land-ice still fast and preventing his approach. While contending with the loose packs, and struggling to advance to the westward, a strong gale of wind on the 1st of September suddenly closed the ice around the ships, wherein they remained helplessly beset until the 25th, by which time they had drifted out of Lancaster Strait, and were off Pond's Bay. As the season was now far advanced, further search that year was thus frustrated by an accident, often experienced in the navigation of the Arctic Seas; and all harbors in that vicinity being closed for the winter, Sir James reluctantly gave the signal to bear up for England.

While Sir James C. Ross was still engaged in the ice on the west side of Baffin's Bay, Mr. James Saunders, Master and Commander of the "North Star," having been sent out with supplies in the spring of 1849, was working up on the east side, with imminent danger to his ship. Owing to the unusual quantity of ice in the bay that summer, and the frosts which glued the floes into one impenetrable mass, he was unable to cross over to Lancaster Sound, and his ship becoming involved in the ice about the same date that the "Enterprise" and "Investigator" were caught in the pack, drifted with it the whole of September, until on the last day of that month she was providentially driven into Wolstenholme Sound, where there being a pool of open water she was at length extricated. There the ship wintered in lat. 76° 33′ N., long. 68° 56¼′ W., being the most northerly position in which any vessel has been known to have been laid up. February was the coldest month, and the thermometer on two occasions marked 63½°, and once 64½°, of Fahrenheit below zero.

On the 1st of August, 1850, the "North Star" was hauled out of the cove in which she had remained ten months, and on the 8th she had crossed over to Possession Bay, which was examined. Mr. Saunders next proceeded to Whaler Point, Port Bowen, Jackson's Inlet, and Port Neill; but being prevented from landing his provisions at any of these places by the heavy land-floes of old ice, he bore up for Pond's Bay, and succeeded in depositing his cargo on Wollaston Island.

Meanwhile the search was proceeding from the quarter of Beering's Straits. Captain Kellett in the "Herald," on July 25th, 1849, after examining Wainwright's Inlet, dispatched Lieutenant Pullen to the Mackenzie; and afterward, in standing along the margin of the ice, discovered a group of high islands on the Asiatic coast in lat. 71° 20′ N.; long. 175° 16′ W., with extensive and very high land to the north of them deeply seated in the ice.* Commander Moore, also, in the "Plover," made several attempts to penetrate to the eastward at this time, and not succeeding, returned first to Kotzebue Sound, and subsequently to Norton Sound, where he wintered.

Lieutenant (now Commander) Pullen, accompanied by Mr. Hooper, mate, and twelve men performed the coasting voyage to the Mackenzie in two 27-foot whale-boats. He was convoyed past

* Captain Smyth suggests that this land may be that reached by Sergeant Andreyer in 1762, which he reported to be inhabited by a people named Kraïhaï.

Point Barrow by the "Herald's" pinnace named the "Owen," and the Royal Thames Yacht-Club schooner, the "Nancy Dawson." The latter was owned and commanded by Mr. Shedden, a mate of the Royal Navy, who had come thus far with his small craft, solely at his own expense, to prosecute his search for the discovery ships, and who, though he was in the last stage of consumption, was not prevented by the languor of the disease, which carried him off two months afterward, from giving most efficient aid to Lieutenant Pullen.

On Sir James C. Ross's return to England in 1849, the Admiralty resolved that a still more vigorous search should be organzed, and accordingly the "Enterprise" and "Investigator" were again fitted out and dispatched to Beering's Straits, the former under the command of Captain Collinson, C. B., and the latter of Commander McClure. These ships having separated in the Pacific, the "Investigator," which was the dullest sailer, through a fortunate choice of route, reached the Straits first, succeeded in passing Point Barrow, and was last seen on the 4th of August, 1850. The "Enterprise," having been unable to penetrate the barrier of ice, went to Hong Kong to refit in the winter, and is to make another attempt this summer of 1851.

The preparations for the search on the side of Lancaster Sound were on a large scale. The "Resolute" was commissioned by Captain Horatio T. Austin, and the "Assistance," Captain Erasmus Ommaney, was put under his orders, together with the "Pioneer" and "Intrepid," steam tenders to the two vessels. Captain William Penny, an experienced whale-fisher, was also engaged for the search, and placed in command of the "Lady Franklin" and "Sophia." In addition to these expeditions fitted out by the Admiralty, others furnished from private sources showed the interest that was widely and deeply felt in the cause. Captain Sir John Ross, notwithstanding his advanced years, sailed in the "Felix" schooner; and by the munificence of Mr. Henry Grinnell, a New York merchant, the United States sent forth the "Advance" and "Rescue" on the same humane quest, under the command of Lieutenant De Haven, U. S. N., and Mr S. P. Griffin. Lady Franklin likewise, with that untiring energy and conjugal devotion which has marked her conduct throughout, dispatched the "Prince Albert" under the orders of Commander Forsyth of the Royal Navy.

This squadron was assembled in Lancaster Sound in the month of August, 1850, at which time the "North Star" was also there, forming in all a fleet of ten vessels. On the return of the last-named ship and of the "Prince Albert," we received intelligence from Barrow's Straits as late as the 25th of August, 1850. By that time both sides of Lancaster Sound had been thoroughly searched as far as Cape Riley on the north side, and Port Leopold on the south, also Prince Regent's Inlet down to Port Neill and Fury Beach. On the above-mentioned date, Mr. Snow, of the "Prince Albert," went ashore at Point Riley to examine a flag-post which had been erected by Captain Ommaney; and found a note from that officer, stating that he had landed with the officers of the "Assistance" and "Intrepid" on the cape on the 23d, had found *found traces of an encampment and collected the remains of materials, which evidently proved that some party belonging to her Majesty's ships had been detained there. Traces of the same party were found on Beechey Island.* The note concludes by the announcement that Captain Ommaney proceeds to Capes Hotham and Walker, in search of further traces of Sir John Franklin's Expedition. No mention is made of the nature of the materials collected; but the tenor of Captain Ommaney's note indicates that he had no doubt of Captain Franklin having been off Cape Riley. Lieutenant De Haven, of the "Advance," landed on the cape on the morning of the 25th, and erected a second signal post, but seems to have carried nothing away. Mr. Snow gathered and brought off five pieces of beef, mutton, and pork bones, together with a bit of rope, a small rag of canvas, and a chip of wood cut by an ax. From a careful examination of the beef bones, I came to the conclusion that they had belonged to pieces of salt-beef ordinarily supplied to the Navy, and that probably they and the other bones had been exposed to the atmosphere and to friction in rivulets of melted snow for four or five summers. The rope was proved by the ropemaker who examined it to have been made at Chatham, of Hungarian hemp, subsequent to 1841. The fragment of canvas which seemed to have been part of a boat's swab, had the Queen's broad arrow painted on it; and the chip of wood was of ash, a tree which does not grow on the banks of any river that falls into the Arctic Sea. It had, however, been long exposed to the weather, and was likely to have been cut from a piece of drift-timber found lying on the spot, as the mark of the ax was recent compared to the surface of the wood, which might have been exposed to the weather for a century.* Mr. Snow counted five rings of stones with two or three slabs in the centre of each circle, which he took for fire-places, but on which there were no traces of smoke nor any remains of burnt wood. As tent-pegs could not be driven into the shingly beach, the stones had been evidently used in the erection of as many tents as there were circles, and the slabs in the centre were likely to have served as stands for magnetic instruments. Colonel Sabine remarked that four tents would be needed in using the instruments supplied to Sir John Franklin's expedition, and a fifth for the protection of the observers. If the ships were stopped in that locality about the time of the monthly term-day, the officers would almost certainly make the term observations, which last for twenty-four hours, and in that case each ship would select a separate place of observation. The term-day in August, 1845, was the 29th; and we may conclude, from the information which we at present possess, that on that day, or about a month after they were last seen, the discovery ships were off Cape Riley. It is ascertained that the bones and rope were not left by any party from other ships of the Royal Navy that have visited Barrow's Straits; and had the "Erebus" and "Terror" wintered there, a cairn, with memoranda, and many other evidences of that fact, would undoubtedly have been found. The ships must have been temporarily arrested by a barrier of ice; and Sir John Franklin, having good anchorage between Beechey Island and Cape Riley, turned the delay to the best account by making the term observations. With the prospect of soon passing the Straits, he evidently had not thought it necessary to erect a cairn, or signal-post, at the threshold, as it were, of the enterprise.

These first traces of the Expedition are exceedingly interesting, and they lead directly to the inference that it pursued its course to Cape Walker. Had Sir John, finding the Strait barred, gone up Wellington Inlet, he would undoubtedly have left a memorandum at Cape Riley, assigning reasons for departing from his instructions.

On the 25th of August, 1850, the "Assistance" and "Intrepid"

* The grounds of these conclusions were fully stated in a report made to the Admiralty by Sir W. Edward Parry, myself, and other officers, which has been published with other parliamentary returns.

were well over toward Cornwallis Island, and a little to the north of Cape Hotham. Captain Penny with his consort was standing up the west side of Wellington Channel, Sir John Ross was making for Cape Hotham, the "Rescue" was near the head of Wellington Sound on the east side, land being visible from Cape Riley, crossing the Sound just beyond her, and the "Advance" was lying under Beechey Island, having grounded on a shoal; but Lieutenant De Haven, expecting to get his vessel off without difficulty, declined the assistance from the "Albert" which was offered. Captain Austin, with the "Resolute" and "Pioneer," was at that time examining the south side of Barrow's Straits. He had visited Possession Bay on the 18th, and then intended to look into Pond's Bay. He has not been seen since, and it is probable that he passed the "North Star" and "Albert" without being descried, in the thick weather that prevailed when they were coming out of Lancaster Sound.

Such was the state of the search in August, 1850. As Sir John Ross intended to return in 1851, after landing his stores on Melville Island,* we may expect that he at least will bring further intelligence in October or November next.

* Sir John Ross took with him four carrier pigeons belonging to a lady residing in Ayrshire, intending to liberate two of them when the state of the ice rendered it necessary for him to lay his vessel up for the winter, and the other two when he discovered Sir John Franklin. A pigeon made its appearance at the dovecot in Ayrshire, on the 13th of October, which the lady recognized by marks and circumstances that left no doubt on her mind of its being one of the younger pair presented by her to Sir John. It carried no billet; but there were indications, in the loss of feathers on the breast, of one having been torn from under its wing. Though it is known that the speed of pigeons is equal to one hundred miles an hour, the distance from Melville Island to Ayrshire being in a direct line about 2400 miles is so great, that evidence of the bird having been sent off about the 10th of October must be had, before that we can well believe that no mistake was made in the identification of the individual that came to the dovecot. Sir John's letters from Lancaster Sound mention that when he wrote he had the pigeons on board.

POSTSCRIPT.

SINCE the preceding sheets were printed, we have received information of the result of the last year's search for the lost Expedition. The first traces of the missing ships, discovered on the south side of Beechey Island and on Cape Riley, as mentioned in p. 327, were followed up by the discovery of seven hundred empty meat-tins, and other remains, which furnish undoubted proof of Franklin's ships having wintered, in 1845–6, on the inside of the above-named Island. The tombs of three men, with head-boards bearing their names and the dates of their deaths, were erected on the east side of the Island, not far from the site of the armorer's forge, an observatory or storehouse, and other inclosures opposite to the anchorage. One of these men belonged to the "Terror," and two to the "Erebus," which is sufficient evidence of the presence of both ships; and the latest death supplies us with the date of 3d April, 1846. The mortality does not exceed that of previous expeditions; and we may therefore conclude, that the Expedition was in highly effective order when it left that anchorage, with only a moderate inroad into its stock of preserved meats, the seven hundred empty tins found on the island forming but a small proportion of the 24,000 canisters with which the ships were supplied.

Captain Penny and his officers, who examined Beechey Island and the neighborhood very carefully and minutely, believe that the Expedition did not quit its winter anchorage till the end of August or beginning of September, 1846, founding their opinion mainly on the lateness at which the ice breaks up; that much of the summer was passed there, they consider as proved by the deep sledge ruts in the shingle, which must have been made after the snow had partially disappeared, and by small patches of garden ground bordered with purple saxifrage, and planted in compartments with the native plants.

It is also the opinion of several officers of the searching party that Franklin's ships left their wintering station suddenly. The reasons assigned for this belief are, that several articles which might have been useful were left behind, and that at a look-out or fowling station, on Cape Spencer, a long day's journey from the anchorage, the lines for securing the covering of a circular inclosure, formed by a low wall of stones, had been cut, instead of having been deliberately untied, when the covering was removed, leaving the ends of line attached to the stones. The absence also of any memorandum of past efforts or future intentions, either at the stone cairn erected on the south side of Beechey Island, at the pile of canisters, or in the neighborhood of the kitchen, forge, and other marked localities opposite the anchorage, is thought by some to be an indication of the sudden departure of the Expedition. The value of the articles left behind is too trifling to support such an inference,* and the absence of the diligently searched-for memorandum does not seem to be sufficiently accounted for by such a supposition. The time required for calling in the parties from Cape Spencer, Caswall's Tower in Radstock Bay, and other points where they have been traced, and for embarking the instruments and utensils from the observatories and kitchen, might have sufficed for the planting of a copper cylinder or bottle, with a memorandum.—That the ships drifted out unexpectedly in a floe of ice is not considered by the nautical men who have examined the anchorage to be possible. The north point of Beechey Island being connected to North Devon by a shingle beach, covered by only two or three feet of water, no pressure of ice can operate on the harbor from that direction so as to drive out vessels by the southeastern and only navigable entrance, and it is almost certain that Franklin's ships must have made their exit by the tedious and laborious operation of sawing out.

The absence of a memorandum at the wintering station is remarkable, and, in my opinion, wholly unexplained by any suggestion that has hitherto been given by the many writers who have made

* These were an armorer's wooden stand, used when laid on its side for the support of an anvil, and when standing on its end for the insertion of a vice; several coal bags, two of them containing coal dust mixed with a small proportion of small cinders and ashes, some pieces of rope, and scraps of old canvas, and a small piece of oaken fire-wood, besides many fragments of worn clothing utterly worthless. An iron stove that had been made on board ship was also found at a fowling station near the east corner of the island, but, it is stated to have been not worth carrying on board. The birds' bones remaining in the vicinity of the stone inclosure on Cape Spencer show that the sportsmen encamped there had been tolerably successful; and much small shot was found scattered among the stones with which the inclosure was paved. In the interstices of the stone wall there were many pieces of newspapers, also two bits of paper of much interest to the friends of two of the missing officers—one being inscribed with the name of Mr. M'Donald the surgeon; the other containing part of a memorandum in the handwriting of Captain Fitzjames, giving directions as to the times of recording certain meteorological observations.

their opinions known, through the medium of the periodical press. From Sir John Franklin's well-known anxiety to act up to the tenor of his instructions, combined with the expressed desire of the Admiralty, that he should embrace every opportunity of forwarding accounts of the progress of the Expedition to England, I should have thought that he would certainly have left a record of his doings at a winter station, which he knew to be within reach of the whalers, before he commenced his voyage of the second season, in the hope of penetrating either to the southwest or northward, where he knew there would be little or no chance of finding a channel of communication, unless he succeeded in overcoming all obstacles, and pushing his way through that archipelago, which has hitherto proved a barrier to successive expeditions. And should he, as some suppose, but contrary as I think to all likelihood, have cut his way out of Beechey Harbor merely to turn his face to England, still I think he would have left some authentic record on the spot, mentioning his labors, and the cause of his return.

As there are no natives on the north side of Lancaster Strait to disturb any memorial or flag-post that may be erected, Sir John Franklin would probably not think it necessary to bury the copper cylinder or bottle containing his memorandum, but would rather suspend it in the most conspicuous way he could devise. Now, I have learnt, by experience, that the wolverene* will ascend trees to cut down a package hung to a branch; and that bears have similar habits was fully ascertained by Captain Austin's sledge parties. A dépôt formed by Lieutenant M'Clintock on Griffith Island was entirely eaten by bears, the tin cases proving to be but a poor defense against the tusks of these omnivorous animals, who expressed their approval of preserved potatoes by the way in which they cleared out the canisters. That they would relish the pemican, which was part of their spoil, might have been predicted. They did not respect even the sign-post, but overthrew it, and bit off the end of the metal cylinder containing the record.

The want of this memorial leaves us totally in the dark as to Franklin's intended course, which would in all probability have been decided upon before he left the harbor; for, from his position, he had the means of ascertaining the state of the ice both in Barrow's Strait and in Wellington Channel. If the former was open, his course would be to Cape Walker and the southwest, agreeable to his instructions; but if Barrow's Strait was closed, as he had found it to be the preceding year, and Wellington Channel open, then he would gladly follow the latter, which one at least of his intelligent officers considered to be the most promising route of all, and which

* The wolverene inhabits the islands north of Lancaster Strait, and its recent footmarks were often seen by Lieutenant M'Clintock.

the spirit of his instructions permitted him to take, if shut out from the west or southwest.

The well-planned and thoroughly organized traveling parties of the searching squadron, though they traced with extraordinary perseverance extensive portions of insular coast, failed in detecting any further decisive vestiges of Franklin's course. Captain Austin's two ships, with their tenders, wintered at the southwest end of Cornwallis Island, under the shelter of Griffith Island. From thence Lieutenant M'Clintock, who made the longest journey of all the pedestrian parties, setting out in spring, rounded the west end of Melville Island in longitude 114° W., and, passing over the extreme discoveries of Sir W. Edward Parry, saw distant land extending beyond the 116th meridian. The intermediate passages and bays were explored by Lieutenant Aldrich, Mr. Bradford, and Mr. M'Dougall. On the south side of Barrow's Strait, Cape Walker, and the adjoining coasts, were traced by Captain Ommaney and Lieutenants Osborne, Meecham, and Browne; Lieutenant Osborne having carried his researches nearly to the 72d parallel on the 104th meridian. This was the most southerly point attained. It lies within 180 miles of the south shore of Victoria Land, and is perhaps part of the same island. Throughout the whole of the great extent of coast-line closely examined by these officers, on both sides of the strait, no traces whatever of Sir John Franklin's ships were discovered, though Lieutenant M'Clintock found the wheels of a cart used by Sir W. E. Parry in 1820, and other traces of that officer's traveling parties. The signal-posts planted by the latter were thrown down by wind or animals.

Captains Penny and Stewart in the Lady Franklin and Sophia, wintered in Assistance Harbor, in company with Rear-Admiral Sir John Ross, of the Felix. The spring journeys of the two former, and of their officers, were directed to the examination of Wellington Sound. Captain Stewart and Dr. Sutherland explored the west and north sides of this inlet, their most northern points being in latitude 76° 24′ N. Messrs. Goodsir and Marshall traced its south and west sides to the 99th meridian; and both parties, from their most westerly stations, saw a navigable sea extending northward and westward, to the utmost limits of their vision. Wellington Strait, closed to the eastward and northward, opens into this westerly passage by three channels, separated from one another by Baillie Hamilton's and Deans Dundas Islands. Baring's Island lies more to the westward, opposite the middle channel. Its shores, and those of the two other principal islands, were examined by Captain Penny, who crossed over to the point of Sir Robert Inglis Bay on the northern shore, which has been named Albert Land; and from whence he had the melancholy prospect of boundless open water, which he had not the means of navigating. A boat was transported over the ice toward it with much labor; but, the provisions of the crew running short, it was abandoned. Mr. Goodsir found a spar of American spruce, untrimmed, with its bark worn off, and broken at both ends, twelve feet long, and as thick as a man's ankle, on the shore facing the open water; also many smaller pieces of the same kind of drift-wood, while none was picked up by Captain Stewart in Wellington Sound. From this fact these officers inferred, that the drift-wood had come from the westward. The currents or tides among the islands at the western outlet of Wellington Strait, were at times, according to Captain Penny's judgment, not less than four knots; and the general opinion of his officers was that the principal set of the stream came from the westward, and the prevailing winds from the northwest.

Animal life was abundant in the open water, and on its coasts. Walruses were seen repeatedly in the several channels, north and south of Baillie Hamilton's Island; and polar bears were numerous and bold, so as to be dangerous to parties not well armed. Several of the bears were killed, and one of them contained an entire seal in its stomach, the practice of these voracious animals being to swallow their prey without mastication when it is not too large to pass their gullets. The walrus can not exist except when it has access to open water; nor is the polar bear usually found at a distance from it, except in its passage from one sheet of water to another. The travelers also saw polar hares, wolves, foxes, herds of reindeer, vast flocks of king and eider ducks, brent geese, and many gulls and other water-fowl of less utility to man. Musk oxen were seen only on Melville Island, where Lieutenant M'Clintock killed four, and might have procured more had he wished to do so.

On the 5th of September, 1850, a floe of ice at least two years old, and upward thirty miles in width, filled the lower part of Wellington Strait, and remained fast, though diminished in breath, when last visited on the 24th of July, 1851. Captain Penny is of opinion that open water existed beyond it all the winter.

With respect to traces of Sir John Franklin's Expedition, beyond Cape Spencer none whatever were observed by Captain Penny's traveling parties, except a small piece of drift-wood, which had been recently charred, and had been exposed to little or no friction subsequent to the operation of fire.* This was found by Mr. Goodsir in Disappointment Bay, in latitude 75° 36′ N., longitude 96° W.; and I con-

* A piece of elm board that had been originally coated on one side with mineral pitch or tar, and after long exposure to the weather split by an ax, was too much weathered even on the most recent surface to come within the date of Sir John Franklin's Expedition. It was found on Baillie Hamilton's Island, and must have drifted a very long way.

sider it to be certainly a relic of Sir John Franklin's Expedition, as these coasts are not now visited by natives, and this piece of charred wood could not have been water-borne from any great distance. It must have traveled however, some short way subsequent to its having been exposed to the action of fire; for if it had been the remains of a fire kindled on the spot, other fragments of charcoal would have been found lying beside it. Franklin would, undoubtedly, during the spring passed in Beechey Bay, send out a party up Wellington Sound as he never never would let the opportunity escape of examining, as far as he was able to do, a route that might influence his future movements; and as the course to the westward within the reach of pedestrian parties was known, the resources of the two ships would be turned to the undiscovered way, commencing in their vicinity. That such exploring party went beyond the limits of Captain Penny's researches, I infer from neither post nor cairn marking the limit of its journey having been seen. If the same expanse of open water was visible, in 1846, from Baillie Hamilton's Island, which Captain Penny saw in 1851, we may readily conceive the efforts that would be made to carry the Erebus and Terror into it by any practicable extent of ice sawing, particularly if Barrow's Strait remained closed. The age of a floe of ice filling a strait does not indicate with certainty the length of time that the strait has been blocked up, for drift ice, loaded with the remains of several years' snow, may be carried into a narrow passage, so as to shut it up, and as suddenly removed again on a favorable concurrence of winds and tides. One navigator, therefore, may be able to sail, as Sir W. E. Parry did, nearly quite through that northern archipelago in one season, while his successors may find impassable barriers thrown across the path which he pursued, and new avenues opened. It would be unsafe, therefore, to argue that Wellington Strait is always closed, because it was choked by a floe of some age in 1850 and 1851.

By the efforts of the searching parties, which have just returned, combined with those of preceding years, all the accessible parts of the continental coast of America have been explored, and both sides of Barrow's Strait, to the further side of Melville Island, and the land beyond Cape Walker. Land has also been traced, though only by distant view, round the bottom of Jones's Sound. This has narrowed the lines of search to two distinct points—that is, to the southwest of Cape Walker, which, from its being the direction in which Sir John was instructed to go, seemed to be especially the one in which he was to be sought; and the newly-found channel opening out to the westward from Wellington Strait. It is greatly to be desired that this one may be pursued by new efforts.

Mr. Rae, in April last, was on the eve of setting out from Great Bear Lake, in the hope of crossing on the ice to Victoria Land,

and of continuing his search in a boat as soon as the navigation opened. Though he may not actually attain Lieutenant Osborn's furthest, he may, under favorable circumstances, approach so near to the scene of that officer's search, or of Lieutenant M'Clintock's, as to prove, should he find no traces of the ships, that the intervening space is too confined for the seclusion of living men. Captain M'Clure, who passed to the eastward of Point Barrow last season, if he found the sea as open as the more sanguine believe it to be, may have reached the west side of Parry's Archipelago, and have spent the winter not far from the supposed outlet of Victoria Channel; and this season Captain Collinson may be sailing eastward in the same direction. It is from Beering's Straits, then, that we are next to look for tidings of great interest to the civilized world, which sympathizes so universally with the efforts made to trace and relieve so many gallant victims to science.*

20th October, 1851.

* With reference to Sir John Ross's pigeons, mentioned in a note on page 329, it appears that he dispatched the youngest pair on the 6th or 7th of October, 1850, in a basket suspended to a balloon, during a W.N.W. gale. By the contrivance of a slow-match the birds were to be liberated at the end of twenty-four hours.

APPENDIX.

No. I.

PHYSICAL GEOGRAPHY.

CHAPTER I.

GENERAL VIEW.—Rocky Mountains.—Their Length.—Their Height.—Glaciers.—Parallelism to the Pacific Coast.—Continental Slopes.—Russian America.—Eastern Slope.—Prairies.—Mississippi Valley.—Its Slope.—Fundamental Rocks of the Basin.—Silurian Strata.—Carboniferous Series.—Tertiary Beds.—Lignite Formations.—Series of Lake Basins.—Transverse Valleys.—Intermediate Belt of Primitive, Hypogenous, or Metamorphic Rocks.—Its Rivers mere Chains ot Lakes.—Its Breadth.—Altitude.—Sources of three great River Systems.—Great Fish River.—The Yukon or Kwichpack.—Basins of Excavation.—Glacial Action.—Active Volcanoes.

VALLEY OF THE ST. LAWRENCE.—Altitudes of the Lakes above the Sea.—Lake Superior.—Lake Michigan.—Lake Huron.—Lake Erie.—Lake Ontario.—Lake Champlain.—Northern Brim of the St. Lawrence Basin.—Its Geological Structure.—North Shore of Lake Superior.—Structure of the Country at the Sources of the Mississippi.

WINIPEG OR SASKATCHEWAN VALLEY.—Height of Lake Winipeg.—Sea River.—Katchewan River.—Thousand Lakes.—Portages.—River Winipeg.—Red River.—Saskatchewan River.

MISSINIPI VALLEY.—Its Lakes.—Frog Portage.

MACKENZIE RIVER VALLEY.—Methy Portage.—Athabasca, Elk, or Red-deer River.—Lesser Slave Lake.—Peace River.—Slave River.—River of the Mountains.—Noh'hanne Bute.—Great Bear Lake.

YUKON VALLEY.—Yukon or Kwichpack.—Volcanic Chain of Alaska.—Coal.—Fossil Bones.

IN drawing up the following Appendix, my object has been to record facts in the light in which they appeared to me. When treating of districts which I did not visit, I have borrowed from every work concerning them to which I had access. The most important sources of information are generally named or expressly alluded to; but I thought that it would give too much formality to so slight a sketch were I to parade every authority for the statements it contains. Where practiced geologists have examined the country, their report has been chosen in preference to my own observations; and this is the case on the route of the expedition up to the 49th parallel. Beyond Lake Winipeg no geologist has yet penetrated, and the descriptions of the rocks occurring within the space of twenty degrees of latitude that lie to the north of that sheet of water are, with all their imperfections, entirely my own. It would be true economy in the Imperial Government, or in the Hudson's Bay Company, who are the virtual sovereigns of the vast territory which spreads northward from Lake Superior, to ascertain without delay the mineral treasures it contains. I have little doubt of many of the accessible districts abounding in metallic wealth of far greater value than all the returns which the fur trade can ever yield.

The Rocky Mountain chain, which is the northern prolongation of the Andes, has a general course of north 26° W. for 2700 geographical miles, from the 30th parallel of latitude up to the shores of the Arctic Sea. Its higher peaks rise from 12,000 to 15,000 feet above the ocean, and enter the region of perpetual snow; but the northern part of the chain, which touches on the Mackenzie, is so much lower, that even its summits* are denuded during the short summer of that district, and perennial patches of snow exist there only in shady crevices which have a northern aspect. I have not been able to discover, after many inquiries, that glaciers which flow through mountain gorges into the lower country are formed in any part of North America, though travelers who have crossed to California, Oregon, and New Caledonia, speak of hills clothed with perpetual snow; and the Copper River, which joins the sea opposite to the peninsula of Alaska, is said by Bäer, on the authority of Klimowskij, to issue from a solid mass of ice. Several passes which traverse the chain do not rise more than 6000 feet above the sea level, and being free from snow in summer may be crossed in that season by pack-horses and even by wagons. The more northern of these passes have long been known to the fur-traders; the southern ones have lately been explored and used by the multitudes who have hurried from the United States to California in search of gold.

Up to the 60th degree of latitude the chain runs nearly parallel to the coast of the Pacific, and not far distant from it; the descent to the level of the sea is consequently rapid on the west—a configuration which M. Guyot† has noted as peculiar to the New World, while in the Old Continent the short slopes are turned to the south, and the long ones toward the north. A large triangular corner, which belongs to the Empire of Russia, and extends westward to Beering's Straits, has a different physical character, in the existence of a transverse series of active volcanoes, as we shall

* Supposed to be at least 3000 feet high, in the 62d parallel.

† Phys. Geogr., p. 50.

hereafter have occasion to notice; at present my remarks will be confined to the continent lying eastward of the mountains.

The width of the chain is stated at from forty to one hundred miles, and the central parts and peaks are said to consist of granite and other igneous rocks.

The eastern slope toward the Atlantic commences by a belt, formed mostly of sandstone, 150 miles in width, which rests on the shoulder of the chain, with an inclination of about 37 feet in the mile, in its descent from 8000 feet above the sea, to 2500. The more gradual slope of the great prairies, beginning at the last-named elevation, has a breadth of 700 or 800 miles, and retains in its descent the prairie character of a treeless, sandy, and moderately undulated, or, as it is locally named, "rolling" plain.

Most of the streams which cross the prairie flow through deep furrows, sunk abruptly below the general level; nevertheless the Mississippi, Missouri, and some of their larger tributaries have wider valleys, skirted by successive terraces and alluvial deposits. On the banks of all the rivers there are belts of woodland, and clumps of trees that encroach on the prairie, intercepting grassy lawns, and producing remarkably fine park scenery, which is often enlivened by small lakes. In the interior of the prairie, however, water is scarce, and there is such a total want of wood, that for days together the traveler can find no other fuel than the dung of the bison. Near the mountains the soil is coarsely sandy, strewn with boulders, and sterile; further eastward the sand is finer, and the boulders disappear, but they recur in numbers on the lower border of the prairie: they are also scarce or wholly absent over very extensive tracts of the rich alluvial deposits of the valley of the Mississippi, south of the Ohio.

The Mississippi drains the entire space between the Rocky Mountains and the Alleghanies or Apalachian chain, embracing thirty degrees of longitude. The whole of this vast water-shed may be considered as one valley, whose bottom, indicated by the channel of the river, has a southerly course, inclining slightly to the eastward. The length of the river, from its source in Itasca Lake, at an altitude of 1490 feet above the sea, to Balize, on the Gulf of Mexico, has been estimated by Schoolcraft at 3160 statute miles, following its windings, and may be stated in round numbers at 2400 geographical miles in a straight line. The moderate hills and eminences of the country in which the river has its origin do not rise above the plane of the general eastern slope of the continent in the same parallel; and the general longitudinal descent of the great valley from Lake Itasca is at the rate of 10 inches in the mile for the upper half of the river, and for the other half of the way, or from St. Louis downward, of only two inches and a half per mile.

As St. Louis is 500 feet above the sea, and about 600 geographical miles distant transversely from the western summit of the prairie slope (which has been taken at 2500 feet above the sea), the lateral descent of the valley to the channel of the river in that parallel is 40 inches in the mile; but the sinuosities of the Missouri, a mightier stream than the one in which it loses its name, and of the other grand affluents of the Mississippi that drain the prairies, give a gentler inclination to their beds. The Illinois, Ohio, and the minor streams which come in from the other side of the valley, flow at a lower level than the western feeders; their upper branches being subordinate to the general slope, which has there descended considerably.*

From the Apalachian chain to the Rocky Mountains the fundamental rocks would appear to be silurian † overlaid in large tracts

* The following facts, ascertained at the Navy Yard of Memphis, in Tennessee, by R. A. Marr, Esq., are interesting points in the history of the Mississippi. The quantity of water passing through the channel of the river at that place in 1849 was sufficient to cover an area of 100,000 square miles to the depth of seven feet and a half, and the quantity of silt it carried down would make a bed of earth one mile square and seventy-six feet deep. The current in the central area of the river generally exceeded three miles an hour, and was less toward the sides.—*Proceed. Amer. Assoc.*, p. 340.

† As the silurian rocks are most extensively developed in North America, and are of constant recurrence throughout the route of the Expedition, I shall introduce here, for reference, a tabular enumeration in the order of their superposition, as ascertained by the United States geologists, beginning with the lowest.

A. Champlain Division; supposed to underlie three-fourths of the territory of the United States, and to occupy the surface of one half.

1. *Potsdam sandstone;* a quartzose rock, generally gray, often striped, sometimes partially or wholly red: is supposed to be the lowest sedimentary fossiliferous rock. The *Taconic rocks* of Dr. Emmons, consisting of lamellar white limestone, with specular iron ore, are supposed to be the earliest deposits of this period, modified by metamorphic agency. Footmarks of reptiles have been found by Mr. Hunt in this sandstone.
2. *Calciferous sandrock;* a deposit of calcareous and earthy matter variously mixed.
3. *Black river or Chazy and Birds-eye limestone;* a brittle limestone, having a smooth, flat-conchoidal fracture. This limestone, characterized by its peculiar fossils, is of frequent recurrence northward, up to the islands of the Arctic Sea.
4. *Trenton limestone;* a dark-colored limestone, interlaminated with dark shaly matter. Often metalliferous.
5. *Utica slate;* dark-colored carbonaceous slates, which readily disintegrate.
6. *Hudson River group, Loraine shales, Frankfort slate, and Rubble-*

by the old red sandstone and the carboniferous series. The sandstone, which rises on the flanks of the Rocky Mountains to the height of 8000 feet above the sea, is referred by some geologists

stone; mostly dark slates and shales, and gray thick-bedded grit stones.

B. Ontario Division, includes a series of limestones, shales, and sandstones, which pass insensibly into each other.

7. *Gray sandstone* of Oswego.
8. *Medina sandstone;* red or slightly variegated, and of every degree of coherence. Originates many brine springs.
9. *Oneida conglomerate;* a variable intermixture of sand and quartz pebbles.
10. *Clinton Group;* consisting of deposits of various characters, such as thin shaly sandstones, shales, conglomerates, thin-bedded impure limestones, with iron ores.
11. *Niagara group;* consisting of dark-bluish shale and dark limestone, and taking its name from the cataract, where a section of both its members is exposed.

C. Helderberg Division.

12. *Onondaga salt group*, is an immense development of argillaceous shales and marls with shaly limestones, veins and beds of gypsum, giving origin to copious and very rich salt springs. This formation re-appears near Slave River, on the 60th parallel of latitude, and also, I believe, on the shores of the Arctic Sea.
13. *Water-lime group*, consists generally of two layers of drab-colored water-limestone, with an intervening layer of blue lime-rock.
14. *Pentamerus limestone;* named from its characteristic fossil, and rarely a pure limestone, being more or less mixed with black shale.
15. *Delthyris or Catskill shaly limestone;* composed of beds of subcrystalline gray limestone, slaty limestone, and slaty argillaceo-siliceous limestone.
16. *Oriskany sandstone;* a whitish sandstone, composed of sand derived from granitic rocks or mica schist.
17. *Caudagalli grit;* named from the feathery forms in which it abounds. It is a drab-colored or brownish calcareous and argillaceous sandstone.
18. *Schoharie grit;* a fine-grained calcareous sandstone, from which the calcareous matter may be washed away by long exposure, leaving the siliceous skeleton of the rock.
19. *Onondaga limestone;* recognizable by its crystalline structure, toughness, and numerous organic remains.
20. *Corniferous limestone;* a fine-grained compact limestone, which is very durable, and produces cascades where the smaller streams traverse it. It is the uppermost of the important limestone beds of the New York or silurian system, being succeeded by shales in which the limestone fossils give place to others of a different character.

D. Erie Division, referred by most English geologists to the Devonian or to the carboniferous series.

21. *Marcellus shales;* black, slaty, bituminous shales, containing septaria, with occasional thin bands of limestone. Similar in lithological characters to Genesee slate.

of the United States to the triassic system; but its exact geological position is not yet satisfactorily determined. Cretaceous beds, known by their organic remains (but not containing white chalk with flints), occur extensively along the Missouri, and spread widely on both sides of the Mississippi, below the influx of the Ohio. They exist also near the Rocky Mountains, high above the sea, in the vicinity of the sources of the Arkansas, Platte, and Gila;*

22. *Hamilton group;* an immense deposit of dull-olive calcareous shales, which change to light ash-gray in weathering. It contains septaria.
23. *Tully limestone;* usually thick-bedded, blue, or nearly black, limestone, often divided by seams into irregular fragments.
24. *Genesee slate;* is a great mass of argillaceous, black, fissile slate, which rapidly exfoliates and falls down. Fluid bitumen is of common occurrence. Either this deposit or No. 21 exists on the north side of Methy Portage, on the Elk or Athabasca River.
25. *Portage group;* a vast deposit of shale, flag-stones, and thick-bedded sandstone, rising with a slope or abruptly from the shales on which it reposes.
26. *Chemung or Ithaca group;* a highly fossiliferous series of shales and thin-bedded sandstones.

To the last succeed the Old Red Sandstone and the Coal Measures.

The lower part of the Champlain division has been considered as the equivalent of the Cambrian system, the *Utica slate* being parallel to the *Landeilo flags;* the *Hudson River group* and Ontario division, up to the *Niagara group*, is thought to be equivalent to the *Caradoc sandstone;* the *Niagara group* and the whole Helderberg division is supposed to be co-equal with the *Wenlock rocks;* and the Erie division as equivalent to the *Upper* and *Lower Ludlow rocks*, including the Devonian system of Phillips. —*Vide Hall, Geol. Report of New York*, p. 518.

* "The cretaceous formations terminate in the Atlantic regions of New York, before they have reached the city of New York, so that their limit scarcely touches the 40th degree of latitude, or 16° lower than in Europe. In Kentucky and Tennessee it remains below 37°, but it is very different far up the Missouri; this great river flows uninterruptedly from the foot of the Rocky Mountains for 1400 miles through strata of chalk, at least as far as the Sioux River. This is the result of the researches of the Prince of Neuwied, and of the reports of the celebrated astronomer, Nicollet. In these western parts of America the chalk rises to 50° of latitude. There, also, it shows a continuous extension, greater than that of any other formation on the globe. Captain Fremont saw chalk strata and fields covered with *Inoceramus crispii*, on the River Platte; Lieutenant Abert found them on the Arkansas, and Dr. Wizlizenus also beyond the Rio del Norte, near Monterey and Laredo. The Rocky Mountains, and their continuation beyond Santa Fé, have entirely cut off this cretaceous sea. No trace of chalk was discovered either by Captain Frémont on the Columbia River or on the Humboldt, in that wonderfully great basin which dips to the Pacific; or yet by the observant Captains Cooke and Johnstone along the River Gila, in Sonora, or California." "The whole of this vastly extended chalk formation consists only of the upper beds. After very

while toward the mouth of the Mississippi there are very extensive recent tertiary and alluvial deposits, which, skirting the Gulf of Mexico, run into Texas on the one side, and along the Florida coast on the other.

Tertiary coal, containing dicotyledonous leaves, exists in the Raton Pass, between the sources of the Red River and the Arkansas, at an elevation of 4600 feet. Coal of the same description, associated with similar leaves, occurs on the Mackenzie in latitude 65° N.; and at various intermediate parts on the flanks of the Rocky Mountains beds of lignite are known to exist, which are probably also of the tertiary era.*

In the whole width of the Mississippi basin, from the falls of St. Anthony downward, no primitive (or hypogenous) rocks appear, except in the low ridges of the Ozark Mountains, which have a hilly prominence, owing to the excavation of the valley; but their summits scarcely rise above the plane of the general slope, supposing that it were extended with an even descent to the Atlantic. They range from Red River to St. Louis, parallel to the Alleghanies, and consequently make an angle with the Rocky Mountain chain. Coal measures crop out on their flanks.

Of the forty degrees of latitude which intervene between the Gulf of Mexico and the Arctic Sea, the valley of the Mississippi occupies, as we have seen, about one half; and the whole drainage of that portion west of the Alleghanies is accumulated in one great channel, which is directed southward and a little eastward. From the head of Lake Superior northward, there is a series of great transverse excavations, occurring in succession, on to Great Bear Lake, which lies under the Arctic circle; and it is remarkable that nearly all the lake basins † of these valleys commence in the silurian strata, and are continued into or entirely across a belt of primitive, or hypogenous, and metamorphic rocks,‡ which extends from Lake

careful and accurate investigation, Sir Charles Lyell decided, that in the whole of North America, chalk strata, from the Maestricht beds down to the gault, alone occurred; and Mr. Ferdinand Römer, as the result of his highly valuable and accurate researches in Texas, goes the length of considering all the strata in that region, already so far removed from the Atlantic coast, as entirely of the upper division, and not even touching on the gault. This peculiarity is, however, singularly enough, limited to North America alone. Even in Mexico deeper chalk beds occur; and Darwin saw cretaceous shells in abundance 2000 feet above the sea, near Port Famine, in 53° south latitude."—*Silliman's Journal*, Sept. 1850.

* Vide Journ. p. 194.

† This grand series of lakes forms the line of canoe navigation from Canada northward; and the fact of its position in the fracture between limestone and granite, was perceived and recorded by Sir Alexander Mackenzie.

‡ My imperfect acquaintance with the science of geology renders me in-

Superior to the shores and islands of the Arctic Sea. The western limit or strike of this formation, which I have traced for more than 1400 geographical miles, has a general course of north 30° W., its rhumb line, consequently, inclines slightly toward the axis of the Rocky Mountains, and the intercepted space grows narrower toward the north.

On the other hand, the Apalachian chain, running parallel to the Atlantic coast for a thousand miles up to its termination in the Shickshock and Notre Dame Mountains of the promontory of Gaspé, and having a direction of north 46° E., diverges from the Rocky Mountain axis at an angle of 72°. These are the three chief pyrogenous systems of North America, the Ozark Hills being of lesser account, and coinciding, as has been said above, in direction with the Alleghanies.

The middle belt of primitive rocks may, both from its position and diagonal direction, be distinguished as the *intermediate primitive or hypogenous* formation. Its altitude nowhere entitles it to the appellation of a mountain chain. Its hypogenous rocks, which are chiefly granite and gneiss associated with trap, scarcely rise above the mean eastern slope, and do not present acute peaks or continuous elevated ridges. They exhibit generally rounded or dome-shaped summits, or form oblong eminences, which are separated from one another by narrow inclined valleys. Most of these valleys, and the larger ones without exception, are occupied by lakes, which are often deep; and the proportion of water in the district is very great, probably considerably exceeding that of the dry land.

The rivers that traverse the intermediate primitive belt (and for ten degrees of latitude all the rivers that originate high on the prairie slope do so) form, on entering it, lake-like expansions, which are studded with rocky islets, and send long winding arms into all the neighboring valleys. These dilatations have little or no current, but they are connected with each other by one or more straits, in which the stream is turbulent and rapid, and the overfall frequently great enough to produce a cascade. The tortuous arms of such expansions often wind for many miles through the country; and the Indian, by making short portages from one string of lakes to another, may travel with his canoe in every direction, as far as the formation extends. Sometimes a river forks in this rocky dis-

competent fully to appreciate the worth of the several systems that profess to explain the mode in which the beds forming the crust of the earth have been formed; neither have I exclusively adopted any of the current opinions: I would therefore be understood to use the terms "primitive," "hypogenous," and "metamorphic," as designations of the rocks so called by geologists, and not as exponents of theories.

trict, and its branches running far apart, just as they would in an alluvial delta, unite again, intercepting a considerable tract of country of the prevailing character, that is, having a predominance of water surface. Examples of this peculiarity occur in the River Winipeg, which conveys the waters of the Lake of the Woods to Lake Winipeg; also in the discharge of the latter sheet of water by Nelson River and Play-Green Lake; as well as in some of the expansions of the Missinipi or Churchill River. Instances also of lakes having more than one outlet are not rare in this formation; and now and then, though comparatively very seldom, these outlets lead to distinct river systems.

On the east side of Lake Winipeg the width of the primitive belt is about two hundred geographical miles; and from the summit of an eminence which rises only a few hundred feet above the general level, but yet is sufficiently conspicuous to have obtained the distinctive appellation of "the Hill," thirty-six lakes may be counted. In many localities, where the knolls are denuded of soil, the surfaces of the rocks are evenly ground down, and are sometimes smooth, polished, and striated.

By a rough measurement, the centre of this formation on the 53d parallel of latitude is found to be between 700 and 800 geographical miles from the Rocky Mountains; the Great Canadian lake district is of equal width; and Labrador on to Newfoundland and the eastern shores of Nova Scotia occupy a similar space in the map. Now assuming, as we have done, and as the observations of the topographical surveyors of the United States entitle us to do, that the height at which the gentle eastern slope of the continent commences is 2500 feet, and supposing the descent to be equable, we should have an altitude above the sea in the country from whence the sources of the Mississippi proper, the St. Lawrence, and the Red River of Lake Winipeg issue, of 1800 feet. The actual elevation of that district is between 1400 and 1500 feet,* and the only marked hilly eminence in the district, which is named the *hauteur des terres*, and is said to consist of drift-sand and boulders, does not appear, from the descriptions we have of it, to rise more than 300 feet beyond the general level. The summits, therefore, of this tract of land, distinguished though it be by

* Schoolcraft estimates the height of Itasca Lake, from which the Mississippi issues, at 1490 feet above the Gulf of Mexico. Major Long, who ascended the St. Peter's, a head branch of the Mississippi, reckons the altitude of the short portage, which separates its sources from those of Red River, at 1400 feet; and my own barometrical observations and estimates place the summit of the water route between Lakes Superior and Winipeg, traversed by the Expedition, at 1460 feet above the tide level of the St. Lawrence.

shedding its waters into three separate river systems and as many different seas, are also subordinate to the general eastern slope of the continent.

Before naming more particularly the transverse basins which cross the intermediate belt of primitive rocks, I may state that the Mackenzie, inferior indeed to the Mississippi, but yet a river of the first class, running in an opposite direction, drains seventeen degrees of latitude into the Arctic Sea, taking its course through a valley which differs in its character from that of the Mississippi, as the details of progress of the expedition through it have already shown. In this place it will be sufficient to recall to mind that from Methy Portage (*Portage la Loche*) to the sea, a distance of 1400 geographical miles, the fall is about 900 feet, the successive portions of the river being designated the Washacummow, Elk or Athabasca, Slave, and Mackenzie rivers.

Two other rivers of magnitude cross the Arctic circle, viz., Back's Great Fish River, which, originating near Great Slave Lake at an altitude of 150 feet above its surface, runs east-northeast into the Arctic Sea, draining the northeastern corner of the continent; and the Yukon, which, rising to the westward of the Rocky Mountains, not far from the union of the Francis and Lewis, which form the Pelly, flows first to the north, and after receiving a large tributary named the Porcupine, to the westward, falling into Beering's Sea, where it is known to the Russians by the name of the Kwichpack.

A glance at the map will show, that on the eastern side of the continent the water basins generally maintain the northeasterly inclination of the Alleghanies, while further to the westward the basins of the two great rivers assume a parallelism to the Rocky Mountains; and that the influence of the intermediate hypogenous formation has been of a different character, the rivers winding their way across it, sometimes with a southerly, sometimes with a northerly inclination, seemingly indicating the obstruction offered by the harder rocks to the agent by which the river channels were excavated. On emerging from the belt, the lower parts of the rivers generally incline toward the northeast, with a considerable degree of parallelism to each other before they fall into Hudson's Bay.*

* A study of the map will show, that the lake basins north of the St. Lawrence have generally their long axes across the river courses to which they respectively belong, and that many assume a greater or less degree of parallelism to the intermediate primitive belt. Perhaps movements of elevation or depression had occasioned an extensive disruption of strata along the western border of the hypogene rocks, previous to the removal of the silurian beds on the excavation of the lake valleys.

The peculiar configuration of the continent which I have endeavored to sketch must be duly considered by any one who endeavors to detect the agency by which the river valleys and lake basins were excavated. It is not, however, my purpose to enter upon the discussion of this question, or to speak of the partial and often repeated elevations and depressions by which the lacustrine and fluviatile terraces have been accounted for; nor would a summary of this kind admit of the necessary elucidations. I shall merely say that, adopting the opinion of the United States' geologists, that they are basins of excavation, I consider them all to be of the same epoch, and that the currents or waves of translation, if such they were, must have had an easterly direction in the middle latitudes, and gained strength as they rolled toward the Atlantic, when they swept away wholly or partially the fossiliferous deposits that once covered the primitive rocks of Hudson's Bay, Canada, and the eastern parts of the United States; the former extent of the newer rocks being indicated by the patches which remain. By a singular coincidence of a political with a natural limit, the northern boundary of the United States, or the 49th parallel, marks the line on the great prairie slope, where the current took a southerly direction, to excavate the wide and magnificent valley of the Mississippi. A similar diversion of the excavating force northwards would produce the basin of the Mackenzie, commencing on the 53d parallel, but in a district narrowed and disturbed by the approximation of the intermediate primitive rocks to the Rocky Mountains.

Supposing the continent to have retained its present form since the era of these excavations, it seems scarcely possible to reconcile the existence of extensive glacial action with any modification of climate; yet the smoothened surfaces, streaks, and furrows referred to that action, whether in the form of glaciers or of drift ice, are of no rare occurrence, wherever durable rocks show themselves, between the St. Lawrence and the Arctic Sea.

In connection with the excavations of the North American continent, the fact may be mentioned of the great indentations of the coast line, including Hudson's Bay, the Gulfs of St. Lawrence and Mexico, and the Caribbean Sea being on the east side; while breaks of the west shore, to the south of the peninsula of Alaska, are comparatively small, and both coasts of South America are nearly entire.

The geologists of the New York Survey consider that the present continent of North America was constructed from the debris of land lying more to the eastward. Mr. Hall, speaking of the strata exposed between the Hudson and the Mississippi, states that they contain organisms which must have lived in the bed of the ocean, and that the chief source of the sedimentary deposits lay to the east and southeast. To the westward, the sedimentary rocks are of a finer grain, and at the same time diminish in quantity, while the carbonate of lime increases, indicating, in conjunction with the contained fossils, the bottom of an ocean of greater depth and more quiet condition. The cretaceous and tertiary deposits of the western prairies show, according to the geologist just quoted, that the eastern part of the continent was first elevated, and that the older rocks in the west were subsequently overlaid by the new deposits. However that may be, the occurrence of the chalk fossils and tertiary deposits in their present position and altitude clearly indicates that the elevation of the Rocky Mountain chain was one of the latest of the great movements that have occurred in this continent. It is to the westward of these mountains only, along the Pacific coast, and in the peninsula of Alaska,* and the Aleutian chain of islands, that recent volcanic action can be traced.

The existence of coal measures, containing ferns and other plants of a tropical character, in Jameson's Land and Melville Island, in the high latitudes of 71° and 75°, is a curious fact, to be accounted for by those who theorize on the ancient condition of the surface of the earth; and the vast accumulations formed at a later epoch in the Siberian Sea and Kotzebue's Sound, of fossil bones of mammoths, rhinoceri, and other animals, which do not exist in Arctic regions at the present day, and the preservation to this date of some of their undecomposed carcases, are equally interesting facts, which need explanation.

VALLEY OF THE ST. LAWRENCE.

The first in order of the great transverse excavations, and the grandest, is the basin of the St. Lawrence, which has a length equal to the whole course of the Mackenzie, and contains by far the greatest accumulation of fresh water in the world. It has no connection with the drainage of the prairie slopes in the same parallels, which is performed by the Missouri and its numerous affluents on one side, and on the other by the Saskatchewan and its tributaries, aided by the upper feeders of the Mackenzie. It differs from the three great lacustrine basins which succeed it, in its head lying within or to the eastward of the termination or elbow of the "intermediate primitive rocks." The position of this head, or as it is well named from its elevation above the other members of the basin, Superior Lake, is midway between the Gulf of Mexico and the Arctic Sea, and its water-surface is 641 feet above the level

* Spelt thus, and also *Alashka* in Cook's Third Voyage. The French and Russian authors write the word *Aliaschka*.

of the tide.* The other great lakes descend successively in the following order of their heights above the sea: Lakes Huron and Michigan, 600 feet; Erie, 565 feet; Ontario, 492 feet; and Lake Champlain, 93 feet.

At the west end of Lake Superior, and on its northern shores, several promontories, having an altitude of from 800 to 1000 feet above the water, give a mountainous character to the coast when seen from the surface of the lake, but which it is far from possessing when viewed in relation to the country lying behind it. It is such as would result from the excavation of the basin by the removal of the softer rocks which have covered the granites, porphyries, and traps of these eminences. The silurian beds, not having been so extensively broken up to the westward of the *Fon du lac*, envelop the pyrogene nucleus so as, in conjunction with recent arenaceous deposits and drift, to cover it on that flank almost to the summit. From this locality, which is rather a plateau than a mountainous district, issue the feeding streams of the three several river systems—of the Mississippi, St. Lawrence, and Saskatchewan—as has been mentioned in a preceding page.

If we trace the south side of the St. Lawrence basin from this quarter, we find that already on the upper lake it assumes a different aspect, being composed chiefly of sandstone, and having less elevation than the north bank. Lake Michigan runs far to the southward among the silurian strata, entering as it were into the valley of the Mississippi. So small is the elevation of the ground between the Fox River which falls into Green Bay on the west side of that lake, and the Wisconsin, a tributary of the Upper Mississippi, that in times of flood a barge may float readily from one stream into the other.† A very moderate elevation in like manner separates the south end of Lake Michigan from the Illinois, another affluent of the Mississippi, so that a depression of 600 feet would produce a communication between the waters of the Gulf of the St. Lawrence and those of the Gulf of Mexico, through the Illinois valley.‡ The basin of Lake Huron is excavated in the

* Some discrepancy exists between the heights assigned to this lake by different authors. We have taken that deduced by Captain Lefroy from barometrical measurements made in connection with the observatory at Toronto. Dr. Houghton, the Michigan geological surveyor, estimates its height at 641 feet; but he makes the descent from it to Lake Huron 45 feet, while Professor Henry reckons this descent at only 18 feet, which must be under-estimated. Mr. Logan, in 1847, sets down the height of Lake Superior as 597 feet, having adopted for the height of Lake Michigan 578 feet, from Professor Henry. In the height of Erie and the inferior lakes authors are generally agreed.

† Featherstonhaugh.

‡ A depression of thrice that amount would carry the ocean to the west-

silurian strata, and the great promontory which divides it from Lake Michigan is said to be a deposit of old red sandstone inclosing the extensive coal measures of Saginaw. The lake shores are bold, but not mountainous.

Lake Erie has lower shores, and is the shallowest of the series. Its bed and much of its northern margin is formed of the corniferous limestone, one of the upper members of the silurian rocks. Mr. Hall remarks that, had the eroding agency removed this bed, and penetrated to the soft strata of the Onondaga salt-group, Erie would probably have been the deepest of the lakes.* The southern brim of its basin is so low, that an easy canal communication has been opened to the Ohio, a tributary of the Mississippi; and other water connections might be made with facility. Lake Ontario occupies a hollow in the silurian rocks inferior to those of the Erie basin; † and the country which lies to the south of it has in general a level character, though the Clinton and Niagara groups of rocks rise in places in high escarpments.

For a full account of the heights of the southern border, the whole valley of the great lakes and the St. Lawrence, and for detailed descriptions of the rock formations, the reader may consult the several geological surveys published under the authority of the legislatures of New York, Michigan, and Ohio: the preceding brief notices have been extracted therefrom for the purpose of showing the general character of the country, and the lowness of the barrier which separates the valleys of the Mississippi and St. Lawrence. I may add, as a further exemplification of the passes on the south side of the St. Lawrence, that a subsidence of 400 feet would cause the waters of Lake Ontario to flow through the valleys of the Mohawk and Hudson into the Atlantic, and at the same time convert Lake Champlain into a maritime strait, thereby forming islands of the State of New York, the New England States, and of the British colonies of New Brunswick and Nova Scotia.

Before we proceed to trace the northern bank of the valley of the St. Lawrence, it will be convenient to notice more particularly Lake Champlain, since it is there that we have an approximation of the Apalachian chain to the primitive rocks which form the

ern border of the prairie islands, leaving, as insular ranges, the summits of the Alleghanies and their continuation in Vermont, New Hampshire, and Gaspé, with a few peaks in the hilly region of New York, which lies between Lakes Ontario and Champlain; while the primitive masses on the north of Lake Superior would be mere reefs, over the highest of which an agitated sea would dash its spray.

* Hall, p. 408.

† In the Medina sandstone, gray sandstone, the Hudson River group of shales, and, toward its eastern extremity, in the Trenton limestone.—*Hall.*

northern brim of the St. Lawrence basin, and unite with the intermediate belt on the north shore of Lake Superior.

Dr. Emmons* estimates the length of the valley of Lake Champlain at 180 miles, and its average width at about twenty. Its bed is most depressed between West Point, Burlington, and Port Kent, where its soundings reach 600 feet, or 500 below the surface of the ocean.† It is in fact a deep chasm with a very abrupt slope on the western side, and a more gradual one on the eastern bank. The direction of the lake is north and south; it opens into the St. Lawrence basin on the north, and the valley of the Mohawk crosses its axis at some distance on the south. The summit level of the canal which connects it to the Hudson is only 147 feet above the tide; and a depression to that amount would cause the waters of the ocean to flow through it from New York Sound into the Gulf of St. Lawrence.

The New York highlands, bounded by Lake Champlain, the St. Lawrence, and Lake Ontario, are formed of primitive granite, hypersthene, and limestone rocks, which constitute many striking and picturesque groups of conical peaks. Among these Mount Marcy rises 5467 feet above the tide; Mount M'Intyre, 5183; Mount Seward, 5100; Mounts Martin and Santanoni, each 5000; and Whiteface, Taylor's Mountain, and Nipple-top are a few hundred feet lower. These highlands extend into Canada, where they form a mountain belt twenty-five or thirty miles wide, along the sources of the Chaudiere and St. Francis.‡ They are separated from the neighboring districts of the State of New York by the river and lake valleys named above; and, being impassable for an army, Lake Champlain was the only route by which the Atlantic States could be assailed from Canada East, or *vice versâ*; hence the

* New York Geol. Survey.

† The lake lies, as has been stated in a preceding page, 93 feet above the tidal waters of the St. Lawrence: hence the fall of the St. John or Richelieu River, which discharges its waters just above the tidal level, may be estimated at 10 inches in the mile; and the descent of the St. Lawrence from Lake Ontario, excluding the comparatively currentless expansions of the Thousand Islands' Lake and Lake St. Francis, is nearly in the same ratio. These facts may aid in the calculation of the inclination of the beds of similar rivers. With the same view, I may add that the fall of the St. Lawrence between Kingston, at the outlet of Lake Ontario, and Montreal is 220 feet, in a distance of 160 geographical miles in a straight line, of which a considerable part is lake-way, which gives an average of 16 inches in the mile.

‡ Orford and Sutton Mountains are each reckoned at more than 4000 feet high. The latter is the summit of a wide hilly tract, composed of chloritic and micaceous schists and gneiss.—*Hunt*.

celebrity of its defiles in the annals of colonial and revolutionary warfare.

The valley of the lake is excavated in the lowest group of the silurian rocks. The Potsdam sandstone, together with a greenish-white marble, and a trap-rock, are quarried extensively at Whitehall, situated at the upper extremity of the lake, for the double purpose of clearing sites for houses, and procuring building stones. In various parts of the lake shores, the silurian rocks are covered with beds of clay and sand, in which there have been detected about twenty species of marine shells, which exist in a living state on the coast of the Atlantic at the present time. The general character of the scenery, and especially of the upper half of the lake, is bold, hilly, and picturesque, often rocky, but occasionally cliffs of clay and sand 100 feet high border on the water. The shores are low and shelving only in the bays, which are formed by short lateral valleys.

The State of Vermont lies along the east side of the Champlain Valley; and the country, as it recedes from the lake, rises gradually into the acclivities of the Green Mountains, which are a continuation of the Alleghanies, and are prolonged into the Shick-Shock and Nôtre Dame Mountains of the promontory of Gaspé on the Gulf of St. Lawrence.*

* Mr. Hunt, of the Canada geological survey, says that "the whole of the Green Mountain Rocks belong to the Hudson River group, with the possible addition of a part of the Shawagunk conglomerates. The fossiliferous rocks of the St. Francis valley are referrible to the Niagara limestones of the upper silurian beds: a similar formation exists at Gaspé, and has been traced 150 miles southwest (in the direction of the Green Mountains); and from the similarity of Nôtre Dame (Gaspé) to the Green Mountains, and the fact that the Hudson River rocks flank the St. Lawrence to the Cape Rosiere (Gaspé), we may conclude that the upper silurian rocks will be found to be nearly continuous throughout. Resting upon this formation, in Gaspé, is a body of arenaceous rocks, 7000 feet thick, which apparently correspond to the Chemung and Portage group of New York, with the old red sandstones. As this formation is found extending quite to the Mississippi, it is probable that it will accompany the silurian rocks through New England, surrounding the coal-fields of New Brunswick, and of eastern Massachusetts, and Rhode Island. To this may be referred, in part, the rocks of the White Mountains, which may sweep around the western border of Massachusetts' anthracite formation, until lost under the super-carboniferous rocks of the Connecticut River. The limestones of western New England seem to be no other than the metamorphic Trenton limestones of Phillipsburg; while the chlorito-epidotic rocks and serpentines of Sutton valley appear again in the rocks of southern Connecticut, between these limestones and the new red sandstone. With such a key to the structure of the metamorphic rocks of New England, and of the great Apalachian chain of which these form a part, we may regard the difficulties

The predominating rocks on this slope are an alternation of argillaceous slates, with slaty and fine grained sandstones, and shaley grits belonging to the Hudson River group. In breaking down, the shales produce a cold, clayey, retentive soil, much less favorable for agricultural purposes than the "birds'-eye" and other limestones which crop out nearer the shores of the lake.*

With respect to the northern slope of the St. Lawrence valley, a reference to the map will show that the brim of the basin, where its feeding streams have their source, is generally about 150 geographical miles from the centre of the lake-way from Ontario upward, and considerably further off from the river channel lower down. The bays of the great lakes, in many places, curtail the breadth of the slope; and it is every where exceeded in breadth by the northern slope toward Lake Winipeg, James's Bay, and Hudson's Straits. It will also be perceived that the valley, with reference to its northern bank, makes an acute bend of which the west end of Lake Erie forms the elbow; the lower or eastern arm being parallel to the Alleghany range, while the upper one takes more the direction of the axis of the Rocky Mountains. The angular form of the basin is in conformity with the course of the primitive rocks from Labrador to Lake Superior, where they blend with the "intermediate belt."† The general level of the ridge formed by these rocks does not exceed the height of 700 feet above the river or lake surface; but, partaking of the general eastern slope of the continent, it attains a height of about 1400 feet above the sea on the north side of Lake Superior.‡ The course of the St. Law-

that have long environed the subject as in a great degree removed, and the bold conjectures as to their metamorphic origin, which have been from time to time put forth, fully vindicated."—*Hunt, Proceed. Am. Assoc. at Cambr.* 1849, p. 333.

* During our voyage through Lake Champlain, I was informed by a fellow-passenger that the agriculture of Vermont was very superior, especially near Burlington, where there are many large orchards, and sheep farming is extensively pursued. Cleared land, he told me, sells currently at forty dollars (£8) an acre, and fifty dollars are thought to be a high price. Two hundred acres form a good-sized farm in the opinion of the neighborhood, and eighteen hundred acres a very large property. Many steamers are constantly employed in summer in visiting the various bays of the lake, and carrying the produce to Whitehall, whence it is transferred by canal to the Hudson.

† The whole belt from the Labrador coast, along the valley of the St Lawrence, and northward to the Arctic Sea, seems to be a segment of the border of a great basin of which Hudson's Bay is the centre, and fragments of its eastern brim may be found on the shores of Hudson's Straits, and in the islands to the north.

‡ The following ascertained points may be noticed: Lake Temiscam-

rence through Canada East is conformable to the general strike of the beds in which the channel is excavated.

The geological structure of the north side of the St. Lawrence basin, as ascertained by the Canadian state survey under Mr. Logan, has been summed up as follows by his assistant, Mr. Hunt: "A formation of syenitic gneiss, often passing into mica-schist, and interstratified with crystalline limestone, forms a ridge of high land extending from the coast of Labrador along the north side of the St. Lawrence, at a short distance of from twelve to twenty miles from the shore, until it crosses the Ottawa, near Bytown, and thence is traced across Lake Simcoe to the shores of Lake Huron, where its northern limit is observed near the mouth of the French River, while it again appears at the southeastern extremity of the lake in Matchedash Bay. Resting upon this is a series of rocks forming the whole north coast of the lake, and numerous small islands. It is made up of sandstones, often coarse-grained, and sometimes becoming conglomerate from the presence of red jasper pebbles. These beds are associated with slates and one or more bands of limestone. . . . The formation is much cut by greenstone dikes, and exhibits very frequently interstratified beds of greenstone, often of great thickness. Both these and the sedimentary rocks contain metalliferous quartz veins, of which the copper mines of this region are examples. Resting unconformably on the tilted edges of this formation, and in other places directly upon the southern limit of the syenitic gneiss, appear the silurian rocks, identical with those which are found in New York, and covering the peninsula between Lake Huron and Lake Ontario. Beginning with the *Potsdam sandstone*, we have, upon the Manatoulin Islands and the coast between Matchedash Bay and Sarnia, a complete exposure of those formations known as the *Trenton limestone*, *Utica slates*, *Loraine shales*, *Medina sandstones*, and the *Niagara limestones*, with the rocks of the *Clinton group*. All these are well characterized by their respective fossils, and are spread out quite undisturbed at a very gentle dip of about thirty-five feet in a mile.

"Passing to the east, we find that the syenitic rocks have divided where they cross the Ottawa, and, taking a southward course, are spread over a considerable extent of country between the Ottawa and the St. Lawrence. Crossing this river below Kingston, they

ing, which is high up on the Ottawa, and near the line dividing the watershed of that river from the Abitibbe and Moose River, which falls into Hudson's Bay, was ascertained by Mr. Logan to be 612 feet above the tide. The dividing portage between the Ottawa and Lake Nipissing is 696 feet above the sea, the lake itself being, according to Mr. Murray, 647 feet high. Lake Simcoe was ascertained, by the same observer, to have an altitude of 704 feet above the tide.—*Logan's Geol. Rep. for* 1848.

constitute the greater part of the Thousand Isles,* and are extensively developed in the northern counties of New York."

Mr. Logan gives a more particular account of the north shore of Lake Huron, of which I have made the following abstract: "The north shore of Lake Huron, on which twenty-two mining locations have been claimed of Government, presents an undulating country rising into hills which sometimes attain the height of from 400 to 700 feet above the lake. These occasionally exhibit rugged escarpments and naked rocky surfaces; but, in general, their summits are rather rounded, and their flanks, with the valleys separating one range from another, are most frequently well clothed with hard and soft wood, often of large growth, and of such species as are valuable in commerce, in many places giving promise of a good arable soil. Many of the slopes are gentle, and many of the valleys wide. The Thessalon, of the reported length of 200 miles, and the Spanish River, of 120 miles, flow through this country, with three other rivers of from 50 to 60 miles' length each." With respect to the rocks occupying this country, he says, that "for 120 miles from the upper end of the lake one great formation, having a breadth in some places of ten and in others exceeding twenty miles, exists. It is composed of sandstones, conglomerates, slates, and limestones; the sandstones often vitreous, and presenting the character of a perfect quartz rock. These are associated with greenstone trap and other igneous rocks, in the form of overflows, dykes, and veins, and with amygdaloidal trap in layers. The whole reposes on granite, which is the metalliferous rock of the district, and lies beneath the lowest known fossiliferous beds. The Potsdam sandstone, Trenton limestone, Utica slates, and Loraine shales are exposed in successive deposits, resting on the tilted beds of the quartz rock in a nearly horizontal, unconformable position in the Grand Manatoulin, La Cloche, Snake, Thessalon, Sulphur, and other islands, and at the east end of the lake. Medina sandstone and Niagara limestones exist in certain localities of the promontory of "Cabot's Head."

Of the mining places above alluded to by Mr. Logan, the Bruce mines, situated on the main shore behind the island of St. Joseph, are the principal; and, in 1849, when I was returning to Canada in the month of September, and had an opportunity of visiting

* In the Lake of the Thousand Isles, as the funnel-shaped outlet of Lake Ontario is denominated, many of the round-backed hummocks of granite which form the innumerable islets exhibit the parallel furrows, streaks, and smooth surfaces attributed by some geologists to glacial action. This expansion, in fact, has exactly the aspect of many of the dilatations of the northern rivers which flow through the "intermediate primitive district."

them, 160 people were there employed, forming, with their families, a considerable village population. Mr. Logan calculated that 250 tons of dressed ore might be raised monthly, yielding, at an average, about 15 per cent. of copper. At present, fuel for mining purposes is obtained from the Pennsylvania coal-field, through the port of Cleveland, on Lake Erie; but ere long, the coal district in Saginaw Bay, on the south side of Lake Huron, will become the smelting place for both the Huron and Lake Superior minerals.

To complete the outline of the north bank of the St. Lawrence valley, we may state that the primitive rocks continued from Lake Huron to the outlet of Lake Superior, form there the bold promontory of Big Cape (*Gros Cap*), which is a mass of flesh-colored granite and porphyry rising 700 feet above the water. Point Iroquois, on the opposite side, three miles distant, is 600 feet high, has a more table-shaped summit, and from its base a line of low sandy beach stretches away on the south side, sending out a projecting tongue named "White-Fish Point." The north shore, as seen from Big Cape, presents a grand and varied coast line, deeply indented by Goulais and Batchewaung Bays, with the promontory of Mamainse, composed of rugged and crumbling amygdaloid, dipping into the wide expanse of waters on the northwest. The granitic ridge which skirts the bottoms of these bays comes out in bold cliffs on both sides of Michipicoten Bay, at the Otter's Head, and at the bottoms of Nipigon and Thunder Bays, projecting also in various capes between them. From the last named bay it stretches across the northern bend of the Kamenistikwoya River, by which stream the canoe navigation to the interior is carried on.

Mr. Logan considers the granite, which frequently passes into gneiss, as the base of the series of rocks composing this bank of the lake. 2. To this succeeds gneiss, and both are traversed by dykes and veins of granite. 3. The next in order are dark green talcose slates, and a pebbly and slaty conglomerate. 4. Resting unconformably on these, is a series of bluish shales, interstratified with trap. 5. Lastly, white or spotted sandstones, indurated marls, and conglomerates interstratified with trap. Trap dykes in vast numbers traverse all the beds down to the granite. Veins containing copper, lead, zinc, and silver, belonging to two systems—one coincident with the rock masses, the other parallel to them—occur in very many places on the north shore. The courses of the veins vary in different bays, and Professor Agassiz has shown that the outline of the lake has a close connection with the directions of the trap dykes, of which he describes six different systems, each of them associated with one of the great curvatures or bays.

A granite porphyry, which is very durable, forms a considerable portion of the boldest and most barren parts of the north shore of the lake; the projections of amygdaloid, being more perishable, assume the most picturesque shapes; and some of the loftiest headlands are thickly capped with greenstone, basalt, and other trappean rocks. The low, flat, well-wooded islands are mostly sandstone. The general elevation of the northern brim of the Lake Superior basin may be stated at between 800 and 900 feet above the water surface, and the distance of its crest from the margin of the lake at from 20 to 50 miles.

Many mining stations have been granted to adventurers by the Canadian legislature, but workings are carried on in three only; viz., one in Pigeon Bay, one among the islands of Nipigon Bay, and the third in Mica Bay. At the last named 100 miners were employed in 1849, when the establishment was broken up by a foray of Chippeways, who thought that their territorial rights were invaded.* But little ore has as yet been shipped from the Canadian side of the lake. The export of native copper from the United States shore is considerable. In 1849, 1114 tons of pure copper were transhipped at Saut Ste. Mary.†

Our limits will not allow us to dwell longer on the St. Lawrence basin; ‡ but with respect to the general character of the ridge which divides it from the Winipeg excavation the aspect of the country traversed in pursuing the canoe route may be considered as a type of the whole. The surface of that tract is hilly, the granite rising in rounded and sometimes in rugged knolls abruptly from lakes or swamps, but only to small heights above the general level. Here and there, but particularly toward the summit of the ridge, there are considerable deposits of sand, gravel, and loam, with many boulders. The term ridge is used with reference to

* Narrative, p. 53.

† The Detroit Free Press states that, in 1850, the shipments exceeded 4000 tons, and it is calculated that they will equal the whole consumption of the United States, which is 6000 tons, in 1851. There are twenty-two copper companies in operation on the Michigan shores, employing 800 men. The masses of native copper on Kewawoonan Point are enormous; but, from their very purity, they can neither be blasted nor hewn out in the ordinary way. The method adopted at present is to use long iron chisels, which are turned slightly by the hand at each blow of the hammer. In this way large slabs weighing two tons or more are cut out, to which not above five per cent. of quartz rock and other impurities adhere. On this point there are said to be parallel ridges of trap rising through beds of sandstone. Among these the native copper lies in walls or veins which have two directions, one running across the trap ridges in N.W. 6 N. $\frac{1}{2}$N. direction, the other parallel to the ridges and strike of the sandstone.

‡ Professor Agassiz has devoted a large volume to the natural history of Lake Superior alone, which is full of interesting facts and comprehensive general views.

its being a height separating two depressions; but its summit is a marshy plateau of some extent, across which narrow winding lakes afford a canoe navigation in a variety of directions.*

This summit of the water-shed, which, level as it is with respect to its water communications, is rendered very uneven by the protrusion of numerous granitic masses to various but moderate heights, lies much nearer to Lake Superior than to Lake Winipeg; and in descending toward the north, the same rocks appear in succession which have been noticed as forming the bank of the St. Lawrence valley, the "bird's-eye limestone" of the Champlain division of the New York system being the newest deposit on the Winipeg Lake.

If we trace the water-shed to the southwest, beyond the head of Lake Superior, by the sources of the St. Louis and head waters of the Mississippi, the uneven marshy surface gradually merges in the sandy prairie lands of the Red River, the Saskatchewan, and Missouri, where the naked rocks disappear, or are to be found only in the deep river channels.

The manner in which an elbow of the "intermediate primitive rocks," which form the nucleus of the water-shed here spoken of, incloses the head of Lake Superior, and extends along part of its south coast until lost under newer deposits, may be learned from the subjoined extract of a "Report by Dr. Owen, geologist to the States of Iowa and Wisconsin." †

* Vide map of the district in Franklin's Second Overland Journey.

† "The protozoic strata form sections on the Mississippi for an average distance, in a direct line, into the interior of Wisconsin, of from fifty to seventy-five miles, or up to the low falls of the principal eastern tributaries of that river, where the crystalline rocks first appear. In this part of the country the igneous ranges do not rise into elevated mountains, but, on the contrary, they are seldom seen except in the immediate cuts of the streams, being, for the most part, covered with drift. The character of the country generally, toward the summit levels leading to Lake Superior, is a succession of terraces of moderate elevation, chiefly composed of drift, having a nucleus, no doubt, of granite, syenite, or hornblende rocks; but these protrude only occasionally. At intervals the streams are ruffled into rapids, being filled with boulders which materially obstruct their navigation. A portion of these boulders may have been transported from great distances, but the greater part appear not to be far removed from their parent rock. It is a matter of surprise that so large an area of the interior of this district, and indeed of the sources of the Mississippi generally, should be level tamarack and cedar swamps, since, on approaching a great water-shed that gives rise to one of the greatest rivers in the world, one is led to anticipate a country with physical features of quite a different character. Interposed between the crystalline and igneous rocks of the interior of the district and the lowest sandstones, some green and red schistose beds have been observed at different localities. These appear to have been derived from the decomposition and detritus of the more easily

VALLEY OF THE WINIPEG OR SASKATCHEWAN.

The next great transverse excavation includes the whole valley of the Saskatchewan and Nelson Rivers; and from the sources of the former near Mount Hooker, one of the Rocky Mountain peaks said to be 15,700 feet high, to the mouth of the latter in Hudson's Bay, the axis of the valley runs in a direct line about east-north-east for 660 geographical miles. Lake Winipeg, the principal lake basin in connection with the river, lies at right angles to that axis, is nearly parallel to the Rocky Mountain chain, and forms one of a series of great lakes succeeding one another in a north-northwest direction. Their names are Lake Superior, Lake Winipeg, Deer Lake, Wollaston Lake, Athabasca Lake, Great Slave Lake, Marten Lake, and Great Bear Lake; the northern coast line being moreover indented on the same bearing by Liverpool and Franklin Bays.

Lake Winipeg itself is 230 geographical miles long and about 40 wide, but its width would be increased to 120 miles, if Moose Lake, Muddy Lake, Winepegoos and Manitoba Lakes, which differ very slightly from it in level, and are evidently component parts of the same lake basin, were included in the measurement. Into this great lake the Saskatchewan, by its two diverging branches, gathers a wide extent of prairie drainage, its northern tributaries being conterminous with the affluents of the Elk or Athabasca River, and its southern ones with those of the Missouri. The Assinaboyn also traverses much prairie land, one of its branches originating on the banks of the great southerly bend of the Missouri; and the Red River, which the Assinaboyn joins, rises, as has been already mentioned, on the same level and in the close neighborhood of the sources of the Mississippi and St. Lawrence. The waters of the lake wash, on the east side of the basin, the "intermediate primitive rocks," and find their way through them by anastomosing

disintegrating felspathic granites. The lower beds of sandstone adjacent to the igneous outburst are not unfrequently changed to hard quartzite." . . . "The highest ranges of the Wisconsin side of Lake Superior, situated from ten to sixteen miles from its south shore, are estimated to be near 1000 feet above the lake, and are formed of hornblende rocks, metamorphic slates, syenite, and trap. No organic remains have been detected in the great mass of sandstone bounding this part of the lake whereby its geological era may be determined. On the west side of the Mississippi, north of the Winebago reserve (Minesota), and as far north as St. Peter's river, limestone, with underlying sandstone, prevails to the extent of half a degree of longitude."

channels named Sea River, Katchewan, and, in its lower part, Nelson River. After it has passed through the intermediate belt, it takes its way over silurian limestones, and finally enters Hudson's Bay through alluvial deposits of some extent.

The surface of the lake has been calculated to be 853 feet* above the sea, and its basin is excavated in the silurian beds. Along the whole eastern shore the granite, gneiss, and trap rocks are every where exposed, the first named rock being the most extensive; and nowhere do these masses rise to the altitude of hills. On the north and west the birds' eye limestone is the prevailing rock, and forms low cliffs, in a country otherwise every where flat; and toward the south end of the lake, and in the narrows, arenaceous deposits appear in the immediate vicinity of granite, trap-rocks, and chlorite slates, having a close resemblance to those of Pigeon Bay of Lake Superior, where argentiferous veins occur. It is, therefore, an interesting quarter for exploration by the practical miner.

In ascending from Lake Superior by the Kamenistikwoya, and its upper branch named the Dog River, to Thousand Lakes (*Milles Lacs*), forty portages are made, in which the whole or part of the cargo is carried, there being besides some long rapids where it is not necessary to unload; and in descending to Lake Winipeg the portages are about fifty, their number and even the route varying with the height of water. Thousand Lakes † (Journ. p. 62) is

* Captain Lefroy's observations:—"In the Geological Appendix to Sir John Franklin's Second Journey, I estimated the height of Lake Winipeg above Hudson's bay at 800 feet, which I considered to be a rough approximation. Major Long places it at the same level with Lake Superior, and the dividing ridge between the two basins at 600 feet higher; but, by the best estimates I have been able to form, he makes the summit of the ridge 230 feet too low."

† The height of this lake above Lake Superior was ascertained approximately as follows:

	Feet.
Foot of Mountain Portage estimated by Captain Lefroy..........	48
Upper end of ditto (bar. meas., Rich.)........................	127
Ascent of river to Dog Portage (estimated, Rich.)...............	150
Upper end of Dog Portage above the lower (bar. meas., Rich.).....	332
Ascent of Dog River to Thousand Lakes (estimate).............	160
	817
Lake Superior above the sea..................................	641
Thousand Lakes above the sea..................................	1458

Owing to the obstructions in the upper part of the Dog River, the canoe route diverges from that stream at Cold-water Lake, and passes by the Prairie, Middle, and Savannah Portages, and a small stream which flows an extremely irregular piece of water, having many extensive arms, some of which are very shallow. Multitudes of islands well covered with birch, aspen, arbor-vitæ (*Thuya occidentalis*), and the various pine trees of the region, render the scenery pleasing. A few granite knolls and mural precipices show among the trees; but many of the islets appear to be formed of sand, of which sections twenty feet high occur. The lake not only branches into the neighborhood, but water communications diverge from it in every direction, as is customary in the "middle primitive belt."

From the Thousand Lakes the canoe route keeps on the border of the primitive rocks, touching on silurian deposits, when it bends to the southward or westward. At first it is flanked on both sides by granite. In Rainy Lake there is much mica-slate; and at its

from the latter into Thousand Lakes. The heights of this part of the route are as follows:

	Feet.
Dog Lake, above Lake Superior	657
Ascent of Dog River	14
Portage to Cold-water Lake	2
West end of Prairie Portage and Middle Portage	161
	834
Lake Superior above the sea	641
Height of Prairie or of Middle Portage above the sea	1475

The difference of height between the ends of the Mountain Portage was carefully ascertained by me with Delcro's barometer in 1849. Major Delafield estimates the difference at 125 feet, and Lieutenants Scott and Denny, who accompanied Major Long in his voyage up the St Peter's, at 130 feet. Mr. Murray, of the Canadian Geological Survey, measured the actual height of the falls here (named *Kakkabikka*) from the smooth water at their summit to the base, and found it to be 119 feet; but at least 8 feet must be added to give the difference between the ends of the portage, which terminates some way above the brow of the cascade. Captain Lefroy makes the height, by barometrical measurement, only 115 feet.

In 1849 the height of the upper end of Dog Portage was ascertained by me with Delcro's barometer; in the previous season the aneroid barometer gave 328 feet as the height, which was a greater degree of accordance between the instruments than I generally found. Major Long estimates the water-shed between Lake Winipeg and Superior at 1200 feet above the tide; Major Delafield calculates the height of Cold-water Lake at 505 feet. to which, if 161 be added for the Prairie Portage, and 641 for Lake Superior, we have 1307 feet for the height of Prairie Portage over the sea. Captain Lefroy, by barometrical measurements made in connection with the observatory at Toronto, makes the west end of Prairie Portage 1361 feet above the sea; but the distance between the two places of observation renders the result liable to some error.

outlet the stream, falling over gneiss rocks, produces the cascade of *La Chaudière*. The greater fertility of the country about Nemican Lake and Rainy River show the vicinity of newer formations. In the Lake of the Woods Dr. Bigsby found the *Pentamerus Knightii*, a fossil characteristic of the upper silurian rocks; but granite is the chief constituent of the islands and shores of the lake. The River Winipeg flows wholly within the granite district, and has the lake-like dilatations and other characteristics of the streams which traverse the "intermediate primitive rocks." The wide extent of land which its reuniting arms inclose is remarkable. One of its affluents, named English River, issues from Lake Sal, which lies near the water-shed, dividing the Winipeg basin from that of James's Bay. The Berens River which falls into Lake Winipeg, issuing from the common brim of the same basins further north, in the vicinity of the sources of the Severn River, affords a canoe route to Hudson's Bay. From my own observations in 1819, and Mr. Barnston's at a later period, it has been ascertained that limestones of the silurian epoch occur on the northern flank of the "intermediate primitive belt," as well as in the basin of Lake Winipeg. The Red River, which has been already repeatedly referred to, and which to prevent confusion with its southern namesake it would be well to call Osnaboyna, lies wholly to the westward of the "intermediate belt," and has a direct course from Lake Travers of 300 geographical miles. The sandy ridge named *Côteau des Prairies*, or *Hauteur des Terres*, separates its upper part from the Missouri valley; but the metamorphic rocks which present themselves around the sources of the Mississippi and its tributary the St. Peter, are also visible near Lake Travers. In the lower part of Red River limestone crops out in one place only, and is quarried by the settlers. Elsewhere the rocks are concealed by the sandy deposit forming the soil of the prairies, along whose eastern border the river flows. Major Long enumerates thirty affluents of the Red River, and its western branch, the Assinaboyn.

The Saskatchewan, which is to be considered as the main feeder of the Winipeg basin, flows from a considerable distance above Pine Island Lake down to Lake Winipeg over horizontal beds of limestone, through so flat a country that the river forks as it would in an alluvial delta. A rich mud is deposited in parts, particularly between Pine Island Lake, and the main stream, and round Moose and Muddy Lakes. In Beaver Lake, which lies immediately to the north of Pine Island lake, the silurian strata are again seen covering the flanks of the primitive rocks; while to the southward an eminence named Basquiau,* lying at the distance of nearly a de-

* See Journal, p. 50.

gree, separates the river valley from the Red Deer Lake and Swan River. As powerful salt springs exist on this eminence, we may conjecture that it belongs to the Onondaga salt group.

I have been more particular in the topographical and geological remarks on the St. Lawrence and Saskatchewan basins, because economy of space made it expedient to omit the details of the voyage of the Expedition through them. The remaining districts will be more briefly mentioned here, since the narrative included many facts relating to them.

VALLEY OF THE MISSINIPI.

The next river basin that we have to notice is that of the Churchill, English River, or Missinipi,* the latter, or Cree appellation, being nearly synonymous with Mississippi. This basin, in crossing the intermediate primitive rocks, lies nearly parallel to the Nelson River or Katchewan, which is the lower part of the Saskatchewan. Deer and Wollaston Lakes, which discharge their waters into the Churchill River, lie in the line of lake basins mentioned above as running northward from Lake Superior; but their axes do not cross the river valley so nearly at a right angle as Lake Winipeg does the valley with which it is connected. Further up the Missinipi the Methy, Buffalo, Clear, and Isle à la Crosse Lakes, which are situated just to the westward of the primitive rocks, taken in the aggregate, lie more in the plane of Lake Winipeg. On the eastern flank of the intermediate primitive ridge lie Big Indian and Weskayow-washgow Lakes, belonging also to the Missinipi river system.

The Missinipi, by its principal feeder, the Beaver River, has its source lower down the eastern slope than the Saskatchewan, and drains a comparatively small extent of prairie lands. At the Frog Portage and elsewhere the two basins are divided from each other by rocks only a few feet high, over which, in times of flood, the waters pour; so that the two may be viewed as one great valley through which two large rivers flow, their trunks running parallel to each other.

VALLEY OF THE MACKENZIE.

Further to the north lies the great valley of the Mackenzie, extending to the Arctic Sea, but having also its subordinate transverse lake basins, which differ from the southern ones in their heads merely entering the western border of the intermediate primitive locks, and their discharging streams taking an opposite direction through the newer deposits.

* *Misi* or *mitchi*, in Cree, signifies "much or great," and *nipi* "water," while *sipi* means "river."

Methy Portage forms the dividing brim between the Missinipi basin and the Mackenzie River valley, at the place where it is crossed in the usual canoe route; and, though the country be wooded, it may be considered as a partial extension of the prairie slope. The strata are bituminous shale* resting on silurian limestone, and covered with a thick arenaceous deposit. This is deeply furrowed by the channels of the Elk River, and its tributary the Washacummow; but the lake basins which mark out the border of the intermediate primitive rocks must be sought for further to the eastward. It is probable that this border touches a straight line drawn from Knee Lake, across the outlet of Athabasca Lake, to the deep northern arm of Great Slave Lake, and onward by Marten Lake, across the two eastern arms of Great Bear Lake, to the Copper Mountains. That portion of the line which lies between Athabasca Lake and Methy Portage is little known, because the water route lies to the westward of it. Wollaston and Deer Lakes already mentioned, will, if this line be correctly drawn, be situated considerably within the border of the primitive rocks; and an illustration of the manner in which the waters occupying the minor valleys of that district communicate with each other is afforded by Wollaston Lake sending a stream from its north end into Athabasca Lake, and one from its south end into the different river system of the Missinipi. This fact, which, as reported on the authority of the fur-traders, is expressed in Arrowsmith's map, may be considered as proving that Wollaston Lake is considerably elevated. Indeed, it seems to be high up on the water-shed which separates the Mackenzie valley from the basin of Hudson's Bay.

According to Captain Lefroy's measurements and estimates, Methy Lake is about 1540 feet above the sea; and I ascertained the summit of Methy Portage road to be 188 feet higher than the lake; while the Washacummow, or Clear-water River, on the north side of the portage, is 590 feet below it, and by my calculations 910 feet above the sea. From Methy Portage westward, the country, though deeply furrowed by river courses and ravines, and more or less thickly wooded, partakes so much of a prairie character that horsemen may travel over it to Lesser Slave Lake and the Saskatchewan. At the junction of the Washacummow with the Elk, the channel of the river is sunk 925 feet below the summit of the portage, which may be assumed as near the mean level of the district, and is 1688 feet above the sea.†

The Athabasca, Red Deer (*La Biche*), or Elk River, the most

* It is doubtful whether this shale be referable to the Marcellus shale or not.

† See Narrative, p. 75. Captain Lefroy's estimate of the height of Methy Lake exceeds mine by 40 feet.

southerly feeder of the Mackenzie, originates in the Rocky Mountains near Mount Browne, which is said to rise 15,000 feet above the sea, and, flowing through prairie lands, receives the waters of Lesser Slave Lake, whose axis crosses its general course, and afterward those of Red Deer Lake. Its bed is in many places deeply cut beneath the level of the prairie plateau, which is not separated by any marked ridge from the Saskatchewan prairie country. Further north, the wide-spreading sources of the Peace River drain the Rocky Mountain chain for four degrees of latitude, and the trunk formed by their union curves across the slope to join the Athabasca a little below the transverse basin of the Lake of the Hills. The Channel of the Peace River is cut into silurian or Devonian strata; but the Lake of the Hills, or Athabasca Lake, runs eastward among the "intermediate primitive rocks," and, as has been said above, receives a tributary stream from Wollaston Lake, situated near the water-shed which divides the basin from Hudson's Bay. The conjoined stream of the Peace and Athabasca Rivers assumes the name of Slave River, which flows in the fracture between the silurian and primitive rocks. The junction of the western arm of the river with Great Slave Lake marks the western border of the primitive rocks, which is also indicated on the northern side of the basin by the Fort Providence Inlet. Aylmer and Artillery Lakes lie 150 feet higher at the east end of the lake, into which they send their waters; and in their immediate vicinity, on the same plateau, are the sources of the Great Fish or Back's River, which falls into the Arctic Sea, as has been already stated. The streams that run from the westward into Chesterfield Inlet must come from near the same place; and this inlet, from its transverse direction, and east-northeast bearing from Great Slave Lake, has an evident connection with that excavation, their axes being nearly at right angles to the Rocky Mountain chain.

In Great Slave Lake the Mackenzie is deflected from the intermediate primitive belt, and flows first westward then northward, in a channel scooped out of the upper silurian strata and still newer deposits, for 1000 statute miles of river course, or nearly 600 geographical miles in a direct line; neither granite, gneiss, nor mica slate are seen on its banks, and even trap rocks are rare, if any actually occur.

The River of the Mountains, embracing by its feeders a more northern part of the Rocky Mountain chain, after disengaging itself from the rugged hills from which it draws its supplies, makes a northerly bend nearly parallel to Slave River, and then joins the Mackenzie at Fort Simpson. Of this mountain stream I have already given a slight notice in the Narrative (see pp. 105, 106); and I may add here that for twenty-five miles upward from its mouth it flows through sand and shale, with limestone occasionally cropping out. At the end of that distance there is a rapid, above which low wooded points exist, with at intervals mountain bluffs coming down to the banks of the river. The most remarkable of these stands at the influx of the Noh'hannè River, and is named the *Noh'hannè Bute.* It is the highest hill in that quarter, and is about seventy-five miles from Fort Simpson. Perhaps it is a member of a range whose prolongation is seen indistinctly in descending the Mackenzie about half-way between Hare Skin River and Fort Simpson. Messieurs M'Pherson and Bell ascended it, and the latter gentleman was seized with nausea and vertigo before reaching its summit; so that its altitude is probably considerable, but the snow disappears from it in summer. *On its top* there is a salt spring, having a basin fifteen feet in diameter, which is never dry. For this notice I am indebted to Mr. M'Pherson, who brought from it some fragments of limestone that were similar in lithological character to those procured at the Rock by the River's Side described in page 113.

Great Bear Lake, the most northerly of the transverse freshwater lakes, lies about 150 feet above the channel of the Mackenzie, and crosses the Arctic circle on the line where the hypogenous and silurian rocks meet. Coronation Gulf is also excavated on the same line, which has a general parallelism to the Great Slave Lake series of excavations enumerated above.

The northerly trending of the coast line west of the Mackenzie is evidently due to the prolongation of the Rocky Mountains, whose successive ridges or spurs come out *en échelon*, diminishing in altitude as they approach the shore.

The short western slope of the continent from the Rocky Mountains to the Pacific differs from the eastern one in its configuration, its river valleys being all more or less transverse. The peculiar wing-like projection in the north, toward Asia, is evidently due to the volcanic chain of Alaska, which runs at right angles to the Rocky Mountains. The great transverse river valley of the Yukon or Kwichpack lies to the north of it. The western sea-coast from Cook's Inlet to Beering's Straits is delineated in a map attached to Bäer's "Nachrichten über den Nordwestküste von Amerika," published in 1839, though it seems to have been hitherto neglected by our topographers: and the interior of Russian America has remained a blank in our maps.*

* The newest work on this district of America is, "Beitrage zur Kenntniss der orographischen und geognostischen Beschaffenheit der Nordwestküste Amerikas, von Dr. C. Grewingk, St. Petersburg, 1850; but this I have not been able to procure.

For the following information respecting the Hudson's Bay Company's land to the west of the Mackenzie, and for some account of Russian America gathered from the natives, I am indebted to Chief Factor M'Pherson, Chief Trader Bell, and to the letters of Mr. Alex. H. Murray, alluded to in a preceding page. By ascending the River of the Mountains, and tracing its northwest branch to Lake Frances, a very elevated mountainous country is reached. In this the Lewis, which flows toward the Pacific, takes its rise, its sources springing to the eastward of those of the River of the Mountains; so that here, as well as in other parts of the Rocky Mountain chain, the rivers falling into opposite seas interlock at their origin. In lat. 61° 30′ N., long. 130° W., the Lewis is joined by the River Francis, which has no connection with the lake of a similar name, but comes from Russian Lake, a sheet of water lying more to the south. At the junction of the Francis and Lewis, Mr. Roderick Campbell has built a trading post, named "Pelly Banks," which, according to his experiments with the boiling water thermometer, is elevated 1400 feet above the ocean. The united streams take the name of the Pelly, which falls into the Pacific, probably into Tchilikat or Lynn Canal, but the exact situation of its mouth has not been ascertained. Native traders come from the head of Tchilikat to Pelly Banks in a fortnight.

From the same elevated district in which the Lewis or Pelly and the northwest branch of the River of the Mountains take their source, the Yukon, a river of great magnitude, issues; and for a considerable part of its course flows to the north, through a country which, as far as I can judge from the descriptive notices of it which I have collected, closely resembles the valley of the Mackenzie. Mr. Murray was disposed at one time to identify this river with the Colville, which falls into the Arctic Sea about 120 miles to the east of Point Barrow; but more recent and full native information leads him now to conclude that it flows toward Norton Sound; and one of the officers of the Enterprise, in a private letter which has been published, states that the Russians have established its identity with the Kwichpack, which falls into Beering's Sea between Cape Stephens and Cape Romanzoff.*

* The following is an extract of a letter to me from Mr. Murray, dated May, 1850, on the Yukon :—"My account of the course of this river, also a sort of chart I made of it from the description given by the Indians, might, perhaps, lead you to have a wrong impression respecting the mouth of the river. I am now convinced that it is not the same with the Colville, and I have for some years suspected that its mouth lay to the west. The Russians have come up the lower part of the river regularly for some seasons: I was at first informed that they entered it from another river, but I am now told positively by Indians who went down

In 1847, Mr. Bell, having heard of the Yukon from the Kutchin who visited the fort on Peel's River, set out in quest of it, accompanied by a native guide. He first crossed the mountains to a stream termed the Rat River, on which an outpost named La Pierre's House has since been built. This post is about sixty miles distant from Bell's Fort on the Peel, and is about ten miles to the southward of it. Shortly after embarking in a canoe on the Rat River, Mr. Bell came to one of much larger size to which it is tributary, and which is named the Porcupine. Three days' descent* of this carried him into the Yukon, which it enters at right angles in the 66th parallel, and in the supposed longitude of 147¼° west.† At this place the Yukon is 1¼ mile wide, and is full of well-wooded islands, with a very strong current in the channels which separate them. After issuing from the mountainous district north of Pelly Banks, the course of the Yukon is, according to the Indians, to the westward of north; and in one place where it cuts a spur of the Big Beaver Mountains, it passes between high limestone cliffs resembling the "ramparts" on the Mackenzie. Its current is every where more rapid than that of the river just named. The first important tributary which it receives is the Red Island River, that flows in from the mountains on its eastern bank, and is divided at its source from the head of the Peel by a single ridge of land. Between it and the Lewis there is a barren plateau, which the Indians cross in four days, but on which they find no water.‡ Another tributary from the east comes in lower down, and below that Deep River enters from the west. The Russians appeared on this stream with a skin boat (baidar), coming overland, it was sup-

and met them last summer that they come into it direct from the sea. By one of these Indians I received a letter from the Russians which, being in their own language, is unintelligible to me. Salmon and hooked-nosed trout (*Salmo scouleri*) ascend the river, but are not found in the Mackenzie, or rivers falling into the Arctic Sea. Again, I have made frequent inquiries of the 'Gens du large,' or the northern Indians, who visit the Arctic Sea coast, and find that they are unacquainted with the mouth of the river. For two winter days' walking below the Porcupine, the Yukon trends to the west and southwest, and the natives say that it flows on in the same direction. I am therefore inclined to believe that the Colville is a smaller river, and that the Yukon empties its waters into Norton Sound."

* In returning from the Yukon against the stream, by tracking the canoe, nine days were required to reach Rat River.

† By Mr. Murray's courses and distances.

‡ There is a district in Siberia on which neither snow nor rain is said to fall. This may be a similar one; for it can scarcely be, in so rigorous a climate, that melting snow, if it exists, should not leave pools of water all the summer.

posed, from the Copper River or Atna, which joins the Pacific in Comptroller's Bay. From below the "ramparts," which are reckoned to be about seventy miles above the influx of the Porcupine, the Yukon flows to the northwest through a flat country; but twenty or thirty miles below the mouth of that tributary it again cuts a spur of the Beaver Mountains, trending at the same time to the southwestward and westward, and finally issuing in Beering's Sea, under the Eskimo appellation of Kwichpack, as has been mentioned above. Below the Big Beaver Mountains it receives from the south the "River of the Mountain Men," which runs parallel to the main stream for some distance before entering it. From thence to the sea the Kwichpack flows, according to the natives, through a low, marshy, and very sparingly-wooded country.

Mr. Murray, in his letter to me quoted above, mentions his intention of exploring the Yukon as far as he had leisure to do so in the summer of 1850; and it was probably the report of his party having been seen which induced Captain Collinson to land Lieutenant Barnard and Mr. Adams at Fort Michaelowsky, that they might ascertain who the white men were.

On the Pacific coast only of North America does volcanic action exist at the present time. Many peaks throwing out fire and smoke are mentioned by Bäer. They lie in a line running west-southwest from the north side of Kanai or Cook's Inlet, through the peninsula of Alaska and the Aleutian Islands. Many of their summits rise into the region of perpetual snow. Among these Ilämän has a height of 13,151 feet, and Winkelhöhe of 11,270, both lying on the north bank of Cook's Inlet. In Alaska there is the Peak of St. Paul, and in the islands Schischaldin, which is 8953 feet high, besides Schimaldin, Makuschkin, Korowenskische, and many others. The line in which these volcanic peaks lie, when prolonged to the eastward, strikes the Big Beaver Mountains on the Yukon.

Mounts Edgecumbe, Fairweather, and St. Elias are, I believe, extinct volcanoes, which form, with those of Alaska, nearly the segment of a circle.

On the side of the Atlantic, modern volcanic rocks occur in Jan Mayen's Island only, whose principal mountain, Beerenberg, rises 6870 feet above the sea. In the coal of Jameson's Land, lying in north latitude 71° on the east side of Greenland, and in that of Melville Island, in latitude 75° north, Professor Jameson found plants resembling fossils of the coal fields of Britain. This fact is sufficient of itself to raise a world of conjecture respecting the condition of the earth when these ancient fossils were living plants. If the great coal measures, containing similar fossil vegetable forms, were deposited at the same epoch in distant localities, there must have existed, when that deposition took place, a similarity of condition of the North American continent from latitude 75° down to 40°.*

In concluding this sketch I shall advert to the distribution of the fossil remains of mammalia, and especially to those of the mammoth. Teeth of this animal have been discovered on the banks of several rivers in Russian America, north of Mount St. Elias; and there is a celebrated locality in Kotzebue Sound where the thawing and waste of frozen cliffs is continually exposing the bones and tusks of mammoths and other quadrupeds. Dr. Buckland, in his interesting account of the specimens collected at this place, on Captain Beechey's voyage, enumerates fragments of bones of mammoths, and of the urus, the leg-bone of a large deer, and a cervical vertebra of some unknown animal, different from any that now inhabit Arctic America. Along with these there were found also the skull of a musk ox and some bones of the reindeer in a more recent condition than the others. In Asia, from the Don to Beering's Straits, on the banks of all the great rivers, bones and teeth are still more plentiful, the local deposits being richer and more extensive the more we advance to the north. They are especially abundant on the islands north of Sviatoi Noss (Sacred Cape), lying between the 73d and 76th parallels; and some of these islands seem to consist in a great part of organic remains, which fill more space than the matrix in which they lie. The soil containing them is frozen as hard as a rock; and, as it thaws annually, the bones drop out or are quarried by the natives. From time immemorial this has been proceeding, and the ivory hunters have been obtaining their annual supply without any sensible diminution of the store they draw from. Some tusks found in New Siberia weigh 480 lbs., and from that island alone the merchants of Yakutzk, in 1821, obtained 20,000 lbs. weight of this fossil ivory. Though the tusks of the mammoth (*Elephas primigenus*) are chiefly sought for, bones and teeth of other animals are found, and among the rest those of the *Rhinoceros tichorinus*, of which the skulls, flesh, and skin with its hair, have been procured. The remarkable discovery, in 1799, by a Tungusian, at the mouth of the Lena, of the entire carcass of a mammoth, in excellent preservation, is well known. The existence of these numerous testimonials of an ancient fauna is suggestive of many curious speculations, and geologists seem hitherto to have failed in explaining the circumstances under which accumulations so vast could occur in such high latitudes. The difficulty is increased when we consider that these bones have not been detected to the east of the Rocky Mountains in the northern

* The coal of Vancouver's Island and Oregon belongs, it is said, to the great coal measures.

latitudes. None have hitherto been found in Rupert's Land, though the annual waste of the banks of the large rivers and the frequent land slips would have revealed them to the natives or fur traders had they existed even in small numbers. They are rare also, or altogether wanting, in Canada, but in the valley of the Mississippi the bone-licks are well known as most extensive, and furnishing the remains of a different series of extinct quadrupeds.

The existence of the skull of a musk ox in Kotzebue Sound is of much interest. This relic is preserved in the British Museum, and the naturalists who have compared it with recent skulls of the animal brought from Melville Island or Churchill, perceive no difference. But, as the species is not known to frequent any district to the westward of the Mackenzie, the transport of any of its bones in modern times to Kotzebue Sound can not be readily accounted for.

In a preceding page I have alluded, in a general way, to the distribution of boulders on the eastern prairie slope and in the valley of the Mississippi. The journal of the voyage shows that these are every where present in the north. The surface soil, the beds of rivers, and sea-shore abound in them. I noticed them also in various places accumulated in clusters, forming small eminences of from 10 to 100 yards in diameter, and from 8 to 20 feet high. These may be ice-borne boulders. The usually circular form of the heaps militates against their being glacier moraines. Such collections are of frequent occurrence on the borders of Great Bear Lake, and in the valleys that separate the spurs of the Rocky Mountains—several hundred feet above the present levels of the lakes and rivers. They also occur in more southern localities.

No. II.

CLIMATOLOGY.

Snow line—Ground Ice—Thermometrical Observations in the Valley of the St. Lawrence—Comparative Temperature of the two sides of the Continent—Phenomena of the Seasons at Penetanguishene: at Fort William: at Fort Vancouver.—Thermometrical Observations in the Valley of the Saskatchewan—On the east and west sides of the Continent in that Parallel—Phenomena of the Seasons at Cumberland House: at Carlton House: at Marten's Falls on Albany River—Thermometrical Observations on the Missinipi, and in the same parallels on the east and west sides of the Continent—Thermometrical Observations in the Valleys of the Mackenzie, Yukon, and Pelly—Progress of the Seasons at Fort Franklin—Thermometrical Observations on the Arctic Seas—General Remarks—Nocturnal Radiation.

THE remarks which follow on the climate of North America have reference especially to the districts through which the Expedition traveled, some general facts being, however, illustrated by observations made in other quarters.

The course of the snow line has always engaged the attention of travelers, who have ascended high mountain ranges, or penetrated into Polar regions. Nowhere on the route of the Expedition is the snow permanent; not even on the summits of that part of the Rocky Mountains which skirts the Mackenzie. Snow may indeed be occasionally found in summer which has drifted into some narrow ravine, or under a high cliff with a northern exposure; but these patches are of small extent, and have no general influence on the temperature of the district in which they occur.

In lat. 65° N., the snow remains continually on the ground from the middle of October till the beginning of May, at which time the soil begins to appear after it has been covered up for 200 days. The thickness of the snowy covering materially affects the depth to which the low winter temperature can be traced into the subsoil. In places of small extent, sheltered from cold winds, and having a good drainage, with a southern aspect, the vernal rays of the sun assist in removing the snow early; while a retentive clay, even in lower and more southern localities, produces a tardy summer and late frosts. Such are the hypogenous or primitive districts, which from their cold climate and ungenial soil have been justly named "barren grounds," and are analogous to the *terra* *damnata* of Lapland, described by Linnæus; or the *tundras* of Arctic Siberia; in such tracts the snow falls early and remains long. The active vegetation of forest lands reacts on the soil; excited by the sun's rays, the trees are roused from their winter's sleep, while the soil is still as hard as a rock, and the snow disappears sooner from over their roots than elsewhere. At the Equator the permanent snow line is said to vary from 15,000 feet of altitude to 20,000, to sink to 3800 on the 60th parallel of latitude, and to be one foot high on the 75th parallel.* The latter assumption does not, however, accord with observations in the northern hemisphere. In no arctic district to which man has yet penetrated, is there a permanent covering of snow through any wide extent of low country. Even at Spitzbergen, only nine degrees from the Pole, there is a summer in which vegetation proceeds, of which we have witnesses in the flora and fauna. The well-fed herds of reindeer, which that hyperborean land maintains, must find grass and lichens, whereon they fatten.†

On the north side of Lake Superior, the duration of the winter's snow is less by at least sixty days than it is in lat. 65° N., that being the difference due to about twenty-seven degrees of latitude. In favorable seasons, at Melville Island, in lat. 74¾° N. and long. 110° W., the snow toward the end of June lies only in the valleys where it had drifted deeply, and the level meadows remain uncovered for seventy days, or till the beginning of September. In Regent's Inlet, which is more to the eastward, the snow dissolves less rapidly; and at Igloolik, still further east, notwithstanding a difference of 5½° of southing, the soil is uncovered for a still shorter period, furnishing an illustration of the depression of temperature on the eastern coast, on which some observations will be made in a subsequent page. The subjoined tables will furnish the dates of the disappearance of the snow in the spring, of various more southern localities.

The opinion held at no distant date by eminent meteorologists, that a mean annual temperature of the freezing point of water coincides with the snow line, has been satisfactorily disproved by

* Von Buch states the height of the snow line at the North Cape (Europe) at 2275 Prussian feet.—*Meyen, Geog. of Plants.*

† The Herald Island, discovered by Captain Kellett in lat. 71° 13′ N., long. 175° 23′ W., to the north of Beering's Straits, is 1200 feet high, and consists of granite precipices to the height of 900 feet, and then a succession of terraces, on which there grows a turfy vegetation. Eight species of plants were gathered from these terraces, viz. a *Hepatica*, *Poa arctica* and another grass, *Artemisia borealis*, *Cochleria fenestrata*, *Saxifraga laurentiniana*, a moss and a red lichen, which covered the rocks.

observation. Every where to the eastward of the Rocky Mountains, the isothermal of 32° F. is below the fifty-seventh parallel of latitude.

When the rigor of the climate of Arctic America is considered, the under limit of permanent snow on the hills appears to be very elevated. The true laws of the altitude of this line have not yet been ascertained. Correct measurements have shown that within the tropics the snow line varies several thousand feet, even in the same range of mountains, according to its different aspects. The active radiation of the sun in the continuous day of an arctic summer, in conjunction with the comparatively small winter deposit, must tend to elevate the snow line; while within the tropics, the effect of a vertical sun is compensated by nocturnal terrestrial radiation, and the deposition that attends the sudden cooling of an atmosphere charged with moisture. The same or similar causes must tend to vary the breadth of the mountain zone, comprehended between the summer and winter snow line. The east coast climate every where north of Canada has an analogy to this zone, and not to its upper limit or the permanent snow line.

Another phenomenon intimately connected with the mean temperature of a district, is the "ground ice" or "permanently frozen subsoil." The lateral extent of this substratum, its southern limits, and its thickness, are interesting subjects of inquiry. It is well known that the warmth excited by the sun's rays is conducted slowly and progressively into the earth, the effects of seasons and years following like tides, but becoming less sensible and less distinct until they are blended, and at a certain depth vanish altogether. Professor Forbes says, that "the decrease of the annual range is common to the strata of the air above the surface of the earth and to those of the soil beneath; both ultimately, no doubt, exhibit a limit, first where the diurnal variations disappear, then the annual." The cause, however, he states is different in the two cases; the one being chiefly the result of the radiation, and the other of the conduction of heat. The limit here spoken of, or the depth to which the periodical changes of summer and winter are felt, is influenced by a variety of circumstances, and differs in different localities. The temperature of that limit would be, it is supposed, that of the mean of a number of years (fourteen or fifteen perhaps) forming a complete cycle of the annual variations.

The limit or commencement of permanently frozen soil, or ground ice, is coincident, according to Bäer, with the isothermal line of 32° F.; and its thickness increases in proportion as the mean temperature of the locality falls below that degree, its unlimited descent being checked by the interior heat of the earth.

Observations oft he temperatures of mines and of artesian wells have established the fact that the temperature rises as we penetrate into the crust of the earth; but the rate of increment has not yet been satisfactorily determined. The temperatures of mines in the same district, and of different parts of the same mine at equal depths, vary greatly. Some authors fix the increment at one degree of Fahrenheit for every forty-five feet of descent, after the superficial stratum which is directly influenced by the solar heat has been passed.

At Yakutzk, in Siberia, on the sixty-second parallel, with a mean annual temperature of 14° F., a well dug to the depth of 382 feet just penetrated the frozen earth, and the resulting increment of heat there is one degree for twenty-eight feet of descent. At Fort Simpson, on the Mackenzie, very nearly in the parallel of Yakutzk, but having a mean annual temperatue of 25° F., the frozen substratum was found to terminate at the depth of seventeen feet from the surface, the underlying bed being loose and sandy. The surface soil there was thawed at the close of summer (19th October, 1837), to the depth of nearly eleven feet, so that the ground ice was only six feet thick.

At York Factory, on Hudson's bay, in lat. 57°, in October, 1835, recent frosts had penetrated eight inches into the soil; the thaw due to the summer heat extended twenty-eight inches beyond this, beneath which a frozen bed seventeen and a half feet thick reposed on thawed mud which had a temperature of 33 F. The mean annual heat of this place is 25½ F., being equal to that of Fort Simpson, which lies five degrees further north.

At Seven outpost, exactly one degree of latitude to the south of York Factory, and on the same coast of Hudson's Bay, the surface had thawed at the close of the summer of 1835 nearly to the depth of six feet, and the frozen substratum was dug into seven and a quarter feet, being thirteen feet in all, but not passed through, so that its thickness was not ascertained.

I have no information respecting the ground ice of the Peace River or Saskatchewan prairie districts.

At Rupert's House, on James's Bay, near the level of the sea, in lat. 51° 26′ N., long. 78° 40′ W., the soil in an exposed situation in the month of April was frozen to the depth of seven feet; but under a snow-drift the frost had penetrated only thirteen inches into the earth.

At East Main, situated on the opposite side of the same bay, in lat. 52° 15′ N., long. 78° 40′ W., the ground under a snow-drift eight feet thick was frozen to the depth of only ten inches. The pits at these places were not dug deep enough to give any information respecting the existence of a permanently frozen substratum.

On Michiskum Lake, which lies in lat. 49° N., long. 78° W., on the north side of the water-shed that divides the Abitibbe, or eastern branch of Moose River, from Lake Temiscaming and the Ottawa, and which is about 700 feet above the sea, the frost toward the close of winter had penetrated three and a half feet in a cultivated field, while in the woods not above six inches of soil were frozen. No search was made for a frozen substratum. These are the chief observations* I have been able to collect, having reference to this subject. It is plain that much more extensive researches are required to enable us to form any general conclusions which can be relied on.

Professor Bäer informs us that in Europe and Siberia, the farther we go east, the more southerly do we find the limit of perpetual ground ice to be. In the environs of Lake Baikal, ice remains at least in one locality all the year. No permanent ice was found at Tobolsk in lat. 58° N., but Humboldt discovered small pieces at the depth of six feet in summer in the elevated district of Boguslowsk, near the foot of the Ural Mountains, in lat. 59° 45′ N. It would appear from these instances that the ground ice has a more southerly limit on the shores of Hudson's Bay than in Siberia; and that in America, as far as observations go, that limit follows the course of the isothermal lines which dip to the south as they proceed to the eastward.

THERMOMETRICAL OBSERVATIONS IN THE VALLEY OF THE ST. LAWRENCE.

The subsequent tables of temperature and notices of the progress of the seasons are arranged so as to convey, as far as my materials go, a view of the climates of the successive lake basins from the St. Lawrence northward. *Table II.* is intended also to illustrate the fact that the climate, on the west side of the continent, is milder than in the eastern States. Fort Vancouver being upward of forty miles from the coast in a direct line, its climate can not be called maritime, and its temperatures may be conveniently contrasted with those of Montreal, which is also situated on the banks of a first-rate river, and not under the influence of sea-breezes.

There is very little difference in the latitudes of the two places, or in their altitudes above the sea; but the direct distance between

* All the experiments made in Rupert's Land, in 1835–36, are detailed in the Ed. New Philos. Journ. for January, 1841.

them exceeds 2200 geographical miles. The difference of their mean annual temperatures is, at least, eight degrees of Fahrenheit in favor of the Pacific coast, and is equal to what would be produced by a diminution of four or five degrees of latitude on the same meridian. The greater mean heat of the Pacific side of the continent has been long known, and a reference to the table will show that it is mainly due to the milder winters; the mean difference between the summer and winter temperatures being twice as great at Montreal as on the Oregon. On the other hand, the summers are sensibly warmer at Montreal. The notices of the progress of the seasons on the two sides of the continent will illustrate these facts equally well. The early spring at Vancouver, with the two inches of snow and a rainy winter, contrasts strongly with the long dry winter and three feet of snow at Penetanguishene, where the ice does not disappear from the lake till past the middle of April.

Columns 1 and 2 of the same table refer to places immediately on the coast, and Franklin Malone is at an altitude above Fort Vancouver, which is considered to be equivalent to two degrees of mean annual temperature. If we compare these columns with column 4, we perceive that with little change of mean annual temperature, places on the coast have a more equable climate, the three winter months being comparatively milder, and the three summer ones less warm.

Table I. shows that as we advance into the interior, the heat of the summer is kept up, or even augments, notwithstanding the elevation above the sea attained by ascending the successive stages of the St. Lawrence basin.

The nature of the rock formations has a considerable influence on the climate of a district. In the primitive country, such as has been described in a preceding page as abounding in lakes and swamps, the climate is extreme, the winters being not only longer, but also more severe, the dissolution of the ice in such districts absorbing much heat. A marked difference occurs when we pass from the "intermediate primitive range" to the prairie districts lying to the westward; for, notwithstanding the greater elevation of the latter, the winters are milder, the snow less deep and less durable, the rivers break up earlier, and the sap flows sooner in the trees. A corresponding difference in the vegetation occurs; the prairie plants have much less of an arctic aspect than those of the primitive districts. Professor Agassiz, in his work on Lake Superior, has instituted a very interesting comparison between the vegetation of that basin and the lower and middle subalpine zones of the higher tracts of the Jura, proving their very great similarity. With the prairie districts, the analogy of the Jura is very much less strong. Many of the plants which give the peculiar character to the prairies south of the Missouri, range northward to the branches of the Peace River, a main affluent of the Mackenzie; and several prairie plants enter the silurian wooded tracts which lie to the westward of the "intermediate primitive range," though they have not been discovered in the more eastern parts of the continent.

TABLE I.

Temperature of various Places in Lakes Ontario, Erie, Huron, and Michigan.

PERIODS.	*Lake Ontario.* North Side Toronto. Lat. 43° 40′ N. Long. 79° 22′ W. Alt. 340 Feet.	*Lake Huron,* East Side Penetanguishene. Lat. 44° 48′ N. Long. 80° 40′ W. Alt. 626 Feet.	*Lake Michigan.* Green Bay, Fort Howard. Lat. 44° 40′ N. Long. 87° 00′ W. Alt. 600 Feet.	*Lake Erie,* East End Buffalo. Lat. 42° 55′ N Long. 78° 55′ W. Alt. 575 Feet.	*Lake Erie* and R. S. Clair, Detroit. Lat. 42° 19′ N. Long. 82° 58′ W. Alt. 575 Feet.
	Hourly Obs. for 5 Years at Magn. Obs.	8 and 8, by C. Todd, Surgeon R. N. corrected by Dove's Form.	7, 2, 9, corrected by Dove's Form.	Computed by the Form. N. Y. Dove. 2 Yrs. Obs-	8 and 2, 3½ Years, Corrected by Dove's Form.
	(1.)*	(2.)	(3.)	(4.)	(5.)
January	25·34	23·42	17·76	23·41	20·26
February	23·71	22·60	20·05	21·14	31·32
March	30·37	31·90	30·87	35·49	39·88
April	43·38	37·82	42·83	40·70	49·68
May	53·28	55·36	56·57	55·29	57·07
June	60·88	67·99	67·86	67·45	68·02
July	66·28	73·31	71·71	71·55	70·83
August	66·73	69·40	68·67	69·99	68·87
September	58·71	56·03	56·91	59·89	61·68
October	45·69	50·11	46·74	48·75	54·28
November	36·03	38·86	33·97	37·22	38·05
December	27·24	25·24	20·95	27·80	29·32
Year	44·81	45·74	44·57	45·54	49·10
Winter	25·43	23·75	19·59	24·12	26·97
Spring	42·34	35·02	43·43	43·83	48·87
Summer	64·43	70·24	69·41	69·66	69·24
Autumn	46·81	48·34	45·87	48·62	51·34
Difference of hottest and coldest months	48·81	49·99	53·95	50·41	50·57
Difference of summer and winter	39·20	46·49	49·82	45·54	42·27

* No. 1 is taken from the volume of observations made at the observatory at Toronto, published by Government. No. 2 from Franklin's Second Journey. Nos 3, 4 and 5, from Dove's Tables in the Report of the British Association for 1847.

TABLE

COMPARATIVE TABLE OF TEMPERATURES ON THE EAST AND

PERIODS.	East Side.			
	Atlantic Coast.		*St. Lawrence Valley.*	
	Maine, Eastport, Lat. 44° 54′ N. Long. 66° 56′ W. Alt. 15 Feet.	Nova Scotia, Halifax, Lat. 44° 39′ N. Long. 63° 38′ W. Alt. 15? Feet.	North N. York, Fr. Malone, Lat. 44° 50′ N. Long. 74° 23′ W. Alt. 645 Feet.	Canada East, Montreal, Lat. 45° 31′ N. Long. 73° 35′ W. Alt. 60 Feet.
	(1.)*	(2.)	(3.)	(4.)
January	19·18	20·00	28·24	13·03
February	22·71	18·00	26·15	17·68
March	29·85	25·00	31·42	30·89
April	38·72	30·00	45·08	46·80
May	48·99	40·00	52·67	53·97
June	56·31	50·00	60·22	59·73
July	63·28	63·00	66·90	68·90
August	63·24	70·00	65·45	68·04
September	57·15	51·00	55·17	57·61
October	46·33	51·00	46·92	46·50
November	36·18	38·00	32·85	31·58
December	24·55	25·00	21·22	19·66
Year	42·41	40·08	43·52	42·12
3 Winter months	22·15	21·00	21·87	16·95
3 Spring months	39·19	31·67	43·06	39·03
3 Summer months	60·94	61·00	64·19	67·26
3 Autumn months	46·55	46·67	44·98	45·18
Difference of hottest and coldest months	44·10	52·00	48·66	55·87
Difference of summer and winter	38·79	40·00	42·32	50·31

* Columns 1, 2, and 3 are extracted from Dove's paper, published in the Report of the British Association for 1847, being corrected for the diurnal variation computed for Toronto by Dove, in the same paper. No. 4 is abstracted from a paper by J. S. M'Cord, Esquire, printed at Montreal, and discussing observations of temperature for 1839 and 1840, and 1840–41, 12 times daily at the even hours.

Column 5 is also from Dove's paper, and is reduced by his formula.

Columns 6 and 8 are abstracted from a year's observations, by G. B. Roberts, Esquire, of the Hudson's Bay Company, with a thermometer of Newman's construction. Column 6 is reduced for the hours of observation, according to Dove's formula for Toronto. Column 8 is transcribed without correction. Column 7 is an abstract of observations by the Rev. S. Parker, also corrected by Dove's formula.

The subjoined is an abstract of observations for seven years, made by Henry Poole, Esquire, at Albion Mines, near Pictou, in Nova Scotia, and on the coast of the Gulf of St. Lawrence. Mr. Poole ascertained the minima in the night by a self-registering thermometer, and the maxima between 1 and 2 P.M., by a thermometer hung on the north side of the house, and sheltered from the north winds by a plantation of firs at the distance of a few yards. The ob-

II.

West Sides of the Continent near the 45th Parallel.

West Side. Oregon State. Fort Vancouver, on the Columbia, 50 m. from the Sea. Lat. 45° 37′ N. Long. 122° 45′ W.				PERIODS.
Hours 7, 2, 9.	Hours 8 and 8.	Hours 7–1.	Hour 2 p.m.	
(5.)	(6.)	(7.)	(8.)	
37·62	36·58	37·10	40·68	January.
42·89	35·37	38·98	44·07	February
43·68	44·05	47·47	55·06	March.
45·55	50·22	..	62·50	April
53·44	58·43	..	74·61	May.
62·48	58·72	..	73·17	June.
65·46	61·76	..	75·71	July.
65·84	63·05	..	79·74	August.
60·30	61·10	..	74·27	September.
53·23	52·62	48·27	68·16	October.
42·68	38·83	39·23	48·27	November.
42·95	34·49	38·60	42·06	December.
51·34	49·73	..	61·91	Year.
41·15	35·49	38·23	42·21	3 Winter months.
47·55	50·90	..	64·07	3 Spring months.
64·67	61·30	..	76·14	3 Summer months.
52·00	50·92	..	64·71	3 Autumn months.
28·28	·28·56	..	39·06	Difference of hottest and coldest months.
23·52	25·81	..	33·93	Difference of summer and winter.

servations are corrected for the Toronto diurnal variation for an hour before sunrise, and 2 P.M As there appears to have been some local cause producing higher winter temperatures than in other parts of the coast, I did not think it expedient to introduce these observations into the comparative table.

Abstract of Temperatures at Albian Mines, 120 Feet above the Sea, Lat. 45° 34½′ N. Long. 62° 42′ W. Means of Maxima.

Jan. 50·58 F. Feb. 51·85 Dec. 51·11	March 52·53 April 54·51 May 58·64	June 68·86 July 70·64 August 70·61	Sept. 62·42 Oct. 55·10 Nov. 49·98	Year . . . . 56·03 Diff. hot and cold months . . 20·66
Wint. 51·18	Spring 53·23	Summer 69·04	Autumn 55·83	Diff. of summer and winter . . 17·86

TABLE III.

Table of Temperatures on Lake Superior and on the upper Part of the Ottawa River.

PERIODS.	Lake Superior.			L. Temiscaming.
	Fort Brady, Saut St. Mary. Lat. 46° 31′ N. Long. 84° 25′ W. Alt. 660 Feet.	Northeast angle, Michipicoten. Lat. 47° 56′ N. Long. 85° 6′ W. Alt. 660 Feet.	Thunder Bay, Fort William. Lat. 48° 23½′ N. Long. 89° 22′ W. Alt. 660 Feet.	H. Bay Post, on the Ottawa. Lat. 47° 19′ N. Long. 79° 31′ W. Alt. 630 Feet.
	7, 2, 9, Dove Corrected.	8 and 8, by G. Keith, Esq., corrected by Dove's form.	8 and 8, Corrected by Dove's form.	Sunrise, sunset, and noon. by —— Severight, Esq., corrected by Dove's form.
	(1.)	(2.)	(3.)	(4.)
January	+18·30	+10·63	+5·70	+9·23
February	19·69	16·66	8·22	18·44
March	27·04	26·09	22·72	24·41
April	38·05	34·66	31·42	39·04
May	52·00	51·88	48·87	49·35
June	58·61	55·00	58·73	62·75
July	65·36	57·03	62·19	67·28
August	64·36	60·04	58·84	65·58
September	55·55	49·67	48·16	53·39
October	44·75	44·92	41·88	40·83
November	33·59	29·01	23·43	25·97
December	22·23	22·38	18·16	17·68
Year	41·63	38·25	35·90	38·58
3 Winter months	20·07	16·68	10·75	15·02
3 Spring months	39·04	37·57	39·67	37·58
3 Summer months	62·77	57·39	59·94	65·23
3 Autumn months	44·63	41·24	37·80	40·07
Difference of hottest and coldest months	46·06	49·41	56·49	58·05
Difference of summer and winter months	42·70	40·71	49·19	50·21

Column 1 is extracted from Dove's paper, already referred to; columns 2 and 3 are compiled from documents obtained from the Hudson's Bay Company; and column 4 is extracted from Mr. Logan's Geological Report for 1845–6.

*Phenomena illustrative of the Climate of Penetanguishene, on Lake Huron.**

The spring sets in very suddenly. Snow continues until the latter end of April, and longer in the forest than in cleared lands. The weather in March is clear and cloudless, and the ice, which in winter has attained a thickness of sixteen inches, begins to dissolve. Toward the end of the month the sap of the maple flows, and the sugar harvest commences. Flocks of Canada geese and various ducks appear about the same time, and are the harbingers of fine weather. The ice disappears on an average on the 24th of April. Alders and various willows flower about the middle of the month, and the *Hepatica triloba* blossoms on the 25th. Potatoes are planted between the 1st and 20th of May, and cucumbers and melons are usually sown between the 25th and the end of the month. *Viola blanda*, *Xylosteum*, *Leontice*, *Erythronium*, and many other plants, blossom in this month; the forest trees come all into leaf about the 16th, and, about the 19th, the musquitoes begin to be troublesome.

In the month of June the temperature rises to 90° F. in the day, and heavy dews fall in the night. Barley and oats are sown about the 15th, and toward the end of the month garden peas are fully podded, and the male flowers of maize spring up. The *Lilium philadelphicum* blossoms at this time.

In July and August the weather is usually dry and sultry. About the beginning of the former month *Penstemon pubescens*, *Rhus typhinum*, garden melons, and cucumbers blossom; and, toward the middle of August, melons grown without artificial warmth are ripe, and the wheat and oat harvest commences. Maize is fit for pulling about the end of the month.

In September numerous flocks of *Turdus migratorius* and other birds arrive from the north, and remain for a time feeding on the berries of various rasp bushes. Maize ripens about the first of the month, and near its end frost destroys the cucumber and melon vines: forest trees change their hue, and potatoes are dug and stored for winter use.

The forest assumes a variety of autumnal hues in the beginning of October; about the middle of the month, many flocks of geese and ducks pass to the southward, and their appearance precedes a series of cold weather, which strips the leaves from the trees. A fall of snow usually occurs about the 25th.

November is generally calm and pleasant, and about three weeks of peculiar weather, named the "Indian summer," occurs. It is characterized by a fog or haze rising from the earth, through which the sun is seen obscurely, and there is little or no wind. In December, the thermometer sinks a few degrees below zero, and much snow falls. The harbor freezes in the beginning of the month. In January, the thermometer sinks 20 degrees below zero, and rarely to —32°. The snow attains a depth of three feet in the woods, but the ground it covers is not frozen. A great fall of snow takes place in February, and there is usually a temporary thaw about the end of the month, accompanied by heavy rain, and occasionally by thunder.

* By the late C. C. Todd, Esq., Surgeon, R. N.

Phenomena indicating the Progress of the Seasons at Michipicoten, Lake Superior, in the year* 1840.

Jan. 19. Open water in the bay. Mergansers frequenting it.

Feb. 14. Bay again closed by ice.

March 8. Snow-birds departed for the north. 26th. The snow-birds returned again, the weather having been severe. Domestic hens began to lay eggs. Two ducks seen.

Apr. 10. Lake clear of ice in Michipicoten Bay. 12th. *Turdus migratorius* came.

May 2. All the snow gone. 6th. Swallows came. 22d. Potatoes planted.

Sept. 4. Small trout collecting in the rivulets to spawn. On the 23d they ceased spawning. Frosty nights. Potatoes not hurt. 26th. *Corregonus lucidus* spawning in the river.

Nov. 10. Ice beginning to drift in the river. 18th. The large trout and *Corregonus albus* cease to spawn here at this time, though they carry on this operation later in other parts of the lake.

Dec. 3. Ice broken up in the river. Little snow lying on the ground. 13th. River again frozen over. Season mild.

Phenomena indicating the Progress of the Seasons at Fort William, Lake Superior, in the year 1840.

Feb. 29. Thermometer at noon rose to 39° F.

March 1. Temperature 61° in the middle of the day. On the 27th a gray hawk, and on the 31st a barking crow, *Corvus americanus*, were seen.

April 2. The sap of the sugar maple began to run. On the 4th small holes began to perforate the ice. On the 9th the first wild ducks of the season came, and on the 10th butterflies, blue flies, and gulls were noticed. 20th. The general thaw commences at this period. Ground frozen to the depth of 3 feet 9 inches. 21st. *Anser canadensis*, *Anas boschas*, and mergansers frequenting the neighborhood. 28th. Heard a nightingale (*Turdus ?*). 30th. River partially open.

May 2. River free of ice. Bay of the lake full of drift ice. 6th. *Anser hyperboreus* passing in flocks. 8th. Musquitoes seen. 10th. The birch tree and maple budding.

June 15. Swallows building in the outhouses. 17th. Sturgeons spawning in the rapids of the river. 19th. *Catastomi* beginning to descend the river from the rapids. 21st. *Corregonus lucidus* comes to the entrance of the river in shoals.

July 3. The *Corregoni* have left the mouth of the river. 15th. Barley just coming into ear. Potatoes in flower. The *Lepus americanus* having its second litter of young. 31st. Raspberries ripening.

Aug. 8. Red currants and blue berries (*Vaccinium*) perfectly ripe. 10th.

* By chief factor George Keith, Esq.

Reindeer begin to rut. 19th. Barley ripening. 29th. Peas quite ripe. 31st. The swallows have disappeared.

Sept. 2. Reindeer rutting season ends. On the 7th the leaves of the birch and aspen change color. 10th. Small trout begin to spawn. 13th. Potatoes, cabbages, turnips, and cauliflowers nipped by the frost. 14th. A few ducks arriving from the north. 16th. The first stock-ducks arrived from the north this autumn. 20th. Small trout spawning abundantly on the shoals. 23d. The orioles have departed for the south. Canada geese arriving from the north, and going southward. 30th. *Corregonus lucidus* at this date begins to spawn in the rapids of the river.

Oct. 6. The large trout begin to spawn in the lake at the Shaguinah Islands: they cease on the 18th. Thunder. 7th. Leaves of the birch and aspen falling. 10th. The *Corregonus lucidus* has ceased spawning on the rapids. 14th. Thunder. *Anser hyperboreus* arriving from the north. 15th. Passing in large flocks. 20th. Hail, thunder, and lightning. Plovers, divers, snipes, orioles, geese, and ducks in the neighborhood. On the 31st, snow-birds began to arrive from the north.

Nov. 3. The small lakes frozen over. On the 9th, the river (Kamenistikwoya) covered by a sheet of ice, which broke up again. 21st. The spawning season of the *Corregonus albus* terminates.

Dec. 1. Ice driving about in the lake with the wind. On the 17th, the bay was frozen across to the Welcome Islands.

The following *Notices of the Progress of the Seasons at Fort Vancouver in* 1838,* may be contrasted with the state of vegetation at the same times of the year in the valley of the St. Lawrence:

Jan. 2. Short young grass, affording good pasturage for sheep in places that were flooded in summer. 8th. Berry-bearing bushes budding, such as the wild cherry, black currant, &c. Swans abundant, ducks, geese, and cranes scarce. 10th–12th. Snow fell to the depth of an inch and a half, and vegetation was retarded by unfavorable weather till

Feb. 17. when wild gooseberry bushes were observed budding. Between the 26th of this month and

Mar. 16. Thunder and hail showers occurred. *Ribes sanguineum* blossoming. *Trillium grandiflorum*, having the local name in Oregon of "Herb Paris," in full flower. 21st. Apple and pear-trees budding. The wild gooseberry in full leaf, and further advanced than bushes from England cultivated in the garden. 24th. The swallow first noticed. 30th. Humming-birds appearing; strawberries flowering.

April 3. Mahonia in blossom, and, on the 5th, peach-trees flowering. 8th. Potatoes that have lain in the ground all winter, beginning to show. 11th. Dog-wood and elder in blossom. 17th. Several species of violet in flower. 20th. Field iris in flower. 23d. Brambles flowering. On the 25th, clover in bloom; and,

* By G. B. Roberts, Esq., of the Hudson's Bay Company

on the 26th, wild tares flowered. Hail and thunder storms. 28th. Blossoms of the fruit-trees falling.

May 1. Lupins in flower. 7th. Wild rose and eglantine in flower. 12th. Strawberries ripening. 28th. Field-peas in blossom. 30th. Garden-peas brought to table. Thunder.

June 1. Spring barley in ear. 5th. New potatoes raised from tubers, left all winter in the ground, fit for table. 7th. Oats in flower. 10th. Spring wheat coming into ear. 17th. Bramble-berries ripening. 23d. Gooseberries, currants, and raspberries ripe. Thunder. 26th. Blackberries (*Vaccinium*) ripe.

July 4. Humming-birds scarce this season. 19th. Barley fit for the scythe. 22d. Winter wheat ripe. 28th. House-flies numerous.

Aug. 3. Oats ripe. 7th. Field-peas harvested. 9th. Foggy mornings, followed by a clear sky and excessive heat. Dewless mornings succeeded by rain. 12th. Salmon season usually ends. Several cases of ague occurring. 27th. Musquitoes very troublesome. Geese arriving from the north.

Sept. 7. Hail and thunder. 12th. Buckwheat harvested. 13th. Ague prevalent. 19th. Mowed barley that was sowed on the 16th of June—a fair crop. 22d. Peas sown on the 19th of June, on alluvial land, that was flooded in winter, now ripe.

Oct. 27. The Columbia or Oregon River unusually low, the greatest depth opposite the fort being less than fifteen feet.

Nov. 8. Potatoes killed by the frost. 18th. A little drift ice in the river. 20th. Rain and sleet.

Dec. 26. Snow at this time two inches deep, being the greatest fall this season, and less than usual.

TABLE IV.

THERMOMETRICAL OBSERVATIONS.—VALLEY OF THE SASKATCHEWAN.

MONTHS AND SEASONS.	Cumberland House. Lat. 53° 57′ N. Long. 102° 20′ W. Alt. 900 Feet. Mean Temp., 8 A.M., 8 P.M., corrected 1839.	Oxford House. Lat. 54° 55′ N. Long. 96° 28′ W. Alt. 400 Feet? Mean Temp., 7 A.M., Noon, and 8 P.M., corrected, 1833.	York Factory. Lat. 57° 0′ N. Long. 92° 26′ W. Alt. 20 Feet. Mean Temp. Morn., Noon, and Evening, corrected, 1830.	Rupert House. Lat. 51° 21′ N. Long. 83° 40′ W. Alt. 20 Feet? Mean Temp., Sunrise, 1½ P.M., Sunset, corrected, 1839.
June	..	..	47·67	
July	..	..	59·99	
August	62·84	..	54·85	
September	44·50	..	41·90	
October	33·15	17·53	33·43	34·80
November	21·48	13·29	25·17	23·33
December	7·94	—23·06	3·73	15 59
January	—0·89	—22·06	—5·12	—4·09
February	—8·06	— 1·90	—6·60	—0 68
March	18·30	8·57	4·77	7·64
April	27·01	28·62	19·21	21·05
May	52·59	38·01	33·53	41·51
Year	33·20	..	25·63	
Summer	62·62	..	52·07	
Autumn	33·04	..	33·50	
Winter	—0·17	—0·82	—2·53	0·14
Spring	32·70	7·51	19·17	0·78
Difference of hottest and coldest months	72·06	..	66·59	
Difference of summer and winter	61·79	..	54·60	

The observations recorded in the above table were made at Cumberland House, by chief factor John Lee Lewis, between the beginning of August, 1839 and the end of September, 1840, the register being deficient in June and July, when Mr. Lewis was traveling on the business of the Company. The temperatures were measured by a thermometer made by Newman, and sent out at that time with many others, by the Geographical Society, but at the expense of the Hudson's Bay Company. The monthly means of the combined temperatures at 8 A.M. and at 8 P.M. were corrected by Dove's table, calculated for Toronto.—*Rep. Br. Ass.* 1847.

The observations at Oxford House were made between October, 1833 to May, 1834, at 7 A.M., noon, and 8 P.M., and these are also corrected by Dove's formula for Toronto.

The register for York Factory, Hudson's Bay, was kept by chief factor Joseph Charles, from June, 1830 to end of May, 1831, the temperatures being recorded in the morning, noon, and afternoon, but the exact hours of the morning and evening are not specified. They have been corrected by Dove's table, on the supposition that the hours were sunrise and sunset.

The observations at Rupert House, on the east side of James's Bay, were made at sunrise, an hour and a half after noon, and at sunset, and Dove's corrections have also been applied to them.

An examination of the several columns will show clearly the effect of open water in Hudson's and James's Bays, in tempering the atmosphere in the months of October, November, and even December, and of the presence of ice in those seas, keeping down the summer heat.

TABLE V.

COMPARATIVE TABLE OF TEMPERATURE ON EAST AND WEST SIDES OF THE CONTINENT.

PERIODS.	Nain. Lat. 57° 10′ N. Long. 61° 50′ W. Alt. 20 Feet. 3 Years. 8, 12, 4, 8. Dove.	Okak. Lat. 57° 30′ Long. 66°. Alt. 20 Feet. 2 Years. 8, 12, 4, 8. Dove.	Sitka. Lat. 57° 3′ N. Long. 135° 18′ W. Alt. 20 Feet. 2½ Years. Reduced. Baer.
June	42.53	44·65	53·83
July	50·18	51·65	57·11
August	50·99	52·00	57·79
September	44·98	44·45	55·96
October	33·98	31·15	46·63
November	26·51	22·40	42·89
December	6·51	8·45	36·32
January	0·95	2·15	34·18
February	3·51	1·95	33·60
March	7·52	8·25	38·01
April	29·97	29·00	40·64
May	36·23	38·25	48·18
Year	27·82	27·86	45·44
Summer	47·90	49·43	56·24
Autumn	35·16	32·67	48·49
Winter	3·66	4·18	34·74
Spring	24·57	25·17	42·28
Difference of hottest and coldest months	50·04	50·05	24·19
Difference of summer and winter	44·24	45·25	21·50

TABLE VI.

ABSTRACT OF A RECORD OF TEMPERATURES OF THE AIR IN THE SHADE, KEPT BY MR. DRUMMOND, AT EDMONTON HOUSE, LAT. 53° 40′ N., LONG. 113° W., ALT. 1800 FEET.

MONTHS.	Monthly Means.			Extreme Temperatures.	
	Of Maxima.	Of Minima.	Of these.	Highest.	Lowest.
1827.					
January	18·68	3·42	11·05	+42·0	−27·0
February	29·96	3·68	14·32	+47·0	−25·0

TABLE VII.

ABSTRACT OF A JOURNAL OF TEMPERATURES KEPT AT CARLTON HOUSE BY THE AUTHOR, LAT. 52° 51′ N., LONG. 106° 13′ W., ALT. 1100 FEET.

MONTHS.	Monthly Means.			Extreme Temperatures.	
	Of Maxima.	Of Minima.	Of these.	Highest.	Lowest.
Feb., last 10 days.....	12·50	—1·20	5·65	31·0	—29·0
March	23·10	+0·74	11·92	42·0	—26·0
April	40·97	18·53	29·75	59·0	+ 2·0
May	61·90	33·95	47·92	75·0	+22·0

TABLE VIII.

TABLE OF THE EXTREME TEMPERATURES OCCURRING IN EACH MONTH AT SEVERAL PLACES IN THE SASKATCHEWAN VALLEY, AND IN HUDSON'S AND JAMES'S BAYS.

	Cumberland House. Lat. 53° 57′ N. Long. 102° 20′ W. Alt. 900 Feet.		Oxford House. Lat. 54° 55′ N. Long. 96° 28′ W. Alt. 400 Feet.		York Factory. Lat. 57° 00′ N. Long. 92° 26′ W. Alt. 20 Feet.		Rupert House. Lat. 51° 21′ N. Long. 83° 40′ W. Alt. 20 Feet.	
	Highest Temp.	Lowest Temp.	Highest Temp.	Lowest Temp.	Highest Temp.	Lowest Temp.	Highest Temp.	Lowest Temp.
June	87	42	..	..	83	25		
July	98	47	..	..	96	35		
August	90	49	..	..	79	40		
September......	73	30	..	..	70	28		
October	68	5	49	— 3	58	19	61	13
November	38	—10	39	— 6	42	8	36	—10
December	25	—26	32	—29	30	—30	37	— 5
January	25	—32	0	—44	19	—34	16	—32
February	34	—35	30	—33	37	—35	39	—36
March..........	50	— 9	34	—27	47	—25	34	—25
April...........	55	— 3	64	— 8	55	— 9	53	—16
May	93	27	65	15	62	15	71	19

Table VIII. shows that tropical temperatures occur in the Saskatchewan basin, for a day or two, or it may be only for a few hours at a time in summer, yet that the three summer months seldom pass without night frosts. These destroy tender plants, and in untoward seasons injure the growth of *cerealia*. Wheat, however, ripens well in the drier limestone districts, and still better in the prairie country; but it is there subject to periodical ravages of the larvæ of caterpillars, which come like an army of locusts, and eat up all that is green. Were the country more generally cultivated, and rooks and domestic poultry encouraged, this plague might be lessened. Maize ripens well at the Red River and Carlton, and, I believe, at Cumberland House also, though I did not see it in cultivation there. The 54th parallel may be considered as its northern limit, but its profitable culture does not, perhaps, extend beyond the 49th or 50th degree on the east side of the Rocky Mountains.

In the following Table of Phenomena, indicating the *Progress of the Seasons at Cumberland House*, I have combined my own observations in the spring of 1820 with those of chief factor John Lee Lewis, in 1839 and 1840, distinguishing the remarks by the years. The supposed altitude of Cumberland House above the sea is 900 feet, according to Captain Lefroy's calculations.

March 4. Water collecting in pools round the establishment. 1840.
7. Much bare ground visible.
8. The snow which covered the ground to the depth of three feet, was observed to moisten in the sun for the first time this season. 1820.
12. Temperature in the shade rose for the first time to +30° F. The melting snow began to drop from the eaves of the houses.
21. Patches of earth became visible, the season being in respect to the melting of the snow fourteen days later than that of 1840. The river Saskatchewan broke up partially, the melting snow covered with *Poduræ*, as it is also frequently in the autumn.
24. A white-headed eagle was seen, this being almost always the first of the summer birds which arrives; it comes as soon as it can obtain fish. In 1840, the first eagle was seen on the 26th.

April 2. The river Saskatchewan froze over again, after some very cold days.
7. Barking crows (*Corvus americanus*) seen. They were not observed till the 19th in 1840.
8. First snow bunting seen (*Emberiza nivalis*). 1840.
9. A merganser seen. 1820.
10. Willow catkins beginning to burst.
12. Geese and swans seen in 1820. In 1840 they were not seen till the 20th; and pelicans and ducks were observed that year on the 21st.
13. Buds of *Populus balsamifera* bursting. 1820.
17. Plovers, grakles, and orioles seen, and, on the following day, Canadian jays and fly-catchers. Frogs croaking.
20. Coltsfoot, *Nardosmia palmata* flowering.
26. Alder flowering. The sugar harvest, which is collected in this district from the *Negundo fraxinifolium*, commenced in 1820, on the 20th of this month, and lasted till the 10th of May. The flow of the sap is greatly influenced by the direct action of the sun, and is greatest when a smart night's frost is succeeded by a warm sun-shining day. The flow ceases in a cold night.
28. The Saskatchewan thoroughly broken up. The ice on Pine Island Lake did not disappear until nearly a month afterward. Wahlenberg observes that the mean temperature of the air in Lapland must rise to 40° F. before the rivers are completely free. The Saskatchewan opens in this district before the mean heat for ten days rises so high; but its upper part flows from a more southerly and warmer, though a more elevated, country.
30. Commenced plowing. 1840.

May 1. *Anemone patens*, or wind flower, in blossom, its leaves not yet expanded. 1820.

May 2. A fall of snow to the depth of two feet. 1840.
13. Planting potatoes.
14. Sowing barley. 1820. *Negundo fraxinifolium* and gooseberry bushes in flower.
17. Willows, gooseberries, aspens (*Populus tremuloides*) in leaf. Various *Drabæ* in flower. 1820. In 1840 the trees were bursting their buds at this time.
17. Wheat sown on the 8th of this month, above ground to-day, having germinated in nine days. 1840.
21. Barley sown on the 14th above ground, having taken seven days to germinate.
22. Leaves of the trees expanding rapidly.
24. *Ulmus americana* flowered. 1820.
25. Pine Island Lake clear of ice. 28th. *Prunus pennsylvanica*, *P. virginiana*, and Amelanchier in flower. 30th. From the 23d to the 30th of this month in 1840, the temperature in the shade at 2 P.M. varied between 78° F. and 93° F. On the 30th, potatoes planted on the 13th appeared above the ground. 1840.

June 12. All the forest trees in full leaf. 1820.

Aug. 1. Commenced reaping barley. On the 15th, 18th, 19th, and September 1, the thermometer at noon ranged between 80° and 90°, being the hottest days in the month. There was much thunder and hail on these days. 1839.

Sept. 2. Flocks of water-fowl beginning to arrive from the north. 3d. The first fall of snow this autumn. 4th. Vast numbers of water-fowl flying southward. A severe fall of snow and frost in the north causes these birds to hurry to the south. 11th. First hoar-frost. Birch and aspen leaves turning yellow. 14th. Wild-fowl numerous. 20th. Snow; 21st, ditto very heavy. 24th. Thunder and lightning.

Oct. 1. Taking up potatoes. 5th. Leaves all fallen from the deciduous trees. On the 11th, thermometer at 2 P.M., in the shade, 68° F., being unusually high.
14. Water-fowl passing southward in large flocks. 1839.
15. Bays of the lake frozen over. 16th. The ground frozen hard. 17th. Last water-fowl seen this season. 18th. Lake entirely frozen over. In 1839 the Little River was frozen over on the 24th of this month, but broke up again in part, and remained partially open all the winter.
31. Waveys (*Anas hyperborea*) passing. Lake partially opened.

The following are the *Phenomena of the Spring of* 1827 *at Carlton House*, in lat. 52° 51′ N., long. 106° 13′ W., on the eastern limits of the Saskatchewan prairie lands, and at an elevation above the sea of about 1100 feet.*

* This was estimated in 1827 at 1000 feet from the length of the river course between Carlton House and York Factory, making a smaller allowance per mile for the descent, as far as Lake Winipeg, and a considerably greater one for the falls and rapids of Nelson River. It accords sufficiently with Captain Lefroy's observations; and its error does not probably exceed 200 feet at most

Feb. 15. Snow thawing in the sunshine, and on the 17th many sandy hummocks on the plains were bare. This is at least three weeks earlier than the thaw commences in an early season at Cumberland House, which is a degree further north, but is 200 feet lower.

March 6. Trees thawed in fine days, and on the 8th the black earth on the immediate banks of the river was softened to the depth of two inches by the power of the sun's rays. At this place the westerly winds bring mild weather, and the easterly ones are attended by fog and snow.

13. Sparrow-hawks (*Falco sparverius*) arrived from the south, and on the 17th several migratory small birds were noticed.

29. Large flocks of snow-birds (*Emberiza nivalis*) came about the establishment; and, by the 31st, steep banks, which had a southern aspect, were clear of snow.

April 1. Many *Fringillidæ* (birds of the sparrow tribe) were seen. On the 2d, swans arrived, and, by the 3d, much snow had disappeared from the plains.

4. The snow at this time was melting in the shade, and the sap of the maple trees (*Negundo fraxinifolium*) began to flow.

April 6. Geese arrived. Stormy weather, about the middle of the month, retarded the arrival of the summer birds; but the plants continued to grow fast. On the 20th, the Telltale plover (*Charadrius vociferus*) and several small birds came.

22. *Turdus migratorius*, *Pyrrhula ludoviciana*, and *Lanius excubitor* were seen, and the flowers of *Anemone patens* expanded.

27. Ice in the river Saskatchewan gave way. Frogs began to croak.

28. Canada cranes *Grus canadensis* arrived.

May 1. *Sturnus ludovicianus* arrived, and the last flocks of *Emberiza nivalis* departed for the north.

2. On this day, *Icterus phœniceus* and *Scolecophagus ferrugineus* were seen, and most of the water-fowl had by this time arrived. On the 4th, *Phlox hoodii* flowered.

5. *Ranunculus rhomboideus*, *Viola debilis*, *Nardosmia palmata*, and several carices flowered.

6. *Hirundo viridis* and many gulls arrived.

7. On this day the sap of the ash-leaved maple, which had flowed scantily for ten days, ceased to run altogether, and the sugar harvest closed. *Avocetta americana* arrived. *Populus tremuloides* in flower.

9. Crow-blackbirds were first seen. *Corydalis aurea*, *Corylus americana* and *rostrata*, *Hippophäe canadensis*, *Thermopsis rhombifolia*, *Vesicaria arctica*, and *Alnus viridis* flowered. 12th. *Potentilla concinna*, *Townsendia sericea* flowered. 14th. Gooseberry bushes coming into leaf. Ash-leaved maple flowering, seven days after the sap had ceased to flow from wounds in the stem. 16th. The *Picus varius* arrived in considerable numbers, and on the 19th the *Viola nuttalliana* flowered.

The average antecedence of spring phenomena at Carlton House to their occurrence at Cumberland House is between a fortnight and three weeks.

The difference of latitude, which is only one degree, is nearly counterbalanced by 200 feet of greater altitude; but the dry sandy soil of the plains, which are early denuded of snow, gives the spring there a great superiority over that of the lower country, where the ground is almost submerged, and the greater part of it icebound for a month after the river is open.

I obtained no other register of temperatures at Edmonton House, or from the country near the base of the Rocky Mountains in that parallel, than the daily maxima and minima for two winter months observed by Mr. Drummond in 1827. These are included in the subjoined table; and it will be observed that the winter, as far as we can judge from a few isolated observations, is even milder at Edmonton than at Cumberland House, though it is on the same parallel and at a much greater altitude. Edmonton House is in latitude 54° N., and longitude 113° W., and its elevation above the sea is estimated by Captain Lefroy, from his observations on the boiling point of water, at 1800 feet.

Neither have I been able to procure registers of temperatures kept at any post in the southern parts of the Saskatchewan basin. The Red River Colony extends to the boundary line of the United States, or the 49th parallel; and I have been informed that the *Fagus ferruginea*, or American beech, grows within the limits of the settlement, though it does not exist on Lake Superior, but terminates in that direction at Michilimackinac, on the 46th parallel. In the interesting account of the Alps of New Hampshire by Professor Agassiz, the trees which grow in the zone comprised between the elevations of 830 and 1500 feet above the sea are the same kinds which grow on the Red River and forks of the Saskatchewan at nearly as great altitudes, and from five to ten degrees further north; with the addition of the oaks, which find their northern limit on Lake Winipeg, though they do not enter the corresponding New Hampshire zone. The same trees, however, if I understand the passage in the work on Lake Superior (p. 186) to which I here refer, continue to ascend the New Hampshire mountains to the height of 2080 feet; above which, to the height of 4350 feet, the vegetation consists almost entirely of *Abies alba* and *balsamea*, and *Betula excelsa* and *papyracea*. *Betula excelsa* scarcely reaches the Saskatchewan basin, or, if it does, is rare even in the southern parts; but the other trees here named go northward on the Mackenzie till the 69th parallel, and terminate there on coming within the direct influence of the winds blowing from the Arctic Sea. The summits of the New Hampshire Alps, 6280 feet high, present a truly arctic flora, much more so, I believe, than the slope of the Rocky Mountains at that elevation on the 52d parallel. On this point, however, I can not speak with any confidence, as I have no measurements of the heights at which Mr. Drummond gathered his plants, and he is the only authority for the botany of that district.

That I may carry on as complete a view of the climate of the country as I can produce from the observations of others as well as from my own, I introduce here a table* drawn up by a very intelligent officer of the Hudson's Bay Company, chief factor George

* Mr. Barnston having sent me this paper in 1840, I published it in the Edinburgh Philosophical Journal for April, 1844.

Barnston, giving the progress of the seasons at Martin's Falls. This post is situated in lat. 51° 32′ N., long. 86° 39′ W., on Albany River, about 200 geographical miles in a direct line from the coast of Hudson's Bay, an equal distance in a westerly direction from the water-shed between the Albany and Winipeg Rivers, and somewhat less from the beach of Lake Superior, bearing south. According to Mr. Barnston's remark which follows, the country about the falls is similar in its geological character to the west side of Lake Winipeg, and referrible to the silurian epoch. "Our geological position," says that gentleman, "is upon the confines of the great basin of James's Bay, an immense extension of the older calcareous strata. Between the 'falls' and the coast the bed of the river is composed of limestones and clays, both containing extinct genera of shells; while above, toward the interior, little is to be seen but gneiss and greenstone schist, with a mixture here and there of less fissile granite rocks. The fossils I have been able to procure in this neighborhood are principally spirifers, producta, terebratula, and impressions of trilobites. Although in winter we have the cold of Russia, in the months of July and August we enjoy the climate of Germany and the north of France." (Barnston, l. c.) A reference to the map will show that there is a canoe route from Winipeg River, through Lake Sal, and by a portage over the water-shed to Lake Saint Joseph, and Albany River, and also a shorter one from the Peek River on Lake Superior to the southern tributaries of the Albany. In short, as has been already frequently stated, the primitive rocks forming the brims of these several river basins are traversed in every direction by sheets of water occupying more surface than the rocks themselves.

On the Progress of the Seasons at Martin's Falls.

Dec. Jan. Feb. We are frequently visited in these dead winter months by the white owl (*Strix nyctea*) from James's Bay, but the hawk owl (*Strix funerea*) is our most common bird of prey. *Tetrao umbellus*, *T. canadensis*, and *T. phasianellus* are residents in this district the whole year. The *Tetrao saliceti*, or willow ptarmigan, is a winter visitor which comes from the north.

March. Martens (*Mustela martes*) pair, and soon afterward rabbits (*Lepus americanus.*)

15. In the middle of the month the snow often melts in the height of the day, and by the 20th a snow-bird (*Emberiza nivalis*) may be seen occasionally if the season be early.

20. Tops of the higher grasses, which have been concealed beneath the snow, begin to show. A few brown feathers clothe the necks of the willow-grouse, and these birds leave us.

April. A slight crust now forms on the snow, produced by night frosts after thaw in the day. In mild weather, a few insects show in the sunshine.

8. Two species of *Perla* and one of *Nemoura* come up through the crevices of the ice and porous snow, and all proceed straight to the nearest beach. The cold renders them too weak to fly, though most of them have got rid of their nymphine investments before emerging from the ice.

15. Snow-birds have become plentiful, and are now joined by the *Emberiza lapponica* and *Alauda alpestris*. 20th. The flesh-fly is still scarce. The small owl (*Scops*) calls in the warm nights, and the common woodpecker (*Picus pileatus*) drums on the hollow trees. 22d. The Canada goose (*Anser canadensis*) and stock-ducks (*Anas boschas*) sometimes arrive at this date, but are frequently forced to return for want of water, and by northerly blasts.

5. A few spots of ground bare; (more than a month later than the earth begins to appear at Carlton House.) 28th. The red-breasted thrush (*Turdus migratorius*) and cattle blackbird (*Scolecophagus ferrugineus*,) are now arriving, and pick up the benumbed grubs and caterpillars. Goshawks arrive.

May 1. Snow melting rapidly. Ground getting barer. 5th. Wild geese and ducks passing to the northward. Hawks still arriving. 10th. Every fine day brings an accession to the small bush birds, fly-catchers, &c. Food for these is still scarce, and they approach the houses in quest of *Dipteræ*, which rise from the manure and rich earth round the place. Snow-birds have left us, and ermines and hares (*Lepus americanus*) are becoming altogether brown. The ice is now shingly and dangerous, and strong currents and rapids are open. Waveys (*Anser hyperboreus*) and brents (*Anser bernicla*) passing toward James's Bay in large flocks. No weather stops them after this time.

12. The northern diver (*Colymbus glacialis*) and several black ducks (*Anas nigra*, *fusca*, and *perspicillata*) are still scarce, but are sometimes seen. The buds of the balsam poplar, aspen, and of various willows, swell. On the latter may be found the earliest *Tenthredines*, the larvæ being nursed in the tender bud. Two species of butterfly (*Vanessa* and *Argynnis*) sport over the ice and snow, when these are not gone.

15. The larger rivers break up. (The Saskatchewan opens about twenty days earlier at Carlton House.) Fish ascend the small streams to spawn. The pike (*Esox lucius*) and dorè (*Lucioperca*) spawn. Sucking-carp (*Catastomi*) soon follow them. Trout take bait greedily. The cliff-swallow (*Hirundo fulva*, Vieillot) is seen. (On May 28th, 1849, Mr. Rae found this swallow on the banks of the Coppermine, having constructed its clustered nest against the cliffs at the mouth of Kendall River, lat. 67° N., but not yet laid its eggs.) Swamps and stagnant pools are thawed. Frogs begin to croak, and musquitoes to bite.

20. Shells (*Limnæi*) begin to move in the pools along the river. Snails (*Limax*, *Helix*, *Bulimus*, &c.) remove from under stones and fallen timber. The end of the month discloses some species of moths (*Noctualites*).

25. Our only goatsucker (*Chordeiles virginianus*), and the golden-

winged woodpecker (*Colaptes auratus*), the last of the spring birds in this district, arrive. Beavers, otters, and musk-rats have their young.

May 28. The balsam poplar, and aspen expand their leaves. (Not so early as on the Saskatchewan, two or three degrees further north.) The Hudson's Bay reindeer has young.

June 1. Sturgeon begin to frequent falls and rapids, and to spawn. 5th. Insects are busy on warm days; the *Tenthredinidæ* on bushes, the *Sphingides*, *Andrenetæ*, and *Pangoniæ*, on the ground, all attended by a variety of parasite ichneumons. The first flowers blow, and those of the willow are surrounded by *Sylphides* and flower-flies (*Anthomoyæ*).

10. A night-frost will sometimes intervene, even at this late period; and, in the woods, the ground is still frozen solidly at the depth of a foot from the surface. Vegetation, nevertheless, still goes forward. Musquitoes become a torment; the swamps and puddles swarm with their larvæ. Small tadpoles abound in the pools. 13th. The country is now covered with verdure. Birds are nestling; geese and ducks hatching. The natives are occupied with the sturgeon fishery.

15. The latest shrubs are in leaf, and the majority of moths and butterflies are disclosing themselves. The large *Ephemeræ* (*Perlæ* and *Phryganeæ*) issue from the water. 20th. Trout take the fly-hook. White-fish (*Corregonus albus*) rise to the surface. Cattle seek the houses to get rid of the tormenting *Tabani*. In dry seasons rivulets become low, though rivers retain their strength.

July Our warmest month. The river unusually gets low. Sturgeon fishing continues. Cattle become lean, being tormented by flies in the day, and feeding in the night only. 10th—20th. Many genera of *Coleoptera* appear, some of them of forms more characteristic of warmer climes; *Cicindelæ*, *Necrophori*, many *Buprestes*, and a species allied to *Lucanus*. Of those whose larvæ live on wood, the *Serropalpus*, a very fine *Dorcacenis*, *Cerambyx*, *Callidium*, *Lamia*, and numerous species of *Lepturetæ*. *Neuroptera* are abundant on the banks of the river—*Libellulæ*, *Agrion*, &c.; and on the leaves *Hemerobius*, *Panorpes*, *Sialis*. In the other orders, also, there are many genera to keep up the character of the month. It ends with bringing in strawberries, and in sending off sturgeon, which return to the depths.

Aug. The raspberries begin to ripen. Young ducks are well feathered. We have sultry weather for a few days, and then thunder-storms followed by chilly nights. 10th. Pigeons (*Ectopistes migratorius*) are numerous. Young geese (*Anser canadensis*) can fly. Gnats decrease, and sand-flies (*Similium*) replace them.

15. The raspberries, and red and black currants ripen. Grasshoppers are full grown. Trout move about, ascending the river. Sturgeon are very scarce. Grass becomes brown in dry situations. 20th. The noisy lesser tell-tale (*Totanus flavipes*) ap-

pears, and if we have much rain we are visited by a species of snipe. The golden and ring-necked plovers are not uncommon.

Sept. The air is generally cooler and the wind stronger, and frosty nights may be expected.

10. Trout spawn. Many insectivorous birds depart. The passenger pigeon disappears. Hawks and the large horned owl (*Strix virginianus*) are common. Night frosts frequent.

15. Tops of potatoes always blackened. Caterpillars nearly all cased. Trout refuse the fly-hook, but still take bait; they are now poor fish. Canada geese pass to the southward. Ducks abound in the grassy lakes. Leaves turning yellow rapidly. 20th. Pleasant weather at mid-day but cold in the night. The fall-moth (*Phalæna autumnalis*) is now to be seen. Sand-flies bite only in the height of the warmer days. The musquito is utterly defunct. Diving ducks common; the others gone.

Oct. 1. Pools and swamps crusted with ice. White-fish (*Corregonus albus*) begin to spawn. 5th. Sucking-carp (*Catastomi*) and trout desert the small streams. Foliage is yellow, and falls. Deer rut. Instead of rain we have snow, which generally melts, the earth being warmer than the atmosphere.

10. A single blast of northerly wind will now suffice to bare the trees, strip the shrubs, and send all water-fowl to the south. The last of these are weak or lingering flocks of waveys (*Anser hyperboreus*), which may be observed passing. They seldom alight unless when met by adverse winds.

20. The small lakes and rivers sometimes frozen. (Between the 21st and 28th is the the usual period for the freezing of the lake at Cumberland House.) Tullibee (*Coregonus artedi*) spawns. Quadrupeds acquire thick fur. The willow-grouse (*Tetrao saliceti*) arrives from the north. There is usually a little snow on the ground; and the American hare and ermine are becoming white.

Nov. 1. The ground is covered with snow, which in mild weather harbors multitudes of *Poduræ*. 10th. I have frequently observed at this time, and later, a wingless tipula (*Chionea hiemalis*) crawling about. 20th. Large rivers and the lakes often solidly frozen; strong rapids filling with ice, and setting fast. American hares and ermines entirely white. Swamps passable, and winter fairly set in.

(George Barnston.)

TABLE IX.

Temperatures between the Parallels of 52° and $61\frac{1}{2}$° North Latitude.

PERIODS.	Churchill. Lat. 59° 02′ N. Long. 93° 10′ W. Alt. 20 Feet. Year, 1838–39. Morning, Aftern., Evening. Corrected by Dove's formula.	Greenland. Fredericksthal. Lat. 60° N. Long. 45° W. Alt. 20 Feet. Dove.	Pelly Banks. Lat. 61° 30′ N. Long. 130° W. Alt. 1400 Feet. 5 Months. 1848, 1849. Sunrise, 3 P.M., Dusk. Corrected.
	(1.)	(2.)	(3.)
January	—21·21	19·62	—21·95
February	— 7·31	18·72	—14·73
March	— 4·63	22·10	— 0·99
April	16·29	27·50	20·44
May	28·42		
June	44·69		
July	56·80		
August	53·39		
September	36·03		
October	26·50	32·45	
November	3·32	35·15	
December	—14·00	29·75	—13·98
Year	18·19	30·00 ?	
Winter	—14·17	22·70	—16·88
Summer	[illegible]63		
Spring	13·36		
Autumn	21·95		
Difference of hottest and coldest months	78·01		
Difference of summer and winter	65·80		

No. 1. The register at Churchill was kept by Mr. Harding, of the Hudson's Bay Company's Service, from February, 1838, to May, 1839, the observations being made daily in the morning, afternoon, and evening. Column 2 is extracted from Dove's table; and for column 3 I am indebted to Murdock M'Pherson, Esq., who sent me a register kept at Pelly Banks by R. Campbell, Esq.

TABLE X.—Thermometrical Observations in the Valley of the Mackenzie; in Greenland; and on the Yukon.

PERIODS.	Godhaab. Lat. 64° 10′ N. Long. 52° 24′ W. Sea level.	Fort Chepewyan. Lat. 58° 43′ N. Long. 118° 20′ W. Alt. 700 ft.	Fort Resolution. Lat. 61° 10′ N. Long. 113° 51′ W. Alt. 500 ft.	Fort Simpson. Lat. 61° 51′ N. Long. 121° 51′ W. Alt. 400 ft.	Fort Reliance. Lat. 62° 46′ N. Long. 109° 00′ W. Alt. 650 ft.	Fort Enterprise. Lat. 64° 28′ N. Long. 113° 06′ W. Alt. 850 ft.	Fort Franklin. Lat. 65° 12′ N. Long. 123° 13′ W. Alt. 500 ft.	Fort Confidence. Lat. 66° 54′ N. Long. 118° 49′ W. Alt. 500 ft.	Yukon. Lat. 66° N. Long. 147° W. Alt. 200 ft. ?
	(1.)	(2.)	(3.)	(4.)	(5.)	(6.)	(7.)	(8.)	(9.)
January	12·38	—8·76	0·42	—12·54	—25·00	—15·57	—23·33	—21·57	—26·85
February	12·56	—4·01	—25·60	— 9·06	—18·84	—25·88	—16·75	—21·52	—26·44
March	15·60	3·08	9·95	5·55	— 6·14	—13·48	— 5·38	—20·21	—11·16
April	22·01	19·80	12·88	26·28	8·23	5·78	12·35	— 4·71	12·66
May	32·16	45·40	40·14	48·16	36·03	31·20	35·18	..	41·24
June	39·09	55·00	..	63·64	..	..	48·02	..	53·49
July	41·92	63·00	..	60·97	..	..	52·10	..	65·75
August	40·84	58·10	..	53·84	..	..	50·56	..	59·90
September	35·65	43·53	..	49·10	..	31·59	41·00	..	38·66
October	29·84	33·00	26·06	24·28	20·70	21·75	22·47	19·43	21·60
November	21·94	19·13	12·04	8·53	13·44	— 1·70	— 0·11	— 3·68	— 8·28
December	17·49	2·76	— 2·59	— 8·36	—17·07	—30·54	—10·88	—38·69	—18·43
Year	26·79	27·52	21·00 ?	25·75	16·00 ?	17·94	17·75	9·00 ?	14·58
Winter	14·14	—3·34	— 8·09	—10·00	—16·97	—24·00	—17·00	—27·26	—23·80
Summer	40·62	58·70	..	59·48	..	55·00 ?	50·41	..	56·73
Spring	23·26	22·76	20·99	26·66	12·21	23·50	12·69	..	14·60
Autumn	29·14	31·89	..	27·34	..	17·21	21·15	..	17·37
Difference of hot and cold months	29·54	71·76	..	76·18	..	86·88 ?	74·43	..	92·60
Difference of summer and winter.	26·48	62·04	..	69·48	..	79·00 ?	67·41	..	80·53

No. 1, extracted from Dove's table, Seventeenth Rep. Br. Ass. p. 376, gives the temperature of the west coast of Greenland, in that parallel, and is inserted for comparison.—No. 2 is compiled from 3½ years' observation by Messrs. Keith and Stewart in 1825 and 1826, and by another officer in 1838 and 1839. In the latter case, the times of observation were eight and eight, and the correction for these hours, calculated by Dove for Toronto, has been applied. The summer months were not entered in the register on the second occasion. Observations were also made at 2 P.M., and are recorded in another table.—No. 3 is abstracted from a register of observations made at eight and eight, in the winter and spring; Dove's correction has been used.—No. 4 is an abstract of 2½ years' observation by Murdock M'Pherson, Esq., in 1837, 1838, 1839, and 1840, at eight and eight, with Dove's correction for Toronto applied.—No. 5 is compiled from Sir George Back's observations in 1834 (fifteen daily).—No. 6 is abstracted from a register kept at Fort Enterprise in 1820, 1821, and 1822. The summer observations are defective; but the mean summer heats are inferred from observations made in traveling over the barren grounds further north.—No. 7. Two years' observations at Fort Franklin; first year nineteen observations daily, except in summer, during which, and in the second year, six observations daily at equal intervals.—No. 8. Fifteen to seventeen observations daily.—No. 9. Observations at six and six in summer; and in winter as early as the thermometer could be read in the morning, and as late in the evening. Dove's correction for these hours at Toronto has been applied.

TABLE XI.

TABLE OF THE SUMMITS OF THE DAILY OR MONTHLY CURVES OF TEMPERATURE IN THE VALLEYS OF THE MISSINIPI, MACKENZIE, YUKON, AND PELLY, BETWEEN THE PARALLELS OF $58\frac{3}{4}°$ AND 66° 54′ N., EXPRESSED IN DEGREES OF FAHRENHEIT.

PERIODS.	Fort Churchill. Lat. 59° 02′ N. Long. 93° 10′ W. Sea Level. (1.)		Fort Chepewyan. Lat. 58° 43′ N. Long. 118° 20′ W. Alt. 700 ft. (2.)		Fort Resolution. Lat. 61° 10′ N. Long. 113° 51′ W. Alt. 500 ft. (3.)	Fort Simpson. Lat. 61° 51′ N. Long. 121° 51′ W. Alt. 400 ft. (4.)	Fort Enterprise. Lat. 64° 28′ N. Long. 113° 06′ W. Alt. 850 ft. (5.)	
	2. P.M. Mean.	Monthly Maxima.	2 P.M. Mean.	Monthly Maxima.	Monthly Maxima.	2 P.M. Mean.	Means of Maxima.	Monthly Maxima.
January	—16·48	—2	—7.04	22	20°	—17·55	— 9·68	20
February	— 1·18	42	3·21	44	2	— 1·68	—19·10	1
March	8·51	37	16·72	47	37	12·68	— 0·90	24
April	31·66	58	39·12	65	32	47·22	16·40	40
May	37·94	65	33·33	76	64	56·98	42·80	68
June	55·17	82	62·00	97	..	73·40	..	..
July	66·81	86	69·00	78	..	70·61	..	..
August	62·97	83	66·10	72	..	65·87	..	..
September	54·09	84	46·00	65	..	60·00	39·30	53
October	33·32	46	41·06	56	47	30·08	27·90	37
November	8·00	23	16·71	39	31	12·51	2·80	25
December	— 9·77	5	4·95	40	20	— 7·98	—25·80	6
Year	27·59	86	32·61	97	..	33·51	..	..
Winter	— 9·14	..	0·38	44	20	— 9·37	—17·15	20
Summer	61·65	..	65·70	97	..	69·96	..	..
Spring	26·04	..	29·72	76	64	38·89	..	..
Autumn	31·80	..	34·59	65	..	34·15	23·33	..

PERIODS.	Fort Franklin. Lat. 65° 12′ N. Long. 123° 13′ W. Alt. 500 ft. (6.)		Fort Confidence. Lat. 66° 54′ N. Long. 118° 49′ W. Alt. 500 ft. (7.)		Yukon. Lat. 66° N. Long. 147° W. Alt. 200 ft. ? (8.)		Pelly Banks. Lat. 61° 30′ N. Long. 130° W. Alt. 1200 Feet. (9.)	
	Mean Temp. at 2 P.M.	2 Years Monthly Max.	Mean Temp. at 2 P.M.	Monthly Maxima.	Mean at 1 P.M.	Monthly Maxima.	Mean at 3 P.M.	Monthly Maxima.
January	—21·53	11·8	—17·83	26·0	—27·58	17	—16·61	16
February	— 8·24	27·8	—16·49	27·0	—23·55	10	—00·32	29
March	1·71	31·8	— 7·45	12·8	— 0·94	28	10·35	34
April	23·21	49·0	6·70	24·5	19·43	52	36·50	50
May	41·65	69·0	..	..	48·81	70		
June	51·00	75·0	..	..	62·00 ?			
July	59·02	80·0	..	..	74·84	76		
August	56·88	74·0	..	..	70·94	86		
September	44·82	66·0	..	..	52·73	69		
October	23·75	47·0	22·42	32·0	9·49	50		
November	4·81	32·5	0·41	19 0	— 5·40	28		
December	—12·07	27·5	—32·53	—7·0	— 5·39	22	— 0·34	4
Year	22·09	80	..	..	23·89	86		
Winter	—14·14	27·8	—22·28	27	—22·70	22	— 5·72	29
Summer	55·64	80	..	..	69·28	86		
Spring	22·18	69	..	..	22·71	70		
Autumn	24·45	66·5	..	.	26·03	69		

Progress of the Seasons at Fort Franklin, on Great Bear Lake, in lat. 65° 12′ *N., long.* 123° 12′ *N.*

The mean temperature of the three winter months varies comparatively little in different years; but the relative temperatures of these months differ greatly among themselves, so that in one year December is the coldest month, in another February, and in a third January. In some years the temperature of places exposed to the sun rises for a day or two in winter above the freezing point, and the snow moistens on the surface; but in other winters no thaw whatever occurs.

In March the snow is deepest, and averages about three feet, being, however, often drifted to a much greater thickness under cliffs and on the borders of lakes. In the end of March or beginning of April trees begin to thaw, the mean temperature in the shade being about zero Fah.; but the effect of the sun's rays on the blackened bulb of a thermometer being sufficient to raise the mercury to +90° Fah.

About the 10th of April the snow begins to thaw decidedly in the sunshine, and myriads of *Poduræ* are seen at such times moving actively in its cavities. Ptarmigan begin to assume their summer plumage toward the end of the month.

From the 1st to the 6th of May, according to the season, water-fowl arrive. The *Colymbus glacialis* and *arcticus?* arrive occasionally earlier, and frequent a piece of water at the efflux of Bear Lake River from the lake, which remains open all the year.

Swans (*Cygnus buccinator* and *americanus*) are among the early arrivals, the larger species coming first. The *Anas acuta*, *A. crecca*, *Clangula histrionica*, and *Oidemia perspicillata* make their appearance within the first eight days. Gulls come about the 9th or 10th. Singing birds, orioles, and swifts arrive about the middle of the month; the latter varying their time of appearance to a week later, if the spring is tardy. Pools of water and swamps must have been thawed long enough to release a sufficiency of winged insects for the support of the swallow tribe, before they show themselves in a district. About the 10th or 12th of the month small streams break up, the mean temperature of the ten preceding days having risen to 37° Fah. Bear Lake River, which is fed from the depths of the lake with warmer water, breaks its bonds at its efflux earlier. Lower down, this river remains fast till the first or second week of June. Mackenzie River usually opens at Fort Simpson about the 7th of May, and in the more northern quarters in the course of a week later; the boats which follow the flood in its descent taking about that time to go to Fort Good Hope. In 1849 the river broke up on the unprecedently late date of the 23d of May.

At this latter date there is bright light at midnight on Great Bear Lake, and the *Fringilla leucophrys* is employed with other songsters in singing at that hour.

Snow-geese arrive about this time, or a week earlier, and are followed in a fortnight by the laughing-geese. Both kinds wing their way northward in bands of from fifteen to forty individuals, which are passing every few minutes, day and night, for about three weeks. Many go on without halting; others alight in the marshes to feed on the nascent stems of the early *Cyperaceæ*, which are developed with marvelous rapidity after the commencement of the thaw, and, though still wrapped in the dead leaves of last season, have acquired juiciness and a sweet taste by the time that the snow has mostly gone. Toward the end of May, or in the first week in June, according to the earliness of the spring, *Chrysosplenium alternifolium*, *Arbutus alpinus*, *Eriophorum vaginatum* begin to flower, and the *Betula glandulosa* and some willows show their tender foliage and catkins. Early in June the *Potentilla fruticosa*, the *Rhododendron lapponicum*, and several anemones flower. Frogs at this time croak loudly; and by the middle of the month, summer may be considered as fairly established. About the 24th or 26th of July, ripe bleaberries (*Vaccinium uliginosum*) may be gathered. Strawberries are generally a week earlier, and the *Arbutus alpina* and *Rubus chamæmorus*, or cloudberry, are somewhat later.

In the beginning of August stars may be seen at midnight; and in the last week of this month the van flocks of snow-geese are seen going southward, having spent between eighty and ninety days at their breeding stations. The laughing-geese follow in a day or two: but they pass on in autumn without any of the delays that characterize their spring flights, which are necessarily checked as often as a few cold days arrest the melting of the snow on the sea-coast. Drift ice obstructs the navigation of the lake in some seasons till the first or second week of August.

In the last week of August, or in the beginning of September, snow falls, and by the 10th of the month the deciduous leaves begin to drop. By the 18th, most of the birds which breed in the district have migrated southward, a few water-fowl and the winter residents alone remaining. Between the first appearance of vegetation, till the falling of the leaves of deciduous trees, about a hundred days elapse; but although this may be taken as the length of the season for the growth of plants, some of the grasses continue to ripen their seeds till the beginning of October, notwithstanding much severe frost before that date. In ordinary seasons the frost sets in severely before the end of September, and the seeds of carices and grasses, instead of dropping off, are frozen hard in their glumes, and remain hanging on the culm till next spring, when they drop off into a soil prepared by the thaw for their reception. It is on these grass seeds of the preceding year that the graminivorous birds feed on their first arrival from the south. In October, when the soil begins to freeze again, the summer thaw has penetrated about twenty-one inches in the neighborhood of Fort Franklin. The small lakes are covered with ice by the 10th or 12th of the month; and, when that occurs, the last of the water-fowl depart. By the 20th of the month the smaller trees are frozen through, the larger ones remaining soft and moist in the centre. By the end of the month, or early in November, the young ice, filling the bays, puts an end to the navigation of the lake, after it has continued open about sixty days. The centre of the lake does not freeze over till late in December

TABLE XII.—Thermometrical Observations in the Arctic Seas.

PERIODS.	Melville Island. Lat. 74¾° N. September, 1819–August, 1820.	Port Bowen. Lat. 73¾° N. September, 1824–August, 1825.	Igloolik. Lat. 69⅓° N. August, 1822–July, 1822.	Winter Island. Lat. 69⅓° N. August, 1821–June, 1822.	Wolstenholme Sound. Lat. 76½° N. Six times daily. August, 1849–July, 1850.	Fort Hope, Repulse Bay. Lat. 62° 50′ N. Long. 86° 56′ W.
	(1.)	(2.)	(3.)	(4.)	(5.)	(6.)
September	+22·66	+25·90	+25·10	+31·62	+26·76	+28·57
October	− 3·02	+10·84	+13·72	+13·15	+11·32	+12·56
November	−20·85	− 5·00	−18·66	+ 7·80	−18·60	+ 0·68
December	−21·80	−19·05	−28·25	−14·24	−27·05	−19·27
January	−30·64	−28·91	−16·13	−23·13	−25·07	−29·32
February	−32·31	−27·31	−19·58	−24·01	−34·02	−26·68
March	−18·10	−28·36	−19·01	−10·78	−17·47	−28·10
April	− 8·30	− 6·50	− 0·85	+ 6·50	− 3·74	− 3·95
May	+16·74	+17·62	+25·14	+23·31	+25·82	+17·88
June	+36·22	+36·12	+32·16	+33·16	+39·73	+31·38
July	+42·43	+38·87	+38·58	+35·33	+40·52	+41·46
August	+32·63	+35·77	+33·88	+36·88	+33·67	+46·32
Year	+ 1·43	+ 4·33	+ 5·76	+ 9·80	+ 4·54	+ 6·14
Spring	− 3·13	− 5·74	+ 1·79	+ 6·35	+ 1·59	− 4·73
Autumn	− 0·43	+10·58	+ 6·80	+17·53	+ 6·55	+13·81
Summer	+37·09	+36·93	+35·07	+35·15	+37·97	+39·59
Winter	−28·36	−25·02	−21·38	−20·35	−28·53	−25·04
Difference of hot and cold months	74·74	67·78	66·83	60·89	74·54	75·64
Difference of summer and winter	65·45	61·95	56·45	55·50	66·50	64·43

Columns 1, 2, 3, 4, are extracted from a paper which I published in the Ed. Phil. Journ. for 1841. Column 5 is an abstract of a thermometrical register, kept by James Rae, Esq., surgeon of the Royal Navy, on board the North Star. An observation was made every four hours. Column 6 is from observations by John Rae, Esq., in 1846–7. The observations were made eight times in the twenty-four hours, at equal intervals; and the house was situated only a few feet above the sea level.

OBSERVATIONS IN SPITZBERGEN.

	June.	July.	August.	Sum. Quarter.
Hecla Cove, lat. 80° N..	+35·86	+40·17	+38·40	+38·15
Trent cruising, m. lat. 80° N., long. 10° E...	33·73	35·98	33 80	34.52

The preceding pages contain the temperatures of the districts through which the Expedition traveled, wherever I have been able to ascertain them, and also data for extending the lines of mean annual heat (*isothermal*), mean summer heat (*isothæral*), and mean winter heat (*isocheimenal*) across the continent. By comparing the sea-coast temperatures in Table II. and those of the shores of the great lakes in Table I. with those of places in France and Italy lying between the same parallels of 42°–45° north latitude, we perceive that the mean annual heat of Europe is from 8° to 15° Fah. greater than that of America at the same distance from the equator, while the summer heats differ only from 2° to 6°.* The inferior mean heat of America is therefore due principally to excessive winter colds, and this is decidedly the case in the interior. As the summer heats, however, regulate the culture of the *cerealia* and the growth of deciduous plants generally, the severe winters of America do not cause a scanty vegetation. From the 50th parallel northward the trees are frozen to their centres in winter; and, consequently, the development of buds and other vital processes which go on in the temperate climate of England, even in the coldest months, are completely arrested. This hybernation of plants increases in length with the severity and duration of winter which, generally speaking, augment in the interior of America with the latitude. The summer heats do not, however, decrease in the same ratio as we go to the north; on the contrary, the isothæral lines nearly follow the canoe route, and run to the northward and westward. The elevation of the prairie slopes has less influence in depressing the summer heat, than the nature of the soil and other causes have in raising it.

Experiments are still wanting whereby we may ascertain the ratio of the decrease of mean heat in America with the increase of altitude. In Table II. we find that, notwithstanding the eleva-

* Dove's table in the Report of the Brit. Association for 1847 has furnished the means of making this comparison. The places compared were Alais, Arles, Bordeaux, Dax, Manosque, Marseilles, Montpellier, Pau, Puy, Tarascon, Toulon, Toulouse, Perpignan, Alba, Bologna, Cascina, and Lucca. Oleron of Bearn has the same altitude with Lake Superior, and Mount Louis is 4900 feet high. All of them lie within the parallels of 43°–45° except the last which is in 42° 50′ N. The maps of isothermal lines of this author express the general results of the study of the table referred to.

tion of Franklin Malone above Eastport of 645 feet, its mean temperature is greater; its interior position giving it an advantage in summer heat over the sea-coast, which its greater altitude does not destroy. If we refer to Dove's table, and contrast the temperatures of Mount Louis, which lies near the 43d parallel, with those of low country situations enumerated at the foot of the preceding page, we find a mean difference of temperature of one degree of Fahrenheit for 350 feet of altitude. A similar allowance for the elevation of the successive steps of the St. Lawrence basin would place in still greater prominence the rise of the isothermal lines, and more still that of the isothæral ones, as they recede from the Atlantic coast. There is, however, this difference in the climate of the summit of a mountain and of an elevated plateau, that in the former case we approach near the line of invariable temperature, and the summer heat therefore differs less from the mean of the year, and more from that of the plains, than on a plateau where the depression of mean temperature produced by elevation is due chiefly to winter colds, and in a small degree only to defect of summer heat.

From Table II. also we may learn that the mean temperature of the coast districts of the Pacific is greater than that of the Atlantic countries, and, at the same time, more equable; the difference between the hot and cold months being less. We find in it an expression of the general fact, that the west coasts of continents are warmer than the east ones; and as Montreal and Fort Vancouver lie nearly in the same latitudes and at the same altitudes above the sea, and both are far enough removed from the coast to be beyond the direct influence of the sea breezes, columns 4 and 6 furnish the means of eliciting many of the peculiarities of climate on the two sides of the continent. Instead of four or five months of continuous snow and ice which Canada may be said to enjoy, for it is the season of general enjoyment, Oregon has an open, rainy winter, with little frost or snow; but, at the same time, a summer of less power.

Table V. exhibits even greater differences in the Pacific and Atlantic climates in a higher parallel. The course of the ocean currents, and the interposition of the peninsula of Alaska and its prolongation by the Aleutian chain of islands, protect the west coast of America from the masses of drift ice which in the same latitudes encumber and chill the Labrador coast for the most of the year. Even in the polar regions the west coasts have milder climates. Table X. shows, as far as it goes, that the mean temperature of the west coast of Greenland exceeds that of places on the continent, up to the 150th meridian, though the summer on the coast is greatly colder than that of the interior. By the study of Table XII. we learn that in the polar seas the summer heats vary little, as we might expect from the constant presence of ice; but the annual mean seems to decrease generally with the latitude, the only exception being that of Wolstenholme Sound, in which we have a confirmation of the greater mildness of the west Greenland coast. In the high latitudes the mean heat of the three winter months does not differ greatly in different years; but in some seasons one of these months, in others another, is the coldest, the temperature being ruled greatly by the prevailing winds.

Generally speaking, the mean annual temperature of places in the interior of North America falls within a degree or two of the mean heat of the two months of April and October. The mean temperature of the whole surface of the earth is, according to Dove's calculations, 58·2° Fah., being 54° in January and 62° in July. For such a mean annual heat in America we must descend to the 34th parallel of latitude; but the July heat of 62° Fah. extends northward to the Mackenzie.

The intense winter colds in the high latitudes are apparently in a great measure owing to the active nocturnal radiation into the clear blue sky. The observatory, which was a small log building without a fire-place, furnished us with the means of judging how much greater the depression of temperature in the night was in places exposed to the sky, than in those covered in.

The daily curve of atmospheric temperature for the three winter months was a bold and nearly regular hyperbolic curve, of which the mean was —25·2° Fah., the maximum —22·2° Fah., and the minimum —26·7° Fah.* The maximum occurs at 1h. 18m. P.M., and the mean line is crossed by the curve at 9h. 19m. A.M., and 6h. 29m. P.M.; the lowest temperature being reached at 7 A.M. The ascending branch of the curve, therefore, corresponds to an interval of 6h. 18m., and the descending one to 17h. 42m. During 14h. 51m. the temperature is below the mean, and it is only 9h. 9m. above it, which indicates a tolerably bold curve in the day, and a nearly horizontal course in the night.

In the observatory the mean for the same period was only —15·91° Fah., and the range no more than 0·97°. The maximum occurred at 6 P.M., being retarded 4¾ hours; and the minimum at 10 A.M. being delayed 3 hours. For most of the night the temperature was above the mean—such being the effect of the interposition of the building between the thermometer and the blue atmosphere. The walls of the observatory, it is necessary to remark, were by no means air-tight, and the door was opened at least once an hour in the day, and sometimes, especially on term days, much oftener. There was, consequently, a considerable and frequent admission of

* The correction for the error of the thermometer at low temperatures used in Tab. X. col. 9, was not applied to these numbers.

the external air; and, on the other hand, during the experiments on magnetic intensity, the heat of the observer's body had an evident effect in raising the temperature of the room.

I had intended to have instituted a series of observations, with Sir John Herschell's actinometer, on the nocturnal radiation, and also on the momentary intensity of the direct rays of the sun; but the instrument was unfortunately broken on the voyage. The Edinburgh New Philosophical Journal for 1841 contains the results of observations made at Fort Franklin, with the black bulb thermometer, on the heating power of the sun's rays, and I renewed these observations at Fort Confidence; but, as they were not carried on later than April, they furnish no information respecting the power of the sun in the months in which the processes of vegetation are active. As the black bulb thermometer indicates the accumulative effect of the sun's rays, it seems to be a useful instrument for ascertaining the heating power of the sun on the stems and larger branches of trees, at least, if not also on their leaves and on herbaceous plants. The hybernation of trees ceases long before the temperature of the atmosphere is sufficient to restore activity to the vegetative processes, and before the earth, still enveloped in its snowy covering, has felt the influence of returning spring. This is evidently mainly or wholly due to the sun's light direct or reflected; and perhaps its rays as reflected from the pure and glassy surface of the snow, after the days have increased considerably in length, may have the same powerful effect on the forest that, according to Professor Forbes, they have on the black-bulb thermometer. For some time after the trees have begun to thaw by day, they freeze again in the night; and in more southern localities, where the sugar-maple grows, the sugar makers are well acquainted with the fact that a hard frost arrests the flow of the sap in the night. Should a hot day, however, follow such an occurrence, the flow is more abundant than ever, the short rest seemingly increasing the irritability of the organs by which the sap is eliminated and circulated.*

* As I was revising this sheet, Sir William Hooker favored me with an extract from the journal of Mr. Berthold Seeman, botanist of the Herald, part of which follows: "During our stay at Port Clarence, in September, 1850, I made several experiments to ascertain the depth to which the thaw penetrated the soil: the result varied; in some places it did not descend above two feet into the earth, while in sandy places the ground was free from frost to the depth of four or five feet. The season was much colder than in 1849, the sea more loaded with ice, and the terrestrial vegetation less vigorous."

No. III.

ON THE GEOGRAPHICAL DISTRIBUTION OF PLANTS IN THE COUNTRY NORTH OF THE 49TH PARALLEL OF LATITUDE.

Generic and Specific Forms of Plants decrease in Number as the Latitude increases.—Analogy between Altitude and Increase of Latitude.—Culture of the Vine.—Of the *Cerealia*.—Maize.—Wheat.—Oats.—Barley.—Potatoes.—Botanical Districts.—Their Physiognomy.—Woodland District.—Barren Grounds.—Prairies.—Rocky Mountains.—Sitka.—Polar Plants.—Arctic Zone.—Trees and Shrubs.—Table of Distribution of Species in three several Zones.—Carices.

Though the *isothæral* lines, when the term is restricted to the mean temperatures of the three summer months of June, July, and August, run from Lake Superior northward to the Mackenzie, yet the short duration of the summer on the banks of that river, and the occasional frosts in June and August, and in some years even in July, render the climate unsuitable for numbers of vegetables which flourish in the northern districts of the United States. Many trees, shrubs, and perennial roots can be frozen without injury if the frost be continuous throughout the winter; and they acquire so much irritability in their hybernation, that the stimulus of perpetual though less fervid day within the arctic circle causes them to perform the functions of foliation and fructification with a rapidity unknown in more temperate regions. Other plants which need longer time to perfect their fruit or woody fibre, terminate in succession according to their several constitutions as the latitude increases. Their place is only partially supplied by other species, which have in like manner their equatorial and polar limits. These are not, however, so numerous as those which die out, there being no rule more general than the decrease of generic and specific forms in passing from temperate zones to arctic or polar ones.

There is a similarity in many respects between the vegetation of alpine tracts and that of high latitudes, but not an identity, the condition of the two regions differing in some essential particulars. No more apt illustration of this fact is needed than that adduced by Meyen, of Titicaca. This alpine lake, situated on the plateau of Chuquito in southern Peru, at the elevation of 12,700 feet, is surrounded by a rich and beautiful vegetation, which flourishes under a perpetual spring. On its banks a populous community, inhabiting magnificent cities, is supported by a fertile soil, yet trees are wanting in the country; whereas we have seen that the pine forest extends in North America to the 69th parallel, beyond the limits of the *cerealia*. On the shores of Lake Titicaca barley and oats grow, but wheat does not succeed, and maize is raised only by artificial heat. In respect of these products, therefore, its climate agrees with that of Fort Simpson on the 62d parallel. Its summer heat, which, according to Meyen, ranges between 52° and 66° Fah., is more equable than that of Fort Simpson which has a mean daily summer temperature of about 59° Fah., with a mean at noon of 70° Fah.; and a range of from 90° Fah., down to below the freezing point.

It is necessary to remark, however, that the decrease of vegetable forms with an increase of latitude has more analogy to that which is observed on a lofty isolated mountain than on an elevated plateau; and plants actually grow on the summits of the White Mountains of New Hampshire which are not met with again until we reach the shores of the Arctic Sea.

The peculiarities of the climate of Canada and Rupert's Land may be in part shown by reference to a few of the plants usually cultivated for food. The *vine* would thrive with the summer heat of Fort Simpson were the season long enough; but the September and October heats, which are required to ripen its fruit, do not occur in any district of Rupert's Land; and the grape is destroyed by the severe night frosts which are frequent in autumn even in so low a parallel as the north shore of Lake Superior. The conditions essential to the due growth of the vine, mentioned by Meyen,* do not extend in the basin of the St. Lawrence beyond the 43d parallel, while on the Rhine wine is a profitable production up to the 51st.

Maize is a plant which thrives best in the dampest and hottest tropical climates, where it brings forth eight hundred-fold. Its culture extends into temperate regions, but with a greatly diminished yield; and it is cultivated near its northern limit only as a green vegetable, the grain seldom ripening, and being eaten in its milky state. This is its condition in most parts of Great Britain, when reared in the open field. On the western shore of Europe it is not cultivated beyond 46°, though in the valley of the Rhine it extends to 49° north lat. In South America, on the Chili coast, it is

* This author states that the culture of the vine is regulated more by the length of summer than by its high temperature, though the latter is also an element in the proper ripening of the grape. It will succeed, he says, under every tropical heat, provided the atmosphere be not too moist. It thrives well under a mean heat of 60° Fah.; it ripens with a lower mean heat of 48° Fah., and a summer heat of 68° Fah., but the juice contains less sugar and yields less alcohol. (Meyen, Geogr. of Plants.)

planted as low as 40° south lat.; and on the Peruvian plateau, at the height of 12,000 feet, above which it requires artificial shelter and warmth. A profitable return can be obtained from it in Rupert's Land between the 49th and 51st parallels, where, however, the vine does not accompany it, as on the banks of the Rhine. Garden cultivation and shelter from spring frosts would extend its cultivation in Rupert's Land even higher than in England. On the fertile acclivity of Young Street which leads from Toronto to Lake Simcoe, and crosses the 44th parallel of latitude, we may behold heavy crops of maize, and cucumbers and gourds, ripening, in the same field, with but little expenditure of care or labor, though the mean annual heat, being 41° Fah., is inferior to that of the Orkney and Shetland Islands, where barley, one of the most northern of the cereals, grows imperfectly. The summer heat of Young Street, however, exceeds that of any part of the British Isles.

Wheat is the cereal which requires most heat of those usually cultivated in England. Its culture is said to ascend to 62° or 64° north lat. on the west side of the Scandinavian peninsula, but not to be of importance beyond the 60th. On the route of the Expedition it is raised with profit at Fort Liard in lat. 60° 5′ north, long. 122° 31′ west, and having an altitude of between 400 and 500 feet above the sea. This locality, however, being in the vicinity of the Rocky Mountains, is subject to summer frosts; and the grain does not ripen perfectly every year, though in favorable seasons it gives a good return. At Dunvegan, on Peace River, lying in lat. 56° 6′ north, long. 117° 45′ west, and at an altitude of 778 feet, the culture of this grain is said to be equally precarious. It grows, however, freely on the banks of the Saskatchewan, except near Hudson's Bay, where the summer temperature is too low. From Mr. M'Pherson I learnt, that on the west side of the Rocky Mountains good crops of wheat are raised with facility at Alexandria, on Frazer's River, in lat. 52° 30′ north, long. 122° 40′ west, and 300 or 400 feet above the sea; also at Fort George, on the same river, more than a degree further north, and 100 feet higher.

At Fort James, on the borders of Stuart's Lake, in lat. 54½° north, in a mountainous region near the source of Frazer's River, wheat continues to grow, but often suffers from the summer frosts. In these quarters the grain comes to maturity in about four months. In the colony of Red River its growth is luxuriant, though the upper part of that country, which touches the 49th parallel of latitude, is elevated about 1000 feet above the sea. Periodical ravages of grasshoppers, however, frequently destroy the hopes of the husbandman.

At Fort Francis, situated on the banks of Rainy River in lat. 48° 35′ north, long. 93° 28½′ west, wheat is generally sown about the 1st of May, and is reaped in the latter end of August, after an interval of about 120 days. In 1847 multitudes of caterpillars spread like locusts over the neighborhood. They traveled in a straight line, crawling over houses, across rivers, and into large fires kindled to arrest them. Throughout the whole length of Rainy River, on the Lake of the Woods, and on the River Winipeg, they stripped the leaves from the trees, and ate up the herbage. They destroyed the *folle avoine* on Rainy Lake, but left untouched some wheat that was just coming into ear. This was the first time that Fort Francis had experienced such a visitation. When we passed that way in 1848, the still leafless trees were covered with the cocoons of last year, in each of which there remained the hairy skin of a caterpillar.

On the island of Sitka, lying in 57°—58° north latitudes, though the forest, nourished by a comparatively high mean temperature and a very moist atmosphere, is equal to that of the richest woodlands of the northern United States, yet corn does not grow.

In the middle temperate zone of France wheat is cultivated to the height of 5400 feet only. In Mexico its culture commences at the altitude of 2500 or 3000 feet, and ascends to more than 9000 feet. On the plateau of southern Peru, 8000 feet above the sea, its yield is extraordinary; and on the foot of the volcano of Arequipa it succeeds as high as 10,000; but it will not grow in the equable temperature of Lake Titicaca, the heat there not being sufficient to ripen either it or rye. It requires for its growth, says Meyen, the mean annual heat of 56° Fah.; a much inferior heat is, however sufficient in the extreme climate of subarctic America, provided the summer heat for 100 or 120 days be great enough.

Oats are litle cultivated in Rupert's Land; they require longer time than barley to ripen, and are therefore not likely to grow so far north. They have not been tried at Fort Norman, however, which is the most advanced post in that direction where barley is cultivated. Mr. M'Pherson finding some grains of oats accidentally in a barley field, propagated them, and raised some good crops on the River of the Mountains, and I believe also at Fort Simpson. On the Scandinavian peninsula this grain is said to extend to 62½° N. and 65° N., but, even on the latter parallel, falling five degrees short of the latitude which barley reaches. Meyen saw ripe oats at Lake Titicaca.

In good seasons *barley* ripens well at Fort Norman on the 65th parallel, as has been mentioned in the Narrative (page 104.) All Mr. Bell's attempts to raise it at Fort Good Hope, two degrees further north, failed. It reaches, as we have just observed, the 70th degree of latitude on the Scandinavian peninsula, and it is cultivated for green fodder in Peru up the height of 13,800 feet, but seldom ripens its grain higher than 10,000 feet. (Meyen.)

Potatoes, which have been cultivated from time immemorial on the banks of Lake Titicaca, yield abundantly at Fort Liard, and grow, though inferior in quality, at Fort Simpson and Fort Norman. They have not succeeded at Fort Good Hope, near the 67th parallel. At the latter place *turnips* in favorable seasons attained a weight of from two to three pounds, and were generally sown in the last week of May. At Peel's River the trials made to grow culinary vegetables had no success. Nothing grew except a few cresses. Turnips and cabbages came up about an inch above the ground, but withered in the sun, and were blighted by early August frosts.

In the preceding narrative, as well as in the geographical sketch, we have had frequent occasion to allude to several great divisions of North America, each of which has a peculiar physiognomical character in its vegetation.

1st. The eastern woodland country constitutes the first division, in which the forest extends from the Atlantic westward till it meets the great prairies.

2d. The second division lies to the north, of the forest lands, and is appropriately named the "Barren Grounds." This tract has its greatest north and south extension on the eastern coast. On the shores of Hudson's Bay and the Welcome it reaches from the 60th or 61st parallel to the extremity of the continent, but narrows to the westward; since the boundary line of the wood takes a diagonal or northeast direction from the 91st meridian, and, before reaching the 120th, has risen to the 67th parallel. Further to the west the Barren Grounds form a border to the Arctic Sea of greater or less breath according to the northerly prolongation of the continental promontories, since the southern limit is nearly coincident throughout with the arctic circle, on which it approaches Beering's Straits—clumps of spruce fir,* the usual outliers of the forest, having been observed on Buckland or Noatak River which falls into Eschscholtz Bay. The fertile alluvial deposits of the well-sheltered valley of the Mackenzie interrupt the continental continuity of the Barren Grounds by carrying the woods nearly to the sea-shore; but there seems to be no other material indentation of the barren district; and even on the Mackenzie the valley is bridged, as it were, by the naked summits of the alpine ridges.

3d. The prairie slope forms a third physiognomical district of vegetation, which has the greatest transverse expansion on the Missouri, and, narrowing as it goes north, runs out on the 60th parallel,

* The species is doubtful.

having, after passing the Saskatchewan, been much indented by the woods which feather the numerous rivers that drain the declivity. These prairies have much analogy with chalk downs in aspect as well as in mineral constitution.

4th. The Rocky Mountain chain, and the alpine ranges and isolated peaks which rise to the westward of it, may be considered as a fourth district which nourishes some peculiar species of plants.

5th. And the lower woodland country on the Pacific side of the range forms a fifth.

If we trace any one of these districts northward, making due allowance for the varying altitude of the country above the sea, we may ascertain the effect of increase of latitude on the vegetation of that meridian; but, if we compare one district with another, we must keep in view the climatological fact of the rise of the isothermal lines in proceeding westward. The course of the forest boundary is one illustration of this phenomenon; and we have another in the range of certain species or forms constituting that forest. Thus the *Cupressus thyoides* is rare beyond the 49th parallel in the eastern district, and terminates altogether along with the *Thuya occidentalis* on the 53d, while the magnificent *Cupressus* or *Thuya nutkatensis* adorns the forests of Norfolk Sound on the 58th parallel of the Pacific coast. The distribution of the *Pinus inops*, *Abies canadensis*, *Rubus nutkanus*, and of some other conspicuous trees and shrubs, show that the vegetation of the district of Sitka on the north-west coast is equal and similar to that of the eastern States of Wisconsin or Minnesota eight degrees further south.

The physiognomy of the *woodland district* through which the canoe route lies has been incidentally touched upon in the descriptions of several localities that have been introduced into the narrative, yet it will not be out of place to recall its general features here. Of this district, which has a breadth of about 600 geographical miles between the 50th and 55th parallels, the white spruce is the most abundant and characteristic tree; yet up to the 54th parallel it is conjoined, and especially on the banks of the rivers, with other trees which break the monotony of the dark evergreen forest. Beyond the banks of the Saskatchewan the oaks, elms, ashes, maples, bass-wood, white thorns, Virginian clematis, and various other trees and shrubs cease to grow; and the white spruce may be said to cover the face of the country, except on the alluvial borders of rivers and lakes, where the aspen, balsam poplar, balsam fir, alder, and multitudes of willows usurp its place, or on the edges of swamps where the black spruce leads a lingering, unhealthy existence. With the black spruce the larch is often associated; though it is not confined to morasses, yet it is too much isolated in its distribution to produce a difference of tint sufficiently massive to please the eye, except in very few localities. The Banksian pine is more frequently seen in considerable patches, and its appearance is agreeable to the voyager; for, independent of the fact that its spreading branches and general form, resembling that of the Scotch fir, is a rest to the eye wearied with the tapering stiffness of the spruce, it offers the prospect of a dry and comfortable encampment. It always grows in a sandy soil, and is remarkable among the Rupert's Land trees for its freedom from underwood. Not so the white spruce, which admits of a thick undergrowth of willows, cornel bushes, viburnum, roses, brambles, and goosberries; and in the country south of Lake Winipeg, of maples, American yews, and many other shrubs and trees. The willows, especially when conjoined with the falling or inclined stems of forest trees—the growth of bygone centuries—form a barrier to the progress of a white man in the forest; but the slim and agile native glides through the tangled thicket with a noiseless and ghost-like ease, impassive to the annoyance of the musquito clouds that darken the air. The prickly twining *Panax horridum*, which interlaces and arms the brushwood on the north-west coast up to the 58th parallel, has no representative on the east side of the continent, except perhaps the *Aralia hispida*, which, though of the same family, has feeble defenses, and is not a climber. The *Cratægi* have the most offensive weapons of any of the shrubs in Rupert's Land.

Even beyond the Saskatchewan, where the maples, ampelopsis, and some other trees and shrubs whose leaves assume the orange and red tints before they fall, cease to grow, the river banks are enlivened by the bright purplish shoots of the white cornel berry (*osier rouge*) and the gay spires of the *Epilobium angustifolium*, which rise above a man's height in the alluvial deposits, and are varied also by other shrubs that have been noticed in the descriptions interspersed through the preceding pages. These are merely the foreground incidents, however; the sombre spruce every where forms the background.

The agency of man is working a change in the aspect of the forest even in the thinly peopled north. The woods are wasted by extensive fires, kindled accidentally or intentionally, which spread with rapidity over a wide extent of country, and continue to burn until they are extinguished by heavy rains. These conflagrations consume even the soil of the drier tracts; and the bare and whitened rocks testify for centuries to the havoc that has been made. A new growth of timber, however, sooner or later springs up; and the soil, when not wholly consumed, being generally saturated with alkali, gives birth to a thicket of aspens instead of the aboriginal spruce.

The frozen subsoil of the northern portions of the woodland

country does not prevent the timber from attaining a good size, for the roots of the white spruce spread over the icy substratum as they would over smooth rock. As may be expected, however, the growth of trees is slow in the high latitudes. On the borders of Great Bear Lake, four hundred years are required to bring the stem of the white spruce to the thickness of a man's waist. When the tree is exposed to high winds, the fibres of the wood are spirally twisted; but in sheltered places, or in the midst of the forest, the grain is straight and the wood splits freely.

At the limit of the woods the white spruce is every where the most advanced tree, growing either solitarily, with its branches clinging to the ground and its dwarfed top bent from the blast, or in small clumps in some favorable locality. The *Salix speciosa* may indeed be said to pass beyond the spruce; but it does so only on the alluvial points of rivers, and not in its tree form.

Though the species of plants become less numerous as we advance northward through the woody region, there is no falling off in the number of individuals of the species that remain. For not only is the forest crowded, and often almost impenetrably so, when the trees are young, but on the margins of rivers, and other open places, there is a dense herbaceous vegetation, which clothes the ground in Rupert's Land as perfectly as it is covered in a lower latitude, though the vegetation be less rank. On the inundated alluvial flats tall carices grow as closely as they can stand, and furnish an abundance of nutritious hay. There is, however a total absence in the north of the *Lianas*, *Tillandsiæ*, and parasitic *Orchideæ* which impart so peculiar an aspect to the forests in some of the warmer districts of the earth. The great hedge bindweed (*Calystegia*), the Virginia creeper, the hop plant, and the twining herbaceous *Smilacina*, with its grape-like clusters of blackberries, disappear on the south side of Lake Winipeg, and the only aerial parasite in the north is the leafless *Arceuthobium oxycedri*, which seats itself on the branches of the Banksian pine. The graceful *Usneæ* which hang from the branches of the ancient black spruces in long, thread-like hanks, have, it is true, some resemblance to the *Tillandsiæ* which forms an elegant drapery to the evergreen oaks of Georgia and Florida.

In the eastern woodland district, from the St. Lawrence to the Saskatchewan, the *Compositæ* are the most numerous family of plants, and they form between the sixth and seventh of the whole phænogamous vegetation. Next to them come the *Cyperaceæ*, which owing to the great development of the genus *Carex*, constitute more than one-ninth of the *Phanerogamia* of the district.

In the *second*, or *barren ground district*, in places where the soil is formed of the coarse sandy *debris* of granite, and is moderately dry, the surface is covered by a dense carpet of the *Cornicularia tristis, divergens, ochroleuca*, and *pubescens*, mixed in damper spots with *Cetrariæ cucullata* and *islandica*. In more tenacious soils other plants flourish; not, however, to the exclusion of lichens, except in tracts of meadow ground. The *Rhododendron lapponicum, Kalmia glauca, Vaccinium uliginosum, Empetrum nigrum, Ledum palustre, Arbutus uva ursi, Andromeda tetragona*, and several depressed or creeping willows, lie close to the soil, their stems short, twisted, and concealed, with only the summits of the branches showing among mosses or lichens. Here and there, on the moister sides of the hills, there is a gay display of saxifrages, pediculares, or primroses; and a few of the sandy spots on the coast are enlivened by a beautiful dwarf phlox or a handsome dodecatheon. On the alluvial banks of rivers only are willows of erect growth to be found, and of the *Salix speciosa* is the most robust and the handsomest. On the sandy shore of the sea the *Pisum* (or *Lathyrus*) *maritimum*, the *Polemonium cæruleum*, various blue and yellow *Astragali*, and several *Artemisiæ* flourish. Most of these plants also occur, though more sparingly, in the interior. One circumstance which came under my observation, and has been cursorily alluded to in the Narrative (page 192), is the existence of very ancient stumps of trees, either solitary or grouped, in various places of the barren grounds, seemingly the vestiges of the forest, which had spread more widely over the country some centuries ago than in the present day. Further evidence that such was the case may be obtained in the extension of *Pyrolæ*, and some other woodland plants to the coasts of the Arctic Sea. On the sheltered banks of rivers, even in the barren grounds, clumps of living trees occasionally occur; but the stumps I speak of stand often on the exposed side of a hill, and indicate a deterioration of climate, however that may have been produced. We saw no young firs growing up in such situations to leave similar vestiges in a future age.

In many sheltered valleys on the sea-coast, and even in the more elevated interior, especially where a fertile soil has been formed by the decomposition of trap rocks, there is a good growth of grasses, several of which flourish well on lands that are occasionally inundated by the sea. Among these are *Elymus mollis, Spartina cynosuroides, Calamagrostis stricta, Carices stans, compacta, glareosa, membranacea*, and *livida, Colpodium, Deschampsia, Festuca*, and several *Poæ*. In some of the maritime meadows to which the reindeer resort to bring forth their young, there are treacherous mud-banks, which are soft enough and deep enough to swallow up a deer or musk ox, that may rush heedlessly into them when chased by a wolf; but in general the frozen subsoil is so near the surface as to preclude any such accident. The existence of these boggy places, which were seen only on the sea-coast, scarcely affords a satisfactory solution to the problem of the entombment of a living elephant or rhinoceros, and the subsequent preservation of the entire carcass in the frozen soil. But in whatever manner this may have taken place, I should infer, from the economy of the arctic regions, that these animals were migratory, like the reindeer of the present day, and wintered in milder climates.

On approaching the arctic circle the relative proportion of the *Compositæ* greatly decreases, and that of the *Cyperaceæ* increases within the woody tracts, though it falls off on the barren grounds. Taking the whole zone between the arctic circle and the extremity of the continent, which includes much woodland, the *Cyperaceæ* are the most numerous family of plants, and are more than double the *Gramineæ*. The *Cruciferæ* come next to the *Cyperaceæ* in this zone. In the polar regions beyond the continent, the *Cruciferæ* take the first place in respect of number of species, then come the *Gramineæ*, which are closely followed by the *Saxifrageæ*.

The *third*, or *prairie district*, has the prevailing aspect of a grassy plain, the herbage, however, having a considerable intermixture of carices among the true grasses. The herbage grows up rather wiry in the dry summers of that region; but, in consequence of the fires that frequently spread over vast tracts, a young growth takes place, to which the bison and deer resort. On the Arkansas, the "buffalo or bison grass" is the *Sesleria dactyloides*. Whether this species extends to the Saskatchewan or not, I am unable to say: we certainly did not gather it there; but at the time that Mr. Drummond and I visited that part of the prairie, recent fires had made flowering specimens of grasses very rare. Of the phænogamous prairie plants actually collected, the *Gramineæ* form about the eleventh, and the *Cyperaceæ* the sixteenth. On the plains the *Compositæ* are numerous and showy; there is a considerable variety of handsome *Leguminosæ*, with some pretty *Boragineæ;* and the *Artemisiæ*, owing to the quantity of surface they cover, though the species are not numerous, contribute greatly to the hoary aspect of the prairie vegetation. The *Rosaceæ* vie with the *Cyperaceæ* in number of species; but many of them are fruit-bearing shrubs, growing on the banks of the rivers that serpentine among alluvial points, in channels sunk deeply below the general surface of the prairie.

Between the 32d and 33d parallels, on the Gila and Rio del Norte, west of the Rocky Mountain ridge, Colonel Emory gathered many examples of *Cacti*, of which Dr. Engelman has described fifteen species belong to the genera *Mammillaria*, *Echinocactus*, *Opuntia*, and *Cereus*. Among these the *Pitahaya*, or *Cereus giganteus*, is the most remarkable, as it grows in the shape of a candelabra, or

Titanic tuning-fork, with three or four prongs, to the height of sixty feet. Cacti are numerous also on the eastern side of the mountains in the same parallel; and the smaller kinds, chiefly *Opuntiæ*, range northward over the prairies to the 49th parallel, and perhaps still further north. We gathered *Opuntia glomerata*, or the *crapaud verd* of the voyagers, on the Lake of the Woods; and a species of the same section of the genus attains an equally high parallel on the Pacific coast.

With the physiognomy of the vegetation on the Rocky Mountains, and of the district to the west of that range, I have no personal acquaintance, and borrow the following notice of the vegetation of Sitka from Bongard. Sitka is situated in the entrance of Norfolk Sound, on the 57th parallel, near an extinct volcano, named Mount Edgecumbe, which marks the entrance of the sound. The most remarkable mountain in the immediate vicinity of the settlement is Westerwöi, which is 3000 Parisian feet in height, and is clothed to its summit by a dense forest of pines and spruces, some of which acquire a diameter of seven feet, and the prodigious length of 160 feet. The hollow trunk of one of these trees formed into a canoe, is able to contain thirty men, with all their household effects. The climate of Sitka is very much milder than that of Europe on the same parallel. The cold of winter is neither severe nor of long continuance; but the atmosphere is charged with vapors, whose condensation occasions almost constant rains. In the month of July the sun is seldom visible on more than three or four days, and then only for an instant. This humidity gives astonishing vigor to the vegetation, yet corn does not grow there; and, in fact, the want of level surface is an impediment to cultivation. In six weeks the botanists collected 222 species of plants, of which thirty-five were new to science

Of the *Polar plants*, amounting to ninety-one species, which inhabit Melville Island, the shores of Barrow's Straits and Lancaster Sound, and the north coasts of Greenland, between the 73d and 75th parallels of latitude, about seven-ninths range to Greenland, Lapland or Northern Asia. Of the remainder, some have been gathered on the shores of the Arctic Sea from Baffin's Bay to Beering's Straits; and it is probable that if these high latitudes were fully explored, the flora of the entire zone would be found to be uniform. Some of the more local plants will perhaps be ascertained, on further acquaintance, to be mere varieties altered by peculiarities of climate. That the flora as well as the fauna in the high northern latitudes is nearly alike in the several meridians of Europe, Asia, and America, has long been known. And even when we descend to some distance south of the arctic circle, we find that this law is superior to the intrusion of high mountain chains, and is but partially infringed upon. In taking the St. Lawrence basin for instance, if we allow for the rise of the isothermal lines on the west coast, and make our comparisons in an oblique zone, including Sitka and Wisconsin, we shall find that there is much similarity in the floras on the two sides of the continent. The Rocky Mountain ridge is not by any means a boundary to the peculiar vegetable forms of the Pacific coast; on the contrary, many of them cross the ridge to its eastern declivity, though they do not descend into the low country; and there is actually more similarity between the vegetation of the prairies of Oregon, and those of the Missouri and Saskatchewan, than there is between the latter and eastern parts of the United States and Canada. In still more southern latitudes the case may be different; and Ehrenberg has found a totally different group of *Infusoria* in California from that which exists on the east side of the continent; the Rocky Mountains, in his opinion, proving a complete barrier to these organisms.

The families of *Polar plants* which are most rich in species are the *Cruciferæ*, *Gramineæ*, *Saxifrageæ*, *Caryophylleæ*, and *Compositæ*. Of these the *Saxifrageæ* are most characteristic of extreme northern vegetation. All of them that inhabit the 74th parallel in America are found also in Spitzbergen, Lapland, or Siberia; and even the polar species are twice as numerous as those which exist in the wide district which Gray's "Flora of the Northern States" comprehends. If we reckon all that enter the arctic circle, we shall find them to be four times as many as those which Dr. Gray enumerates; and we may add that the plant which Humboldt traced highest on the Andes was a saxifrage. The *Caryophylleæ* and *Crucifereæ*, which vie with the saxifrages in number on the 74th parallel, include many of the doubtful local species above alluded to. Of the most northern *Gramineæ*, about one half are, as far as we yet know, exclusively American; the few species which the other families contain have as extensive a lateral range as the saxifrages.

Arctic zone.—On descending to the main land from the 71st parallel down to the arctic circle, including a zone of four degrees of latitude, we find that the species have increased eight-fold in number, and there is a large addition of generic forms, as might be expected on entering within the limits of the forest.

The Polar families are—

Ranunculaceæ	Compositæ	Polygoneæ
Papaveraceæ	*Cichoraceæ*	Salicaceæ
Cruciferæ	*Eupatoriaceæ*	Junceæ
Caryophylleæ	*Senecionideæ*	Cyperaceæ
Leguminosæ	Campanulaceæ	Gramineæ
Rosaceæ	Ericeæ	Lycopodineæ
Onagrariæ	Polemoniaceæ	Equistacéæ
Saxifrageæ	Scrophularineæ	Cryptogamia

In addition to the above the following enter the *Arctic Circle*—

Sarracenieæ	Caprifoliaceæ	Chenopodieæ
Fumariaceæ	Valerianeæ	Eleagneæ
Violarieæ	Compositæ	Santalaceæ
Droseraceæ	*Asteroideæ*	Empetreæ
Polygaleæ	Vaccinieæ	Urticeæ
Lineæ	Monotropeæ	Betulaceæ
Balsamineæ	Gentianeæ	Coniferæ
Celastrineæ	Diapensiaceæ	Juncagineæ
Halorageæ	Hydrophylleæ	Aroideæ
Ceratophylleæ	Boragineæ	Naiades
Portulaceæ	Orobancheæ	Smilaceæ
Crassulaceæ	Labiatæ	Melanthaceæ
Grossularieæ	Verbenaceæ	Asphodeleæ
Umbelliferæ	Primulaceæ	Orchideæ
Araliaceæ	Plumbagineæ	Irideæ
Corneæ	Plantangineæ	Filices

I made a pretty full collection of lichens and mosses within the arctic circle; but since so many of them are almost cosmopolites, and a still greater number are common to both the temperate and frigid zones, under similar conditions of moisture and exposure, I have avoided swelling the lists with their names. *Fungi* are not wanting in the northern regions, but the difficulty of preserving them prevented me from gathering many. All the families in the above two lists are represented in England, except *Diapensiaceæ*, which is a Lapland form; and *Sarracenieæ* and *Araliaceæ* which are more purely American.

Between the arctic circle and the south side of the Winipeg or Saskatchewan basin on the 50th parallel, embracing the entire width of the continent, the following families make their appearance—

Berberideæ	Terebinthaceæ	Euphorbiaceæ
Nymphæaceæ	Cucurbitaceæ	Ulmaceæ
Capparideæ	Loranthaceæ	Cupuliferæ
Cistineæ	Paronychieæ	Myriceæ
Malvaceæ	Jasmineæ	Liliaceæ
Tiliaceæ	Apocyneæ	Alismaceæ
Hypericineæ	Asclepiaceæ	Pontederiaceæ
Acerineæ	Convolvulaceæ	Restiaceæ
Ampelideæ	Solaneæ	Hydrocharideæ
Geraniaceæ	Amaranthaceæ	Marsiliaceæ
Oxalideæ	Aristolochieæ	Salvinaceæ
Rhamneæ		

The families which reach the St. Lawrence basin, but do not extend northward to the Winipeg valley, or enter the western prolongation of that zone, are—

Menispermaceæ	Hamamelideæ	Saurureæ
Podophylleæ	Compositæ	Juglandaceæ
Limnanthaceæ	*Vernoniaceæ*	Platanaceæ
Oxalideæ	Acanthaceæ	Camelineæ
Rutaceæ	Nyctagineæ	Hypoxideæ
Lythrarieæ	Phytolacceæ	Dioscoreæ
Cacteæ	Laurineæ	

To give a further view of the accession of families in going southward, the following are added from Dr. Gray's "Botany of the Northern States"—

Magnoliaceæ	Hippocastanaceæ	Nyssaceæ
Anonaceæ	Melastomaceæ	Podostomeæ
Cabombaceæ	Hydrangeæ	Balsamifluæ
Resedaceæ	Aquifoliaceæ	Amaryllidaceæ
Elatinaceæ	Ebenaceæ	Hæmodoraceæ
Anacardiaceæ	Bignoniaceæ	Xyridaceæ

In tracing individual species to their northern limits, we did not discover in any one instance that the crest of a water-shed between successive transverse river basins was a boundary to the plant. Many of the more remarkable trees, oaks, &c., flourish in the neighborhood of Rainy Lake and on the upper part of Red River, but die out on approaching the south end of Lake Winipeg. Others go a degree or two further north to the banks of the Saskatchewan, about Cumberland House, and there make their last appearance; among these are the ashes, elms, and maples. Some which are not seen beyond that locality on the canoe route, go three or four degrees further north on the western side of the prairies, in the sheltered valleys of the Rocky Mountains. In these valleys also the lamented Drummond found a considerable number of the species of the Pacific coast, their range not being cut short by the dividing ridge, but being seemingly more effectually limited by the dry prairies. It is unfortunate that the vertical limits of the species gathered by Drummond in the mountains were not noted, as a careful list containing that element, and which no one was more able than he to make, would have conveyed much information with respect to the distribution of plants. The statistical enumeration of the mountain species, collected between 52° and 57°, in the subjoined table evidently contains a mixed flora; some families having an arctic, almost a polar character; others a subarctic, or almost temperate one.

LIST OF TREES AND SHRUBS.*

Ranunculaceæ.—*Clematis virginiana* is common to Oregon, the eastern United States, and Canada, and extends northward to the Saskatchewan.

Berberideæ.—*Berberis vulgaris* has been found in Canada, Newfoundland, and New England, and is considered as having been introduced from England. The pinnate-leaved barberries or *Mahoniæ* are natives of Oregon, and perhaps extend northward into New Caledonia or Vancouver's Island.

Cistineæ.—*Hudsonia tomentosa* grows in New Jersey and Canada, on the borders of all the great lakes, and onward on the canoe route to Clearwater River on the 57th parallel, beyond which it was not observed.

Tiliaceæ.—*Tilia glabra*, the lime tree, white-wood, or bass-wood, is a familiar ornamental and useful tree in the United States and Canada. We observed it as far north as Lake Winipeg, but only as underwood, sending out long flexible branches, which the natives convert into temporary cordage.

Acerineæ.—*Acer montanum (vel spicatum)*, the mountain maple, has a range from Maine, Pictou, Wisconsin, and Minnesota, to the River Winipeg, and, from the beautiful orange and red tints which its leaves assume in decay, is a great ornament to the woods in autumn. *A. circinatum* is confined to the west coast, is common in Oregon, and extends to the British territory on that side of the mountains. It grows in the woody country only, and chiefly in the pine forests, where its pendulous branches, taking root, form almost impenetrable thickets. The close-grained tough wood is used by the natives for making hoops.

A. saccharinum, sugar maple, with the variety, or perhaps species, named *A. nigrum* by Michaux, has been traced by Dr. Asa Gray along the Alleghany Mountains to Georgia. In the low country it scarcely passes to the south of Pennsylvania, but on the west side of the valley of the Mississippi is found as far south as Arkansas. Its northern limit is a short way beyond the 49th parallel on the elevated southern water-shed of Lake Winipeg; but it may, perhaps, attain a greater northern latitude in the lower country of Canada. A little to the south of Rainy Lake it yields abundance of good sugar. The variety named bird's-eye maple grows on one of the islands of the Lake of the Woods, and has been employed for making gun-stocks. Goat Island, at the Falls of Niagara, according to Mr. David Douglas, nourishes some of the largest sugar maples in North America.

A. rubrum, red or swamp maple, ranges southward, according to Dr. Gray, to Florida, and round the whole Gulf of Mexico to northern Texas; but some of the southern forms, he says, would probably be considered by European botanists to be specifically distinct from the northern tree. It grows in Nova Scotia, throughout Canada, westward to Lake Winipeg and the Rocky Mountains on the 52d or 53d parallel, and also crosses that chain to the head waters of the Columbia.

A. pennsylvanicum, striped maple, or moose-wood, comes down along the

* I am indebted to Dr. Asa Gray for some valuable information respecting the range in the United States of some of the trees in the following list. The *Northern States* referred to in the list extend from New England to Wisconsin, and south to Ohio and Pennsylvania, inclusive.

coast to Boston, and follows the mountains from Pennsylvania to the borders of Georgia, to which Dr. Gray has traced it. It grows also in Kentucky, and was seen by us on the banks of the Winipeg, where it has more the character of a flexible willow than of a tree. *A. macrophyllum* is confined to the mountainous country on the Pacific up to the 50th parallel, and is one of the most graceful trees, rising to the height of ninety feet, with a circumference of sixteen. *A. dasycarpum*, white or silver maple, is a fine large ornamental tree, well known in the United States. It is found on Lake Huron, but does not appear to rise northward out of the St. Lawrence basin. Good sugar is made from the juice of this tree.

Negundo fraxinifolium or *aceroides*, ash-leaved maple, does not, to Dr. Gray's knowledge, grow wild in New England. It abounds in Pennsylvania, and extends westward to western Texas and the Rocky Mountains, growing at the high elevation of 6000 or 7000 feet near Santa Fé and the Pawnee Fork, according to Lieut. Abert. It terminates northward about the 54th parallel on the banks of the Saskatchewan, and is the tree which yields most of the sugar made in Rupert's Land. Though this product varies much with the skill of the operator, the kind obtained from the juice of this tree is generally of a darker color than that which the true sugar maple yields.

AMPELIDEÆ.—*Ampelopsis quinquefolia*, Virginia creeper, or American ivy, extends northward to Lake Winipeg, and is a great ornament to the protruding rocks over which it creeps. It is a familiar shrub in the Northern States, but I have not been able to ascertain its southern limit. The *Vitis cordifolia* or *riparia*, frost grape, grows, on the evidence of collections made on my former journeys, as far north as the south end of Lake Winipeg, on the 50th parallel. I did not observe it on my late voyage, in which, indeed, I had very little leisure to search for plants; and if it actually grows in so high a latitude, it does not produce edible fruit so as to attract the attention of the residents, who could give me no information concerning it. Together with the *Vitis æstivalis*, or summer grape, it is common in Wisconsin and Minesota. Some of the native American vines are cultivated in the Eastern States; and the Isabella grape, a variety of the *Vitis labrusca*, has an agreeable though peculiar flavor.

ZANTHOLACEÆ.—*Zanthoxylum americanum*, northern prickly ash, and *Ptelea trifoliata*, the shrubby trefoil, grow in Canada and Wisconsin, where they seem to find their northern limit.

CELASTRINEÆ.—*Staphylea trifolia*, bladder nut; *Euonymus atropurpureus*, burning bush; *E. americanus*, strawberry bush; and *Celastrus scandens*, inhabit Canada and Wisconsin, but were not observed to the north of Lake Superior. *E. atropurpureus* crosses the continent to Oregon.

RHAMNEÆ.—*Rhamnus alnifolius*, alder-leaved buckthorn, grows from Maine and Michigan northward, to about the 58th parallel. It is a low shrub, and is applied to no economical purpose. *R. purshianus* is an Oregon plant, which extends to Vancouver's Island and New Caledonia. *Ceanothus americanus*, New Jersey tea, ranges from Maine, Michigan, and Wisconsin, to Canada West, but was not gathered by us to the north of Lake Superior. The *C. lævigatus*, a west-coast species, which extends from Oregon to Vancouver's Island, seems to be the only member of the genus that enters the British territory. *C. sanguineus* is common in the valley of the Columbia, and crosses the mountains to the upper tributaries of the Missouri, forming one of the many instances of west-coast plants traversing the dividing ridge to the eastern prairies, but not extending to the eastern woodland districts.

TEREBINTHACEÆ, or ANACARDIACEÆ.—*Rhus radicans* or *toxicodendron*, the poison oak; *R. aromatica*, the fragrant sumach; and *R. glabra*, the smooth sumach, reach the banks of the Saskatchewan, or latitude 45°. *R. typhina* and *R. venenata* extend to Canada, but have not been discovered north of Lake Superior.

LEGUMINOSÆ.—*Amorpha fruticosa*, false indigo; *A. canescens*, lead plant; and *A. nana*, grow abundantly on the prairies of Osnaboya, and are the only shrubby leguminous plants which extend to Rupert's Land. *Robinia pseudacacia* is plentiful in Canada East.

CÆSALPINEÆ.—*Gymnocladus canadensis*, the Kentucky coffee-bean tree; *Cercis canadensis*, red bud; and *Cassia chamæcrista*, partridge pea, have their northern limits in Canada or Wisconsin.

ROSACEÆ.—*Prunus americana*, the wild yellow plum, seems to reach its northern boundary on the River Winipeg, not having been observed by me beyond the 50th parallel. Lieut. Abert gathered its fruit as far south as the banks of the Canadian and Pawnee forks of the Arkansas, on the 40th parallel. It is a common bush on the river banks in the Northern States. The American plums and cherries require further investigation, as the number of the species and their distinctive characters are imperfectly known. This one grows to the height of ten or fifteen feet on the Winipeg, producing in the woods long flexible branches, armed with a few slender sharp thorns. Its ripe fruit is fleshy and well flavored, but rather mealy, of a yellowish color inclining to orange. It has an ovoid shape, with a shallow groove on one side like a peach, is nearly an inch in diameter, and its stone is so much compressed that its thickness is less than half its width; while its length, being 0·63 inch, exceeds the width by a fourth part. The nut is oblique, with convex valves, being circumscribed by two unequal curves. One edge is acute, with a groove on each side of it; the other edge is occupied by a narrow groove. I have been thus particular in the description of the northern fruit, that it may be compared with plums growing in other districts. The fruit is the *Puckĕsāminan* of the Crees; and *La Prune*, or the plum, of the white residents.

The *Nekā-u-mina* of the Chippeways, or *Thekā-u-mina* of the Crees, and sand cherry of the residents on Rainy River and Lake Winipeg, is a bush or small shrubby tree, a foot and a half high, which grows on sand hills. The bark of its annotinous and biennial shoots is reddish, and the older twigs are brownish, with small warty specks. When in fruit (in which state only I examined it in September, 1849) the fruitstalks are solitary, and spring from the base of the summer's growth; they are rather more than half an inch long, or about equal to the diameter of the fruit, which is black and rather austere, but edible. The stone is 0·38 inch long, almost regularly elliptical, and acute at the ends, but more so at one end than the other. Its valves are very convex, so that its width exceeds its thickness very little. The sides are not acute-edged: one suture is depressed, forming a shallow groove; on the other side, which is very obtuse and almost flattened, there is a furrow above and below the suture, and rather remote from it. The annotinous shoots are smooth, angular, and generally flexuose, with the leaves springing alternately at the curves

The leaves measure two inches and a half, the footstalk forming about one fifth part of this length. The lamina is lanceolato-elliptical; that is, nearly regularly elliptical, with an acute end, and a gradual tapering into the footstalk; it is serrated by acute, appressed teeth at the upper end, and is entire toward the footstalk; its under surface is pale and somewhat glaucous, the upper one dark green, and both sides are perfectly smooth. The footstalks are edged by the decurrent lamina for more than half their length, and the deciduous linear-lanceolate stipulæ are incisco-pinnate inferiorly. This cherry is probably the *Cerasus pumila* of Michaux and later American botanists. It was not traced by me beyond the 50th parallel.

Another small shrubby cherry grows on moist sandy soil, by the banks of rivers and lakes, from Lake Superior to the Elk River on the 57th parallel. Its fruit is scarcely half the size of the preceding, but is, like it, black, and hangs generally on solitary footstalks, though the flowers grow by twos or threes in short racemes. This shrub lies close to the ground; makes no approach to the tree form; seldom exceeds a foot in height; and so much resembles the *Salix myrsinites* and some other depressed willows, that, on looking for catkins, I have not been undeceived until I found the footstalks of the last year's cherries. The fruit of this sand cherry is sweeter than that of the preceding one. Whether it be the *Cerasus depressa* of some botanists I can not determine; nor do I pretend to clear up the confusion that exists in botanical works respecting *C. pumila* and *depressa*.

C. pennsylvanica, wild red cherry, the *Pāsis-so-wey-minan* of the Chippeways, and *Pāsi-ā-wey-minan* of the Crees, produces a small sour red fruit, which grows in a many-flowered raceme on long slender footstalks. Its equatorial limit, according to the United States botanists, is the New England States and Pennsylvania, where it is a slender tree 20 or 25 feet high. Its polar limit is within the Saskatchewan basin, which it ascends toward the base of the Rocky Mountains, nearly to the height of 2000 feet above the sea. *C. virginiana*, choke-cherry, is named by the Crees *Ta-kwoy-minan*, and by the Dog-ribs *Ki-e-dunnè-yerrè*. It was found by Lieut. Abert on the Kansas and Arkansas, and on Purgatory Creek; and is, in northern latitudes, a shrub with long branches. At Fort Liard, on the 61st parallel, it is 20 feet high, and on the confines of the arctic circle, where it terminates, it does not exceed four or five feet. The fruit can scarcely be said to be edible by itself; but it is often pounded, stones and all, and mixed with pemican. *C. serotina*, wild black cherry, is a general inhabitant of Rupert's Land, extending westward to the valleys of the Rocky Mountains on the Pacific side, where, however, it is generally dwarfed; and northward to near Great Slave Lake. It is said by Dr. Gray to be a fine large tree in the Northern States, with purplish-black fruit, having a pleasant vinous flavor. Besides these I gathered specimens of a cherry-tree, not in flower, on Athabasca and Slave Rivers, which Sir William Hooker is inclined to consider as the *C. mollis*, discovered by the unfortunate David Douglas, on the banks of the Columbia, growing on subalpine hills to the height of from 12 to 25 feet.

Purshia tridentata inhabits the Rocky Mountain prairies near the head waters of the Missouri and Columbia, and extends northward to the 49th parallel. *Spiræa opulifolia*, nine-bark, ranges from Maine, Canada, and Wisconsin, westward to the valley of Oregon, in which it is found from the sources of the Columbia downward. It is common on the low islands of Lake Superior, and has its polar limit in the colony of Red River. *S. chamædrifolia* inhabits the northwest coast up to Sledge Island in Beering's Straits, and Chamisso Island in Kotzebue Sound. It does not cross the Rocky Mountains, nor does any other *Spiræa* go so far north on the east side of the continent. *S. betulifolia* is another western species which inhabits the Blue Mountains and Mount Hood, and crosses the Rocky Mountains to their eastern valleys between latitudes 52° and 54°, but does not descend to the lower eastern country. *S. tomentosa*, hard-hack or steeple-bush, is common in the meadows and low grounds of New England, and spreads through Nova Scotia, Canada, and Rupert's Land to Lake Winipeg. *S. douglasii* is an Oregon species resembling the preceding, which extends to the Straits of Da Fuca. *S. ariæfolia* forms part of the underwood in forests on the Pacific coast, on the Kooskoosky, Spokan, Flathead, Salmon, and M'Gillivray Rivers, up to the 49th parallel. *S. salicifolia* is very abundant on the banks of every lake and river in the St. Lawrence and Saskatchewan basins, and northward to Slave River. It is often associated with the *Myrica gale*, growing in the water. In its northern range it approaches the *S. chamædrifolia* of the west coast, but does not attain so high a latitude, owing to the greater severity of the climate on the east side of the mountains.

Rubus occidentalis, black raspberry or thimbleberry, extends from the Northern States to the Saskatchewan basin, and also to the Pacific coast. *R. strigosus*, wild red raspberry, is also found on both sides of the continent; on the east side it inhabits the United States, Newfoundland, and Canada, and may be traced in the interior canoe route throughout the Saskatchewan basin. *R. nutkanus*, white flowering raspberry, was discovered in Queen Charlotte's Sound by Mr. Menzies in lat. 51° on the Pacific coast, since which time it has been found in Norfolk Sound, lat. 57°, and traced down to Cape Orford in lat. 43°, and to the head waters of the Columbia in 52°. On the eastern declivity of the Rocky Mountains it grows between latitudes 52° and 54°, and on the River Winipeg, Lake Superior, and Upper Michigan. Near the Pacific it is ten feet high, and forms the underwood on the island of Sitka; but in the passes of the Rocky Mountains dwindles down to a foot or eighteen inches. In thickets on the Winipeg its leaves attain remarkable dimensions: the fruit is inedible. *R. odoratus*, purple flowering raspberry, is a native of the Northern States, Canada, and the country between Lake Superior and the Saskatchewan. *R. spectabilis* is a prickly shrub, ten feet high, inhabiting the Pacific coast from Oregon to Unalashka. *R. suberectus*, bramble. This species, which is also European, is an inhabitant of Newfoundland, and of the country between Lake Superior and the Saskatchewan, where it was found both in 1825 and 1848. *R. villosus* is common in the Northern States, and is found also in Nova Scotia and Canada West up to Lake Huron and Wisconsin. It is included in Elliott's "Flora Carolina," but I have not ascertained its equatorial limit. *R. hispidus vel obovalis*, running swamp blackberry, is common in the Northern States, and extends through Canada to Lake Superior. *R. canadensis* (L.), *vel trivialis* (Pursh), low blackberry or dewberry, has a similar range with the preceding species. *R. nivalis* (Douglas), is an alpine shrub, found on the snowy ridges of the Rocky Mountains, and not growing more than six inches high. There are also several herbaceous species of this genus; as *R. triflorus*, dwarf raspberry, which is common in the Northern States and throughout Rupert's Land, northward to Slave and Mackenzie Rivers. The Dog-ribs name it *Tăsillè-ki-eh*. Its northern limit is about lat. 68°. *R. chamæmorus*, cloudberry, is found on the White Mountains of New Hampshire near the limit of trees; also in Maine and Nova Scotia, Newfoundland and Labrador. In lat. 54°, and more to the north, it crosses the continent, and is found on the summits of the Rocky Mountains between latitudes 52° and 56°; in Unalashka, on the shores of Beering's Straits, and on the most northern promontories of the continent. Near the Arctic Sea it is a common plant on mossy plains, but produces fruit there only in fine seasons. The fruit, which has a rich honey flavor, perishes with the early frosts. It is perhaps the most delicious of the arctic berries, when in perfection, but cloys if eaten in quantity. *R. stellatus*, resembling the preceding, has been found only at Foggy Harbor on the northwest coast. *R. arcticus* and *R. acaulis* inhabit the shores of Hudson's Bay, Labrador, and the country westward to Kotzebue Sound. Their southern limit seems to be in the Saskatchewan basin, in about lat. 53°. On many parts of the flat beaches of Slave and Mackenzie Rivers the lively red flowers of *R. acaulis* cover large patches of ground which are partially flooded by small rivulets. In woods the last-mentioned species has a stouter growth, and emits long flagelli which run among the mosses.

Potentilla fruticosa, the shrubby cinquefoil, grows abundantly from the Northern States to the Arctic Sea, by river banks as well as in the most exposed and elevated situations. It occurs in the high valleys of the Rocky Mountains, at Pelly Banks on the west side of that range, and in Kotzebue Sound. On the Coppermine River near the sea it is almost herbaceous, the woody stem being extremely short and subterranean.

Rosa woodsii, *R. carolina*, *R. blanda*, *R. cinnamomea*, *R. majalis*, and *R. stricta* grow in the wooded districts; but, from their similarity to each other, their respective limits have not been ascertained. *R. blanda* was found flowering freely near the mouth of the Mackenzie on the 69th parallel. This species and *R. cinnamomea* cross to the Pacific coast. *R. woodsii* and *majalis* have been traced as far north as the Mackenzie. *R. nitida* and *lucida* grow in Newfoundland and in the New England States. *R. fraxinifolia* is confined to the Pacific coast; and *R. lævigata* has not been found beyond Lake Huron. *R. setigera*, a fine climbing rose, grows from Ohio to Wisconsin, but has not been detected to the north of the great lakes.

Cratægus punctata, dotted thorn, is found every where in the Northern States, extends northward to Wisconsin, and crosses the continent to the coast of the Pacific; but has not, so far as I have learnt, been found within the British territory. *C. glandulosa* occurs in Canada, and northward to the south side of the Saskatchewan basin; probably also on the Pacific coast. *C. coccinea*, scarlet-fruited thorn, is a common low tree in the Northern States; was found by Lieut. Abert, as far south as Stranger Creek, in lat. 39°; and extends to Wisconsin and the great lakes. *C. cordata*, Washington thorn, is supposed by Dr. Asa Gray to have been introduced into New England, but to grow wild in Pennsylvania and the more southern States. It is found from Canada to the Saskatchewan and the valleys of the Rocky Mountains, and about the sources of the Columbia, in between lat. 52° and 54°. The *Cratægi* flourish on the banks of

Rainy and Winipeg Rivers; but are scarce further north. Mertens found several in the forest lands of Sitka. *Amelanchier canadensis* (*botryapium et ovalis*,) shad-bush and service-berry, is *La Poire* of the voyagers, the *Misass-ku-tu-mina* of the Crees, and the *Tchè-ki-eh* of the Dog-ribs. This shrub extends along the banks of rivers nearly as far northward as the woods go, and produces fruit up to the 65th parallel on the Mackenzie. It is common in the Northern States, in Nova Scotia, Newfoundland, and Labrador, and westward to the Pacific. The black fruit is aboût the size of a pea, is well tasted, dries well, and in that state is mixed with pemican, or used for making puddings: for which purpose it nearly equals the Zante currant. Its wood, being tough, is used by the natives for making arrows and pipe-stems, and has obtained on that account the name of *bois de flèche* from the voyagers; but in the United States the name of arrow wood is given to a different tree. The variety or species named *A. sanguinea* was traced up to the 60th parallel.

Pyrus rivularis, Powitch tree, inhabits Oregon and Vancouver's Island. Its fruit is edible, and its wood, which is hard enough to take a fine polish, is used for wedges. *P. americana*, the American mountain ash, is found on the southern parts of the Alleghanies, and more commonly in the swamps and mountain woods of the Northern States. It is frequent on the shores of Lake Huron and Superior; but is seldom seen on the canoe route beyond Lake Winipeg. On the acclivities and in the valleys of the Rocky Mountains, however, it ranges northward to Fort Liard, near the 60th parallel. It has been observed as high as the 56th degree of north latitude on the Pacific coast, from whence it extends southward through the subalpine regions of Oregon. *P. arbutifolia*, choke-berry, is common in the damp thickets of the Northern States, in Newfoundland, in Canada, and onward to the Saskatchewan basin; but was not observed so far north as the immediate banks of that stream.

Grossularieæ.—The species of this family seem to attain their maximum number to the north of the United States. *Ribes oxyacanthoides*, sharp-thorned gooseberry, inhabits Newfoundland, Canada, and the canoe route northward to the 62d parallel, or perhaps further. *R. cynosbati*, prickly gooseberry is common in the rocky woods of the Northern States, and accompanies the preceding species northward to Slave Lake. *R. saxosum* inhabits New England, the shores of Lake Huron, and the valley of the Saskatchewan, extending also probably to Oregon. *R. hirtellum*, short-stalked wild gooseberry, the most common species in New England, extends to Canada, Wisconsin, and northward to Great Slave Lake. The preceding one seems to be considered by Dr. Asa Gray to be a variety of this species. *R. lacustre*, swamp gooseberry, is common in the most northern parts of the United States and Nova Scotia; crosses the mountains to North California and Oregon; and extends northward along the Mackenzie, nearly or quite to its delta. It is the *Tagossay-ki-eh* of the Hare Indians, which name is common to several kinds of gooseberry in the Dog-rib country, where there is a greater variety of species. *R. divaricatum* is common near Indian villages on the northwest coast, from 45° to 52° north lat. *R. rotundifolium*, Michx. (*triflorum*, Willd.,) is a rare inhabitant of the mountainous districts of Oregon, and inhabits the Northern States from Massachusetts to Michigan and Wisconsin, but has not been found on the north side of Lakes Superior or Huron. *R. rubrum*, the common red currant, native both of Europe and America, extends from the Northern States very nearly to the shores of the Arctic Sea, having been gathered beyond the 69th parallel; and it ranges westward to Kotzebue's Sound. It is the *Ki-eh-eth-lule-āzè* of the Dog-ribs and Hare Indians. *R. prostratum*, fetid currant, inhabits cold damp woods from Nova Scotia and the Northern States northward to the Athabasca, and westward to the Rocky Mountains and Oregon. The fruit is produced in copious racemes; but, in common with the foliage, it has an unpleasant odor, and a strong taste of turpentine. *R. hudsonianum*. This is the *Nut-sinnè* of the Dog-ribs, and is a common gooseberry from Hudson's Bay to the Rocky Mountains and subalpine districts of Columbia; also in a northerly direction on the Mackenzie to lat. 67°. *R. floridum*, wild black currant, resembles the preceding, and is a common species in the Northern States, westward to Wisconsin, and ranges northward to lat. 54°. *R. sanguineum*, which has become so common an ornament of our gardens, is a native of the Pacific coast only, where it ranges from 38° north lat. to 52°. There are several other very handsome species in Oregon, and, among others, the rich *P. aureum;* but they have not been traced beyond the 49th parallel.

ARALIACEÆ.—*Panax horridum*, prickly ash-leaved panax, a twining shrub common in California, Oregon, and New Caledonia, as far north as 57° or 58°, crosses the Rocky Mountain ridge to the upper tributaries of the Saskatchewan, but does not descend to the eastward. *Aralia hispida*, bristly sarsparilla, may be considered as the eastern representative of the preceding, though it is scarcely shrubby, having merely a very short, tough stem, almost buried in the crevices of the rocks from which it springs.

CORNEÆ.—*Cornus alba vel stolonifera*, red osier cornel. This willow-like shrub, which is the *osier rouge* of the voyagers, ornaments the river strands from the Northern States, Nova Scotia, and Newfoundland, northward to near the mouth of the Mackenzie, and westward to the shores of the Pacific. It is named by the Crees, on account of the bright red color of its twigs, *Mithwka-pè-min-àhtik* (red stick,) and its fruit *Muskwamina* (bear-berry,) because the bears eat it. The Dog-ribs call this berry, *Kai-gossai-ki-eh*. A warm decoction of the bark and twigs is used by the natives for bathing their limbs when swelled by fatigue. *C. alternifolia*, *C. paniculata*, *C. sericea*, and *C. circinata*, which are inhabitants of the Northern States, are said to extend to Canada; but except the last named, which occurs on Lake Superior, none of them were gathered by us on the canoe route. *C. cericea* and *C. florida*, also Canadian species, cross the continent to Oregon, but do not occur north of the great lakes. The herbaceous *C. canadensis* reaches the shores of the Arctic Sea, crossing the continent from east to west; and the *P. suecica* a European plant, is found in the Gulf of St. Lawrence, and on the west coasts of arctic America as high as Kotzebue Sound, and southward to Oregon; but has not been detected in the interior districts.

LORANTHACEÆ.—*Arceuthobium oxycedri*, this leafless parasitical shrub, is common to Europe, Central Asia, and North America, where it grows on cedars and pine trees. On the eastern declivities of the Rocky Mountains it ranges from lat. 52° to 57° north, and also eastward to Hudson's Bay, growing on the *Pinus banksiana*. On the western side of the mountains, from the Spokan River, in 47° north lat., to near the sources of the Columbia, it infests the *Pinus ponderosa*.

Caprifoliaceæ.—*Sambucus canadensis*, black-fruited elder, was gathreed by Lieut. Abert on the Cottonwood Creek of the Neosha, in lat. 38½° north, at an altitude of about 1400 feet. It has its northern limit in the Sascatchewan basin, and ranges westward from Nova Scotia across the prairies. *S. racemosa vel pubens*, the red-fruited elder, is common in the Northern States on the shores of Lake Superior, going northward to the Saskatchewan, and westward to Oregon. Its polar limit, as far as ascertained, is on the eastern declivity of the Rocky Mountains between 52° and 59° north lat. *V. prunifolium*, black haw, or sloe-leaved viburnum, reaches the north shore of Lake Huron; but is more common in New York and Ohio. *V. lentago*, sweet viburnum, is a handsome tree in the Northern States, grows in Nova Scotia and Wisconsin, and extends northward to the south side of the Saskatchewan basin. *V. nudum*, withe-rod, is more common from New Jersey southward than toward the great lakes; but occurs as far northward as the last-named species. *V. dentatum*, arrow wood, is common in the low grounds of the Northern States up to Wisconsin, and is said by Pursh to extend to Canada; but it seems to be rare in that country. *V. acerifolium*, maple-leaved arrow-wood, is a more northern species; and, in common with other shrubs that approach the arctic circle, it crosses the Rocky Mountains to the valley of Oregon, and also to Sitka. It has been traced as far north as Great Slave Lake, occurs also in Newfoundland, and is common in the rocky woods of the Northern States. *V. opulus vel oxycoccus*. Sir William Hooker is inclined to consider the European and American shrubs known by these names—and of which the handsome snow-ball tree, or guelder rose, is a cultivated variety—to be one species, and Dr. Asa Gray unites them. In America the shrub extends from the Northern States and Nova Scotia to lat. 68° on the Mackenzie, and perhaps very nearly to the verge of the woods. It also crosses to the Pacific coast, having been found in the valley of Oregon. Its fruit, of a bright pinkish red color, has a sharp acid taste, and is the *Mongsö-a minā* (moose-berry) of the Crees, and the *Dunnè-ki-e* or Indian-berry of the Dog-ribs and Hare Indians. The fruit being sometimes used as a poor substitute for cranberries, has obtained for the bush the name of cranberry tree in the Northern States. *V. edule*, the *pembina* of the voyagers, was traced by us northward to the Elk River. It is much less common than the preceding, and has a more fleshy and less acid fruit, of an orange-red color. The voyagers relish this fruit; and it has given name to many of the rivers of Rupert's Land. It is the *Nipi-minan* (water-berry) of the Crees. Michaux, Dr. Asa Gray, and other authors consider it to be scarcely a variety of *V. opulus*. I have found, however, its foliage retaining pretty constantly its peculiar character. In *V. oxycoccus* the lobes of the leaves are separated by acute sinuses, and have long, tapering, jagged, or deeply serrated points. In *V. edule* the sinuses between the lobes are rounded, and the lobes themselves are shorter, though the lamina of the leaf is cut to within a short distance of its base. The European *V. opulus* has generally obtuse sinuses, and a less deeply cut lamina, but the lobes also short. Pursh, who separated *V. oxycoccus* from *edule*, described the bases of the leaves of the one as acute and of the other obtuse; but, as there seems to be no difference in that respect, it is probable that he meant the sinuses.

Diervilla trifida vel canadensis, bush honeysuckle, has a herbaceous aspect, and is one of the most common underwoods on the portages. It oc curs in all the woody districts of the Saskatchewan basin up to the acclivities of the Rocky Mountains, but is rare to the North of Cumberland House. It grows also on Lake Superior, in Wisconsin, Nova Scotia, and the Northern States. *Lonicera parviflora*, small honeysuckle, has a conterminous range with the preceding. *L. douglasii* gathered on Saskatchewan is considered by Dr. A. Gray to be merely a variety produced by cultivation. *L. hirsuta*, hairy honeysuckle, is a coarse-leaved climber, common in moist rocky woods of the Northern States and Canada as far as Lake Huron. *L. ciliati*, fly honeysuckle, grows at Pictou, on the Catskill Mountains, in Ohio, Wisconsin, generally throughout the rocky woods of the Northern States; also on Lake Superior, and northward along the whole Saskatchewan basin. *L. cærulea*, the mountain fly honeysuckle, extends northward to the arctic circle; it likewise ranges from the Labrador coast and Newfoundland to the Rocky Mountains, and we should suppose also to the Pacific coast, since it is both a European and a Siberian species; but it is not named by Mertens or Bongard among the Sitka plants, nor does it appear to have been found by Douglas, Tolmie, or Scouler in Oregon. It grows in Wisconsin, New Hampshire, Massachusetts, and New York.

Symphoricarpus racemosus, snow-berry, and *S. occidentalis*, wolf-berry, range from Vermont, Michigan, and Wisconsin, over the St. Lawrence and Saskatchewan basins, to the 60th parallel on the Mackenzie. They also occur in the Oregon valley, Vancouver's Island, and doubtless much further along that coast.

Rubiaceæ.—*Cephalanthus occidentalis*, button bush. This shrub, which belongs to the sub-family of *Cinchoneæ*, occurs in thickets of the Northern States and Canada, but does not extend to Lake Superior.

Compositæ.—Of this large family no shrub has been detected in the canoe route north of Lake Superior; though the *Crinitaria viscidiflora* grows as high as the 55th parallel on the banks of the Salmon River, west of the Rocky Mountains, and on the upper branches of the Columbia above the Kettle Falls. A small annual herb was found on the Saskatchewan, which Sir William Hooker placed next this species; but, from the imperfect specimens, he could not ascertain its genus satisfactorily.

Vaccineæ.—*Gaylussacia resinosa*, black huckleberry, is common in the Northern States westward to Wisconsin, and extends northward to the Saskatchewan. *Vaccinium corymbosum*, common swamp blueberry, extends from the Northern States to Newfoundland and Canada, as far north as Quebec, but has not been gathered to the westward of Lake Superior. *V. pennsylvanicum*, low shining-leaved blueberry, is very common in the dry rocky woods of the Northern States, Canada, and the country between Lakes Superior and Winipeg. *V. canadense*, downy-leaved blueberry, is the most abundant species by the sides of streams and in thickets, from Maine and Michigan to the shores of Hudson's Bay, and northward in the woody districts to the arctic circle. It extends also westward across the mountains to the upper feeders of the Columbia. *V. uliginosum*, bog bilberry, occurs on the summits of the New Hampshire Alps; on the Green Mountains of Vermont, and on Essex county mountains of New York; on the Newfoundland, Labrador, and Greenland coasts; also from Lake Superior northward to the Arctic Sea. On the west side of the Rocky Mountains it has been gathered on Sitka, Unalashka, and Kotzebue Sound. In Europe it grows in the forests of the higher Jura, in England, and the Scandinavian peninsula. Beyond the arctic circle its fruit is not abundant every year; but in good seasons it is plentiful to an extraordinary degree, and is of a finer quality than in more southern localities. It then affords food to the bears and large flocks of geese, which fatten on it, and acquire a fine flavor. The berries, when frozen by the autumnal frosts, remain hanging on the bushes until the snow melts in the following June, and may be then gathered in a very juicy but tender condition. *V. salicinum*, willow-leaved bilberry, is an inhabitant of Unalashka. *V. myrtillus*, myrtle-leaved bilberry, was gathered by Mr. Drummond on the summit of the pass between the head waters of the Saskatchewan and Columbia, but has not been detected further to the east, though it is a European plant. Chamisso found a *Vaccinium* on Unalashka, which he was inclined to refer to this species; but his specimens were imperfect. Bongard, however, enumerates it as existing among the plants gathered by Mertens on Sitka. *V. myrtilloides vel angustifolium* is found in Canada, and from Hudson's Bay to the woody declivities of the Rocky Mountains, between the 52d and 54th parallels. It crosses the dividing ridge also to the alpine valleys of Oregon, and to the sea-coast further north, where the purplish-brown fruit is eaten with relish by the natives. *V. cæspitosum*, dwarf bilberry, grows on the Alps of New Hampshire, the shores of Lake Superior near James's Bay, and northward to the valley of the Saskatchewan and the Rocky Mountains, between 52° and 57° north; also in the Oregon valley. *V. ovalifolium* grows in Oregon from the mouth of the Columbia up to the Portage River, near the crest of the Rocky Mountains, on the 50th parallel, and also on the island of Sitka at lat. 57° north. *V. vitis-idæa*, cow-berry, or alpine cranberry, is the *Wi-să-gù-mīnă* of the Crees, and the cranberry most plentiful and most used throughout Rupert s Land. This berry is excellent for every purpose to which a cranberry can be applied; and though inferior to the *V. oxycoccus* in flavor in autumn, is far superior to it after the frosts; and, as it may be gathered in abundance in a most juicy condition when the snow melts in June, it is then a great resource to the Dog-ribs and Hare Indians, as well as to the immense flocks of water-fowl that are migrating to their breeding places at that date. It grows in perfection, in the most exposed situations, round a boulder or granite rock, over whose face its branches may spread, and where it can have at one time both moisture and the reflected heat of the sun's rays. It is found at Danvers, in Massachusetts, in Maine, and the higher mountains of New England, where its fruit is reported by Dr. Asa Gray to be barely edible, bitter, and mealy. In the parallel of Lake Superior it spreads from the Atlantic to the Pacific (being absent, however, on the prairies.) In a higher latitude it crosses the continent also from Churchill Fort to Sitka and Kotzebue Sound, and it extends in the middle districts to the Arctic Sea in latitude 71°. In Sitka its leaves are said to be small. In Rupert's Land they vary in size, according as the plant is exposed or under shade. *V. ovatum* is common in Oregon and rocky places of the west coast northward to the 49th parallel. *V. oxycoccus*, dotted cranberry, is, like the preceding, common to the New and Old World. It grows in peat bogs from New England and Wisconsin, northward to the arctic circle, and from Newfoundland and Labrador to the Rocky Mountains, between 52° and 57° north lat.; in Sitka on the latter parallel, and in Kotzebue Sound. *V. macrocarpum*, American cranberry, is common in the peat bogs of the Northern States, and has its limit in the Saskatchewan basin. It crosses the continent from Newfoundland to Oregon; and the natives near the mouth of the Columbia eat its fruit, when boiled, under the name of *Su-labich*. *Chiogenes hispidula*, creeping snowberry, is common in the Northern States, where it grows under evergreens in turfy places. It extends across the continent from Newfoundland to the sources of the Columbia, and northward along the Rocky Mountains to the 55th parallel.

Ericeæ.—*Gaultheria procumbens*, creeping winter-green tea-berry, checker-berry, partridge-berry, or box-berry. This fragrant creeping shrub is a great ornament of the woods north of Lake Superior. It inhabits moist woods in the Northern States, grows at Pictou and on Lakes Huron and Superior, and was traced by us northward to the Lake of the Woods, or near the 50th parallel. *G.? myrsinites* has hitherto been found on the declivities of the Rocky Mountains only between the 52d and 57th parallels. Mr. Drummond says that its small berries have a delicious pine-apple flavor. The plant was cultivated in the Botanic Garden at Glasgow, but I have not heard that it produced fruit there. *G. shallon* is an Oregon plant growing between Cape Mendocino and Puget Sound, but not extending inland more than a hundred miles from the sea-coast. *Epigæa repens*, ground laurel, or trailing *Arbutus*, inhabits sandy and rocky woods in the Northern States, Canada, Nova Scotia, Newfoundland, and Rupert's Land, as far north as the Saskatchewan. *Arbutus menziesii* and *A. tomentosa* inhabit Oregon northward to Puget Sound; but no true *Arbutus* has been detected on the east side of the Rocky Mountains. *Arctostaphylos uva-ursi*, bear-berry, is common to Europe and America, and descends from the Arctic Sea coast to Rainy Lake and the rocks and hills of the Northern States. It crosses the continent to the valley of Oregon, where the Chenook Indians mix its dried leaves with tobacco. It is used for the same purpose by the Crees, who call it *Tchakashè-pukk;* by the Chepewyans, who name it *Klèh;* and by the Eskimos north of Churchill, by whom it is termed *At-tung-ă-wi-at*. On account of the Hudson's Bay officers carrying it in bags for a like use, the voyagers gave it the appellation of *Sac-a-commis*. On the northwest coast, Mertens found it at Sitka, and it doubtless extends along the whole coast. Its dry farinaceous berry is utterly inedible. *A. alpina*, alpine bear-berry, though a herbaceous plant, may be mentioned with the others: it is also European. In the United States the only habitat given is the Alps of New Hampshire; but it grows at a much lower altitude in Newfoundland and Canada. It was found by Drummond on the Rocky Mountain ridge, and is very common on the barren grounds beyond the woody district, and along the whole arctic coast to Kotzebue Sound. There are two varieties, one with bright red and more juicy fruit; the other, having a dark purplish-black berry, of more fleshy consistence, and a stronger peculiar flavor. Both are eaten in the autumn; and, though not equal to some of the other native fruits, are not unpleasant. The two kinds are exactly alike in foliage. *Andromeda hypnoides*, moss-like andromeda, an inhabitant of the Alps of New Hampshire, Mount Marcy in New York, Labrador, and the northwest coast, was not detected by us on the interior canoe route. *A. lycopodioides*, a Kamtschatka plant, was found by Chamisso on Unalashka. *A. cupressina* inhabits the Rocky Mountains in lat. 56° north. *A. mertensiana* and *A. stelleriana*, so named by Bongard, were discovered on Sitka by Mertens. *A. tetragona* is one of the most northern plants, being an inhabitant of the north end of Spitzbergen. It occurs on all the islands and coasts of the Arctic Sea, from Greenland to Kotzebue Sound, at Sitka, and as far south as Mount Hood on the 45th parallel. It is also a Lapland and Siberian plant. Like the two preceding species, it is rather a wiry herb than a shrub. The withered leaves of past years remain attached to the thread-like stem, and may be used as fuel, a fact which Mr. Rae so fully demonstrated, as we have mentioned, in a preceding page. *A. polifolia*, rosemary andromeda, inhabits the Alps of New Hampshire and New York, Wisconsin, Lake Superior, and the country northward to the Arctic Sea; also the whole breadth of the continent from Newfoundland and Labrador to Sitka and Kotzebue Sound, with the exception of the prairies. It is an inhabitant also of the higher Jura. *A. calyculata*, rusty-leaved andromeda, grows in sphagnous bogs and on the flooded strands of clear streams in the Northern States and Rupert's Land as far as the upper part of the Mackenzie, and also on the shores of Beering's Sea. *A. racemosa*, cluster-bearing andromeda, grows in the moist copses of Canada, Massachusetts, and New Jersey near the coast, extending from thence southward. *A. ligustrina*, privet andromeda, a common shrub of the Northern States, extends northward to the Saskatchewan basin. In this genus and in *Arbutus* it may be noticed, that the more herbaceous species have generally the highest range.

Phyllodoce taxifolia, or *Menziesia cærulea*. This English plant grows on the New Hampshire Alps, and has been found on the Labrador coast. Steller is also said to have gathered it on the American coast and islands opposite Kamtschatka. *M. ferruginea* and *M. aleutica* were found by Mertens at Sitka; the former, which is one of Menzies's discoveries, has since been gathered by Seeman on the coast of Beering's Sea, and the latter was previously found by Chamisso on Unalashka. *Menziesia glanduliflora* is one of Mr. Drummond's discoveries on the Smoking River, an elevated tributary of Peace River, on the 55th parallel. It is remarkable for its gracefully drooping yellow flowers. *M. empetriformis* inhabits Vancouver's Island and the Alpine districts of Oregon. *M. grahamii* and *M. intermedia* grow on the Rocky Mountains, in lat. 55° eastward of their crest. *M. globularis* inhabits the same districts on the Smoking River northward to 56°; and, according to Pursh, it occurs also on the high mountains of Carolina, and on the Cacapon Mountains, near Winchester, in Virginia. None of the *Menziesiæ* are mentioned by Dr. Asa Gray as existing in the Northern States; and it would appear that many of the species are very local, particularly the alpine ones.

Kalmia latifolia, calico bush ,mountain laurel, or spoon-wood, forms dense thickets on the mountains of Carolina and Pennsylvania, and is common northward from Maine to Ohio and Canada, where it is a much humbler shrub. It was not observed by us on the north side of the St. Lawrence basin. *K. glauca* inhabits moorish places from the Northern States to the Arctic Sea, and crosses the continent to Sitka. *K. angustifolia*, sheep laurel, is common in the Northern States and Canada, to James's Bay and Newfoundland. We did not observe it on the canoe route north of Lake Superior. *Azalea viscosa* inhabits the Northern and Eastern States and Canada, but was not seen by us beyond the St. Lawrence basin. *A. nudiflora*, purple azalea, or pinxter flower, a common shrub in the Northern States, extends to Canada. The *Rhododendron maximum*, which is common on the mountains of Carolina and Pennsylvania, and is more rare in the Northern States and Canada, grows also in Oregon, on the subalpine range of Mount Hood, and more to the north on the high mountains near the "Rapids" of the Columbia. *R. lapponicum*, Lapland rose-bay, is another arctic plant which is found isolated on the peaks of the White Mountains of New Hampshire, and on Mount Marcy in the north corner of New York. It has been gathered as far south on the coast as the Labrador peninsula and the shore between York Factory and Churchill River, and grows on the summits of the Rocky Mountains on the 56th parallel, and throughout the whole extent of the barren grounds from Repulse Bay to Norton Sound, and northward to the Arctic Sea. An infusion of the leaves and flowering tops was drunk by us instead of tea, but it makes a less grateful beverage than the *Ledum palustre*. It is a Scandinavian plant. *R. kamtschaticum* is an inhabitant of the northwest coast in lat. 53°, and of Unalashka, as well as of the Asiatic shore. *R. albiflorum*, an elegant and ornamental plant, was discovered by Drummond on the Rocky Mountains between 52° and 57° of north latitude, where alone it has been found.

Loiseleuria vel azalea procumbens inhabits the Alps of New Hampshire, and the coasts of Newfoundland, Labrador, Hudson's Bay, and the Arctic Sea; also the northwest coast at Mount Edgecumbe, Sitka, and Kotzebue Sound. *Ledum palustre*, narrow-leaved Labrador tea, the *Kā-ki-ki-pukwā* (perennial leaves), or the *maskègo-pukwā* (medicine leaves), of the Crees, is an inhabitant of the colder parts of Canada, the coasts of Newfoundland and Labrador, and the whole of Rupert's Land to the Arctic Sea, on whose shores it grows from Repulse Bay and the mouth of the Thlewee-choh to Kotzebue Sound. It is also found at Sitka; but Dr. Asa Gray has seen no specimens gathered south of the United States boundary line. It is frequently used as a substitute for tea. *L. latifolium* grows in the woody districts of Rupert's Land, often in the immediate vicinity of the other species; but extends further south, being common in cold boggy grounds in the Northern States.

MONOTROPEÆ.—*Cladothamus pyrolifolius* (*Tolmiea*, Hook.) inhabits Norfolk Sound on the Pacific coast in lat. 57°, and the country southward to Puget Sound. *Chimaphila umbellata*, Prince's pine, Pipsissewa, goes northward to 53° on the Rocky Mountains, but does not pass the 50th parallel in the much lower country through which the canoe route lies. It crosses the dividing range, descends to the mouth of the Columbia, and is common in the Northern States. The Chippeways, in whose country it grows abundantly, do not appear to have discovered its admirable diuretic qualities. *C. maculata* is a more southern species: it was gathered by us on the great lakes, but is not common north of the Middle States.

JASMINÆ.—*Fraxinus sambucifolia*, black ash, is said to grow in Virginia, and by Dr. Asa Gray to range from Maine to Wisconsin. It also inhabits New Brunswick, Nova Scotia, and Canada. *F. americana*, white ash, was found by Lieut. Abert on the Arkansas, high up on the western slope of the Mississippi valley. It is a large forest tree in the Northern States. It grows at Pictou, also on Lake Superior, Rainy Lake, the River Winipeg, and the banks of the Saskatchewan, to latitude 54° north, where it is still a tree. *F. pubescens*, red ash, does not grow thicker than a man's thigh on Rainy River, where it terminates near the 49th parallel. It extends southward to the Middle States, and also across the mountains to Oregon.

ELEAGNACEÆ.—*Eleagnus argentea*, silver-berry, is a very common shrub on the banks of rivers throughout the basins of the Saskatchewan and Mackenzie, up to the 68th parallel of latitude. Its dry husky berries are covered with the same silvery epidermis that gives the hoary appearance to the leaves, and are used by the Kutchin to ornament their dresses. This apparently sapless fruit is often found in the stomachs of geese on their northerly migrations. It is the *wapow-muskwa-minan*, or "white-bear berry," of the Crees; and the branches, which harden in drying, are used by the natives for making pipe-stems. It ascends the Saskatchewan, and occurs in Canada, but does not find a place an Gray's "Flora of the Northern States." *Shepherdia canadensis* grows from Vermont and Wisconsin northward to beyond the arctic circle, and is very common on the Mackenzie. Its small, red, juicy, very bitter, and slightly acid berry is useful for making an extempore beer, which ferments in twenty-four hours, and is an agreeable beverage in hot weather. *S. argentea* is a prairie shrub common to the plains of the Missouri and Saskatchewan, but which does not grow in the eastern districts. It is the *Mith-yŭ-mìnă* or blood-red berry of the Crees.

THYMELEÆ.—*Dirca palustris*, leather-wood, is common in the Northern States, and extends to the north side of Lake Superior, disappearing about the Lake of the Woods.

EMPETREÆ.—*Empetrum nigrum* occurs on the Alps of New Hampshire and New York, and is found throughout the whole extent of Rupert's Land up to the Arctic Sea, along which it ranges to Kotzebue Sound, descending the western coast to Sitka, and perhaps lower. It is absent only on the prairies. In the more sandy tracts of the barren grounds it covers the surface with its prostrate branches, that are loaded with fruit in favorable seasons. The snow-geese feed and fatten on the berries, which, after the fresh frosts, become very juicy, and are highly refreshing to the weary and thirsty traveler.

ULMACEÆ.—*Ulmus americana*, white elm, was found by Lieut. Abert on the Pawnee Fork in latitude 38° 10′ N., at an elevation of 1658 feet; and Dr. Asa Gray informs me that it descends to Texas. It is a majestic tree in the Northern States, much prized for its rapid growth and the beauty of its form. Its wood is in requisition there for the use of wheelwrights. On the north banks of the Saskatchewan, in about latitude 54°, which is its polar limit, it grows only in rich alluvial soil, and, being crowded among balsam poplars and other trees which inhabit such places, does not exhibit its handsome outline so as to strike the eye. Its timber there is often decayed at heart, and, even when sound, is so porous that we found it to be unfit for planking boats. It is probable that the *U. fulva*, slippery or red elm, known by the corky and angled bark of its branches, has an equal northerly range; but we did not trace it, though two kinds extend to the Saskatchewan, and we gathered their flowers. It is perhaps an elm of which I heard, but did not see in leaf, which inabits wet places on the banks of Rainy River, and produces a wood that is considered there to be of no value. Dr. Asa Gray has not traced *U. fulva* on the Atlantic side of the Alleghanies further south than Maryland. *Celtis occidentalis*, sugar-berry or hackberry, is common in the Northern States, and extends to

Wisconsin and to the Oregon, but was not seen by us on the north side of Lake Superior.

JUGLANDINEÆ.—*Juglans cinerea*, butter-nut; *J. nigra*, black walnut; *Carya alba*, shell-bark or shag-bark hickory; *C. amara*, bitter-nut or swamp hickory; and *C. glabra*, pig nut or broom hickory, reach Wisconsin, and the basin of the St. Lawrence, but were not seen by us north of Lake Superior.

CUPULIFERÆ.—*Quercus obtusiloba vel stellata*, post oak, abounds in Texas as far as San Antonio de Bexar, wherever any hard wood grows. It was traced with *Q. rubra*, red oak, as far as the River Winipeg. Of the latter, Dr. Asa Gray says that he does not know whether it extends to Texas or not. It is a good-sized tree in the Northern States, and common in rocky woods. *Q. alba*, white oak. This, a most valuable forest tree in the Northern States, ranges northward to Lake Winipeg, where it has a crooked and rather unsightly growth of 20 feet. Michaux states that its southern limit is in Florida; but I do not find it in the lists of plants gathered by Messrs. Emory and Abert in their journeys from Fort Leavenworth on the Missouri to North Mexico and California. *Q. bannisteri; Q. tinctoria*, quercitron or black oak; *Q. macrocarpa*, burr oak; *Q. bicolor*, swamp oak; *Q. prinos*, swamp chestnut oak; and *Q. palustris*, pin oak, grow in Canada or Wisconsin, but were not detected by us to the north of the great lakes, with the exception perhaps of *Q. macrocarpa*, since I gathered the immature acorns of an oak resembling this on Rainy River. *Q. garryana*, inhabits Oregon, northward to Puget Sound, is a tree which reaches 80 feet in height, and is well adapted for shipbuilding.

Fagus ferruginea et sylvestris, American beech: is said by Pursh to range southward to Florida; it ceases at Mackinac on Lake Huron, and does not grow on Lake Superior, but reappears further to the northwest on the Red River of Lake Winipeg, beyond which it was not seen. *Carpinus americana*, hornbeam, blue or water beach, called also iron-wood, inhabits Canada, Wisconsin, and the other Northern States, but was not seen by us on the canoe route. *Ostrya virginica*, hop hornbeam. This tree, which has also the trivial name of iron-wood, grows as far north as the River Winipeg, and is plentiful on Rainy and Red Rivers. Its southern range, according to authors, is to Carolina and Georgia.

Corylus americana, hazel nut, and *C. rostrata*, beaked hazel nut, range northward to the Saskatchewan, the former also crossing the continent to the Pacific coast.

MYRICACEÆ.—*Myrica gale*, sweet gale or Dutch myrtle, is also a European shrub. It is common in North America, on the stony margins of lakes, and in peat bogs, from Virginia to the arctic circle, and in the island of Sitka. The native population of Rupert's Land use the buds as a material for dyeing. *Comptonia asplenifolia*, sweet fern, is common in the Northern States, and terminates on the northern slope of the Saskatchewan basin; it ranges southward along the mountains of Carolina and Georgia.

BETULACEÆ.—*Betula papyracea*, paper or canoe birch, is an invaluable tree to the population of Rupert's Land. Its bark is indispensable for the construction of their canoes, and also serves for the covering of tents, in localities where the skins of large animals are scarce. Neatly sewed and ornamented with porcupine quills, it is moulded into baskets, bags, dishes, plates, and drinking vessels; in short it is the material of which most of the light and easily transported household furniture of the Crees is formed. The ruder 'Tinnè use it, but dispense with many of the forms into which it is worked by their southern neighbors. The wood serves for paddles, the framework of snow-shoes, sledges, hatchet helves, and occasionally for gun-stocks; and in spring the sap forms a pleasant sweet drink, from which a syrup may be manufactured by boiling. Beyond the arctic circle it is a scarce and crooked tree, but occurs of a small size as high as the 69th parallel. It grows in perfection on the north shore of Lake Superior, in the neighborhood of Fort William, where, owing to the ample supply of good bark, a manufactory of canoes for the use of the Hudson's Bay Company has been established. As the Kolushes north of Sitka use birch-bark canoes, I infer that this tree extends to the Pacific; but I have not seen it in the lists of plants of that coast. Pursh mentions Hudson River as its southern limit; and Gray states its range as extending from New England to Wisconsin, but chiefly through the northern parts of that district. It grows in Nova Scotia, Newfoundland, and Labrador. *B. occidentalis* occurs in Oregon, along the Rocky Mountains, northward to the straits of Da Fuca, and crosses the ridge to the vicinity of Edmonton House, on the 54th parallel. *B. excelsa*, yellow birch, was not traced by us beyond the banks of the Kamenistikwoya, which falls into Lake Superior. In the Northern States it is a stately tree, 60 feet in height. *B. lenta*. cherry or sweet birch, is a rather large tree, which is common in the Northern States, Nova Scotia, Canada, and Newfoundland; but does not appear to go far westward, as it was not found by us nor by Agassiz on Lake Superior. *B. pumila vel glandulosa*, little birch, is rare in New England, but grows in bogs of the northern parts of Pennsylvania, Ohio, Michigan, and Wisconsin; also in Nova Scotia, Canada, Newfoundland, and Labrador. It goes considerably beyond the arctic circle, being found on the banks of the Thlewee-chow, the Coppermine River, and other arctic streams, and also on the Mackenzie to about the 68th parallel. It is very like the following, but has a more erect and slender growth, which may be perhaps owing to locality. The leaves are generally longer. *B. nana*, dwarf birch, exists on the summits of the White Mountains of New Hampshire, and of the Essex Mountains of New York. It grows also in the higher parts of Labrador and Canada, along the shores of the Arctic Sea, from Davies' Straits to Kotzebue Sound; and generally throughout the Barren Grounds. *Alnus viridis*, green or mountain alder, and *A. incana*, speckled or hoary alder, range northward to the delta of the Mackenzie on the 68th parallel, and from Newfoundland and Labrador to Kotzebue Sound; the first species being also found on Sitka. They are common bushes in New England and Wisconsin; but I have not seen their southern limits mentioned. *A. rubra*, red alder, is an inhabitant of Sitka and Norfolk Sound, and extends southward to Oregon. It seems to be the western representative of the *A. serrulata* of the southern parts of New England.

SALICACEÆ.—Willows are numerous on the east side of the Rocky Mountains, and seem to attain the maximum development of species in the southern parts of Rupert's Land, but to be less abundant on the Pacific coast, except to the north of the peninsula of Alashka. From Lake Superior to the Arctic Sea they form dense thickets on the shores of every river and lake. It is scarcely possible to note the range of willows, or to collect satisfactory specimens of the species on a rapid journey, as many of them which perfect their catkins before the evolution of their leaves, remain undetermined in the herbarium. Of twenty-two species described by Dr. Asa Gray as inhabitants of the Northern States, only *S. tristis*, Aiton; *S. humilis*, Marshall; *S. sericea*, Marshall; *S. alba*, L.; and *S. angustata*, Pursh, were not collected by us on our northern voyages. *S. uva ursi*, *S. repens vel fusca*, and *S. herbacea*, which grow on the Alps of New Hampshire, extend beyond the arctic circle on both sides of the continent, the latter being one of the most northern plants, as it grows on the north end of Spitzbergen. The other fourteen named by Dr. Gray reach one or more of the northern basins.

The following were traced from Lake Superior to the arctic circle, or beyond it: *S. villosa*, *S. rostrata*, *S. discolor*, *S. viminalis*, *S. lucida*, *S. longifolia*, *S. cordata*, *S. rigida*, *S. planifolia*, and *S. pedicellaris*. Some others were not gathered higher than the valley of the Saskatchewan, but their southward range included the St. Lawrence basin: as *S. candida*; *S. petiolaris*, *S. rosmarinifolia*, *S. purpurea*, and *S. fragilis*.

S. drummondii, *S. barattiana*, and *S. cordifolia* were gathered near the elevated sources of the Saskatchewan only, though the last named has been detected by other collectors on the Labrador coast. *Salix sitchensis* is known only as an inhabitant of the island from whence it derives its name. *Salix richardsonii* and *S. acutifolia* are common to the Saskatchewan and Mackenzie River basins, the former being also an inhabitant of the coast between York Factory and Churchill.

The following are specially arctic in their habitats: *S. myrsinites*, *S. vestita*, *S. speciosa*, *S. reticulata*, and *S. nivalis*, which grow on the peaks of the Rocky Mountains between 52° and 57° north, and within the arctic circle, some of them reaching very high latitudes. *S. reticulata* grows on the coast between York Factory and Churchill. *S. speciosa* inhabits the Arctic Sea coasts from Coronation Gulf to Kotzebue Sound, and ranges southward on the Mackenzie to about the 60th or 61st parallel. It is perhaps the handsomest of the genus, having an agreeable growth, and very large leaves, which are of a silvery whiteness beneath, and when bruised have a rather pleasant odor. On the Mackenzie it grows to the height of 15 feet, in form of a bush, with very stout and long yearly shoots, which distinguish it from all the other willows of the same localities. On the coasts of the Arctic Sea, wherever the rivers afford a suitable point of alluvial soil, a thicket of this willow may be expected, as tall as a man. Mr. Seeman observed it in the tree form on the northwest coast, where it is from 18 to 20 feet in height, and having a stem five inches in diameter. It resembles *S. lapponum* in its habit. *S. stuartiana* and *S. retusa* grow on the more northern banks of the Mackenzie and in Kotzebue Sound, and have not as yet been detected south of the arctic circle. *S rostrata*, *S. speciosa*, *S. lucida*, *S. longifolia*, *S. depressa*, *S. reticulata*, *S. arctica*, and *S. polaris*, have been enumerated by authors as crossing to the Pacific side of the Rocky Mountains in their respective zones. *Salix glauca* was found by Seeman on the shores of Beering's Sea.

The following are not confined to the American continent, but range to either Europe or Asia, as well as to Rupert's Land, or the arctic coasts: *S. petiolaris*, *S. rosmarinifolia*, *S. viminalis*, *S. purpurea*, *S. fragilis*, *S. acutifolia*, *S. fusca*, *S. myrsinites*, *S. stuartiana*, *S. reticulata*, *S. herbacea*, *S. polaris* and *S. ammaniana*.

The most common in Rupert's Land are the *S. rostrata*, which extends southward to New England, and in the north forms almost impenetrable thickets 20 feet or more in height, in which the old twisted and sordid gray stems spread in all directions. *S. longifolia*, which has the growth of an osier, covers the new-formed sandbanks of the rivers up to the 68th parallel, its flexible, densely growing young stems serving to arrest the mud, and speedily to raise the bank above the ordinary level of the water. In drier spots, by river banks in the Saskatchewan basin, it forms bushes from 20 to 25 feet in height. Even in the passes of the Rocky Mountains, as at Jasper Lake, it grows freely on drifting sands. Lieut. Abert found it growing at Council Grove, and Hundred-and-ten-mile Creek, at the height of 1200 feet above the sea, between the 38th and 39th parallels. It inhabits also the banks of the Susquehanna, and all the Northern States. The soft pliable twigs are a favorite food of the moose-deer, and might, indeed, as they grow on the flooded sandbanks, be mowed like hay. *Populus balsamifera*, balsam poplar, or tacamahac, was found growing on the banks of the Mackenzie up to lat. 59°, where it makes a very slender tree. In the southern part of the delta of that river it forms groups of healthy young trees, and from thence to the United States it flourishes on rich alluvial and occasionally flooded banks of rivers to the exclusion, on such spots, of most other trees: its trunk attains a greater circumference than any other member of the northern forest, but its wood is of no value, except for fuel; and, when old, the tree is unsightly from having very generally lost its top. Its growth is rapid, and its decay apparently equally so. I measured some drift logs of this tree which were floating down the Mackenzie, and found them to be about 15 feet in circumference, with a very moderate tapering upward. The Crees name it *Mathèh-mètus*, or ugly poplar. Dr. Asa Gray gives as its southern limit New England, Wisconsin, and perhaps Pennsylvania, but not further. It crosses Beering's Sea to Kamtschatka; and on the rivers of Oregon it grows, according to Douglas, to the height of 140 feet, and 20 feet in diameter. *P. candicans*, balm of Gilead, which greatly resembles the preceding, has not been detected north of Wisconsin: it is the common balsam poplar of Pennsylvania, New York, and New England. *P. monilifera, lævigata vel canadensis* (Mx.), is a more southern species, being rare in New England, but taking the place of *P. candicans* in Western Pennsylvania, Ohio, and Kentucky. It grows on the banks of the Arkansas and other southern tributaries of the Missouri. *P. tremuloides*, aspen. Dr. Gray believes that, south of Pennsylvania and Kentucky, this tree is confined to the Alleghanies, and even on these mountains it is rare. In the Northern States it is common, and varies in height from 20 to 50 feet. It abounds in Rupert's Land in the more fertile soils, and very generally springs up in place of the white spruce, when that tree has been destroyed by fire. Its range is co-extensive with the forest land; but toward the Arctic Sea, and in lat. 69° on the Mackenzie, it is a slender willow-like tree. It is the best fire-wood in the country, but is applied to no other economical purpose, except that its ashes are collected on account of the abundance of potash they contain. I do not know whether it inhabits the Pacific coasts or not. *P. grandidenta*, big-toothed aspen, is common in the Northern States, and reaches New Brunswick and Canada, but did not come under our notice on the canoe route.

Platanaceæ.—*Platanus occidentalis*, American plane-tree, resembles the well-known *P. orientalis* in the way that its exterior bark falls off in thin plates. It extends northward to Canada, but does not appear north of Lake Superior.

Coniferæ.—*Pinus banksiana*, gray pine, the *Cyprès* of the voyagers, grows from the arctic circle on the Mackenzie, down to the great Canada lakes, south of which, Dr. Gray has scarcely seen it, but has heard that it is found in the northern districts of Maine; and it occurs in the list of Wisconsin plants published by the American Association. It crosses the Rocky Mountains to the Spokan River in latitude 47° north. This would be an ornamental tree on many sandy and otherwise unproductive wastes. *P. resinosa*, red pine, has its southern limit, according to Emerson, at Wilkesbarre, in Pennsylvania (latitude 41½° north). I have traced it to 56½° of latitude on Methy River, and it crosses the Rocky Mountains to latitude 43° in Oregon. Dr. Gray says that its height in the Northern States is from 60 to 80 feet, and Emerson relates that a few years ago it was not uncommon to find trees of this species in the southern parts of Maine exceeding 100 feet in height, with a stem four feet in diameter. *P. inops*, which was not seen by us on the canoe route to the north of the United States boundary, extends on the northwest coast from Oregon to Sitka, and ascends Mount Rainier to near the snow limit. *P. strobus*, white or Weymouth pine, has its equatorial limit on the Alleghanies of Virginia or North Carolina, and it ranges northward to the south end of Lake Winipeg. In the Middle States this tree has a shaft of 100 feet; and Emerson has collected instances of trees formerly existing which had the extraordinary length of from 220 to 260 feet. Even near its northern termination it is still a stately tree.

Abies balsamea, balsam fir, was not traced beyond the 62d parallel on the canoe route. It is *Le Sapin* of the voyagers, who prefer its spray to that of any other tree for laying the floor of a tent or winter bivouac. Dr. Gray traced it on the Alleghanies only to Pennsylvania. In the latitude of Norfolk Sound (57°) it crosses the Rocky Mountains to the Pacific. In Virginia, North Carolina, and Georgia, *Pinus fraseri*, or the small fruited balsam fir, occupies the Alleghanies to the exclusion of the preceding. It does not reach the great lakes. *A. canadensis*, hemlock spruce, was observed on the Kamenistikwoya, but not further north than the 49th parallel; though Mr. Tolmie traced it up to the 57th degree of latitude on the shores of the Pacific, and it was observed by Mertens on Sitka. In Maryland this species is found on the Alleghanies only; and Dr. Gray thinks that it ceases to grow in North Carolina and Tennessee. *A. alba*, white spruce. Of this species we have had frequent occasion to speak in the preceding pages, as it is especially the forest tree in Rupert's Land. It is *L'epinette blanche* of the voyagers, and the *Mina-hik* of the Crees. Within the arctic circle it seldom exceeds 40 or 50 feet in height; though in ravines, where it is well-sheltered, and has a suitable soil, it attains twice that altitude. Its age in these high latitudes exceeds 400 years before it shows signs of decay. It most probably has a range from one side of the continent to the other, but has not yet been detected on the west coast. From the 69th parallel on the Mackenzie, it crosses obliquely to the 61st or 60th on the coast of Hudson's Bay; and it is the common spruce in Canada, Nova Scotia, New Brunswick, and New England, but its southern limits are unknown to Dr. Gray. In Canada the sweet cedar is much used for the thin hoops (*varandes*) and lining of the bark canoes, being a straight-grained light wood; but in more northern districts the white spruce supplies its place. It is also exclusively used north of Lake Winipeg, for building purposes, sawing into deals, and boat-building. With its tough roots split to a convenient thickness, and used under the Cree appellation of *Watap*, the pieces of canoe bark are sewn together; and, in districts where birch bark is scarce, a rude canoe is formed of the bark of a spruce fir. A well grown tree, with 30 feet or so free from branches, is chosen; an incision made down to the wood along one side; and the bark, being skillfully raised in one piece, receives the canoe shape by the two ends being skewered together and stuffed with a few branches to add stiffness. The cargo is then placed in the middle, and two or three Indians will descend a rapid river in this extempore vessel. Before many days, however, it becomes water-logged, and, losing its stiffness, spreads out flatly almost to the level of the water, so as to be nearly useless as well as dangerous. Pieces of the bark are sometimes used for covering the roofs of houses.

A. nigra, black spruce, falls little short in its northern range of the preceding, but in the higher latitudes it is a much inferior tree in numbers, beauty, and utility, and is almost confined to swamps and bogs. According to Emerson, it is in perfection in the northern parts of Maine, or about the 46th parallel, and is less flourishing in more southern localities. It is found on the higher mountains of North Carolina and Tennessee. Up to the Saskatchewan it retains a vigorous growth, beyond which it becomes visibly inferior to the white spruce, its branches being short, irregular, and overgrown with *usneæ* and other parasitic lichens. *A. mertensiana* and *A. sitchensis* grow in the forests of Norfolk Sound on the northwest coast in lat. 57°; Mr. Seeman found the latter extending northward to the coasts of Beering's Sea; and a spruce which grows on the banks of the Niatok or Buckland River is thought by Sir William J. Hooker to be one of these Sitka species, and decidedly different from *A. alba*, to which Mr. Seeman at first referred it.

Larix americana, American larch, tamarack or hackmatack, *L'epinette rouge* of the voyagers, and the *Wagginā-gan* or "tree that bends" of the Crees, ranges northward to the arctic circle, and from Newfoundland and Labrador across the continent to the Pacific. It grows in the swamps of the Northern States, and extends southward to Virginia, where it is confined to the mountains. In high latitudes this tree yields a very heavy wood, so much twisted in the grain as not to be readily worked, but it is tough and very durable. It is a tree of no great importance, and is generally thinly scattered through the forest, and if it is any where grouped in numbers it is on the borders of swamps, where it never attains much height.

Cupressus thyoides, white or sweet cedar, extends from North Carolina to the south side of the Saskatchewan basin. Clumps of it grow on the west side of Rainy Lake, and solitary trees range northward to the vicinity of Cumberland House in latitude 54°, where a specimen was gathered by Mr. Drummond. *C. nutkatensis vel Thuja excelsa* inhabits the Pacific coast from Norfolk Sound down to Observatory Inlet and Vancouver's Island. *Thuja occidentalis*, American arbor vitæ, also called white cedar, has its northern limit on the east side of the Rocky Mountains at *Lac Bourbon* or Cedar Lake, a dilatation of the Saskatchewan lying between the 53d and

54th parallels. Michaux mentions the mountains of Virginia as its southern limit. It is a handsome ornament to the banks of Rainy River and the River Winipeg, where it overhangs the water in a picturesque manner; but, as it commonly grows on the occasionally inundated points of lakes and in swamps unmixed with other trees, it has a sombre aspect; and its stems are generally inclined, crooked, and even contracted. *T. gigantea*, the *Wyeth* of the Wallamet Indians, grows in the valley of Oregon from the Rocky Mountains to the sea, and northward to Vancouver's Island. Mr. Douglas found it growing to the height of 170 feet, with a trunk 40 feet in circumference.

Juniperus communis extends from the vicinity of the Arctic Sea to the New England States and Newfoundland. It produces berries freely on elevated grounds within the arctic circle. *J. prostrata* (Persoon), *repens, vel humilis aliorum*, is considered by Sir William Hooker to be a variety of *J. sabina*, which includes the *J. virginiana*. It has always the prostrate form in Rupert's Land, and was observed within the arctic circle, 1000 feet above the sea, associated with the preceding, and bearing fruit. Dr. Gray informs me, that it is not found in this prostrate flagelliform condition south of New York and Northern Pennsylvania. The ordinary *J. virginiana*, red cedar or savin, ranges to the furthest limits of Texas, and to the country about Santa Fé and Tampas Creek, which is elevated from 3000 to 5000 feet above the sea. Col. Emory found it on the 35th parallel, at an altitude of from 6000 to 7000 feet, in form of a large tree. *Taxus canadensis*, American yew or ground hemlock, grows in Massachusetts and Newfoundland, and on the borders of the great lakes northward to the southern slope of the Saskatchewan basin. It is a very pleasing underwood, with an almost herbaceous aspect, which grows thickly under the shade of many kinds of trees. On the canoe route it never assumes the tree form; but in the valley of Oregon Mr. Douglas found yew trees as large as those of Europe.

The following table is founded on Sir William J. Hooker's *Flora Boreali-Americana*. Since the publication of that work, Sir George Back's voyage down the Thlewee-choh, Messrs. Dease and Simpson's through the Arctic Sea, Mr. Rae's from York factory to Repulse Bay, the voyage detailed in the preceding pages, a list of Nova Scotia plants contributed by Mr. Dawson of Pictou, a collection of plants gathered by Mr. Campbell at Pelly Banks, and Dr. Asa Gray's important Botany of the Northern States, have contributed to our knowledge of the distribution of species, and have been severally had recourse to.

The first zone extends on the eastern side of the continent from latitude 45° to 55°, or it comprehends the St. Lawrence and Saskatchewan basins: it rises obliquely, in accordance with the course of the isothermal lines, in going westward, and on the Pacific coast it includes the 49th and 58th parallels, or Vancouver's and Sitka Islands. It is subdivided into three districts; viz., the eastern forest country the eastern prairies, and the country west of the crest of the Rocky Mountains.

The second zone comprehends all the country lying between the arctic circle and the extremities of the continent in latitude 72°. It was not found practicable, owing to the way in which the herbaria were formed, to separate the barren ground species from those growing in the woody country, and the zone has been made to include three districts, one of which is Kotzebue Sound, where 273 species have been collected by Chamisso and others.*

The Rocky Mountain ridge, between the 52d and 57th parallels, has been made a second district of this zone, as many arctic species go southward along the elevated crest of the ridge. I have not been able, however, to separate the species collected by Drummond in the lower valleys of the ridge from those gathered high up on the peaks. From the untiring diligence and unrivaled quickness of eye of this celebrated collector, we may consider the district as well explored; and had the vertical ranges of the species been noted, there would be nothing more required for the present investigation.

The third district of the zone comprises the entire arctic country from Barrow's Point to Davis' Straits.

The third zone lies to the north of the 73d parallel, and extends from Melville Island to Spitzbergen, including a few species gathered above that parallel in Baffin's Bay, and some collected by Dr. Scoresby on the east coast of Greenland. No American land to the westward of Melville Island in so high a latitude has been discovered; and I have met with no flora of the polar Asiatic islands; but it is probable that, so far north, a nearly uniform vegetation encircles the earth.

* Mr. Seeman, who was employed as botanist, in the Herald's late voyages to that quarter, has made a much more ample herbarium of the northwest coast; but as he has just arrived in England as these pages are passing through the press, I can avail myself only partially of his researches for the improvement of the table.

STATISTICAL ENUMERATION OF THE NUMBERS OF SPECIES BY THEIR FAMILIES AND GENERA IN DIFFERENT ZONES.

[For the plants marked * see corrections on p. 462.]

FAMILIES.	Total Species in Three Zones.	First Zone. Between Lat. 45° to 55° on East side, and Lat. 49° to 58° West side.				Second Zone. From Arctic Circle northward to 72° N.				Third or Polar Zone. Lying North of 73° Lat.				.
		Species in First Zone.	On Pacific Coast.	On Eastern Prairies.	In East. Woody Distr.	Species in Second Zone.	Kotzebue Sound.	Rocky Mts. 52° to 57°.	Eastern Arctic Distr.	Species in Third Zone.	Lancaster Sound, &c.	North Greenland.	Spitzbergen.	Europe, Asia, and Africa.
DICOTYLEDONEÆ*	1718	1499	498	391	1130	551	230	340	376	70	57	40	29	403
I. RANUNCULACEÆ*	72	59	24	19	47	42	12	31	23	5	5	1	1	22
Clematis	3	3	3	1	1	1		1						
Thalictrum	5	4	2	2	3	2	1		2					1
Anemone	11	9	4	4	8	8	2	7	4					6
Hepatica	1	1	1	1	1									1
Hydrastis	1	1			1									
Adonis	1	1			1									
Ranunculus	30	23	6	8	20	21	7	17	13	4	4	1	1	13
Caltha	4	3	2	2	2	4		3	2	1	1			1
Trollius	1					1		1						
Isopyrum	1	1			1									
Coptis	2	2	2		1									
Aquilegia	2	2	1		2	2			2					
Delphinium	3	2	1		1	2	1	1						
Aconitum	2	2	1		1	1	1	1						
Actæa	2	2		1	2									
Cimicifuga	1	1			1									
Hydrastis	1	1			1									
Pæonia brownii	1	1	1											
II. MENISPERMACEÆ	1	1			1									
Menispermum	1	1			1									
III. BERBERIDEÆ	5	5	3	1	2									
Berberis	3	3	2	1	1									1
Leontice	1	1			1									
Epimedium	1	1	1											
IV. PODOPHYLLEÆ	3	3			3									
Jeffersonia	1	1			1									
Podophyllum	1	1			1									
Hydropeltis	1	1			1									
V. NYMPHÆACEÆ	4	4	1		4									1
Nymphæa	1	1			1									
Nuphar	2	2	1		2									1
Nelumbium	1	1			1									
VI. SARRACENIEÆ	1	1		1	1	1		1	1					
Sarracenia	1	1		1	1	1		1	1					
VII. PAPAVERACEÆ	3	2			2	1	1	1	1	1	1	1	1	1
Papaver	1						1	1	1	1	1	1	1	1
Stylophorum	1	1			1									
Sanguinaria	1	1			1									

FAMILIES.	Total Species in Three Zones.	First Zone. Between Lat. 45° to 55° on East side, and Lat. 49° to 58° West side.				Second Zone. From Arctic Circle northward to 72° N.				Third or Polar Zone. Lying North of 73° Lat.				Europe, Asia, and Africa.
		Species in First Zone.	On Pacific Coast.	On Eastern Prairies.	In East. Woody Distr.	Species in Second Zone.	Kotzebue Sound.	Rocky Mts. 52° to 57°.	Eastern Arctic Distr.	Species in Third Zone.	Lancaster Sound, &c.	North Greenland.	Spitzbergen.	
VIII. Fumariaceæ	9	7	3	1	5	4	2	1	2					2
Dielytra	4	3	3	1	1	1	1							
Adlumia	1	1			1									
Corydalis	3	2			2	3	1	1	2					1
Fumaria	1	1			1									1
IX. Cruciferæ	104	65	31	21	48	66	16	34	53	16	14	7	5	50
Cheiranthus	1	1	1											
Nasturtium	4	4	2	1	3	1	1	1	1					3
Barbarea	3	3	2	2	2	1		1	1					2
Turritis	7	5	2	3	4	6		5	5					1
Arabis	7	7	3	2	6	2	1	2	2					4
Cardamine	7	5	2	2	5	6	4	3	5	1	1	1	1	4
Dentaria	2	2			2									
Parrya	2					2	1		2	1	1		1	1
Vesicaria	3	3	1	3	1	2		2	1					
Draba	25	8	5	1	5	22	5	9	16	7	6	2	2	15
Erophila	1	1	1		1									1
Cochlearia	8	4	3		1	3	1		3	4	3	3		4
Thlaspi	3	3			3	1			1					3
Hutchinsia	1					1	1	1						1
Cakile	1	1			1									
Hesperis	2	1			1	1	1							2
Sisymbrium	10	9	6	5	7	7	1	5	7					3
Camelina	2	1			1	1								2
Braya	4					4		1	4					
Platypetalum	2					1		1	1	2	2	1	1	
Eutrema	2					2			2	1	1			
Oreas	1	1	1											
Sinapis	2	2			2									2
Lepidium	3	3	2	2	2	2		2	1					1
Capsella	1	1			1	1		1	1					1
X. Capparideæ	2	2	2											
Cleome	1	1	1											
Polanisia	1	1	1											
XI. Cistineæ	5	5			5									
Helianthemum	1	1			1									
Lechea	3	3			3									
Hudsonia	1	1			1									
XII. Violarieæ	18	18	4	5	15	2		2	2					1
Viola	18	18	4	5	15	2		2	2					1
XIII. Droseraceæ	9	7	3	3	6	6	3	6	3					4
Drosera	4	4	2	2	4	2	1	2	1					3
Parnassia	5	3	1	1	2	4	2	4	2					1
XIV. Polygaleæ	7	7			7	1			1					
Polygala	7	7			7	1			1					
XV. Caryophylleæ*	66	48	26	13	33	35	15	19	25	11	7	7	7	29
Dianthus	1					1	1							1

FAMILIES.	Total Species in Three Zones.	First Zone. Between Lat. 45° to 55° on East side, and Lat. 49° to 58° West side.				Second Zone. From Arctic Circle northward. to 72° N.				Third or Polar Zone. Lying North of 73° Lat.				Europe, Asia, and Africa.
		Species in First Zone.	On Pacific Coast.	On Eastern Prairies.	In East. Woody Distr.	Species in Second Zone.	Kotzebue Sound.	Rocky Mts. 52° to 57°.	Eastern Arctic Distr.	Species in Third Zone.	Lancaster Sound, &c.	North Greenland.	Spitzbergen.	
CARYOPHYLLEÆ—(*continued.*)														
Saponaria	1	1			1									1
Silene	10	10	3	1	7	2	1	1	2	1	1	1	1	3
Lychnis	3	2			2	1	1	1	1	2	1	2	1	2
Agrostemma	1	1			1									1
Mollugo	1	1	1		1									
Spergula	4	3	2	2	2	3	1	2	1	1			1	4
Larbræa	1	1	1			1		1	1					1
Stellaria	14	10	7	3	5	9	4	5	8	3	1	2	2	3
Arenariæ	19	11	7	5	9	13	4	6	11	3	3	1	1	9
Merckia	1					1	1							
Cerastium	10	8	5	2	5	4	2	3	1	1	1	1	1	4
XVI. LINEÆ	3	3	1	2	3	1	1	1	1					1
Linum	3	3	1	2	3	1	1	1	1					1
XVII. MALVACEÆ	5	5	1	2	3									1
Malva	3	3	1	1	2									1
Hibiscus	1	1			1									
Sida	1	1		1										
XVIII. TILIACEÆ	2	2			2									
Tilia	2	2			2									
XIX. HYPERICINEÆ	8	8			8									1
Hypericum	8	8			8									1
XX. ACERINEÆ	8	8	2	1	6									
Acer	7	7	2		5									
Negundo	1	1		1	1									
XXI. AMPELIDEÆ	4	4			4									
Ampelopsis	1	1			1									
Vitis	3	3			3									
XXII. GERANIACEÆ	6	6	3	1	3									4
Geranium	5	5	2	1	3									3
Erodium	1	1	1											1
XXIII. BALSAMINEÆ	2	2	1		2	1		1	1					
Impatiens	2	2	1		2	1		1	1					
XXIV. LIMNANTHACEÆ	1	1			1									
Floërkia	1	1			1									
XXV. OXALIDEÆ	5	5	3	1	3									1
Oxalis	5	5	3	1	3									1
XXVI. RUTACEÆ	4	4			4									
Xanthoxylum	3	3			3									
Ptelea	1	1			1									
XXVII. CELASTRINEÆ	8	8	2	1	7	1		1						
Staphylea	1	1			1									

FAMILIES.	Total Species in Three Zones.	First Zone. Between Lat. 45° to 55° on East side, and Lat. 49° to 58° West side.				Second Zone. From Arctic Circle northward to 72° N.				Third or Polar Zone. Lying North of 73° Lat.				Europe, Asia, and Africa.
		Species in First Zone.	On Pacific Coast.	On Eastern Prairies.	In East. Woody Distr.	Species in Second Zone.	Kotzebue Sound.	Rocky Mts. 52° to 57°.	Eastern Arctic Distr.	Species in Third Zone.	Lancaster Sound, &c.	North Greenland.	Spitzbergen.	
CELASTRINEÆ—(*continued.*)														
Euonymus	2	2	1		2									
Celastrus	1	1			1									
Myginda	1	1	1	1		1		1						
Ilex	1	1			1									
Prinos	1	1			1									
Nemopanthes	1	1			1									
XXVIII. RHAMNEÆ	6	6	3		3									
Rhamnus	2	2	1		1									
Ceanothus	4	4	2		2									
XXIX. TEREBINTHACEÆ	6	6	1		6									
Rhus	6	6	1		6									
XXX. LEGUMINOSÆ*	98	88	27	34	51	19	11	11	14	2	2	1		19
Thermopsis	2	2	1	1										1
Baptisia	4	4			4									
Medicago	2	2			2									2
Trifolium	7	7	6	2	3									3
Psoralea	5	5	3	3										
Petalostemum	2	2		2	2									
Glycirrhiza	1	1	1	1	1									
Tephrosia	1	1			1									
Amorpha	3	3		3	2									
Robinia	1	1			1									1
Phaca	12	10	2	7	2	4	2	2	4	1	1			2
Oxytropis	9	4		4	2	6	4	4	4	1	1	1		5
Astragalus	12	10	4	7	2	2		1	1					2
Desmodium	5	5			5									
Hedysarum	2	1			1	2	1	1	1					
Lespedeza	5	5			5									
Vicia	5	5	3		4	1	1	1	1					2
Ervum	1	1			1									
Lathyrus	7	7	2	3	6	2	1	1	2					1
Amphicarpea	1	1			1									
Apios	1	1			1									
Phaseolus	1	1			1									
Lupinus	6	6	5	1	1	2	2	1	1					
Gymnocladus	1	1			1									
Cassia	1	1			1									
Cercis	1	1			1									
XXXI. ROSACEÆ*	124	106	37	33	78	43	20	25	27	5	4	4	1	19
Cerasus	8	8	1	4	8	1			1					
Prunus	1	1			1									
Purshia	1	1	1											
Lütkea	1	1	1											
Spiræa	8	8	6	3	3	2	1	1						1
Gillenia	1	1			1									
Dryas	3	2	1	1	2	3	2	2	3	1	1	1	1	1
Geum	4	4	1	2	3	1		1						1
Sieversia	5	2	1	1	1	3	3		1	1	1			
Comaropsis	2	2	1		1									1

FAMILIES	Total Species in Three Zones.	First Zone. Between Lat. 45° to 55° on East side. and Lat. 49° to 58° West side.				Second Zone. From Arctic Circle northward to 72° N.				Third or Polar Zone. Lying North of 73° Lat.				Europe, Asia, and Africa.
		Species in First Zone.	On Pacific Coast.	On Eastern Prairies.	In East. Woody Distr.	Species in Second Zone.	Kotzebue Sound.	Rocky Mts. 52° to 57°.	Eastern Arctic Distr.	Species in Third Zone.	Lancaster Sound, &c.	North Greenland.	Spitzbergen.	
ROSACEÆ—(*continued.*)														
Rubus	18	16	8	3	12	5	3	3	4					2
Dalibarda	1	1			1									
Fragaria	3	3	1		2	1		1	1					
Potentilla	36	24	6	14	16	19	10	13	12	3	2	3		7
Sibbaldia	1	1			1	1		1						
Chamærhodos	1	1		1		1		1						1
Agrimonia	1	1			1									1
Alchemilla	1	1			1									1
Sanguisorba	2	2	2		2	1	1		1					
Poterium	1	1			1									1
Rosa	9	9	1	2	8	3		1	3					2
Cratægus	8	8	1	2	8	1		1						
Amelanchier	2	2	1		2	1			1					
Pyrus	6	6	4		3									
XXXII. ONAGRARIÆ	28	25	9	11	11	7	5	6	4	2	1	2		10
Epilobium	12	9	5	2	4	7	5	6	4	2	1	2		6
Gaura	4	4		4										
Œnothera	7	7	2	5	3									1
Clarkia	1	1	1											
Isnardia	2	2			2									1
Circæa	2	2	1		2									2
XXXIII. HALORAGIÆ	10	8	2	4	7	5	3	2	4					5
Proserpinaca	1	1			1									
Myriophyllum	3	3		1	3	2		1	2					2
Callitriche	3	3	1	2	2									2
Hippuris	3	1	1	1	1	3	3	1	2					1
XXXIV. CERATOPHYLLEÆ	1	1		1	1	1		1	1					1
Ceratophyllum	1	1		1	1	1		1	1					1
XXXV. LYTHRARIÆ	3	3			3									1
Lythrum	2	2			2									1
Decodon	1	1			1									
XXXVI. CUCURBITACEÆ	2	2		1	2									
Sicyos	1	1			1									
Echinocistus	1	1		1	1									
XXXVII. PORTULACEÆ*	13	11	9	1	3	3	3	1						1
Montia	1	1	1											1
Portulaca	1	1			1									
Lewisia	1	1	1											
Talinum	1	1			1									
Claytonia	9	7	7	1	1	3	3	1						
XXXVIII. PARONYCHIEÆ	2	2		1	1									
Anychia	1	1			1									
Paronychia	1	1		1										
XXXIX. CRASSULACEÆ	3	3		1	2	1	1		1					1
Sedum	2	2		1	1	1	1		1					1
Penthorum	1	1			1									

FAMILIES.	Total Species in Three Zones.	First Zone. Between Lat. 45° to 55° on East side, and Lat. 49° to 58° West side.				Second Zone. From Arctic Circle northward to 72° N.				Third or Polar Zone. Lying North of 73° Lat.				Europe, Asia, and Africa.
		Species in First Zone.	On Pacific Coast.	On Eastern Prairies.	In East. Woody Distr.	Species in Second Zone.	Kotzebue Sound.	Rocky Mts. 52° to 57°.	Eastern Arctic Distr.	Species in Third Zone.	Lancaster Sound, &c.	North Greenland.	Spitzbergen.	
XL. CACTEÆ	2	2	1	1	1									
Cactus	2	2	1	1	1									
XLI. GROSSULARIEÆ	16	16	8	3	10	4	1	2	4					1
Ribes	16	16	8	3	10	4	1	2	4					1
XLII. SAXIFRAGEÆ	56	32	13	4	19	35	23	21	18	12	11	7	7	18
Heuchera	6	6	3	1	3	1		1	1					
Tiarella	3	3	2		1									
Tellima	1	1	1											
Mitella	4	3		1	2	3		3	1					
Chrysosplenium	2	1			1	1	1	1	1	1	1			1
Saxifraga	38	17	7	2	11	28	20	15	15	11	10	7	7	17
Eriogyna	1	1			1	1	1	1						
Leptarrhena	1					1	1							
XLIII. UMBELLIFERÆ	39	36	13	6	29	10	6	2	5					10
Hydrocotyle	2	2			2									1
Eryngium	1	1			1									
Sanicula	1	1	1		1									
Cicuta	3	3	1		3	2		1	2					1
Zizia	3	3		1	3	1		1	1					
Ammi	1	1			1									
Carum	1	1			1									1
Cryptotænia	1	1			1									
Sium	2	2	2	1	2									1
Bupleurum	2	1			1	1	1							1
Seseli	1	1		1		1			1					
Cnidium	1	1			1									
Thaspium	1	1			1									
Ligusticum	1	1	1		1	1	1							1
Conioselinum	2	2	1		1	1	1		1					1
Angelica	2	2			2									
Archangelica	1					2	2							1
Pleurospermum	1	1	1											
Færula	2	2	1	2										
Imperatoria	1	1			1									1
Pastinaca	1	1	1		1									1
Heracleum	1	1	1		1									
Laserpitium	1					1	1							
Polytænia	1	1		1										
Daucus	1	1	1											
Osmorrhiza	2	2	2		2									
Conium	1	1			1									
Erigeneia	1	1			1									
XLIV. ARALIACEÆ	7	7	1	1	6	1		1	1					1
Adoxa	1	1			1	1		1	1					1
Panax	3	3	1	1	2									
Aralia	3	3			3									
XLV. HAMAMELIDEÆ	1	1			1									
Hamamelis	1	1			1									

FAMILIES.	Total Species in Three Zones.	First Zone. Between Lat. 45° to 55° on East side, and Lat. 49° to 58° West side.				Second Zone. From Arctic Circle northward to 72° N.				Third or Polar Zone. Lying North of 73° Lat.				Europe, Asia, and Africa.
		Species in First Zone.	On Pacific Coast.	On Eastern Prairies.	In East. Woody Distr.	Species in Second Zone.	Kotzebue Sound.	Rocky Mts. 52° to 57°.	Eastern Arctic Distr.	Species in Third Zone.	Lancaster Sound, &c.	North Greenland.	Spitzbergen.	
XLVI. CORNEÆ	7	7	4	2	7	2	2	1	1					1
Cornus	7	7	4	2	7	2	2	1	1					1
XLVII. LORANTHACEÆ	1	1	1		1	1	1	1						1
Arceuthobium	1	1	1		1	1	1	1						1
XLVIII. CAPRIFOLIACEÆ	24	24	7	4	23	6	2	6	3					2
Sambucus	2	2	1	1	2									
Viburnum	9	9	2	2	9	1		1	1					
Diervilla	1	1			1	1		1						
Lonicera	8	8	1		7	3	1	3	1					1
Triosteum	1	1			1									
Symphoricarpus	2	2	2		2									
Linnæa	1	1	1	1	1	1	1	1	1					1
XLIX. RUBIACEÆ	15	15	5	2	14	3	3	2	2					2
Hedyotis	3	3			3									
Mitchella	1	1			1									
Cephalanthus	1	1			1									
Galium	9	9	5	2	8	3	3	2	2					2
Spigelia	1	1			1									
L. VALERIANEÆ	6	5	2	1	3	3	1	2	1					1
Valeriana	5	4	2	1	2	3	1	2	1					1
Fedia	1	1			1									
LI. COMPOSITÆ*	321	285	66	80	218	85	31	60	57	7	6	3	1	46
1. *Cichoraceæ*	41	36	7	9	31	11	2	10	6	1	1	1	1	12
Sonchus	7	6	3	2	5	1		1	1					4
Nabalus	6	6	1		5									
Lygodesma	2	1			1	1		1						
Leontodon	2	2	2	2	2	2	2	2	2	1	1	1	1	2
Apargia	2	2	1		1									1
Cynthia	1	1			1									
Lapsana	1	1			1									1
Cichorium	1	1			1									1
Crepis	3	3		1	2	2		2	1					1
Hieracium	12	10		3	10	3		2	1					2
Troximon	3	2		1	1	2		2	1					
Krigia	1	1			1									
2. *Cynareæ*	13	10	1	1	10	5	1	4	2					5
Centaurea	1	1			1									1
Arctium	1	1			1									1
Carduus	6	5	1	1	5	3		3						2
Cirsium	3	3			3									
Saussurea	2					2	1	1	2					1
3. *Vernoniaceæ*	2	2		1	2									
Vernonia	2	2		1	2									
4. *Eupatoriaceæ*	15	14	2	4	14	4	2	2	3	2	2			2
Kuhnia	1	1			1									
Eupatorium	4	4			4									1

FAMILIES.	Total Species in Three Zones.	First Zone. Between Lat. 45° to 55° on East side, and Lat. 49° to 58° West side.				Second Zone. From Arctic Circle northward to 72° N.				Third or Polar Zone. Lying North of 73° Lat.				Europe, Asia, and Africa.
		Species in First Zone.	On Pacific Coast.	On Eastern Prairies.	In East. Woody Distr.	Species in Second Zone.	Kotzebue Sound.	Rocky Mts. 52° to 57°.	Eastern Arctic Distr.	Species in Third Zone.	Lancaster Sound, &c.	North Greenland.	Spitzbergen.	
COMPOSITÆ, *Eupatoriaceæ*—(*continued.*)														
Mikania	1	1			1									
Liatris	5	5		2	5									
Nardosmia	3	2	1	1	2	3	1	1	3	2	2			1
Adenocaulon	1	1	1	1	1	1	1	1						
5. *Senecionideæ*	116	104	36	29	74	35	21	24	25	4	3	2		23
Silphium	6	6		1	6									
Xanthium	2	2			2									
Ambrosia	5	5	2	1	4									
Parthenium	1	1			1									
Iva	1	1	1	1										
Heliopsis	1	1		1	1									
Echinacea	1	1			1									
Rudbeckia	3	3		2	2									
Lepachys	1	1			1									
Coreopsis	3	3			3									
Helianthus	12	12	2	3	9									
Bidens	6	6	1		6									
Gaillardia	2	2	2	2	1									
Trichophyllum	2	2	2											
Hymenopappus	1	1	1											
Picradenia	1	1		1										
Helenium	1	1	1	1	1	1	1	1	1					
Anthemis	1	1			1									1
Marruta	1	1			1									
Achillea	3	3	2	1	2	2	1	2	2					1
Chrysanthemum	4	3	2		1	3	3	1	2	1	1			2
Pyrethrum	1	1			1	1	1	1	1					1
Cotula	1	1	1											1
Omalanthus	1	1	1		1	1			1					
Tanacetum	3	3	1		2									1
Artemisia	17	12	6	8	5	11	7	7	8					5
Gnaphalium	6	6	1		5									4
Antennaria	6	5	3	3	5	4	1	4	2	1	1	1		4
Arnica	5	4	4			3	1	3	1	1	1	1		1
Senecio	16	12	3	3	10	9	6	5	7	1				2
Cacalia	2	2		1	2									
6. *Asteroideæ*	134	119	20	36	87	30	5	20	24					4
Solidago	31	31	4	5	30	3	1	3	2					1
Aster	59	49	7	10	37	11	3	5	9					1
Eurybia	2	2			2									
Seriocarpus	1	1			1									
Tripolium	2	2		2		1			1					
Galatella	2	2			2									
Townsendia	1	1		1		1		1						
Erigeron	17	12	6	6	9	12	1	10	8					2
Dipplopappus	9	9	1	6	4									
Boltonia	1	1			1									
Brachyris	1	1		1										
Polymnia	1	1			1									
Madia	1	1		1										

FAMILIES.	Total Species in Three Zones.	First Zone. Between Lat. 45° to 55° on East side, and Lat. 49° to 58° West side.				Second Zone. From Arctic Circle northward to 72° N.				Third or Polar Zone. Lying North of 73° Lat.				Europe, Asia, and Africa.
		Species in First Zone.	On Pacific Coast.	On Eastern Prairies.	In East. Woody Distr.	Species in Second Zone.	Kotzebue Sound.	Rocky Mts. 52° to 57°.	Eastern Arctic Distr.	Species in Third Zone.	Lancaster Sound, &c.	North Greenland.	Spitzbergen.	
COMPOSITÆ, *Asteroideæ*—(*continued.*)														
Crinitaria	2	2	1	1										
Donia	4	4	1	3		2		1	1					
LII. CAMPANULACEÆ	14	11	3	1	9	4	3	2	2	1		1	1	3
Campanula	8	5	3	1	3	4	3	2	2	1		1	1	2
Lobelia	6	6			6									1
LIII. VACCINEÆ	16	16	13	5	10	6	2	5	4					4
Vaccinium	16	16	13	5	10	6	2	5	4					4
LIV. ERICEÆ*	40	33	19	3	23	18	8	13	10	2	1	1	2	10
Gaultheria	4	3	2	1	2	2		2						
Arbutus	4	4	3	1	2	2	2	2	2					2
Andromeda	10	9	6	1	6	4	2	2	3	2	1	1	2	2
Menziesia	8	4	3		1	4		4						1
Kalmia	3	3	1		3	1		1	1					
Epigæa	1	1			1									
Rhodora	1	1			1									
Rhododendron	6	5	1		4	2	1	2	1					2
Azalea	1	1	1		1	1	1		1					1
Ledum	2	2	2		2	2	2		2					2
LV. MONOTROPEÆ	16	15	8	4	10	6	2	5	5					5
Cladothamnus	1	1	1											
Pyrola	10	9	6	3	5	6	2	5	5					5
Pterospora	1	1			1									
Monotropa	2	2			2									
Chimaphila	2	2	1	1	2									
LVI. JASMINEÆ	3	3	1	1	3									
Fraxinus	3	3	1	1	3									
LVII. APOCYNEÆ	4	4		1	3									1
Apocynum	4	4		1	3									1
LVIII. ASCLEPIADEÆ	11	11		2	11									1
Asclepias	11	11		2	11									1
LIX. GENTIANEÆ	34	27	10	2	19	13	8	9	8					5
Gentiana	23	16	6		11	10	6	7	6					2
Pleurogyne	1	1			1	1	1		1					1
Swertia	1	1	1											1
Halenia	3	3		1	3	1		1						
Sabattia	2	2			2									
Menyanthes	1	1	1	1	1	1	1	1	1					1
Villarsia	2	2	2											
Limnanthemum	1	1			1									
LX. POLEMONIACEÆ	13	11	4	2	8	4	2	2	2	1			1	2
Polemonium	2	2		1	2	1	1	1	1	1			1	1
Phlox	9	7	2		5	3	1	1	1					1
Collomia	2	2	2	1	1									

FAMILIES.	Total Species in Three Zones.	First Zone. Between Lat. 45° to 55° on East side, and Lat. 49° to 58° West side.				Second Zone. From Arctic Circle northward to 72° N.				Third or Polar Zone. Lying North of 73° Lat.				Europe, Asia, and Africa.
		Species in First Zone.	On Pacific Coast.	On Eastern Prairies.	In East. Woody Distr.	Species in Second Zone.	Kotzebue Sound.	Rocky Mts. 52° to 57°.	Eastern Arctic Distr.	Species in Third Zone.	Lancaster Sound, &c.	North Greenland.	Spitzbergen.	
LXI. DIAPENSIACEÆ*	1					1			1					1
Diapensia	1					1			1					1
LXII. CONVOLVULACEÆ	6	6	2	2	6									2
Convolvulus	2	2			2									1
Calystegia	2	2			2									1
Cuscuta	2	2	2	2	2									
LXIII. HYDROPHYLLEÆ	5	4		1	4	2	1	2	1					
Hydrophyllum	3	3			3									
Eutoca	2	1		1	1	2	1	2	1					
LXIV. BORAGINEÆ*	27	25	7	6	20	9	4	5	6					6
Myosotis	4	3	2	1	1	1	1	1	1					3
Echinospermum	7	7		1	7	2		1	1					2
Lithospermum	14	13	4	3	11	6	3	3	4					1
Onosmodium	1	1		1	1									
Echium	1	1	1											
LXV. SOLANEÆ	8	8	4	2	3									
Solanum	2	2		1	1									
Datura	1	1			1									
Nicandra	1	1			1									
Nicotiana	2	2	2											
Physalis	2	2	2	1										
LXVI. OROBANCHEÆ	6	6	3	3	4	1	1		1					1
Orobanche	4	4	3	3	2	1	1		1					1
Conophilus	1	1			1									
Epiphegus	1	1			1									
LXVII. ACANTHACEÆ	2	2			2									
Dipteracanthus	1	1			1									
Dianthera	1	1			1									
LXVIII. SCROPHULARINEÆ*	74	69	27	22	46	20	10	13	14	2	2	1	1	21
Verbascum	3	3			3									2
Linaria	2	2			2									1
Collinsia	4	4		1	3									
Chelone	1	1			1									
Penstemon	8	8	5	4	1									1
Mimulus	4	4	2	1	2									
Gratiola	1	1	1	1	1									
Synthyris	1	1			1									
Limosella	1	1			1									1
Veronica	11	11	6	4	8	3	1	3	1					6
Gymnandra	2	1	1			1	1		1					
Romanzoffia	2	2	2			1	1							
Gerardia	6	6		1	6									
Orthocarpus	3	3	3	1	1									
Castilleja	5	5	3	3	4	3	2	3	3					
Euphrasia	1	1			1									1
Bartsia	1	1			1									1
Rhinanthus	1	1	1	1	1	1		1	1					1

FAMILIES.	Total Species in Three Zones.	First Zone. Between Lat. 45° to 55° on East side, and Lat. 49° to 58° West side.				Second Zone. From Arctic Circle northward to 72° N.				Third or Polar Zone. Lying North of 73° Lat.				Europe, Asia, and Africa.
		Species in First Zone.	On Pacific Coast.	On Eastern Prairies.	In East. Woody Distr.	Species in Second Zone.	Kotzebue Sound.	Rocky Mts. 52° to 57°.	Eastern Arctic Distr.	Species in Third Zone.	Lancaster Sound, &c.	North Greenland.	Spitzbergen.	
SCROPHULARINEÆ, (*continued.*)														
Melampyrum	1	1		1	1									1
Pedicularis	16	12	3	4	8	11	5	6	8	2	2	1	1	6
LXIX. LABIATÆ	40	40	8	10	37	2		2	2					8
Mentha	1	1	1	1	1									
Lycopus	4	4	1	2	2									
Salvia	1	1			1									
Monarda	3	3			3									
Blephila	1	1			1									
Pycnanthemum	3	3			3									
Collinsonia	1	1			1									
Cunila	1	1			1									
Hedeoma	1	1			1									
Micromeria	1	1			1									
Melissa	1	1			1									
Prunella	1	1	1	1	1									1
Scutellaria	6	6	3	2	5									1
Lophanthus	3	3		1	3									
Nepeta	1	1			1									1
Dracocephalum	1	1		1	1	1		1	1					
Physostegia	1	1	1	1	1									
Lamium	1	1			1									1
Leonurus	1	1			1									1
Galeopsis	1	1			1									1
Stachys	3	3	1	1	3	1		1	1					1
Marrubium	1	1			1									1
Teucrium	1	1			1									
Trichostema	1	1			1									
LXX. VERBENACEÆ*	15	15	5	5	13	4		4	2					6
Verbena	6	6	2	2	5									1
Phryma	1	1			1									
Utricularia	6	6	1	1	6	2		2	1					3
Pinguicula	2	2	2	2	1	2		2	1					2
LXXI. PRIMULACEÆ	23	17	9	7	13	11	7	4	8					10
Dodecatheon	2	2	2	1	1	2	2		1					
Androsace	2					2	2	2	2					2
Douglasia	2	1		1		1			1					
Primula	6	3	1	1	3	5	3	2	3					4
Trientalis	3	3	2		1									1
Lysimachia	6	6	2	2	6	1			1					1
Glaux	1	1	1	1	1									1
Samolus	1	1	1	1	1									1
LXXII. PLUMBAGINEÆ	2	2	1		2	1	1		1					1
Statice	2	2	1		2	1	1		1					1
LXXIII. PLANTAGINEÆ	5	5	3	1	4	2			2					3
Plantago	5	5	3	1	4	2			2					3
LXXIV. NYCTAGINEÆ	3	3	1		2									
Oxybaphus	1	1	1											
Allionia	2	2			2									

FAMILIES.	Total Species in Three Zones.	First Zone. Between Lat. 45° to 55° on East side, and Lat. 49° to 58° West side.				Second Zone. From Arctic Circle northward to 72° N.				Third or Polar Zone. Lying North of 73° Lat.				Europe, Asia, and Africa.
		Species in First Zone.	On Pacific Coast.	On Eastern Prairies.	In East. Woody Distr.	Species in Second Zone.	Kotzebue Sound.	Rocky Mts. 52° to 57°.	Eastern Arctic Distr.	Species in Third Zone.	Lancaster Sound, &c.	North Greenland.	Spitzbergen.	
LXXV. AMARANTHACEÆ.....	6	6	1	2	6									
Amaranthus............	6	6	1	2	6									
LXXVI. CHENOPODEÆ.......	20	19	6	11	12	4	3		1					14
Salicornia..............	2	2	1	1	1									2
Salsola..................	1	1			1									1
Corispermum...........	1	1	1	1	1									1
Eurotia..................	1	1		1										1
Blitum..................	2	2		1	1									1
Chenopodium...........	8	7	3	6	4	1			1					5
Ambrina...............	1	1			1									
Atriplex................	4	4	1	1	3	3	3							3
LXXVII. PHYTOLACEÆ......	1	1			1									
Phytolacca.............	1	1			1									
LXXVIII. POLYGONEÆ.......	34	32	16	8	22	11	6	6	9	2	2	2	2	17
Konigia.....	1					1	1	1						1
Oxyria..................	1	1	1			1	1	1	1	1	1	1	1	1
Rumex...................	10	10	5	4	9	5	1	2	4					8
Polygonum.............	18	17	7	2	13	4	3	1	4	1	1	1	1	7
Eriogonum.............	4	4	3	2										
LXXIX. LAURINEÆ..........	2	2			2									
Benzoin................	1	1			1									
Sassafras....	1	1			1									
LXXX. ELEAGNEÆ..........	3	3		2	2	2		2	2					
Eleagnus...............	1	1		1	1	1		1	1					
Shepherdia.............	2	2		1	1	1		1	1					
LXXXI. THYMELEÆ.........	1	1			1									
Dirca..................	1	1			1									
LXXXII. SANTALACEÆ......	2	2		2	2	1		1	1					
Comandra..............	2			2	2	1		1	1					
LXXXIII. ARISTOLOCHIÆ....	1	1			1									
Asarum................	1	1			1									
LXXXIV. EMPETREÆ.......	1	1			1	1	1	1	1					1
Empetrum..............	1	1			1	1	1	1	1					1
LXXXV. EUPHORBIACEÆ....	8	8	2	2	8									1
Euphorbia..............	7	7	2	2	7									1
Acalypha...............	1	1			1									
LXXXVI. URTICEÆ.........	10	10	1	1	10	1			1					4
Urtica..................	5	5	1	1	5	1			1					2
Pilea...................	1	1			1									
Parietaria..............	1	1			1									
Boehmeria.............	1	1			1									
Humulus...............	1	1			1									1
Cannabis...............	1	1			1									1

FAMILIES.	Total Species in Three Zones.	First Zone. Between Lat. 45° to 55° on East side, and Lat. 49° to 58° West side.				Second Zone. From Arctic Circle northward to 72° N.				Third or Polar Zone. Lying North of 73° Lat.				Europe, Asia, and Africa.
		Species in First Zone.	On Pacific Coast.	On Eastern Prairies.	In East. Woody Distr.	Species in Second Zone.	Kotzebue Sound.	Rocky Mts. 52° to 57°.	Eastern Arctic Distr.	Species in Third Zone.	Lancaster Sound, &c.	North Greenland.	Spitzbergen.	
LXXXVII. ULMACEÆ	3	3	1		3									
Ulmus	2	2			2									
Celtis	1	1	1		1									
LXXXVIII. SAURUREÆ	1	1			1									
Saururus	1	1			1									
LXXXIX. JUGLANDINEÆ	4	4			4									
Juglans	2	2			2									
Carya	2	2			2									
XC. SALICACEÆ	48	39	8	6	36	27	8	14	24	3	1	3	1	14
Salix	44	35	7	5	32	25	7	12	22	3	1	3	1	13
Populus	4	4	1	1	4	2	1	2	2					1
XCI. BETULACEÆ	11	11	4	4	9	6	3	6	5					2
Betula	8	8	2	2	7	4	1	4	3					1
Alnus	3	3	2	2	2	2	2	2	2					1
XCII. PLATANACEÆ	1	1			1									
Platanus	1	1			1									
XCIII. CUPULIFERÆ	15	15	2	2	14									
Quercus	10	10	1		9									
Fagus	1	1			1									
Carpinus	1	1			1									
Ostrya	1	1			1									
Corylus	2	2	1	2	2									
XCIV. MYRICACEÆ	3	3	2	1	2	1		1	1					1
Myrica	2	2	2	1	1	1		1	1					1
Comptonia	1	1			1									
XCV. CONIFERÆ	20	20	9	6	14	7	1	4	6					2
Pinus	5	5	2	1	4	1			1					
Abies	6	6	3	2	4	3	1	2	2					
Larix	1	1			1	1			1					
Cupressus	2	2	1		1									
Thuja	2	2	1	1	1									
Juniperus	3	3	2	2	2	2		2	2					2
Taxus	1	1			1									
MONOCOTYLEDONES	554	493	162	122	399	198	53	120	146	21	20	9	9	188
XCVI. COMELINEÆ	1	1			1									
Tradescantia	1	1			1									
XCVII. ALISMACEÆ	3	3	2	2	3									3
Sagittaria	1	1	1	1	1									1
Alisma	2	2	1	1	2									2
XCVIII. JUNCAGINEÆ	4	4	1	3	4	2		2	1					3
Scheuchzeria	1	1		1	1	1		1						1
Triglochin	3	3	1	2	3	1		1	1					2

FAMILIES.		First Zone. Between Lat. 45° to 55° on East side, and Lat. 49° to 58° West side.				Second Zone. From Arctic Circle northward to 72° N.				Third or Polar Zone. Lying North of 73° Lat.				
	Total Species in Three Zones.	Species in First Zone.	On Pacific Coast.	On Eastern Prairies.	In East. Woody Distr.	Species in Second Zone.	Kotzebue Sound.	Rocky Mts. 52° to 57°.	Eastern Arctic Distr.	Species in Third Zone.	Lancaster Sound, &c.	North Greenland.	Spitzbergen.	Europe, Asia, and Africa.
XCIX. AROIDEÆ	13	12	2	3	11	3	1	3	2					10
Acorus	1	1		1	1									1
Orontium	1	1			1									
Arum	1	1			1									
Calla	1	1			1									
Symplocarpus	2	2	2		1									2
Lemna	3	3			3									3
Sparganium	3	2		1	2	2	1	2	2					3
Typha	1	1		1	1	1		1	1					1
C. NAIADES	14	14	3		14	1		1	1					9
Zannichellia	1	1	1		1									1
Naias	1	1			1									
Ruppia	1	1	1		1									1
Potamogeton	11	11	1		11	1		1	1					7
CI. SMILACEÆ	19	19	6	8	16	3		3	2					4
Smilax	3	3		1	3									
Streptopus	3	3	3	2	2									2
Uvularia	6	6	1	1	5									
Smilacina	6	6	2	4	5	3		3	2					1
Polygonatum	1	1			1									1
CII. MELANTHACEÆ	16	16	7	4	13	4	2	3	3					3
Limnanthemum	1	1			1									
Zigadenus	1	1	1	1	1	1	1	1	1					
Xerophyllum	1	1	1											
Helonias	1	1	1											
Veratrum	1	1	1	1	1	1		1						
Tofieldia	3	3	2	2	3	2	1	1	2					2
Medeola	1	1			1									
Trillium	7	7	1		6									1
CIII. LILIACEÆ	9	9	6	1	5									1
Lilium	3	3	2	1	3									
Fritillaria	2	2	2											1
Erythronium	3	3	1		2									
Calochortus	1	1	1											
CIV. ASPHODELEÆ	12	11	7	3	4	2	1		2					1
Anthericum	1					1	1		1					
Allium	8	8	4	3	4	1			1					1
Camassia	1	1	1											
Brodiæa	2	2	2											
CV. PONTEDERIACEÆ	2	2			2									
Pontederia	1	1			1									
Leptanthus	1	1			1									
CVI. RESTIACEÆ	1	1			1									
Eriocaulon	1	1			1									1
CVII. JUNCEÆ	23	20	10	9	15	17	8	14	13	3	2	2	2	14
Luzula	7	6	4	2	4	5	4	5	3	2	1	1	2	4
Juncus	16	14	6	7	11	12	4	9	10	1	1	1		10

FAMILIES.	Total Species in Three Zones.	First Zone. Between Lat. 45° to 55° on East side, and Lat. 49° to 58° West side.				Second Zone. From Arctic Circle northward to 72° N.				Third or Polar Zone. Lying North of 73° Lat.				Europe, Asia, and Africa.
		Species in First Zone.	On Pacific Coast.	On Eastern Prairies.	In East. Woody Distr.	Species in Second Zone.	Kotzebue Sound.	Rocky Mts. 52° to 57°.	Eastern Arctic Distr.	Species in Third Zone.	Lancaster Sound, &c.	North Greenland.	Spitzbergen.	
CVIII. HYDROCHARIDEÆ	2	2			2									1
Valisneria	1	1			1									1
Udora	1	1			1									
CIX. ORCHIDEÆ	54	52	17	9	43	10	3	8	8					1
Microstylis	3	3	1		2									
Liparis	2	2			2									
Corallorrhiza	3	3	2	1	2	1	1							
Amplectrum	1	1			1									
Calypso	1	1	1	1	1	1		1	1					
Orchis	2	2	1		1									
Gymnadenia	1	1			1									
Platanthera	20	20	7	5	15	3	1	3	3					
Peristylus	2	2	1	1	1									
Arethusa	1	1			1									
Pogonia	2	2			2									
Calopogon	1	1			1									
Spiranthes	3	3	1	1	3	2		2	1					
Goodyera	2	1			1	1		1	1					
Listera	2	2	2		2									
Cypripedium	8	7	1		7	2	1	1	2					
CX. IRIDEÆ	8	7	2	1	7	3	2		1					
Iris	7	6	1	1	6	2	2							
Sisyrinchium	1	1	1		1	1			1					
CXI. HYPOXIDEÆ	1	1			1									
Hypoxis	1	1			1									
CXII. DIOSCOREÆ	1	1			1									
Dioscorea	1	1			1									
CXIII. CYPERACEÆ	218	184	51	33	160	102	12	54	77	5	5	1	2	81
Carex	183	153	39	24	131	91	9	45	68	3	3	1	2	65
Elyna	2					2		2	1					
Eleocharis	7	6	4	3	5	3		2	2					5
Scirpus	10	10	3	5	10									4
Eriophorum	8	7	4		7	6	3	5	6	2	2			6
Dulichium	1	1			1									
Cyperus	5	5		1	4									
Rhynchospora	2	2	1		2									1
CXIV. GRAMINEÆ	153	134	48	46	96	51	24	32	36	13	13	5	6	56
Leersia	2	2			2									
Hydropyrum	1	1			1									
Alopecurus	4	3	2	1	3	2	1	2	2	1	1	1		3
Phleum	2	2	2	1	1	1		1	1					2
Phalaris	1	1	1	1	1	1			1					1
Hierochloe	3	2		1	2	3	2	1	2	2	2	1	1	2
Anthoxanthum	1	1			1									1
Milium	1	1			1									1
Panicum	8	8	1	3	8									
Holcus	1	1			1									1
Oplismenus	1	1	1	1	1									1

FAMILIES.	Total Species in Three Zones.	First Zone. Between Lat. 45° to 55° on East side, and Lat. 49° to 58° West side.				Second Zone. From Arctic Circle northward to 72° N.				Third or Polar Zone. Lying North of 73° Lat.				Europe, Asia, and Africa.
		Species in First Zone.	On Pacific Coast.	On Eastern Prairies.	In East. Woody Distr.	Species in Second Zone.	Kotzebue Sound.	Rocky Mts. 52° to 57°.	Eastern Arctic Distr.	Species in Third Zone.	Lancaster Sound, &c.	North Greenland.	Spitzbergen.	
GRAMINEÆ—(*continued.*)														
Setaria	1	1			1									
Cenchrus	1	1			1									
Oryzopsis	3	3		2	3									
Stipa	4	4		3	1									
Muhlenbergia	6	6	2	1	5									
Phippsia	2					2	2			1	1		1	1
Colpodium	3	1			1	2	2	1	2	1	1			
Vilfa	1	1		1										
Agrostis	6	6	6	2	3	3		3	1					4
Calamagrostis	10	9	3	2	7	3	2	3	3					2
Ammophila	1	1			1									1
Graphephorum	1	1			1									
Phragmites	1	1	1	1	1									1
Spartina	1	1			1	1			1					
Eutriana	1	1		1		1		1						
Deschampsia	2	1	1	1	1	1	1		1	1	1			1
Dupontia	1	1			1	1	1		1	1	1		1	1
Boutelouia	2	2			2									
Aira	3	2	1		2	1		1			3			1
Trisetum	2	2	1		1	1	1	1	1	1		1		1
Avena	2	1		1	1	2		1	1		1			1
Danthonia	1	1			1									
Poa	26	23	10	6	15	11	7	5	9	3		2	2	13
Eragrostis	1	1			1									
Glyceria	9	8	3	3	5	1			1					2
Pleuropogon	1									1	1			
Reboulea	2	2		1	2	1		1						
Catabrosa	1	1			1									1
Koeleria	1	1	1	1	1									1
Festuca	9	6	2	3	2	5	3	3	4	1	1		1	4
Bromus	3	3	1		2	1	1	1	1					
Ceratochloa	1	1	1											
Brizopyrum	1	1	1	1										
Triticum	3	3	2	2	1	2		2	1					3
Elymus	8	8	3	2	6	3	1	3	2					5
Asprella	1	1			1									
Hordeum	2	2	2	1	1	1			1					1
Andropogon	4	3		3	1	1		1						
ACROGENES	71	67	26	18	57	30	6	21	24					26
CXV. EQUISETACEÆ	9	9	3	2	8	6	2	4	5					8
Equisetum	9	9	3	2	8	6	2	4	5					8
CXVI. FILICES	47	43	16	11	35	20	3	13	16					20
Polypodium	3	3	3		3	3		1	3					3
Woodsia	3	3		1	3	2		1	2					2
Cistopteris	3	2		1	2	2	1	2	1					2
Aspidium	11	10	5	2	9	4	2	3	4					6
Onoclea	1	1			1									
Struthiopteris	1	1			1									
Athyrium	1	1	1	1	1	1		1	1					
Asplenium	7	5			5	2		1	1					3
Blechnum	1	1	1			1			1					1

FAMILIES.		First Zone. Between Lat. 45° to 55° on East side, and Lat. 49° to 58° West side.				Second Zone. From Arctic Circle northward to 72° N.				Third or Polar Zone. Lying North of 73° Lat.				
	Total Species in Three Zones.	Species in First Zone.	On Pacific Coast.	On Eastern Prairies.	In East. Woody Distr.	Species in Second Zone.	Kotzebue Sound.	Rocky Mts. 52° to 57°.	Eastern Arctic Distr.	Species in Third Zone.	Lancaster Sound, &c.	North Greenland.	Spitzbergen.	Europe, Asia, and Africa.
FILICES—(*continued.*)														
Pteris	3	3	1	1	3	1			1					1
Cryptogramma	1	1	1	1		1		1	1					
Adiantum	1	1	1		1									
Cheilanthes	1	1	1			1		1						
Dicksonia	1	1			1									
Osmunda	3	3			3									1
Schizæa	1	1			1									
Botrychium	5	5	2	4	1	2		2	1					1
CXVII. LYCOPODINEÆ	12	12	7	5	11	4	2	4	3					7
Lycopodium	11	11	7	5	10	4	2	4	3					7
Selaginella	1	1			1									
CXVIII. HYDROPTERIDES	3	3			3									1
Isöetes	1	1			1									1
Salvinia	1	1			1									
Azolla	1	1			1									
DICOTYLEDONES	1725	1499	498	391	1130	568	271	340	377	69	57	40	29	403
MONOCOTYEDONES	554	493	162	122	399	198	53	120	146	21	20	9	9	188
	2279	1992	660	513	1529	766	324	460	523	90	77	49	38	591
ACROGENES	71	67	26	18	57	30	6	21	24					26

Obs. In the preceding table, and in that which follows, species that range to several zones are enumerated in each. The proportionate numbers of the second table are found by dividing the whole *Phanerogamæ* of a district by the numbers of each family in that district, and they may, therefore, be considered as denominators of fractions having 1 for a numerator.

The proportions vary remarkably in different districts. The predominance of Compound Flowers, Leguminous and Rosaceous plants in the Prairies, combined with the paucity of Saxifrages, Gentians, and Ericaceous plants, affect the proportions of the other families materially. The Grasses, as might be expected, are more numerous in the Prairies than elsewhere, with the remarkable exception of the Polar Zone, in which the *Gramineæ* form one-seventh of the species, and in conjunction with the *Cruciferæ*, *Caryophylleæ*, and *Saxifrageæ*, constitute more than half the *Phanerogamæ*. The small numbers of Asters, Willows, and Carices on the Pacific coast, modify the numbers of that district.

PROPORTIONATE NUMBERS OF SOME OF THE PRINCIPAL FAMILIES TO THE WHOLE PHANEROGAMOUS VEGETATION OF THE SEVERAL DISTRICTS OF THE THREE ZONES.

FAMILIES.	First Zone.			Second Zone.			
	Pacific Coast.	Eastern Prairies.	East. Woody Distr.	Kotzebue Sound.	Rocky Mountains. 52° to 57°.	Arctic Region.	Polar Zone. North of 73° Lat.
Monocotyledones.........	4·1	4·2	3·8	6·1	3·8	3·6	4·3
Compositæ...............	10·0	6·4	7·0	9·0	7·7	25·1	12·9
Cyperaceæ...............	12·9	15·9	9·5	27·0	8·5	6·8	18·2
Gramineæ	13·8	11·2	15·9	13·1	14·4	14·5	7·0
Rosaceæ.................	17·8	15·6	19·5	12·0	18·4	19·4	18·0
Cruciferæ...............	21·3	24·4	31·9	17·1	13·9	9·8	6·1
Leguminosæ	24·5	15·1	30·0	23·1	41·8	37·3	45·5
Scrophularineæ	24·4	23·3	32·2	27·0	35·4	37·1	45·5
Caryophylleæ	25·4	39·5	46·3	18·0	24·2	20·9	8·3
Ranunculaceæ	27·5	27·0	32·5	23·1	14·8	22·7	18·2
Ericeæ	34·7	171·0	66·5	36·0	35·4	52·2	45·5
Orchideæ	38·8	57·0	35·6	108·0	57·6	65·3	
Saxifrageæ...............	50·8	128·2	80·5	12·5	21·9	29·0	7·5
Umbelliferæ	50·8	85·5	54·5	54·0	230·0	104·4	
Gentianeæ...............	66·0	256·5	80·5	40·5	51·1	65·2	
Coniferæ	73·3	85·5	106·4	324·0	125·0	87·0	
Labiatæ	82·5	51·3	41·3		23·0	26·2	
Salicaceæ................	82·5	85·5	42·5	40·5	32·9	21·7	30·3
Boragineæ	94·3	85·5	76·4	64·8	11·5	87·0	

A list of the plants collected by Mr. Seeman on the coasts of Beering's Sea having been received subsequent to the first part of the preceding tables having passed through the press, some emendations are requisite. The numbers of *Dycotyledones*, in p. 461, are to be substituted for those in p. 445, and the following changes made in the column headed Kotzebue Sound: viz. *Ranunculaceæ*, 14; *Cruciferæ*, 19; *Caryophylleæ*, 18; *Leguminosæ*, 14; *Rosaceæ*, 27; *Portulaceæ*, 5; *Saxifrageæ*, 26; *Compositæ*, 36; *Ericeæ*, 9; *Boragineæ*, 5; *Scrophularineæ*, 12; *Verbenaceæ*, 1; *Diapensiaceæ*, 1. These occasion some slight alterations in the first column of the second zone, and in the total number of plants.

The following Table of the Distribution of the Carices was drawn up by Dr. Boott, whose intimate acquaintance with that genus of *Cyperaceæ* renders it of the highest value to the student of the Geography of Plants.

London, May 6, 1850.

MY DEAR SIR JOHN,

I have examined the Carices you brought from your last excursion to America. They are—

C. scirpoidea *Michx.* Arctic Sea coast.
ursina *Dewey.* Arctic Sea coast.
glareosa *Wahlg.* Arctic Sea coast.
stans *Drejer.* Arctic Sea coast.
saxatilis *L.* Arctic Sea coast.
compacta *Brown.* Arctic Sea coast.
fuliginosa *St. & Hoppe.* Arctic Sea coast.
livida *Willd.* Arctic Sea coast.
Novæ Angliæ *Schwz.* Arctic Sea coast and Methy Portage.
canescens *L.* (*& b.*) Arctic Sea Coast and Methy Portage.
" var. polystachya. Lakes Superior, Rainy and of the Woods.
adjusta *Boott.* Methy Portage.
siccata *Dewey.* Methy Portage, Saskatchewan.
lanuginosa *Michx.* Methy Portage, Saskatchewan.
lenticularis *Michx.* Methy Portage.
Houghtonii *Torrey.* Methy Portage.
Raeana *Boott.* Methy Portage.
utriculata *Boott.* Methy Portage, Lakes Superior, Huron.
aquatilis *Wahlg.* Methy Portage, Lakes Superior, Huron, Fort Simpson and Chepewyan.
oligosperma *Michx.* Lake Superior.
aristata *Br.* Lakes Superior and Huron.
scoparia *Schk.* (*& b.*) Lakes Superior and Huron, Winipeg, Athabasca.
vulgaris *Fries.* Lakes Superior and Huron, Winipeg.
retrorsa *Schwz* Winipeg River, Lake of Woods.
pedunculata *Muhlg.* Winipeg River, Rainy Lake.
intumescens *Rudge.* Lake of Woods, Rainy Lake.
Œderi *Ehrh.* Rainy River and Lake.
Pennsylvanica *Lam.* Saskatschewan, Winipeg, and Cumberland Lakes.
incurva *Light.* Valleys of the Sask. and Mack.

Of the above, C. Raeana is new, and C. stans new to British America.

I find from my notes that the number of Carices in North America is 250; of which 178 are found in all Arctic America, including 97 common to Arctic America and the States, leaving 81 Arctic species.

Of these 81, there are 36 common to Europe, leaving 45 peculiar to Arctic America.

Of the 97 found in Arctic America and the States, 28 are common to Europe, leaving 69 exclusively American.

There are in the States, besides, 72, of which 4 only are European, leaving 68 exclusively American.

The exclusively American species are therefore 182, and 68 common to America and Europe.

American species in Arctic America.....	45		81
European do.		36	
American species in Arctic America and the States..............	69		97
Enropean do. do.		28	
American species in the States	68		72
European do.		4	
	182	68	250

You will find that the large proportion of Carices in the Northern part of America, common to it and to Europe, is in accordance with the observations of Agassiz, made in his late interesting excursion to Lake Superior. He remarks that the farther north we proceed the greater is the uniformity of the plants common to the two continents; and it is remarkable that Leconte, in his list of the Coleoptera of Lake Superior, was struck with the absence of all the groups peculiar to the American continent, the large increase of the species of genera feebly represented in the more temperate regions, and the existence of many genera heretofore regarded as confined to the southern parts of Europe and Asia.

Yours sincerely,
F. BOOTT.

The 97 species found in Arctic America and in the States are—

C. anceps *Muhlg.*
arida *Tor.*
aristata *Brown.*
aurea *Nutt.*
angustata *Boott.*
arctata *Boott.*
adusta *Boott.*
aperta *Boott.*
aquatilis *Wahl.*
atrata *L.*
blanda *Dewey.*
bromoides *Schk.*
Backii *Boott.*
bullata *Schk.*
Buxbaumii *Wahl.*
cephalophora *Muhl.*
conoidea *Schk.*
cristata *Schwz.*
crinita *Lam.*
commutata *Gay.*
capillaris *L.*
canescens *L.*
chordorrhiza *Ehrh.*
capitata *L.*
debilis *Michx.*
Deweyana *Schwz.*

C. digitalis *Willd.*
eburnea *Boott.*
Ehrhartiana *Hoppe.*
festucacea *Schk.*
flexilis *Rudge.*
filiformis *L.*
flava *L.*
fulva *Good.*
grisea *Wahlg.*
gracillima *Schwz.*
granularis *Muhlg.*
gynocrates *Worm.*
hystericina *Muhlg.*
intumescens *Rudge.*
irrigua *Willd.*
lupulina *Muhlg.*
lagopodioides *Schk.*
Liddoni *Boott.*
longirostris *Tor.*
lanuginosa *Michx.*
lacustris *Willd.*
lenticularis *Michx.*
limosa *L.*
livida *Willd.*
Muhlenbergii *Schk.*
miliacea *Muhlg.*

C. monile *Tuckn.*
muricata *L.*
Novæ Angliæ *Schwz.*
oligosperma *Michx.*
Œderi *Ehrh.*
polytrichoides *Muhl.*
pubescens *Muhlg.*
Pennsylvanica *Lam.*
pedunculata *Muhlg.*
plantaginea *Lam.*
pseudocyperus *L.*
pallescens *L.*
pauciflora *Light.*
rosca *Schk.*
Richardsoni *Brown*
retrorsa *Schwz.*
rostrata *Michx.*
retroflexa *Muhlg*
rigida *Good.*
subulata *Michx*
squarrosa *L.*
striata *Michx.*
stipata *Muhlg*
scoparia *Schk.*
straminea *Schk*
scabrata *Schwz.*

C. Schweinitzii *Dewey*.	C. teretiuscula *Good*.	C. vulpinoidea *Michx*.
siccata *Dewey*.	tenuiflora *Wahl*.	vesicaria *L*.
scirpoidea *Michx*.	tenella *Schk*.	vulgaris *Fries*.
stellulata *Good*.	umbellata *Schk*.	vitilis *Fries*.
triceps *Michx*.	utriculata *Boott*.	Willdenowii *Schk*.
tentaculata *Muhlg*.	varia *Muhlg*.	
trisperma *Dewey*.	verticillata *Boott*.	

Of the 97 in Arctic America and the States, 28 are European.

In England.	In Scotland.	In North of Europe.
C. Buxbaumii *Wg*.	C. capilaris *L*.	C. chordorrhiza *Ehrh*.
canescens *L*.	aquatilis *Wahl*.	capitata *L*.
filiformis *L*.	atrata *L*.	gynocrates *Worm*.
flava *L*.	pauciflora *Light*.	tenuiflora *Wahlg*.
fulva *Good*.	rigida *Good*.	livida *Willd*.
irrigua *Willd*.		tenella *Schk*.
limosa *L*.		vitilis *Fries*.
muricata *L*.		Ehrhartiana *Hoppe*.
Œderi *Ehrh*.		
pseudocyperus *L*.		
pallescens *L*.		
stellulata *Good*.		
teretiuscula *Good*.		
vesicaria *L*.		
vulgaris *Fries*.		

Of these 28 species of Europe, 12 are Alpine, or found in high northern latitudes.

C. Ehrhartiana is found in Germany, and is probably a form of C. teretiuscula *Good*.

C. fulva was originally established upon a Newfoundland specimen, and has only been found once near Boston, U.S.A.

The 72 found in the United States are—

C. alopecoidea *Tuckn*.	C. exilis *Dewey*.	C. lucorum *Willd*.
æstivalis *Curtis*.	Elliottii *Tor*.	lupuliformis *Sartw*.
alveata *Boott*.	Floridana *Tor*.	lævigata *Smith*.
Boottiana *Benth*.	Fraseri *Sims*.	mirabilis *Dewey*.
Barrattii *Tor*.	fœnea *Willd*.	Mitchelliana *Curtis*.
Baltzellii *Chapman*.	formosa *Dewey*.	microdonta *Tor*.
Buckleyi *Dewey*.	flaccosperma *Dewey*.	Meadii *Dewey*.
crus-corvi *Shutt*.	folliculata *L*.	mirata *Dewey*.
Careyana *Dewey*.	glaucescens *Eh*.	oligocarpa *Schk*.
Cherokeensis *Schwz*.	Grayii *Carey*.	oxylepis *Tor*.
Crawei *Dewey*.	gigantea *Rudge*.	præcox *Jacq*.
Cooleyi *Dewey*.	hyalina *Boott*.	panicea *L*.
Caroliniana *Buckley*.	Halseyana *Dewey*.	platyphylla *Carey*.
comosa *Baott*.	Hitchcockiana *Dew*.	polymorpha *Muhlg*.
decomposita *Dewey*.	imbricata *Boott*.	planostachys *Kunz*
Davisii *Schwz*.	juncea *Willd*.	refracta *Schk*.
dasycarpa *Muhlg*.	Knieskernii *Dewey*.	retrocurva *Dewey*.

C. sterilis *Willd.*	C. stenolepis *Tor.*	C. turgescens *Tor.*
sparganioides *Muhl.*	Sullivantii *Boott.*	Tuckermani *Boott.*
Sartwellii *Dewey.*	sychnocephala *Car.*	vestita *Willd.*
setacea *Dewey.*	strictior *Dewey.*	venusta *Dewey.*
Shortii *Tor.*	tenax *Chapman.*	virescens *Muhl.*
Steudelii *Kunth.*	tetanica *Schk.*	vulpina *L.*
styloflexa *Buckley.*	torta *Boott.*	Woodii *Dewey*

Of these 72 species, 4 only are common to Europe (England).

C. præcox *Jacq.* (introduced), found only in Salem, Massachusetts.
 lævigata *Smith* (introduced), found once near Boston, Massachusetts.
 panicea *L.*
 vulpina *L.* Doubtful, probably a form of C. stipata. (Ohio, Illinois).

I can offer you little that is satisfactory to myself as to the geographical range of the 97 species that are common to Arctic America and the States, for want of precise data as to the Carices of the Southern and Western States.

A.—I find, from such data as I have, that from lat. 30° to 35°, that is, from N. Orleans through the Carolinas, there are 33 species extending into Arctic America, one of which, C. Novæ Angliæ *Schwz.*, ranging from N. Orleans to the Arctic Sea, maintains an equally vigorous development through 40° of latitude.

B.—From lat. 37° to 41°, Kentucky to Rhode Island, there are 21 species extending northward.

C.—From lat. 42° to 45°, Massachusetts to Wisconsin, there are 43 species extending northward.

Of the 72 species found in the States, 27 are southern species, ranging from Florida to Kentucky.

Of the 81 in Arctic America, there is, in

Newfoundland,	1. C. remota, *L.*
Labrador,	3. C. recta *B.*, nigra *All.*, ustulata *Wg.*
Greenland,	9. C. duriuscula *Mr.* hæmatolepis *Dr.*, reducta *Dr.*, rufina *Dr.*, holostoma *Dr.*, hyperborea *Dr.*, microglochin *Wg.*, pedata *Wg.*, microstachya *Ehrh.*
Canada,	1. C. miliaris *Mx.*
Rocky Mountains,	8. C. petasata *Dy.*, petricosa *Dy.*, filifolia *Nutt.*, Geyeri *B.*, Lyoni *B.*, Jamesii *T.*, dioica *L.*, Pyrenaica *Wg.*
Rocky Mountains and Altai,	1. C. Franklinii *B.* (C. macrogyna *Turczon*).
Northwest coast,	18. C. anthoxantha *Pr.*, anthericoides *Pr.*, Hoodii *B.*, amplifolia *B.*, Gmelini *H.*, circinnata *Mr.*, leiocarpa *Mr.*, marcida *B.*, mi-

	cropoda *Mr.*, macrocephala *W.*, Mertensii *Pres.*, nigella *B.*, Sitchensis *Pres.*, Tolmiei *B.*, elongata *L.*, leporina *L.*, stricta *G.*, physocarpa *Presl.*
Newfoundland to Rocky Mts.,	1. C. ovata *Rudge.*
Greenland to Lake Superior,	1. C. bicolor *All.*
" and Newfoundland to Arctic Sea,	1. C. glareosa *Wg.*
" to Cumberland House,	1. C. subspathacea *Worm.*
" to Arctic Sea,	3. C. stans *Dr.*, vahlii *S.*, ursina *Dy.*
" Slave Lake and Ft. Enterprise,	1. C. rotundata *Wg.*
" and Mackenzie River,	1. C. rariflora *Sm.*
" Bear Lake and Rocky Mountains,	2. C. supina *Wg.* vaginata *Tausch.*
" Bear Lake, Church R. and Sask.,	1. C. ampullacea *G.*
" and Repulse Bay,	1. C. fuliginosa *St. & Hop.*
" Arctic Sea, Rocky Mts., and Northwest coast,	1. C. compacta *Br.*
" to Rocky Mountains,	7. C. festiva *Dy.*, incurva *Light.*, lagopina *Wg.*, nardina *Fr.*, obtusata *Lil.*, rupestris *All.*, saxatilis *L.*
" to Northwest coast (not Rocky Mts.),	3. C. salina *Wg.*, cryptocarpa *Mr.*, stylosa *Mr.*
Arctic Sea,	1. C. marina *Dy.*
Hudson's Bay to Methy Portage,	1. C. Houghtonii *Tor.*
" to Cumberland H.	2. C. heleonastes *Eh.*, maritima *Mull.*
Cumberland House, Mackenzie River, and Rocky Mts.,	1. C. concinna *Br.*
Carlton House,	2. C. Torreyi *Tuck.*, Hookeriana *Dy.*
" to Northwest coast,	1. C. Parryana *Dy.*
Methy Portage,	1. C. Raeana *B.*
Wooded Country,	2. C. affinis *Br.*, podocarpa *Br.*
Rocky Mountains to Northwest coast,	5. C. Douglassii *B.*, Rossii *B.*, nigricans *Mr.*, macrochœta *Mr.*, stenophylla *Wg.*

Of the above 81 species, 36 are European!

England!	Scotland!	North of Europe!
C. remota *L.*	C. ustulata *Wg.*	C. bicolor *All.*
dioica *L.*	vahlii *Schk.*	subspathacea *Worm.*
elongata *L.*	rariflora *Sm.*	fuliginosa *S. & Hop.*
leporina *L.*	vaginata *Tausch.*	nigra *All.*
stricta *G.*	incurva *Light.*	microglochin *Wg.*
ampullacea *G.*	lagopina *Wg.*	microstachya *Ehrh.*
	rupestris *All.*	rufina *Dr.*
	saxatilis *L.*	Pyrenaica *Wg.* (Pyrences!)

England !	Scotland !	North of Europe !
		C. obtusata *Lil.* supina *Wahl.* salina *Wahl.* maritima *Muller.* stenophylla *Wahlg.* heleonastes *Ehrh.* glareosa *Wahl.* festiva *Dy.* nardina *Fries.* pedata *Wahlg.* rotundata *Wahlg.* hyperborea *Dr.* (Lapland). holostoma *Dr.* (Lapland). cryptocarpa *Mr.* (Iceland).

Of these 36 European species found in Arctic America, 30 are Alpine or belonging to high northern latitudes.

C. Pyrenaica *Wahlg.* is confined in Europe to the Pyrenees!

C. remota *L.*, common in England, has been found only in Newfoundland. Royle has found it also on the Himalayas and at Kunawur in the East Indies!

C. festiva *Dewey*, found in Norway, Finmark and Lapland, extends in America from Greenland to Unalashka, and along the Rocky Mountains to the Straits of Magellan!

The 45 American species in Arctic America are:

C. duriuscula *Meyer.* circinnata *Meyer.* leiocarpa *Meyer.* micropoda *Meyer.* stylosa *Meyer.* nigricans *Meyer.* macrochæta *Meyer.* anthoxantha *Presl.* anthericoides *Presl.* physocarpa *Presl.* Mertensii *Prescott.* Sitchensis *Prescott.* miliaris *Michaux.* filifolia *Nuttall.* Jamesii *Terrey.*	C. Houghtonii *Terrey.* reducta *Drejer.* hæmatolepis *Drejer.* stans *Drejer.* petasata *Dewey* petricosa *Dewey.* Hookeriana *Dewey.* marina *Dewey.* ursina *Dewey.* Parryana *Dewey.* Torreyi *Tuckerman.* ovata *Rudge.* Gmelini *Hooker.* macrocephala *Willd.* affinis *Brown.*	C. podocarpa *Brown.* compacta *Brown.* concinna *Brown.* Geyeri *Boott.* Lyoni *Boott.* Hoodii *Boott.* Rossii *Boott.* Tolmiei *Boott.* marcida *Boott.* nigella *Boott.* Douglassii *Boott.* Franklinii *Boott.* Raeana *Boott.* amplifolia *Boott.* recta *Boott.*

A.—New Orleans, Cumberland House, Rocky Mountains, *C. umbellata, debilis.*

" Greenland, Rocky Mountains, Arctic Sea, *C. Novæ Angliæ.*

Texas to Canada, *C. retroflexa, grisea, blanda, triceps.*

Texas to Hudson's Bay, *C. Muhlenbergii.*
" Carlton House, Northwest coast, *C. anceps.*
Georgia to Canada, *C. lupulina, commutata, squarrosa, tentaculata, striata, hystericina, miliacea.*
South Carolina to Canada, *C. cephalophora, varia, granularis, vulpinoidea.*
" " " and Rocky Mountains, *C. bromoides.*
" " Cumberland House, *C. crinita, inumescens.*
" " " " Rocky Mountains, Northwest coast, *C. stellulata.*
" " " " Northwest coast, *C. stipata, scoparia, lagopodioides.*
" " Hudson s Bay, Norway House, *C. polytrichoides.*
South Carolina to Lake Winipeg, *C. lacustris.*
" " Northwest coast, *C. rosea.*
North Carolina Mountains of, Observatory Inlet, Cumberland House, Northwest coast, *C. Buxbaumii.*
" " to Canada, *C. conoidea.*
" " Cumberland House, Rocky Mountains. *C. Pennsylvanica.*

B.—Kentucky to Canada, *C. pubescens, digitalis.*
" Mackenzie River, Rocky Mountains, *C. eburnea.*
Illinois, Cumberland House, Rocky Mountains, Northwest coast, *C. Richardsonii.*
New Jersey to Canada, *C. Schweinitzii.*
" Northwest coast, *C. aperta.*
" to Hudson's Bay, Arctic Sea, Rocky Monntains, Northwest coast, *C. livida.*
Ohio to Cumberland House, *C. arida.*
" Carlton House and Rocky Mountains, *C. Backii.*
Pennsylvania to Canada, *C. subulata, scabrata, bullata.*
" Cumberland House, *C. gracillima, cristata, plantaginea.*
" " " Rocky Mountains, *C. pedunculata, utriculata.*
" Greenland, Northwest coast, *C. vesicaria.*
" Northwest coast, *C. angustata.*
Rhode Island to Bear Lake, *C. monile.*
" Greenland, Cumberland House, Rocky Mountains, Northwest coast, *C. adusta.*

C.—Massachusetts and Newfoundland, *C. fulva.*
" Northwest coast, *C. muricata, verticillata.*
Michigan to Canada, *C. festucacea.*
" Cumberland House, *C. Ehrhartiana.*
" Carlton House, Rocky Mountains, *C. teretiuscula, trisperma.*
" Canada, Northwest coast, *C. straminea.*
" and Northwest coast, *C. Liddoni.*
New York to Canada, *C. arctata.*

New York to Canada, Rocky Mountains, *C. Willdenowii.*
" Cumberland House, *C. pallescens, pseudocyperus, tenuiflora, vitilis, flava, aristata, filiformis, irrigua.*
" " " Rocky Mountains, *C. siccata, longirostris.*
" " " Northwest coast, *C. retrorsa.*
" " " Greenland, Hudson's Bay, *C. Œderi.*
" Mackenzie River, Rocky Mountains, *C. aquatilis.*
" " Northwest coast, *C. lanuginosa.*
" Bear Lake, Rocky Mountains, Northwest coast, *C. limosa.*
" Hudson's Bay, Carlton House, Rocky Mountains, Northwest coast, *C. aurea.*
" Newfoundland, *C. flexilis.*
" " Greenland, Arctic Sea, Rocky Mountains, Northwest coast, *C. vulgaris, canescens.*
" " Rocky Mountains, Northwest coast, *C. pauciflora.*
" Greenland, *C. gynocrates.*
New York Mountains of, to Mackenzie River, *C. lenticularis.*
" " Greenland, Arctic Sea, Rocky Mountains, *C. scirpoidea.*
" " Labrador, Arctic Sea, *C. rigida.*
" White Mts. of New Hampshire, Bear Lake, *C. oligosperma.*
" " " " Northwest coast, *C. rostrata.*
New Hampshire, White Mts. of, Canada, Greenland, Rocky Mountains, *C. atrata.*
" " " Greenland, Hudson's Bay, Rocky Mountains, *C. capitata.*
" " " " Bear Lake, Rocky Mountains, *C. capillaris.*
Wisconsin to Canada, *C. tenella.*
" Cumberland House, Hudson's Bay, *C. chordorrhiza.*
" " " Rocky Mountains, *C. Deweyana.*

The geographical range, as far as I know it, of the 72 species found in the United States is as follows:

Florida, *C. tenax, Baltzellii, oxylepis.*
" to New Orleans, *C. gigantea, Floridana.*
" Georgia, *C. dasycarpa, turgescens.*
" Carolinas, *C. glaucescens, venusta, Elliottii.*
" New England, *C. folliculata, polymorpha.*
Texas, *C. alveata, hyalina, imbricata, microdonta, planostachys.*
" to Alabama, *C. Cherokeensis.*
" Kentucky, *C. stenolepis.*
" New Jersey, *C. flaccosperma.*
" Rhode Island, *C. fœnea.*

New Orleans, *C. Boottiana.*
" to Wisconsin, *C. Meadii.*
Carolina, South, *C. Buckleyi, Caroliniana, Mitchelliana, Fraseri, juncea, styloplexa, lucorum.*
" to Virginia, *C. æstivalis.*
" Massachusetts, *C. comosa.*
Carolina, North, to Ohio, *C. Sullivantii.*
Virginia to Kentucky, *C. Shortii.*
Kentucky to New York, *C. oligocarpa, Hitchcockiana*
" Connecticut, *C. Davisii, virescens.*
Ohio, *C. tetanica, crus-corvi.*
" to Pennsylvania, *C. strictior.*
" New Jersey, *C. Tuckermani.*
" New York, *C. Careyana.*
" Illinois, *C. vulpina.*
Pennsylvania, *C. refracta.*
" to Cherokee, *C. sterilis.*
" Connecticut, *C. torta.*
New Jersey, *C. Barrattii, Kneiskernii.*
" to Rhode Island, *C. Halseyana, platyphylla.*
" Connecticut, *C. vestita.*
New York, *C. alopecoidea, formosa, lupuliformis, mirata, sychnocephala, Woodii.*
" Michigan, *C. crawei, Sartwellii, decomposita, Steudelii.*
" Rhode Island, *C. retrocurva.*
" Massachusetts, *C. exilis.*
" Connecticut, *C. Grayii.*
Massachusetts, *C. setacea, præcox, panicea, lævigata.*
" to Michigan, *C. mirabilis.*
Michigan, *C. Cooleyi.*
Wisconsin to New England, *C. sparganioides.*

For the following notice and list of insects collected on the Expedition, I am indebted to Adam White, Esq., F.L.S., &c., of the British Museum. With respect to the extent of the collection, it is to be observed that no time was devoted to the capture of insects. Such as presented themselves at convenient times were taken, but none were sought for; and the numbers of the list are not, therefore, to be considered as a criterion of the richness of the country in that division of the animal kingdom.

Note on Hymenoptera in Arctic North America.

"Otho Fabricius first, perhaps, recorded any of the *Hymenoptera* of Arctic North America. Doubtless Baffin, Frobisher, and other manly navigators recognized humble bees and other bees during their summer voyages, and *may* have, in print or in manuscript, with sailor-like earnestness, made mention of every such occurrence in their journals. It is delightful to read the notices of flowers and verdure in their accounts of the hurried spring, summer, and autumn two months of a Greenland year, of five-sixths winter. *Where* flowers and verdure abound, even for six weeks or a shorter time, *there* insects must be found;—*there* insects of the order *Hymenoptera*, the order to which this notice is limited, *must* occur. Flowers and *Hymenoptera* must be together.

"O. Fabricius records two species of *Hymenoptera* as being brought by him from Greenland. His book, so admirable a model of a local fauna as to be even now one of the standards of excellence, was published in 1780. The next considerable accession to our acquaintance with the *Hymenoptera* of British America was made by Redman, who collected in Nova Scotia many fine species now in the British Museum. Some of these, such as *Pelecinus*, *Sirices*, *Ichneumonidæ*, &c., were very prominent species, and are now being worked out in the vast collections of the National Museum.

"Sir John Richardson and his brave comrades collected many species, which were lost during their disastrous journey. They still, however, brought many insects to England, and in the '*Fauna Boreali-Americana*' these insects are described by the venerable Kirby. The species of *Hymenoptera* are very few; there are only *thirty-two altogether;* the circumstances attending the journey not admitting of their collection and preservation.

"An eminent man, reasoning on such data as he had, has recorded his belief that it will be found that *Hymenoptera* do not abound in British North America; now it may be remarked in making generalizations on the distribution of animals, especially those of the lower orders, 'that, before generalizing on a collection from any place not often visited or not often explored, attention be paid to the taste or tastes, or, in other words, to the bias or direction of the eye, hand, and mind of the person or persons who collect, supposing such reasoning is recorded as on authentic data.'

"Mr. George Barnston, to whose researches Sir John Richardson directed public attention in the 'Edinburgh New Philosophical Journal' for April, 1841, has published a very admirable summary of the Progress of the Seasons as affecting Animals and Vegetables at Martin's Falls, Albany River, James's Bay, about lat. 51° 30′ N., and in long. 86° 20′ W. In this fresh and refreshing journal, there are *more than indications* that *Hymenoptera*, *Diptera*, and *Neuroptera* abound. In a year or two afterward Mr. Barnston came to London and presented his collection to the British Museum.

"As one instance of his excellence as a collector, I may mention that Mr. Walker, who named and described the species of *Diptera* in the Cabinet of the British Museum, has alluded to or has described nearly 250 species of his dipterous insects from the single station mentioned above; there being only 14 species of these insects recorded in the '*Fauna Boreali-Americana*' of the Rev. Wm. Kirby. Mr. Barnston's researches among the *Neuroptera* also were considerable and very valuable. One insect brought by him, the *Pteronarcys regalis* (although previously found in Canada), afforded Mr. Newport a fit subject for his genius as an accurate anatomist and recorder of facts and reasonings on the insect economy. This gentleman discovered persistent *branchiæ* in the *imago* or perfect state of the *Pteronarcys*, and has recorded his discovery and quoted some observations of Mr. Barnston's in a paper read at the Linnæan Society.

As Mr. Gray's Catalogues of the collections in the British Museum (mines of information to the reasoner and writer on geographical distribution), are published, it will be seen how valuable are Mr. Barnston's and Sir John Richardson's collections to our acquaintance with the articulated animals of British North America, especially in its more northerly parts.

"I have mentioned that Kirby *describes* or alludes to only thirty-two species of *Hymenoptera* in his 'Insects of North America;' while Mr. Barnston *in one spot* found **192** distinct species, exclusive of *Chalcididæ*. I subjoin a comparative list of the families of *Hymenoptera*, the comparison being made with the British species existing in the Museum collection at the time of this record. Mr. B. and myself worked out the *Tenthredinidæ;* my friend and coadjutor Mr. Frederick Smith, an able hymenopterist, determined the other species; so the list may be deemed as correct as the circumstances will admit.

"It must be borne in mind that our British collection of *Hymenoptera* has been accumulating for at least thirty years, was a favorite part of Dr. Leach's collection, and has been made over a wide and variegated country; while Mr. Barnston's was formed in three months, on one spot and under almost unheard-of disadvantages, counterbalanced, however, by an enthusiasm not easily deterred by difficulties.

	British Collection in British Museum.	Collected at Martin's Falls.
Cimbicidæ	10	4
Tenthredinidæ	157	76
Siricidæ, &c.	7	2
Ichneumonidæ	200	47
Chalcididæ	?	?
Chrysididæ	22	1
Formicidæ	11	7
Mutillidæ	5	0
Sapygidæ	2	0
Pompilidæ, &c.	38	2
Crabronidæ	57	16
Vespidæ	17	4
Apidæ	170	33

"A striking proof that the time has not yet come to reason correctly on the distribution of Hymenópterous insects—at least in British North America."

LIST OF INSECTS

TAKEN BY SIR JOHN RICHARDSON AND JOHN RAE, ESQ., IN ARCTIC NORTH AMERICA, DRAWN UP BY ADAM WHITE, ESQ., F.L.S., ETC.

COLEOPTERA.

Cicindela longilabris, *Say*. (C. albilabris, *Kirby*). Shores of Arctic Sea, lat. 70° N.; and at Fort Simpson, lat. 62° N.

Cicindela hirticollis, *Say*. Borders of Mackenzie and Slave Rivers, lat. 59°—62° N.

Dromius nigrinus, *Eschsch.* Great Bear Lake, lat. 66°—67° N.
Carabus — ? n. s. (C. gladiator, *Barnston MS.*). Borders of Mackenzie and Slave Rivers, lat. 50°—65° N.
Carabus Chamissonis, *Eschsch.* Borders of Mackenzie and Slave Rivers; and Cape Krusenstern, lat. 58°—68° N.
Carabus — ? n. s. (C. Hudsonicus ?) Borders of Mackenzie and Slave Rivers, lat. 58°—65° N.
Calosoma calidum, *Auct.* Borders of Mackenzie and Slave Rivers, lat. 58°—65° N.
Loricera pilicornis, *Auct.* Great Bear Lake, lat. 66°—67° N.
Elaphrus intermedius, *Kirby.* Great Bear Lake, lat. 66°—67° N.
Notiophilus sibiricus, *Motchoulsky.* Great Bear Lake, lat. 66°—67° N.
Dicælus — ? n. s. (D. sculptilis ?) Borders of the Mackenzie and Slave Rivers, lat. 58°—65° N.
Agonum melanarium, *Dej.* Great Bear Lake, and district to the south of Lake Winipeg, lat. 49°—68° N.
Argutor brevicornis, *Kirby.* Borders of Mackenzie and Slave Rivers.
Omaseus orinomum, *Leach.* District south of Lake Winipeg, lat. 50°—54° N.
Platysma vitrea, *Eschsch.* Borders of Mackenzie and Slave Rivers.
Pæcilus lucublandus, *Say.* Borders of Mackenzie and Slave Rivers.
Harpalus — ? n. s. (near H. obtusus). Borders of Mackenzie and Slave Rivers.
Stenolophus — ? n. s. Great Bear Lake, lat. 66°—67° N.
Amara — ? sp. (near A. trivialis). South of Lake Winipeg, lat. 49° N.
Amara — ? sp. Great Bear Lake.
Bembidium conicolle, *Motchoulsky* (B. impressum, *Kirby*). Great Bear Lake, and north of Lake Winipeg, lat. 49°—67° N.
Acupalpus — ? n. s. Great Bear Lake, lat. 66°—67° N.
Peryphus — ? sp. Great Bear Lake.
Platytrachelus — ? n. s. Great Bear Lake.
Notaphus nigripes, *Kirby.* Great Bear Lake.
Notaphus variegatus, *Kirby.* South of Lake Winipeg, lat. 49° N.
Dytiscus Harrisii, *Kirby.* South of Lake Winipeg.
Agabus — ? n. s. (near A. arcticus). Cape Krusenstern, lat. 68° N.
Colymbetes — ? sp. Borders of Mackenzie and Slave Rivers.
Hydrophilus picipes, *Auct.* South of Lake Winipeg, lat. 49° N.
Heterocerus — ? n. s. (near H. fossor). Great Bear Lake.
Staphylinus villosus, *Grav.* South of Lake Winipeg, lat. 49° N.
Quedius, n. s. (near Q. molochinus). Great Bear Lake, lat. 66°—67° N.
Omalium, n. s. (near O. rivulare). Shore of Arctic Sea, near mouth of Mackenzie River, lat. 70° N.
Anthophagus, sp. Borders of Mackenzie and Slave Rivers.
Silpha Lapponica, *Auct.* Fort Simpson; borders of Mackenzie and Slave Rivers.
Silpha opaca, *Auct.* Borders of Mackenzie and Slave Rivers.
Silpha, n. s. (near S. Baikalica), *Motchoulsky.* Borders of Mackenzie and Slave Rivers.
Ptinus fur, *Auct.* South of Lake Winipeg. Throughout Rupert's Land.
Byrrhus — ? n. s. Borders of Mackenzie and Slave Rivers, lat. 58°—65° N.

Byrrhus — ? n. s. South of Lake Winipeg.
Rhisotrogus fervens, *Gyll.* South of Lake Winipeg, lat. 49° N.
Platycerus piceus, *Web.* Fort Simpson, on the Mackenzie River, lat. 62° N.
Cyphon fuscipes, *Kirby.* Great Bear Lake.
Elater æripennis, *Kirby.* Shore of Arctic Sea, near Mackenzie River, lat. 70° N.
Elater æneus? Borders of Mackenzie and Slave Rivers.
Elater, n. s. (near E. melancholicus). Borders of Mackenzie and Slave Rivers.
Elater, n. s. (near E. sanguineus). Borders of Mackenzie and Slave Rivers.
Ludius, n. s. (near L. sibiricus). Great Bear Lake, lat. 65°—67° N.
Ampedus, n. s. Great Bear Lake.
Buprestis tenebrica, *Kirby.* Borders of Mackenzie and Slave Rivers.
Chrysobothris, n. s. Fort Simpson, on the Mackenzie River, lat. 62° N.
Trachypteris Drummondi, *Kirby*, var. Fort Simpson, on Mackenzie River, lat. 62° N.
Trachypteris decolorata (Bupr. appendiculata, *Kirby*). Fort Simpson, on Mackenzie River, lat. 62° N.
Ellychnia corrusca, *Auct.* South of Lake Winipeg.
Ragonycha cembricola, *Eschsch.* Great Bear Lake, lat. 65°—67° N.
Thanasimus abdominalis, *Kirby.* Great Bear Lake.
Hydnocera, n. s. Great Bear Lake.
Blapstinus æneus, *Deg.* South of Lake Winipeg.
Upis ceramboides, *Auct.* Fort Simpson, and Borders of Mackenzie and Slave Rivers.
Anthicus — ? n. s. Great Bear Lake.
Formicoma — ? n. s. Great Bear Lake.
Stenotrachelus Roulieri, *Motch.* var. Shores of Arctic Sea, near Macken zie River, lat. 70° N.
Serropalpus — ? sp. Fort Simpson, on Mackenzie River.
Hylobius — ? sp. Borders of Mackenzie and Slave Rivers.
Alophus — ? sp. South of Lake Winipeg.
Alophus — ? sp. Cape Krusenstern, lat. 68° N.
Erirhinus, sp. (near E. tremulæ). South of Lake Winipeg.
Tomicus — ? sp. South of Lake Winipeg.
Asemum striatum, *Auct.* Fort Simpson on the Mackenzie River.
Callidium bifoveolatum. Cape Krusenstern and Arctic Coast, between 67½° and 68°.
Callidium Proteus, *Kirby;* and C. simile, *Kirby*, var. Arctic Coast, between 67½° and 68°; Fort Simpson on the Mackenzie River, lat. 62° N.
Clytus undulatus, *Say.* Shore of Arctic Sea; Mouth of Mackenzie River.
Clytus — ? sp. Shore of Arctic Sea; Mouth of Mackenzie River, lat. 70° N.
Acanthocinus pusillus, *Kirby.* Great Bear Lake, lat. 66°—67° N.
Monochamus resutor, *Kirby.* Fort Simpson, on the Mackenzie River.
Monochamus confusor, *Kirby.* Borders of Mackenzie and Slave Rivers.
Acmæops Proteus (*Kirby*), *Leconte;* Leptura strigilata, var.? Fort Simpson, on the Mackenzie River.
Acmæops strigilata (*Fabr.*), *Lec.* Shore of Arctic Sea (Mouth of Mackenzie).

Pachyta liturata, *Kirby*. Fort Simpson, on the Mackenzie.
Pachyta, n. s. Fort Simpson, on the Mackenzie.
Rhagium lineatum, *Auct.* Fort Simpson, on the Mackenzie.
Syneta carinata, *Eschsch.* About Great Bear Lake.
Galleruca marginella, *Kirby*. Borders of Mackenzie and Slave Rivers; and about Great Bear Lake.
Chrysomela multipunctata, *Say*. Borders of Mackenzie and Slave Rivers.
Phædon Adonidis, *Pall.* Shore of Arctic Sea (Mouth of Mackenzie); Fort Simpson.
Adoxus vitis, *Fabr.* District about Great Bear Lake, and Borders of Mackenzie and Slave Rivers.
Coccinella 13-punctata, *Auct.* Great Bear Lake, lat. 65°—67° N.
Coccinella 5-notata, *Kirby*. Shore of the Arctic Sea; Mouth of Mackenzie, lat. 70° N.
Coccinella ocellata, *Auct.* Borders of Mackenzie and Slave Rivers.

ORTHOPTERA.

Locusta tuberculata, *Pal de Beauv.?* Borders of Mackenzie and Slave Rivers; Fort Simpson.
Locusta, four species. Borders of Mackenzie and Slave Rivers.
Acrydium granulatum, *Kirby*. Borders of Mackenzie and Slave Rivers; Fort Simpson.

NEUROPTERA.

Æschna borealis, *Zetterst.*? Borders of Mackenzie and Slave Rivers.
Libellula — ? sp. Fort Simpson on Mackenzie.
Libellula scotica, *Donov.*? Between Lake Winipeg and Lake Superior.
Agrion cyathigerum, *Charp.* var. Borders of Mackenzie and Slave Rivers.
Ephemera viridescens — ? n. s., *Barnston*. Between Lake Winipeg and Lake Superior, lat. 47°—52°.
Ephemera — ? n. s. South of Lake Winipeg, lat. 49° N.
Pteronarcys regalis, *Newman*. Borders of Mackenzie and Slave Rivers.
Pteronarcys Proteus, *Newman?* Shores of Arctic Sea (Mouth of Mackenzie River), lat. 70°.
Perla — ? (sp. near P. abnormalis, *Newman*). Borders of Mackenzie and Slave Rivers.
Perla — ? (sp. near P. sonans, *Barnston*). Borders of Mackenzie and Slave Rivers.
Semblis — ? n. s. Borders of Mackenzie and Slave Rivers.
Phryganea striata, n. s., *Barnston*. Borders of Mackenzie and Slave Rivers.
Phryganea variegata, n. s., *Barnston*, and two or three other species. Borders of Mackenzie and Slave Rivers.

HYMENOPTERA.

Trichiosoma lucorum, *Auct.* Fort Simpson, on Mackenzie River.
Tenthredo (Nematus). Great Bear Lake.
Tenthredo (Nematus) — ? n. s. South of Lake Winipeg.

Tenthredo integra? About Great Bear Lake.
Tenthredo (Dolerus). South of Lake Winipeg.
Sirex flavicornis, *Fabr.* Cape Krusenstern; Fort Simpson on Mackenzie, and country south of Lake Winipeg.
Ephialtes —? sp. Fort Simpson, on Mackenzie River.
Aspizonus —? sp. Arctic Sea (Mouth of Mackenzie River).
Ichneumon —? sp. Fort Simpson, on the Mackenzie River.
Ichneumon —? sp. Cape Krusenstern, and Fort Simpson on the Mackenzie River.
Cryptus —? sp. South of Lake Winipeg.
Chrysis —? sp. Cape Krusenstern.
Mutilla —? sp. About Great Bear Lake.
Formica herculeana. About Great Bear Lake; borders of Mackenzie and Slave Rivers; Fort Simpson.
Formica sanguinea. South of Lake Winipeg, and Fort Simpson.
Pompilus —? n. s. Borders of Mackenzie and Slave Rivers.
Odynerus —? n. s. Shore of Arctic Sea (Mouth of Mackenzie River.)
Vespa maculata, var. Borders of Mackenzie.
Vespa vulgaris, *Auct.* Borders of Mackenzie and Slave Rivers and Fort Simpson.
Vespa marginata, *Kirby.* Cape Krusenstern.
Halictus —? (n. s. near H. quadricinctus). South of Lake Winipeg.
Halictus, three black species. South of Lake Winipeg.
Megachile Willughbiella? Fort Simpson, on the Mackenzie River.
Bombus arcticus? Arctic Coast between 67½° and 68°; borders of the Mackenzie and Slave Rivers.
Bombus (sp. near B. lapponicus). Arctic Coast between 67½° and 68°.
Bombus —? sp. Shore of Arctic Sea (Mouth of Mackenzie River).
Bombus pratorum. Borders of Mackenzie and Slave Rivers.
Bombus, n. s. (near B. lucorum). Arctic Coast between 67½° and 68°.
Bombus, n. s. Arctic Coast.

Hemiptera.

Acanthosoma boreale, *Hope.* Great Bear Lake.
Acanthosoma nebulosum, *Kirby.* South of Lake Winipeg.
Miris —? sp. Great Bear Lake.
Rhyparochromus, two species. South of Lake Winipeg.
Salda —? sp. (near S. riparia). Cape Krusenstern, lat. 68° N.

Homoptera.

Aphrophora, sp. Great Bear Lake, and to the south of Lake Winipeg.

Lepidoptera.

Papilio Turnus, *L.* Fort Simpson, on Mackenzie River.
Pontia casta, *Kirby.* Arctic Coast between 67½° and 68°.
Pontia, sp. Fort Simpson, on the Mackenzie.
Anthocharis —? n. s. (near A. Simplonia). Arctic Coast between 67½° and 68°.

Colias Palæno, *L.* Fort Simpson, on the Mackenzie River.
Colias Boothii, *Curtis.* Arctic Coast between 67½° and 68°.
Colias Chione, var. ? *C.* Arctic Coast; Cape Krusenstern.
Argynnis Freija (*Thunb.*), var. Melitæa Tarquinius, *Curtis.* Arctic Coast between 67½° and 68°.
Argynnis — ? n. s. Arctic Coast.
Vanessa Milberti, *Godart.* (V. furcillata, *Say*). Fort Simpson, on the Mackenzie River.
Vanessa Progne, *Godart.* (V. C. argenteum, *Kirby*). Fort Simpson, on the Mackenzie River; Arctic Coast between 67½° and 68°.
Nymphalis Artemis, *Auct.* Fort Simpson on Mackenzie River, and Borders of Mackenzie and Slave Rivers.
Chionobas Bore, *Boisd.* ? Arctic Coast between 67½° and 68°.
Hipparchia, n. s. ? (near H. discoidalis), *Kirby.* Arctic Coast between 67½° and 68°.
Hipparchia Rossii, *Curtis.* Arctic Coast between 67½° and 68°.
Polyommatus Franklinii, *Curtis.* Arctic Coast.
Arctia Americana, *Harris*, var. Borders of Mackenzie and Slave Rivers.
Hadena Richardsonii, *Curtis.* Arctic Coast between 67½° and 68°.
Anarta — ? sp. Arctic Coast between 67½° and 68°.
Geometridæ, two species. Arctic Coast between 67½° and 68°.
Tineidæ, three species. Arctic Coast between 67° and 68°.

Diptera.

Culex — ? sp. Borders of Mackenzie and Slave Rivers.
Chironomus, sp. Borders of Mackenzie and Slave Rivers.
Tipula, sp. Arctic Coast between 67½° and 68°.
Tabanus, three species. Borders of Mackenzie and Slave Rivers.
Tabanus, two species. Arctic Sea, Mouth of Mackenzie River.
Eristalis flavipes, *Walker.* Borders of Mackenzie and Slave Rivers, and district to the South of Lake Winipeg, lat. 49°—65° N.
Syrphus, sp. Borders of Mackenzie and Slave Rivers.
Musca, five species. Borders of Mackenzie and Slave Rivers.
Musca — ? sp. South of Lake Winipeg.
Œstrus Tarandi? Arctic Coast between 67½° and 68°.

No. V.

VOCABULARIES.

A. *Eskimo Vocabulary.*

THE Kuskutchewak column of the following vocabulary is extracted from Bäer's work.* To draw up an effective comparative table would require a thorough acquaintance with both dialects, since the names of articles of dress, and implements of art, change with the materials of which they are formed; natural objects are differently designated, according to the circumstances under which they are viewed; and the terms for actions are altered, as the agents, time, place, and other circumstances, vary. Unless, therefore, these facts be known and attended to by one who forms a vocabulary, there may appear to be no resemblance between the dialects of two tribes who mutually understand each other, and converse together with ease. The introduction of the syllabic characters used by the late Rev. Mr. Evans in teaching the Cree Indians would, I believe, remove the difficulties which orthography throws in the way of a European, who endeavors to reduce the native languages of North America to writing.

The column containing Eskimo spoken on the Labrador coast, is extracted from a pretty large vocabulary and grammar, which the Rev. Peter Latrobe had the kindness to procure for my use on the expedition. I have reason to believe that some errors may have crept into this vocabulary, from the similarity of the German written *h* to *s* not being always adverted to by the transcriber, and also from the uncertainty of the proper English equivalent of the German *v*. These are not, however, I trust numerous among the examples I have used. Where the Labrador dictionary was defective, the excellent English and Eskimo vocabulary, drawn up

* BAER, *Statische und ethnographische Nachrichten über die Russischen Besitzungen an der Nordwestkuste von Amerika.* St. Petersburg, 1839. p. 259. The Kuskutchewak words being written in this work in Russian characters, with which I am unacquainted, J. F. von Bach, Esq., of the British Museum, had the kindness to furnish me with a translation. This gentleman drew up carefully columns representing the conventional English equivalents of the Russian characters for each word, and added also the French pronunciation, which want of space compels me reluctantly to omit. I have made some small alterations in words written by him according to the English pronunciation, to suit the plan of orthography which I have followed in the other vocabularies.

by Captain Washington, and published by the Admiralty for the use of the Searching Expeditions, has been referred to. The dialects spoken by the intermediate Eskimo tribes inhabiting the north shores of the continent are seldom quoted, my object having been to identify the language spoken by members of the nation occupying geographical positions the most remote from each other.

In writing out the table it was obvious to me that the Labrador dialect is in general the softer of the two. Instead of the hard *tch* so frequent in the Kuskutchewak tongue, the east coast tribes generally use *s*, and in Coronation Gulf *h* is substituted. The strongly aspirated sound which is heard in the Scottish word "loch" is of frequent occurrence in the Kuskutchewak column of the vocabulary, where it is denoted by *kh*. An Englishman in attempting this sound lets the *k* be heard, which he ought not to do. The difficulty of constructing a correct Eskimo vocabulary is increased by the necessity of previously mastering the exceedingly numerous inflections of the nouns, adjectives, pronouns, and verbs, which supply the place occupied by auxiliary verbs, possessive pronouns, prepositions, and adverbs in the European languages. These inflections are briefly noticed in the introduction to Capt. Washington's vocabulary, and I shall merely add here, that in the Labrador grammar, obtained for us by Mr. Latrobe, there are examples of thirty different terminations of the dual and plural numbers of nouns, which have evidently had their origin in euphuism.

Each noun has six cases in each number, distinguished by their terminations, the vocative being, however, absent in some. The cases are formed by affixes having the power of prepositions, as *mut*, *mik*, *mit*, *me*, and *kut* in the singular, and *nut*, *nik*, *nit*, *ne*, and *gut* in the plural. The nominative is also varied by affixes which perform the functions of possessive pronouns, as *ga*, *go*, *ne*, *ait*, *anga*, *ara*, &c.; as *kivgah*, a servant, *kivganga*, my servant, *kivgane*, his servant; *nuna*, land, *nunaga*, my land; *nelegah*, a master *nelegara*, my master; *tunnusuga*, my nation, &c. *Pagit*, *panga*, or *parma* are affixes employed when the noun is connected with a verb, signifying action or suffering. The noun, when changed by a qualifying affix, is declined in its new form, in the usual way. Besides the ordinary active nominative, each noun has also an intransitive one, which ends in *b*, and is differently declined. Examples are subjoined. The power of the affixes varies according as the noun is used with a transitive or intransitive verb. Nouns may also be varied by affixes expressive of augmentation, diminution, affection, ridicule, humility, or multitude. Some of these terminations are *arsuk*, *arsuit*, diminutives; and *soak*, *sudset*, *sudsek*, augmentatives; *vak* placed before any of them increases their power; and the adverbial *aluk* denoting "very" may be put after them, and is applicable to either good or bad. *Vavak*, signifying an extraordinary number, is placed before *söareluit.**

Adjectives have also their declensions; and likewise comparisons made by the addition of the syllable *nek*, or by verbs. The adjective generally follows the noun, and must agree with it in case and number. If the substantive have an affix, so must the adjective. Nouns may be changed into verbs by the affix *evok* or *ovok*, and the adjective then must take the same termination.

Pronouns are declined like the nouns by affixes, which require much nicety in their due employment. Affixes supply the place of possessive pronouns.

The third person singular of the indicative is considered to be the root of the verb, and may be used as a noun with a change in the termination, "a hunter" being equivalent to "he hunts." The inflections of the verb are extremely numerous, and are expressive of affirmation, negation, interrogation, and of the various circumstances in which the agent or object can be placed with respect to time, place, mood, or possession. The infinitive, formed by the termination *nek*, is used when things are spoken of indefinitely, or when two verbs come together, and is conjugated in the same way as the other moods and tenses, there being a past and future infinitive. Generally the verb in the Labrador Eskimo agrees in its inflections with the Greenland dialect; but there are some special differences, and particularly with regard to the future, which has a threefold construction in the Greenland tongue, but is more simple in the Labrador speech. With the ample means which the regular verb possesses of expressing every mood and tense, the Eskimo has little occasion for auxiliaries, and in fact the structure of the language is very regular and exact.

There are, however, one or two auxiliaries which have an affinity to adverbs—such as *pi-wok*, which is used in a variety of ways, sometimes in immediate relation to a noun, sometimes only as an adjunct to a verb: it occasionally seems to be equivalent to the English "get" or "do." When placed after participles, which is its most common position, it signifies the action of a thing. *Ipsok*, another auxiliary, seems to be equivalent to the Latin *est;* it often increases the meaning of the verb with which it is connected. "To be," or "to have," is denoted by the syllables *gi* or *vi* in composition, as *nunagiva*, "it is his land."

The adverbs are numerous, and have relation to time, place, equality, size, number, order, union, separation, &c.; and also to questioning, denying, affirming, negativing, including, excluding, desiring, admonishing, and distinguishing.

* *Akkatu* is employed by the Eskimo of Churchill in the same way as *söareluit*.

An example of the inflections of a single verb would occupy many pages, and can not be given here; but the preceding short notices will suffice to show that vocabularies of the same language, formed by different people, may have little similarity, and that much care is requisite before we can venture to affirm the distinct origin of two tribes upon such evidence. In a language which is transmitted orally alone, and which is not preserved in its integrity by an appeal to the eye, alterations for the sake of euphony are frequent, and these, which are not uncommon with the Eskimo, vary with the delicacy of the ear of the speaker. Thus when the termination *wangha* does not blend pleasingly with the preceding syllable, *langha* is substituted, and the general pronunciation is more nasal with some small communities, more guttural with others.

Examples of Nouns declined transitively and intransitively.

Tupek, *a tent*.

	Sing.	*Dual.*	*Plural.*
Nom. *tr.*	Tupek }	tuppak	turket.
intr.	turkib }		
Gen.	turkib	tuppak	turket.
Dat.	tuppek	tuppak	turket.
	tuppermut	tuppangnut	tuppernut.
Acc.	tuppak	tuppak	turkinut.
	tuppernik	tuppangnit	turkit.
Voc.	caret.		
Abl.	tuppermit	tuppangnit	tuppermit.
	tuppermut	tuppangnut	turkinnut.

Nelegara, *my master*.

	Sing.	*Dual.*	*Plural.*
Nom. *tr.*	Nelagara	nelegakka	nelekakka.
intr.	nelekama	nelekang-ma	nelekama.
Gen.	nelekama	nelekangma	nelekama.
Dat.	nelegara	nelegakka	nelekakka.
	nelekamnut	nelegamnut	nelekamnut.
Acc.	nelegara	nelegakka	nelekakka.
	nelekamnik	nelegamnik	nelekamnik.
Voc.	nelegara	nelegakka	nelekakka.
Abl.	nelekamnit	nelegamnit	nelekamnit.
	nelekamnut	nelegamnut	nelekamnut.

Nelegane, *his master;* neleganga, *another person's master;* and similar variations of the noun, have, in like manner, their various inflections.

Comparative Table of the Dialects spoken by the Beering's Sea and Labrador Eskimos.

Obs. S. denotes "singular;" D. "dual;" P. "plural." W. points out words taken from Captain Washington's vocabulary. *A* is sounded as in "father;" *ā* as in "law;" *e* as in "there;" *i* as *ee* in "see;" *ĭ* as in "ink," "pin;" *u* as *oo* in "good;" *kh* as *ch* in the Scottish word "loch," or Irish "lough;" *h* after *g* signifies that the latter has the soft pronunciation as in "give;" the hyphen following *g*, gives nearly the same sound.

English.	Kuskutchewak.	Labrador Eskimo.
The only Creator	....	Ping-ortitsi-o-wok.
God (the Creator)....	nuna-lishta........	Nuna, *country*.
Heaven; the firmament	ki-il-yak..........	killĕk.
Earth, land, a country	nuni.............	nuna.
Air	u-i-uchu-yughi-ak...	u-i-a-wak, *west wind*.
Air, wind, also the world and reason	....	silla (sillata, *in the open air*).
Wind	....	anorre; sullu-ar-nèk, *also breath*.
The sun	....	sekkinek (P.-erngit); nai-i-a.
Sun	akkta; pukli-anok ..	akki-suk-pok, *the sun breaks forth*.
Moon	tang-ek..........	takkek.
Month	tang-ak; igal-i-uk.	
Stars	mittit...........	ubluri-ak; ubloriak (D. ubloritsek).
Comet............	ag-i-akhn-akhtak.	
A star surrounded by a halo	....	agsuk (P. aguthet).
Water............	mu-ek...........	immek, *fresh water*.
River	kvak............	ku; kok; kogguk; koggut.
A large river........	....	kokso-ak.
Sea	immakh-pik	immak.
The wide ocean......	....	immarbikso-ak.
Lake	nanvik...........	akker-oktok, a *lake where deer are speared*.
A pond, fresh.......	....	tessek (P. tessit).
Brook	kitchikli-ak.......	kōgak.
A tear............	....	kogve (P. kogvit).
Straits	u-ikakh..........	ikkarasak, W.
Cliff	....	ikkargok (P. ikkarut).
Deeply cleft	....	korok (P. korkut).
A long inlet	....	kang-erdluk.

English.	Kuskutchewak.	Labrador Eskimo.
Gulf or creek........	nang-vagnak	(kang-ŭ-agnak? u-agnak *the west*).
Current............	tchag-vak	sag-vak; sarvak.
Current in the sea ...	....	ingerarnek.
Current in fresh water	....	aksarnak.
Bottom of waters	notu-ik	tung-a-wik, W.
He treads his boot down at the heel	....	tung-mark-pok.
Shore	chna; agavnu-ik....	sǐg-sak.
Mouth or source of rivers, a well	pa-i..............	(pè, *a court-yard;* pa, *the round opening in a kaiyak*).
A bank in the sea, a sunken rock	ithalh-nuk	siorak-so-ak-nuk, *much sand;* ipek-so-ak-nuk, *much mud.*
Stone	tkalhk-uk	ka-ertok, *a rock.*
Deep.............	tuli.	
A sea bird..........	....	tullik.
It is deep (valley or river)	....	ittiwok.
Day	ignu-ik	uvlok, or úblok, *a day.*
In the morning......	....	uvlakut; ublo-tilugo.
The day closes	....	uvlokliwok.
The morning, or the day	....	kau; kauk; kaut.
The front or forehead .	....	ke-uk; kauk (P. karrut).
It is daylight	....	kau-ma-wok.
To-day	....	uvlut; ubluk; ovetsi-ak.
Night	....	u-nu-ak (P. unu-et), unu-ame.
Marsh.............	mag-ik	(mannèk, *moss*, W.)
Dry summer	ki-nu-ig-nu-ik.	
Summer	....	au-i-ak.
Mountain	ing-ik	kakkak (P. kakket).
Lowland	tchu-iv-nu-ik	(su-yuk, *wet, dirty.*)
Shell..............	ammokt	amomio-yok, *an oyster* (P. amomio-yut).
White shell fish	....	ayöarnet.
Mussel	....	uvi-lok.
Snail shell	....	si-ut-terok, W.
The ear............	....	si-ut.
A grain of sand	....	si-orak (D. -kek, P. -ket).
Sand..............	kag-u-i-ak.	
Clay (loam, chalk)...	magai-ak	machak.
Fire	knu-ak	ikoma, *ignis.*
One who fires a gun..	....	kukni-wok.
Wind	a-nuka	annore, W.; an-o-i, W.
Thunder	kalik.	
It thunders.........	....	kalukpok; kalludlarpok.
Lightning	....	kaumarlok (*vide* bright).
Rain	tchali-ali-ak; kitok..	silla-luk (silla, *air*).

English.	Kuskutchewak.	Labrador Eskimo.
Hailstone	kakhulat	(kakkulak, *it is round*, W.)
Any thing sharp		kakilaut; P. lautit.
It is round		angma-la-rik-pok.
Snow	kanikh-chak	kannek, *falling snow* (P. kang-it.
Ice	tchiko	siko.
Storm	anu-gavak	ani-gavak, *an extraordinary quantity of snow* (anio, *a snow-storm*).
Strong wind	(anug-wei, Kotz. S. W.)	akkunak; akkunak-so-ak.
Wind or air		annorre; annorer-ho-ak, *great wind.*
The wind is still		annorre-karung-napok.
Calm	ku-nu-ik	kunigok (D. kunikuk).
Clear	tankikh-tchuk	alla-kak-pok.
A bright sky		alla-ki-wok.
Dark	telkh	tek, *darkness.*
Fog		tek-tuk; tar-tuk (P. tar-tu-it).
It is foggy		niptai-pok, W.
The weather clears		nipter-pok.
Vapor or fog		iseriak (isse, *severe cold*).
A cold		ikkĕ.
Clouded	tali-guk	tali-pok, *it is hidden* (P. -pot).
A cloud		nu-vu-i-a.
Bright, or light	ugakhtok	kauma-wok, *it is bright.*
Coals	khumavit	P. aumakut (S. aumako).
Ashes	agak	arsek, W.
Blue	vitok; minukh-kat	(minnu, *a sea weed*, W.)
Bluish		tungo-i-uktak.
Berry juice		tungo.
Red	kivagok	aupa-luk-tok, *it is red.*
Blood		auk; aggut.
White	ugolh-kak	kaggark-pok, *it is white.*
Night	unuk	u-nu-ak (D. u-nu-ek; P. unu-et).
Smoke	punk	pu-i-ok, *damp smoke, steam.*
It smokes		pu-i-ok-pok.
Smell	nagnak	naimawa, or nai-wok, *he smells something.*
Man (*homo*)	tatchu.	
A shadow		tatchak.
A looking-glass		tatchartut.
Man	nukalhni-ak	(nukak, *a brother.*)
Man (relation) stock		ang-ut (D. ang-u-tek; P. ang-ulit).
His father		ang-uta.
A helm		ang-ut.

English.	Kuskutchewak.	Labrador Eskimo.
An adult	....	ang-uti-marik.
People (Eskimos)....	tagut; yugut	inu-it (S. i-nuk).
A name	....	taggisek.
My nation..........	....	tunnisuga.
Life	....	inusek.
A portrait of a man..	....	inu-i-ak.
White man, European	....	kablunak (P. kablunet).
Eyebrows	....	kablo (D. kabluk; P. -t).
Inland Indian, stranger	alli-a-guk	allani-a-wok (S. allak).
A man's footmark ...	....	allok (D. alluk).
Stranger Eskimos....	....	si-ad-ler-mi-u; he-ad-ler-mi-u.
An unprotected man .	....	sek-sariak.
Husband...........	vi-na.............	wi, *a married man* (P. wi-nit).
One is with another ..	....	una.
She has a husband...	....	wi-ghi-wok.
Wife..............	nuli-ga	nuli-a (nuli-ang-a, *my wife*).
She or he has a brother's son or daughter.	....	nu-a-karpok.
An unmarried woman	....	wi-ga-sok.
A bachelor	....	nule-tok.
Widower...........	....	nuler-tok.
Old man	utchi-nuk; anuli-uvak.	
Male of man or beast.	....	ussuk.
He is old..........	....	itta-wok (itok, W.)
Hindmost	....	ittik.
Woman	aganak	ning-i-yok (aköa, D. akö-ak, *a mother*).
Old woman.........	aganukli-uvak	ning-i-vok.
The oldest of a family	....	ang-ai-i-uklek.
Brother's or son's wife	....	uk-ang-a.
Grandfather	apnugli-uk.	
Grandmother	annugli-u	anenak-si-ak.
They have a mother..	....	mikli-ak-attig-ekout.
Mother	ani	anenak; akko-a.
Mother's milk.......	....	ammak.
She is beautiful	....	enanau-wok.
A relative	tunka.	
Father	atti	attatak (P. attatet).
The father	....	attatu.
An adopted father ...	....	attatak-sak.
Son...............	igni-ak	ergnek (P. ergnerit).
Daughter	panaga; panik	panik (P. paniknit).
Brother....	annak	(anènak, *mother*, *beauty*.)
His elder brother	....	anningna.
Elder brother or sister	....	ango-i-uma.
Younger brother	....	nukak; nukka, *my brother*.

English.	Kuskutchewak.	Labrador Eskimo.
Sister	agna-vu-ik	neya; neyango, *his sister;* neyara, *my sister.*
Two uterine brothers .	....	angu-tauk-attigekpuk.
A twin	....	ikking-ut; karrisarek, *twins.*
A boy	tangoy-ali-uvak	nuka-pi-ak (P. nuka-pitset).
A young woman	....	ni-wi-ark-si-ak.
A girl	nozi-atchak	ni-wi-azi-ak.
Grandchild	tut-khih	
A child belonging to the parents	....	kittorng-ak.
A mother's only child.	....	attung-ektak.
A woman's last child.	....	mikki-erngo-a.
It is my child.......	....	kittom-yarivara.
An orphan	....	ananak-ang-ilak.
A fatherless child....	....	atatai-tok.
An orphan deprived of both parents	....	illi-arksuk (P. illi-arksu-it).
His sister's child	....	u-i-orva.
A child............	....	kittorng-ak; sorusek.
A little, or new born, child	....	nutarak.
Uncle	anahkli-uvan.	
Aunt	annomak.	
His mother's sister...	(*root?* aganak, *a woman.*)	ai-ang-a.
His father's sister....	....	at-sang-a.
Sister-in-law........	....	sak-i-a, W.
A prudent woman ...	....	arnanda.
A robust man.......	....	atsu-illik.
A countryman	....	nuna-kat (P. nuna-kattiget).
A friend, one of two in company	....	illek-sak.
A walking companion	....	tupperkat (P. tupperkatiget).
A traveling companion	....	ing-i-a-ket (D. ingi-akattek; P. ingi-akattiget).
A comrade (housemate)	....	iglo-mokat (P. iglo-mokat-tiget).
I................	khvana	u-wang-a.
Thou.............	lhpu-it	ig-vit.
Thee, *acc.*.........	....	illing-nik.
He	ikum	taim-na.
We	....	u-vag-ut P. (u-va-guk, D.)
Ye, you...........	ilhli-te-pik	ilipsè, P. (illiptik, D.; il-o-wit, *Church. Esk.*)
They.............	....	okköa.
He, *demonstr*.......	....	taimna; taipsoma

English.	Kuskutchewak.	Labrador Eskimo.
This		tamanno.
He, or she, *intransitive*	una	una; tamna.
He, or she, *transitive*		oma; tapsoma.
They	unut	okköa; tapköa.
Of mine		uvango; uwango.
To me	kvinum	uvamnut.
To thee	lhpinum	iling-nut.
To us		uvapting-nut.
To you		illipting-nut.
To him	umu-in; ikumin	omunga; tapsomunga.
To them		okkomunga; tapsomunga.
Of himself		ing-me.
To themselves		ingming-nut.
Whose? what? what kind? of whom?	kai-a; tchambi-a	kina? ki-a, *who* (P. ki-kut).
Who? what thing?		suna? su-ub (P. su-unt)?
What do you say?		suva.
Of or by whom		ki-mit; kikkunit.
With what thing?		sumik?
What company?		kikkut?
Which	ke	kiput.
His	umnia	(*vide* of his)
Of his		oma; tapsoma.
Mine	kho-in-tchati-ka; khvona.	u-wanga.
Thine	ilh-pu-it; lhpu-it-ik	igvit, *of thine*.
This, *masc.*	unakh-wina	una(inung-una,*this man*).
This thing		oma (oma-pung-a, *his thing*).
That		imna.
Self		nang-ninek (P. nang-merngit).
How? what?	tcha-itun	kannak.
The same		ingna.
Thus		tava.
Who are these people? who is the head of the family?		kik-ut?
Head	kamikuk; uksi-u; ni-ba-gun	ni-akko (P. ni-akkut).
Crown of the head		kausek.
Forehead	tchughi-uk	ke-uk; ka-uk (P. karrut).
Eyes	vi-tatu-ik	i-ye (ai-i-ga, W.); isse (P. issit).
Eyebrow	ka-i-ag-mi-ut	kablo, S. (P. kablut.)
Eyelashes		kemerit-set.
Ears	na-i-utu-ik; tchu-u-tu-ik	si-ut (P. si-utik).
Mouth	kanik	kannerk; kaurngit.
Face		kènak.

English.	Kuskutchewak.	Labrador Eskimo.
Nose	nikh	king-ak (P. king-et).
A horn		naksuk (P. naksu-it).
Cheek		ulu-ak.
Muzzle		katang-ak.
Teeth	khu-u-tu-ik	ki-u-tit (S. kig-ut).
Beard	unik	umik (P. umgit), *also a curtain.*
Neck	u-i-anut	u-i-ak (P. u-i-ait), *fore quarter of an animal;* konge-sek.
Hair (human) on the head	nu-i-at	nu-i-ak (P. nutset).
Hair, fur	mu-ilh-kut	merserpok, *also a feather* (merkut, W.)
A needle		merkut.
A skin, general name.		amek (D. amok; P. armgit).
Hands	yagatchu-tu-ik	aggait, *hands and fingers.*
Two, numer.		aggait.
My hands		aggakka; aggaktit, *your hands.*
Foot	ig-uk	ittigak (P. ittiket).
Finger	si-evogat.	
Thumb		kublo (P. kublut).
Forefinger		tikkek (P. tikkerit), *also a thimble.*
A sign to indicate any thing (finger post)		tikkorut.
Middle finger		kettert-lek, W.
The middle		kerka.
He is in the middle		ketterpok.
Third finger		mikilirak, W.
Smallest		mikke (mikkinek, *the least*).
It decreases		mikki-orpok.
The first		mikkledklek.
Little finger		erkekok, W.; mikkillerak (P. mikkillaket).
Hands and feet together		igluktuk.
Belly	aksi-ak	nek, *also the body.*
Tongue	ali-anuk	okak (P. oket).
He licks with his tongue		alluktorpok.
My tongue		okara.
A member of the body		nabgo-ak.
A leg		nabguk-pa (ni-o, *leg or thigh*).
The trunk or body		mimmernet; time (P. timet).
A headless body		kattik.
The back bone		ku-i-a-pigak.

English.	Kuskutchewak.	Labrador Eskimo.
The rump	....	nullok.
Blood	ka-i-unkak	auk; aggut.
To speak	kalkhtu-ik	o-karlune.
He is spoken of	....	kalle-mavok.
To cry	vikhpa-ga-ga.	
The waves roar	....	kadlarpok.
He weeps much	....	kai-u-mi-wok-nudlarpok.
He distorts his face in crying	....	kakkerlu-arpok.
To laugh	nu-inhli-akhta.	
He laughs in mockery	....	i-yorkpork.
He is in a laughing or weeping mood.	....	illapsukpok.
To kill	tchikaliz-gi-u.	
Dead (he is)	tukumak	tokkowok.
A corpse	....	time (*body*) -tokkung-a-yok.
Alive	u-nung-vak	innu-wok.
Life	....	innusek.
A living man	....	innuk.
Bad	tchakli-uk	(sèg-lu-wok, *he lies.*)
He is very bad	....	a-yorpok; yudlarpok.
He becomes bad	....	assilè-wok.
Not good	....	nama-lung-ilak.
Good	knu-ignag-kuk	ai-ung-itok.
Very good, or great	....	ang-i-yok.
He is a good man	....	pillorik-pok; ridlarpok.
Brave	tuvgak	aksut, *also strong*.
Very brave or strong	....	aksorso-ak.
Coward	alantak	erksinadlarpok, *afraid.*
He is terror-struck	....	sakko-arpok.
He is timid	....	innimi-wok.
He loses courage	....	kotso-alavok.
Thick	ukughelghi-a.	
Lean, not stout	kui-migu-ilhnagak.	
He is stout	....	kuini-wok.
Thin, or lean (he is)	....	sallukpok; ludlarpok.
High	yukhtuli	portovok.
Low	yukh-kalhna-gak.	
Warm	kikh-tchatuk	ki-ek-pok; onatomik.
Heat	kalhtok	kivek; onarsivok.
It boils	....	kallapok.
To smell	nagne-chuk	nawok, *he smells.*
To spit	kchigu	(sigguk, *the beak.*)
He spits out	....	oviakpok.
To cough	kuzgh-ga	ko-erlorpok, *he coughs.*
Pain	aknakhtu-a.	
Health	yuguntu-a.	
He is healthy	....	atsu-ili-wok, W.
Angry	wik-nu-i-chuk	ning-akpok; gadlarpok.

English.	Kuskutchewak.	Labrador Eskimo
Quarrel	agu-i-a-uk	akgiwok, *he retaliates.*
Terrible	alu-innakh-kuk	adhei! (*interjection*).
Buy	kupuzg-u	akpang-erpok, *he buys*, W.; pussi-wok, *he buys it.*
Taken	tkhwaka.	
Take	tkhu	pi-uk, W.; pivok, or pi-wok (*auxiliary*).
Give	ta-iz-ghu	ta-ug-si-lugo, *let us barter*, W.; pillata-wok, *he gives in reward;* pittipa, *he gives it to him to hold.*
Sell	kiputna-waka	ni-u-werpok, *he trades.*
Lively, joyous		pio-ri-wok; ku-wi-a-sik-pok, W.
Merry	nuna-nikh-kuk	nunan-ghi-a-suk-pok, *brisk.*
Tedious	nuna-ni-tu.	
He is weak		nuneněpok; sanghe-pok.
Song	i-vagun	iming-arpok.
A song, hymn, or psalm		iming-erut-set.
Dance	kazi-i-achi-kut	okkigenek, W.
Truth	pachikh-pi-ak	(padsitik-sak, *an excuse.*)
Lie	ikli-uk	seglu-wok, *he lies.*
Thief	tu-igli-nak	tiglik-pok, *he steals.*
Forest	nu-i-ku-ig-vakhtut	nappartok, *trees, something erect.*
Place having no trees.		nappartu-itok.
Grass	tchangu-it.	
Straw or grass on the sea shore		i-wik; ibgit.
Berries	nangat	paung-at.
Moss	kumagu-i-tu-it	ting-ang-yak, *a bluish moss;* marnek, W.; ne-kagasek, W.
Sphagnum palustre		orkso (P. orksut).
Fir tree	nu-ikvag-vakh-tugvak	
Firewood		ikko-maksak, W. (ikko-ma, *fire.*)
Birch	ilhgnuk	okpit; kai-volik (P. kai-vogit).
Alder	tchugvagvat.	
Poplar	avgnut.	
Willow	tchagatu-it	okau-jak.
Rock	u-ipnat	u-i-arak, *a stone;* kai-ertok, W.
Vessel, a bark	shunnak	umi-akso-ak.
A small boat of wood.		umi-arak.
Baidare (skin boat)	anh-i-ak	umi-avik.
Baidarka	pukhtan; kai-ak	kai-ak (*for one person*) (P. kai-net).

English.	Kuskutchewak.	Labrador Eskimo.
Flat-bottomed boat ..	anhi-akh-li-uk.	
Dog	anna-kukta	kemmek; king-mek (P. king-mit).
Dog-sled	i-kam-chak	kam-utik, W.
Calls the dogs together	....	kang-marpok.
Tanned sea-cow hide .	amakh-kak.	
Arch	ugli-vu-ik	korok, W., *hollòw.*
A valley	....	korkinek.
A bird arrow	....	nugit.
Arrow.............	ikkh-uk; pickh-tchagak	karksok (P. karksut).
Fish-hook	....	karsuk-sok.
Strap	nuk-tchaklik.	
Hand drum, tambourine	tcha-ul.	
Shaman, sorcerer	tungalhkh; analhkh-tuk	anghe-kok.
House	....	iglo (P. iglut).
Hut (abode of married people).	u-ina.	
Tent	....	tuppek (P. turkit).
Snow-house	....	iglorigak.
Indian tavern	akumgavak.	
To take a vapor bath.	mu-ichtak.	
He bathes, he dips it .	....	missukpok.
Armor.............	annu-i-akh-chutu-it.	
Guest	ali-anik...........	allak; allam-a-wok.
Give for a treat......	tchaktchu.	
He gives a feast.....	....	nerri-marpok.
Eat...............	nuiga	nerri-wok, W., *he eats.*
Dining-hall.........	....	nerriving-me.
To make a present of.	yaguzhgh-ghi; pi-kazhzgh-ghi	pilli-ta-wok.
To sew	minka	mersorpok, *he sews.*
A needle...........	....	merkut (P. merkutit).
To beat...........	pilli-akhku	anauwok, W.
He cuts it in pieces ..	....	pillakpok.
Red fish	nu-i-ku-it.	
Salmo orientalis	taghi-akvak.	
Salmo sanguineus	kak-ki-a.	
Salmo muksun	ka-ukh-tut	ekalluk (P. ekalluktut), *salmon trout.*
Salmo alpinus.......	ankhli-u-gat	ekalli-et, *trout.*
Salmo proteus.......	atakak.	
Chaiko ?	nu-ik-ni-at.	
Syrka	inakh-ping-at.	
Smelt	kpuka-chat.	
Eel pout	managnat.	
Pike	tchukvak.	
Fishing-net.........	kughya.	
A bag, a poke.......	....	pok.

English.	Kuskutchewak.	Labrador Eskimo.
Spawn	mass-i-uk.	
Cup	val-i-uk	korkok, *wide-mouthed cup*; erngusok, *drinking-cup.*
Spoon		alu-pa-ut.
Pot	gant.	
Earthern pot, native		illuterkut, W. (illuli-wok, *he hollows it out*).
Bladder	imangvik.	
Oar (boat's)	anvagun	(pa-ut; P. pa-utit) epat.
Entrails		erchavit P. (S. erchavik.)
Gut	iggzh-u-igli-uk	inelo (P. inelu-it).
Kamlaika (cloth)	iggmagna-tu-ik.	
Woolen cloth		ateg-ek-sai-ah, W.
Parka	atkuk.	
Fur-boots	kamu-ik-si-ak	kamikso-ak.
Breeches	khulik	karlik (P. karlit).
Cap	nachak	ketsivak, akkordlek, *also a jacket.*
Castor oil	alli-ukit-khak.	
Beaver	kini-i-uli	kig-i-ak (P. kig-ilset).
Otter	chvignil'nuk.	
Sable	kakhchichvak	karvi-ait-si-ak.
Gray fox	u-ikh-pu-ikhtuk	arvngasek.
Red fox	kavhiatchak	ka-i-ok.
White fox	ulhi-gu-ik	teri-enniak; P. teriennitsek.
Young fox	pi-i-a-gak	pei-a-raka, *a young quadruped or bird.*
Female bear		akbik.
Bear	unu-valh-iäkh	akhlak.
Polar bear		nennok (P. nennut).
Wolf	ku-isli-unu-ik	amarrok (P. amarkut); ammarwok.
Hare	ka-i-ukh-li	ukalek (P. -lit); ikkingna.
Wolverene	kab-tchak	kablia-ri-u.
Marmot	kalh-ganakh-tuli	sik-sik; ullick.
Muskrat	sig-vak	kiv-galuk, W.
Casan marmot (*Citillus*).	kaninik	ik-ik; sik-sik.
Ermine	nagulhkk-ak	terri-i-a, W.
Lesser otter	amagmi-utak	pammi-oktok.
Tail of an animal		pammi-ok.
Mouse	avilh-nat	awing-ak, W.
Fly	chuvat	nivu-i-wok, *a large fly.*
A spider		assi-wak (P. assi-vait).
Gnat, musquito	ig-tughi-ak; miku-ghi-ak	kiktoriak (P. kiktoritset).
Walrus tusk	tul-i-ak	tógak, W. (tok, *a point.*)
Walrus	azgh-vu-ik	ai-wek (P. ai-werit).
Mammoth tooth	chagu-nu-ek.	

English.	Kuskutchewak.	Labrador Eskimo.
Black-fat of deer	....	tunnuk (tunno, *the back*).
Tallow	anu-ignak	kui-ni-wok, *fat he is.*
Fat	u-ig-nu-ik.	
Reindeer	tun-tu	tuktu (P. tuktut).
A large whale	....	korchak (P. korchetset).
Delphinus leucas	chtvak	(sav-gak, *a water-serpent.*)
A seal, general name .	....	puese (P. pue-sit).
Seal, largest kind	izli-ugvak	uksuk; oguk.
Small seal	....	netsi-arksuk; netsek.
Middle seal	....	kai-rolik.
Young ditto	....	pai-yarak.
Seal lying on ice	....	otok.
Spotted seal	....	kassigi-ak.
A seal with a pointed nose	....	abba (P. abbit).
Unborn seal	....	iblau.
The seal comes up to breathe	....	pu-i-rook.
I will	piyukh-tu-a	pivok.
I will not	piyuk-nak-tu-a	piwak.
Go!	ai-i; ai-aghi	ailer-it, W.
Come!	ikh-tchika	kai-it!
He comes	....	kai-wok.
I come	ta-i-tchika.	
Bring	ta-iski-u	kaitsi-wok, *to bring.*
It is	pitankh-tok.	
Verily	....	ahammarik (*emphatic affir.*)
Yes	....	katz; kassak; kaitsok; aheila; ang-erpok.
Certainly not!	tchata-i-tok	se-i-ovut, aukai-lo.
He says no	....	ang-ing-ilak.
Not	....	naukak, aukok.
Yet	tchali	sulle.
It is so (as you say) ..	....	ahale.
Where?	na-ni?	nane?
Where is it?	....	nauk?
Whither?	nairt?	namut?
Which way? whereby?	....	naukut?
Here	khlonikho	ovane mane; tamane.
Here about, around ..	....	ovona.
There	yani; ung-napi	mâne; tamâne.
Thither	yavu-it.	
Hither	akavu-it	ma-ungo; owunga.
There	khavana	ikkane.
Throw	igazhghi-u	millorpok, *he throws*; egi-pok, *he throws away.*
So	khwatum	taimak; sorlo, *as.*
Now	khwatu-a	mâna.
In the present time ..	....	manakut.

English.	Kuskutchewak.	Labrador Eskimo.
Later, afterward	atakh	king-urgane.
Aforetime		itsak.
Before	tchu-nu-imtpu	sivurnga-gat.
Straight before		miksane.
Before another		ane-taima, *also southward*.
Behind	kanulhklimtiv.	
Backward		knig-o-mut.
Above, upward	kulhma	kollanut; pa-ungo.
Below, beneath	ochi-mi	kanna; sammand.
Num. 1	atu-u-chik	atou-sek.
" 2	a-i-nak; malhkhok	marruk; maggok, W. (agga, *hands*).
" 3	pa-i-na-i-vak	ping-a-sut; ping-ahuk, W.; ping-a-nuk, W.
" 4	tchamik	sittamut.
" 5	tali-mik	(tallek, *the hand*) tedlima; tellimet.
" 6	akhvinok	arvanget; ping-a-su-yok-tut, *twice three*.
" 7	a-i-na-akh-vanam	ping-a-sullo sitta-mello, W., *three and four*.
" 8	pi-na-i-vi-akh-vanam	pina-i-u-ik, W.; ping-a-nuk, W.
" 9	chtami-akh-vanam	tellimella sitta-mello, W., *five and four*.
" 10	tamimi-akh-vanam	tellima-yoktut, W.
" 11		arkang-et.
" 16		arvertanget.
" 20	tzvinnak	igluktut, *hands and feet together*.
" 21		ungna.
Whale	akh-vu-ik	ar-wek.
Bird, in general	tu-in-mi-ak	ting-mi-ak (D. ting-mit-sek).
Eagle	nu-itu-i-gavi-ak	nektoralik.
Raven	kolh-ka-guk	kallu-gak.
Magpie	kalh-ka-gai-ak.	
Hawk	naptak	kiga-wik, W.
Owl	iggi-akhtu-gali	upik, W.; upigu-ak, *great owl*, W.
Goose	nu-ikli-uk	nerlek (P. nerlit).
Swan	kughi-uk	kog-uk, W.
Crane	ghi-na-tuli.	
Duck	tu-in-mu-ik	ting-mi-ak (P. ting-mid-get), *a bird, in general*.
Sinew	ulhi-un; ivali-ut	uli-yut, W.
Glass bead	tu-ikh-lit	(tuè, *the shoulder*); sang. pang-ak, *beads*.
Blue	tchunaizi; tchu-a-gat	tung-a-yuktak (tungo, *berry juice*).

English.	Kuskutchewak.	Labrador Eskimo.
White	katu-ighi-agvak	kaud-luk-pok, W. (kau, *day*.)
It is white		kaggarpok, W.
Red	kivikh-tchitkkhlat	auk-palliki-tak (auk, aggut, *blood*).
Black	tunulhgat	kerngut; kernerpok; kernian-garvok.
Krelle	tchunaglat; anat	(sunak, W., *polar-bear*) (annak, *refuse of animals*).
Ax	kalhk-anak	tukkingai-ok (P. -ut; kuksau-tok).
Adze		nella-yok.
Pickax		tik-lak.
Knife	ulhvak	ulima-ut (P. ulima-utit, *a hatchet*).
Aleutian ax	knun.	
Scissors	ku-ipli-a-unu-ik	kipsaut, W.
He cuts something off		kippiwok.
Needle	minkuk; tchikuk	merkut (sig-uk, *a beak*).
Button	nikht-ku-tu-it	sennero-ak (P. sennerutset).
Mirror	tangh-i-u-guk	takh-artut.
He sees him, or it		takko-wok.
Iron	nu-ilhkh-agak	kikki-ek, *general name, also a wooden or ivory pin*.
Copper	kanukh	kanu-yak, W.
Lead	khu-i-akak	aggiktok, W.
Shirt	tulhpakhak	uvinerok (uvinite *the flesh*).
Linen waistcoat	alkuk	altighigha, *under jacket*, W.
Worsted ditto	tunulhkh-u-i-alkuk.	
Kettle	gantchavak	uk-ku-sik, *stone kettle*.
Dentalium shell	nuinhi-vaghi-ut.	
Ear-rings	aklatu-it	ukla.
Long	tatkhli.	
Short	nanilh-nuk	nai-pok.
Broad	yu-gu-tuli	silikpok, *it is broad and thin*.
Narrow	igu-kink-nuk	nerikipok, W.; amitok, W.
Fresh	milukapak	
Sweet	mi-iknik-kuk	mamakpok, *it tastes nice*.
Sugar		mamamak-sauk.

VOCABULARY OF THE KUTCHIN OF THE YUKON, OR KUTCHI-KUTCHI, DRAWN UP BY MR. M'MURRAY; TO WHICH THE CHEPEWYAN SYNONYMS WERE ADDED BY MR. M'PHERSON.

English.	Kutchin.	Chepewyan.
Animals.		
A bear	so	sasz.
grizzly bear	si-i	tlizè.
beaver	sè	tza; tsha.
red fox	na-kath	na-ghirhè-gossè.
black fox	nakath-barhata-nil-iz-zè	na-ghirhè-sin.
cross fox	nakath-so	na-ghirhè-netlizzè.
white fox (Arctic)	etchi-a-thwi	na-ghirhè-gai.
Canada lynx	ni-itchi	ghisè.
marten	tsu-ko	tha.
mink	tchith-ei	til-chusè; tekh-tusè.
otter	tsu-e	na-pi-ekh.
musquash	tzĕnn	tzĕn; tshĕn.
wolf	zo	yess; nuni-è.
hare (American)	kè	ka.
wolverene	lekh-ethu-e	nakh-ei.
seal	năt-tchuk	(nètsèk; netsi-arksuk, *Esk*)
moose-deer	tin-djukè	dunikh.
rein-deer	bet-zey	bedzi.
goose	krè	tcha.
swan	ta-arr-zyne	kha-goss.
crane	che-a	dhell.
duck	tet-sun	yurrth-tcho.
grouse	akh-tail	dikh (*pintailed gr.*); kas bà (*white gr.*)
fish, a salmon	tleukh-ko	tlu-e-tcho; tlu-e-zanè, *trout.*
white fish (*Coregonus*)	tleukh-ko-tak-hei	thlu.
pike	alle-ti-in	uldai.
blue-fish (grayling)	rsi-tcha	thlu-è-detla.
methy (*Lota*)	che-tlukh	tin-tellei.
Trading Goods.		
An awl	tha	thuth, *a spear.*
An ax	ta-è	thell; thelth.
Beads	nak-kai-e.	
A belt	tho.	
A blanket	tselta	tsurai.
A tobacco-box	tseltrow-ti-ak.	
Buttons	yei-kai-thit-le	bun-eil-lay; pa-il-lay.
A cap	tsa-kol-u	tsa-kŭlay.
A bonnet	tsa-til-ck-ha.	

English.	Kutchin.	Chepewyan.
A capot or coat	ik	ekh.
A duffle coat	chai-ik.	
A chisel	so-itt-se.	
A comb	tcheir-zug.	
A dagger	nil-ei-sho	la-thuth.
A file	kuk-i	hogulth; hok-kelth.
Tape gartering	lekath-at-hai-è	
A looking-glass	mutchai-e-i-a.	
A gun	te-egga	tel-gŭrthè.
A gun-flint	bech-tsi	tlè-tell.
A gun-worm	koggo-te	ko-èdèh.
Gunpowder	tegga-kon	telgŭrre-koun-nè.
A powder-horn	a-ki-itchè.	
A kettle	thi-a	tillè.
A knife	r' si	bèss.
A ring	ilăt-thĕkk.	
A shirt	azu-e-i-ek	tse-tsi-eh; thisitei.
A small shot	tegga-ătsil	teli-thai-è.
A ball	tegga-atcho	tell-gith-tcho.
A fire-steel	il-i-a.	
Cloth	athit-li.	
Thread	athit-li-itchi.	
Tobacco	se' ei-i-ti-it	sel-tu-i.
Trowsers	illei-ik	(karlik, *Eskimo*).
Vermilion	tingi-ta-tseikh.	

Miscellaneous.

English.	Kutchin.	Chepewyan.
A tree	tetch-hau	tsu.
A willow	kai-i	kai-thsinnè.
Grass	tlo	tlo.
The ground	nŭnn	nih.
Water	tchu	tu; to.
A river	han	dessh.
A lake	van	theu-tu-ĭ.
Rain	akh-tsin	dsha.
Warm	konni-etha	etu.
Cold	konni-eka	etdza.
Hungry	sei-ze-kwetsik	seth-ithu.
Fatigued	kei-a-sethelth-krei	ni-nitsau.
Sick	ĕth-ill-seyk	ai-a'.
A mountain	tha	sheth.
A valley	kra-tannè	shegussè.
The sun	r' sey-è	sakh.
The stars	thun	thun.
A rock	tchi	thi 'tsunnè-cho.
A house or fort	izzè.	
A lodge or tent	ni-ti-a	nèballè; nepalle.
A bow	alt-heikh	elthi.
An arrow	ki-e	kah.
A canoe	tri	tsi.

English.	Kutchin.	Chepewyan.
Good	neir-zi	nesu; neso; nazu.
Bad	bets-hè-tè	neso-ulla.
Day	tzin	tzinna.
Night	tatha	hetleghè.
Sleep	nokh-tchi	belkh.
Rest	tuggath-ĭlla-è	thĭlleh.
To sit	tchith-u-ĕtcha	théda.
To walk	ka-whot-ĕl	nathall.
To run	sha-tocha	thebakall.
To shoot	at-ĕl-ke	thelguth.
To kill	beshei-en-i-echa	thega-thul.
A man	tenghi	'dŭnnè; duneh; 'tinnè.
A woman	tren-djo	tshēkwè.
A boy	tse-a	dunne-yazè.
A girl	mitchet-ei	tsekwe-azè.
A dog	tleine	thling; thline.
A sled	latchan-vultl	bet-tchinnai.
Numerals.		
1	tih-lagga	nthlare, D. (en-clai, L.), (sthlagi, C.)
2	nak-hei	nakkhe, D. (nakka, L.), (nakke, C.)
3	thi-eka	khtare, D. (ita-rgha, L.), (takkè, C.)
4	tān-na	tinghe, D. (iting, L.), (tingee, C.)
5	illa-kon-ĕlei	zazunlare, D. (sa-soo-la, L.), (sasulagi, C.)
6	neckhki-ĕt-hei	elcathare, D. (ut-ke-tlai, L.), (alkitakhe, C.)
7	ataitsa-newk-he	nthlazuntinghe, D. (kko-sing-ting, L.), (sthlasi-tingie, C.).
8	nak-hei-etan-na	alcatinghe, D. (elzenting, L.), (alketingie, C.)
9	nuntcha-niko	nthla-otta, D. (kkahooli, L.), (katchine-onnuna, C.)
10	tikh-lagga-chow-et-hi-en	'nthla-una, D. (ito-nan-na, L.), (onnuna, C.)
11	tikh-lagga-mik-ki-tagga	(sthlagi-juthet, C.)
12	nak-hei-mikki-tagga	(nacke-juthet, C.)
13	thi-eka-mikki-tagga	(takhe-juthet, C.)
14	tanna-mikki-tagga	(tingee-juthet, C.)
15	ilakon-ĕlei-mikki-tag-ga.	
20	nak how-chow-ethi-en	(non-nanna, L.), (nackhe-onnuna, C.)

English.	Kutchin.	Chepewyan.
21	nak-how-chow-ithi-en-unsla-tikh-lagga.	(nacke-onnuna, nathetsin sthlage, C.)
30	thi-eka-chow-ethi-en.	(tacke-onnuna, C.)
40	tanna-ha-chow-ethi-en	(tingie-onnuna, C.)
50	atla-konělei-chow-ethi-en.	(sasulagi-onnuna, C.)
60	nikh-ki-at-hei-chow-ethi-en.	(alkitakhe-onnuna, C.)
70	atait-sa.	
80	nich-ki-etanna-chow-ethi-en.	
90	muntcha-niko-chow-ethi-en.	
100	tikh-lagga, chow-ethi-en-chow-ethi-en.	(onnuna-onnuna, C.), *ten tens.*
200	nak-kaggo-chow-ethi-en-chow-ethi-en.	(nacki-onnuna-onnuna, C.)
300	thi-eka-chow-ethi-en-chow-ethi-en.	(takhi-onnuna-onnuna, C.)

Note.—The orthography of the names of numerals inclosed by parentheses is different from that of the other parts of the vocabulary. D. denotes Dog-rib words obtained by myself. L. is Dog-rib recorded by Captain Lefroy. C. denotes Chepewyan words extracted from a list furnished by Mr. M'Pherson, who has adopted the French orthography in part.

The fragment of a vocabulary of the Chepewyan dialect, which follows, was formed entirely from the diction of Mrs. M'Pherson, to whom the language has been familiar from her infancy. It was written in the following manner: Having at hand a pretty full vocabulary of the Cree, drawn up at Carlton House in 1820, in which the words were arranged in alphabetical order, I propounded the Cree expressions to her in succession, assisting her with a French translation when she had any doubt of their meaning. The Chepewyan equivalent was pronounced by her again and again, until my ear caught the sound, and I was able to repeat it after her. I then wrote it down, and read it to her from the manuscript. Such words as I was unable to pronounce to her satisfaction, and they were not few, were left out. The nasal sounds resembling the French final *n* were the most difficult, and they are of frequent occurrence in the language. The Chepewyan tongue also abounds in the burring sound of the letter *r* combined with an aspirate, which I know not how to express in English; and such words have consequently been left out of the vocabulary. The ordinary aspirate, similar to the *och* of the Scottish or Irish, is denoted in the vocabulary by *kh*. The vocabulary, short as it is, took some weeks to produce. It was interrupted by a change in our arrangements in traveling, canoes having been substituted for boats, which made it less convenient for me to receive lessons in Chepewyan. This difficulty would not, however, have prevented the prosecution of the task, especially as Mrs. M'Pherson with much kindness expressed her willingness to proceed until we had gone through the whole Cree vocabulary, of which about nine-tenths remained; but knowing that the language was becoming a written one, under the active superintendence of the Roman Catholic missionaries at Isle à la Crosse, I gave up my intention of endeavoring to ascertain its structure, and contented myself with the following specimen.

Vocabulary of the Chepewyan Tongue, with Cree and English Translations.

A as in "father; *ā* as in "awe" "law;" *è* as in "theme;" *e* as in "dell" "bed;" *i* as in "ravine," or as *ee* in "see;" *ĭ* as in "ill" "ink;" *o* as in "or" "for;" *ŭ* as in "us" "husk;" *ai* as in "aim" "maim;" *au* as *ow* in "how;" *oy* as in "hoy;" *yu* as "you;" *eu* as *ew* in "dew" or *eau* in "beauty;" *kw* as in "awkward;" *ng* as the French nasal *n*; *kh* as *ch* in "loch," *Scottice.*

Cree, of Carlton House.	Chepewyan, of Athabasca	English.
A.		
Abu-ye	tu-a-wĭll	liquor, soup, or drink.
Agatha-shu, or aggai-a-shu	thè-ut-'tinnè	an Englishman.
Aggĭskow, or akkĭskow	el-ka-ti	pin-tailed grouse.
Aggŭsk	sis-thère	a blunt arrow.
Atchak, or akhchak	i-yu-nè	the soul.
Ai-ammi-hè-u	yu-alānè-pallè	a flag.
Akop	tsirrè, or tchirrè	a blanket or covering.
	Tsirrè-kai-cho	a large blanket.
Ai-u-wannis	yu	all kinds of goods.
Akwa-napoy-igan-as-kek	tillè-arakai-ĭnka	a covered kettle.
Akwatĭn	hātkin	frost.
Amĭsk	tza	a beaver.
Amis-kwa-wistè	ekhkè; tza bèkong	a beaver-house.
āmu; amo	klizè; ti-ranna	a bee.
Annèk-kutchass	tli-i; tchillà	a squirrel.
Annèk-kutchassis	tillel-kuzè	small or ground-squirrel.
Apikh-tow-kishi-kow	'tchi-èn-tizè; tchinnè-tan-ni-sè.	mid-day, or half a day.
Apikh-tow-tĭppĭskow	thir-nize	midnight.

Cree, of Carlton House	Chepewyan, of Athabasca.	English.
Apisi-mongsus.......	....	jumping deer.
Apistè-shipis	él-kurrè ; tchikhth-i-a-sè-akhth.	a teal, or small duck.
Apistat-tchèkus......	....	prong-horned antelope.
Appakwa-sun	ni-pallé............	a leathern tent-cover.
Appèk-kusis.........	kleunè.............	a mouse.
Appisk	tannonè-tcho, *big bird*	black or white-headed eagle.
Appistis-kis	kai-yazè ; kai-gusè..	a Hutchins's goose.
Appoyè.............	toth ; tö-a.........	paddle or oar.
Appoy-nask	kès................	a spit.
Appŭssuk, (Pl. appŭss-ye-akhtik, or appussu-yuk).	thai-ye; nepalli-tetch-un (*tent legs*).	tent poles.
Miskahtuk	nepalli	a man's legs.
Akhtai-yè (P. -wuk)..	thè	a fur skin.
Ammiskwa-tai-yè	tzà-thè	beaver skin.
Askik	tĭllè	a kettle.
Aski ; assiski........	kwotlès............	land.
Int'aski.............	ni-tanninnè	my native land.
Kit'aski	na-hinnè...........	your native land.
Int'askinan	....	our native land.
	Bè-anninnè	his land.
Aski-tin-wè-as ; aski-we-as.	bét ; per-elinè	raw or fresh meat ; flesh.
Askow-i	ten-de-ila (*ice, hard, not) ;* ten-nailer.	holes in the ice.
	Kin-the leuk........	ice breaking up.
Assăm (P. assăm-uk)	akhè ; akh ; akhi ...	a snow-shoe.
Ahkik	tĭllè	a kettle, or copper kettle.
	Sampas-tĭllè	a tin kettle.
Assini	'thèkh.	
Assini-uspogan	seltu-yè-thekh ; tchè-tut-thékh	stone pipe, or calumet.
Assini-watche-a	sheth ; thè-she......	Rocky Mountains.
Assini-poyt, or, E-askab	....	a Stone Indian.
Assiskè.............	otlès	mud or earth.
Assiske-pakwè-sigan..	thlès	wheaten flour.
Assiggan (P. assiga-nuk).	tel	a sock, foot stocking.
Assiss-wi	ètlè	an ice chisel (*lit.* a horn).
Astu-thèggum-ik.....	tsi-yè	a shed in which canoes are built.
Astu-tin	tsá	ladies' cap or bonnet (beaver).
	'Tsa-kallé	man's hat.
Astum-astaik........	tsa-ne-tum.........	sunshine.
Atchakht, or atchăk..	thin...............	a star.
Atchappi	eltè	a bow.
Atchappè-kan	klewlghè-elting	a fiddle.

Cree, of Carlton House.	Chepewyan, of Athabasca.	English.
Atchĭmmosis........	thling-yazè.........	a puppy.
Atekh, or attek......	èt-thin	reindeer.
Athappi	tā-bith	a net.
Athabiskow.........	thè-minnè-u-ye	a rocky country.
Atha-wak-kiska-mat-tinow	hokar-ritha.........	a very steep bank.
Athā-wastin	tethi-èl	a calm.
Athik	tsai èllè	a frog (*grenouille*).
	Tsai-el-cho	large frog (*crapaud*).
Athuskan	ta-kallè-chi-a	a raspberry.
Atchak-ash	til-chusè	a mink (*mustela lutreola*).
Attei-gan	yu	trading stock.
Attikh-hameg	thlew	white fish (*coregonus*).
Attim	thling	a dog.
Atuspi	kaithlin-sĭnnè	alder.
ākuu-pusè-wĭn........	thai-i	a platter.
āpètte-kā-hĭggan.....	denti-lita-thil-tillè...	a chest-lock key (properly, but used for keys in general).
Akhàkhk (*a guttural grunt*)	hèkh	yes.
Akā-mik............	nannè	across.
Akā-mik............	yanna.............	on the other side.
Akwa-kukhtin	tit-sa	it is mouldy.
Annutch; attè	tu-hu	now; at the present time.
Attè	kaltunè............	already.
Annutch-kak-kè-sikak	ti-dzinnè...........	this very day.
Annutch kā-tippiskak	terri-kitha	this night.
Annu-watch-gai-as ...	....	rather long ago.
Apikh-tow	tanizè.............	in the middle.
Apatishew	bet-arutha	it is useful.
Appātun............	bèt-taritha	useful.
Apputchiga	....	once on a time.
Askow	athkè	sometimes.
Astum-uspi	ekku-azè	since such a time.
Athi-mun...........	sutu-yè	it is difficult.
Athè-wak	ona-hadzŭn	more.
Athè-wak kishè-wak..	edzun-kuthè........	nearer; very near.
Athe-wak-pètsow	hona-hedza-nitha ...	further; very far.
Eshunila	hulè-ho	he is troublesome; badly disposed.
Ai-ā	nista-ula...........	keep it; have thou it.
Int'ai-an; or int'ai-a-wā-u	se-itza-heila; hunè-zoni	I possess it; it is mine.
Kit'ai-an	netzè..............	it is yours.
Ai-akuski-tè-u	petothè-karth.......	it has a broad bill.
Ai-āmi	yan-ilti	speak thou.
Ai-amew; ai-atchi-mè-u.	yalti	he talks.
Ai-amihin	zedzun-yar-ilti......	speak to me.

Cree, of Carlton House.	Chepewyan, of Athabasca.	English.
Ai-ami-hi-tu-tak	althlai-yalthi (*together let us speak*).	let us talk together.
Ai-ami-hi-tu-wuk	elthney-alti	they talk to one another.
Ai-ami-hè-u	yedzonne-alti	he spoke to him.
Int'ai-amin	è-ăsti	I talk.
Int'ai-ami-ha-u	bedze-asti	I talk to him.
Int'ai-ami-hik	zedzun-alti	he spoke to me.
Kit'ai-ami-hik	nedzunè-alta	he talks to you.
Ekau-witha-atche-mow	zedzun-ye-innè-alti-hila-kula	do not tell it.
Ai-atchĕmow-akwa	nu-hei-lunè	tell us the news; relate thou now.
Ai-apè-tikă-u	peyè-onla-honnè	it is full of partitions.
Aikh-tu-kă-mik	nu-anku	another house.
Waska-iggan	yè	a house.
Wiggi	....	a tent or dwelling.
Ai-ish-ku-shu; ai-ish-ku-tè-u	kalyè-ni-nan-idza	he is tired (with walking).
Ai-ish-kutan-nè-wu	kalyè-ni-tan-idza	they are tired (ditto).
Int'ai ish-kuzin	kalyè-ne ninna-chă	I am tired (ditto).
Int'ai-iskutchè-man	toth-ne-zin-alnilza	I am tired with paddling.
Kitai-iskutchè-man-nă?	teth-ne-ni-nan-ilza-uza?	are you tired with paddling?
Ai-u	nu-a-edzon-illa	he is there.
Akkushew	ey-a-hilla	he is sick or ill.
Int'akushin	ey-a-hèzlè	I am sick.
Akkustemmu	edzil	he is wet.
Int'akkusktemmun	dzedzil	I am wet.
Annĭskutăpan	chăs-inninne-ai	a knot.
Anniskutăpè!	chăs-nos-al!	tie a knot!
Ing'annisku-tăpan	....	I will tie a knot.
Gè-annisku-tă-pè-u	chăs-ninne-al	he has tied a knot.
Apikh-ku-pai-u	tey-kunnè-takh	it has become loose; it is loose.
Ge-appaha	tey-kunnè-arlth; ney-ke-urth.	he has untied it.
Apith-kuna; appaha	pey-kè-urth	loose it (a knot); open it.
Nè-gè-apith-kunain	kalthonna-pey-kè-urth	I have loosened it.
Ne-ge-apa-hain	ey-kè-urrth	I have untied it.
Int'apikh-tă-pă-hă-u	peino-harre-kluk	I gave him a blue eye.
Int'apikh-tă-pă-huk	zunno-arrè-kluck	he gave me a blue eye.
Apisa!	per-il-thilth!	warm it (as a garment at the fire).
Apisum	yi-ěr-il-thilth	he warms it.
Să-sey-int'apisain	kuda-ber-il-thilth	I have already warmed it.
Appi!	thein-'tă!	sit down!
Utè-appi!	ey-er-thein-ta!	sit here! (here sit!)
Appew	nèltă	he sits.
Appè-wé-ŭk	hed-nilthi	they sit.
Int'appin	thi-tă	I sit.

Cree, of Carlton House.	Chepewyan of Athabasca.	English.
Kit'appin-nă?	thin-ta-uzang?	are you sitting?
Ki-wi-appin-nă?	unta-uzang	do you wish to eat?
Ashamin!	bega-van-ilchu!	give me food to eat!
Kiga-ashami-tĭn	ne-a-urchu	I will give you food to eat.
Michèma; hughès....	bet-ho	meat and drink; food; victuals.
Aspun-ishew	ă-ă-ontzun	he is niggardly (of his victuals).
Assitĭnă	èltan-nilè	mingle them; add one to another.
A-sustatin	....	it is hidden.
Kiga-kasustatin......	necha-itus-'i........	I will hide myself from you.
As-swè-tè-u	te-yè-thèlla.........	it is in (a bag).
As-swè-tă-u	te-yè-yèlla	he puts it in.
Int'aswetàn	te-ye-ila	I put it in.
Aswèthim	bega-etu-u-elnè	be on your guard against him.
Aswithi-min	zethè-sekor-u-elnè ...	be on your guard against me.
Ing'aswithi-mow	pa-us-o-èlnè	I will be on my guard against him.
Atchis-chapum-wè-u .	na-seil-hitchè	he gives a side glance to a girl.
Athăg-uskow	ne-etèl	it is broad.
Athin-isew; eythiniseu	hung-ya; huya.....	he is wise or knowing.
Int'sip-gathi-nisĕn ...	hong-she-a	he is wise or prudent.
Athinew	....	he is abstemious.
Atuskè-u	è-hul-ana	he labors.
Int'atuskaim........	e-walasna	I labor.
Atta-wanna! (imp.)..	na-ĭnni!	barter! trade!
Ki-wi-atta-wanna? ..	na-ukh-uneuza?	will you barter?
Atta-wa-gun........	yu	goods for trade.
Atta-thow-ki!	sel-honninnè!	tell a story or fable.
	Honnè............	a story.
Attè-mishi-kă-tè-u ...	necha-ladi-nelthun ..	he grows bigger.
Attè-mi-shè-u.......	....	it grows bigger.
Atikh-tè-u	neuth-lurth	it is ripe or mellow.
Artisum	udedza	she dyes or tinges it.
Int'atisain	uridza	I dye it.
Atima-ow	....	he overtakes.
Int'atima-ow	ne-ni-esha.........	I overtook him.
Ing'atima-ow	ben-nisha-lillè......	I will overtake him.
Int'atimik..........	....	he overtook me.
Int'atimahuk	....	he overtook me (by water).
Atimi-thowuk.......	nar-helteth	they fly from us (birds).
Attohu	tchirr-iltè	he is choking.
Int'attohun.........	tchirr-estè	I am choking.
Ki-wi-au-totè mĕmitĭn	et-te-to-tin-in-ustè ..	I wish to be your friend.
A-wuss; a-wussètè! .	nusè!	keep off! let me alone!

Cree, of Carlton House.	Chepewyan of Athabasca.	English.
E		
è-atchi-inyu-wŭkk ...	et-dunni-'tinnè	Indians of a strange nation.
Ek-kwă	yah	a louse.
èpètchè-kishi-wŭkk...	yelkon	dawn of day.
Eskann-shi-ka-un	edtè-thidzi	a horn comb.
Eskwai-atch-tchi-tchan	tinnè-la-dthaille-dzilla	the last or little finger.
Esputtinow.........	kokkarritha........	high ground ; a bank (*une cote*).
Ethik-kwatin	ne-edja	hoar frost—hoar frost.
Ethiko-pew.........	....	hoar frost—rimy.
Ethikwuk	tchanti	ants.
Etiskew	èkei-ghè	foot-mark or track of an animal.
ĕ-a-hă-u ; or, ya-ha-u !	ey !	ha ! (*interj.*)
E-a-kusin ; thah-kusin	nedtha	light.
	Ned-tarrilla	not heavy.
E-apitch...........	....	still ; quiet.
	Na teillè..........	it stirs not.
èka	hila..............	not.
èpètchè-kishikak	....	as the day was coming.
Ekushi-kak.........	dzithè	by day.
Kishi-kow ; kisgow...	dzinè.............	the day.
èkospi ; èg-guspi	klasing-tingè-vaiyè..	at that time.
è-okwo-pŭkku.......	ashmoh...........	only that.
Eskwai-atch........	no-ontè	last.
Espimmi-sik........	i-yazè-bèkè	a little above.
Espimmik..........	bèkè	above.
ètakkusik	thè-dzini-ghè	yesterday.
Ethipinnè-ok-tapo-an .	oti-a-èlthè	truly.
Etippiskak	hedklèghè	by night.
	Dza-kin	beaver lodge.
è-aske-u	dza-kin-nannelya ...	he breaks up a beaver lodge.
Int'e-askann........	....	I break up a beaver lodge.
Ekau-witha !	....	do not !
Entau-wi	....	go and open it.
E-ukh-tinnè-gatè-u ..	peta-harelta ; peta-haelta	it is opened.
E-ukhte-nammuk	peta-klell	open ye it.
E-ukhte-num	....	he opened it.
Nè-ukhtè-nain	....	I opened it.
ètapoy-ikhta........	bethna-ilkis	mix it ; stir it.
Ethepo-akwow ; nepo-akwow ; athin-ni-sew	huya	he is wise—knowing.
Ey-thin-akhtĕk......	eln	small spruce fir (*Abies balsamea*).

Cree, of Carlton House.	Chepewyan, of Athabasca.	English.
Ey-thin-attu-shiship. .	tchith-tcho	stock duck (*Anas boschas*).
Ey-thinni-kannu-shè-u	ultai-yè.	a pike or jack.
Ey-thinni-mina	. . . .	bilberries.
Ey-thinni-pithey-u . . .	. . . .	Canada grouse.
Ey-thin-yu (P. eythin-yu-wŭkk).	'dtinne	an Indian of the speaker's nation.
Int'ey-apa-huk	. . . .	he made my eye ache by a blow.
Tans-ey-sinikassort ? .	etla-hulyè ?.	what is his name ?
Tans-ey-sini-kassu-yŭn ?	ey-la-hunlye ?.	what is your name ?
Tanna-si-te-kateg-oma ?	. . . .	how do you call this ?
Init. I. sounded as EE.		
I-ā-pit		
Kah-nup-âtè-i-a-pit . .	nakith.	he has an eye on one side.
Tann-ikè ?	etla-djah ?	what is the matter ?
Tann-ikh-tè-an ?	. . . .	what is the matter with you ?
Istè-kwa-nan ; ustĕ-kwan ; mistekwan	edthi	the head.
Int'istèkwanan	zedthi-ey-a.	my head aches.
Nistè-kwan	zedthi	my head.
Uta-pètchè-itotè !	è-o-kŭ-si !	come hither !
Nètè-itotè !	è-o-kŭ-si-nēk-iltkh ! . .	go there or thither !
Tanti-wy-i-tukh-tè-an?	etla-se-nek-ăltkh ?. . .	where are you going ?
TCH.		
Tchakkatinow	shethi-azè	a knoll ; small hill.
Tchi-kè-kum	shith	a wart.
Tchi-tchè	'tinnila-theyl lé (*man's toe*)	a finger.
Tchi-ka-ĕgan	thell ; kong-kwi	a hatchet.
Tchi-ka-ĕgan-akthik . .	thell-tchinnè	a hatchet helve.
Tcheŭk-sa-ĕgan	klell-thelth ; thléh-kon	a gun flint.
Tchè-măn.	tsi ; alle	a canoe.
Tchi-pai (P. -pa-ŭkk)	ethi-a	a dead body ; the deceased.
Tchi-pai-ŭktim	. . . .	you dead dog ! an opprobrious epithet.
Tchi-pai-ŭkk (*dance of the dead*)	nè-èlkai	Aurora borealis.
Petāpan	yel-kon	dawn of day.
O-wanni-wagan.	. . . .	dusk of the evening.
Tchis-ā-wan.	pernatal	a hash, or haggis.
Tchis-kè-pi-son	dza-thulth	a garter.
Tchis-ta-bà-sun.	pan-neyla ; luneylè. .	a button ; an anchor.
Tchista-sè-pŏwin.	pè-o-koyl	a fork.

Cree, of Carlton House.	Chepewyan, of Athabasca.	English.
Tchista-ka-wè-sew ...	ther-onna	a wasp.
Tchista-ka-nan-wi-ship	el-karrè (*pine-leaf duck*)	a teal (*Anas discors*).
Tchistèm-ow	sel-tu-yè	tobacco.
Tchimm-ashĕn	ned-tu-a	it is short.
Tchuk-tchuk-athu; tchuk-tchuk-ai-u	tadzon-zellè	a blackbird (*Scolephagus*).
Tchika-wa-sis	yazè (few)	not many.
Tchi-kima	ta-tu-ahaddè	true, truly; verily.
Tchi-kima-numma? ..	ta-tu?	do you doubt it? it is true.
Tchi-ka-ka-win	nè-o-ka	close to the shore.
'Tchist! tchistè!.....	'tchu!.............	hist! listen! look!
Tchŭppasis..........	pei-yā-thi	below; underneath.
N'tchā-kā-pă-huk	dzènoy-ĭnkè	he poked it (a finger or stick) into my eye.
N'tchā-kā-pi-chi-nin..	dzè-noy-èkè	it has run into my eye (a stick).
Tchèp-wow	kai-intchuthè.......	it tapers.
Tchès-kwă; tchès-kwa-pitta!	karrè!	wait! wait a little!
Tchitchei-mi-kŭskwè-su	pè-'kunne-neltu-yè ..	he or it has short nails.
Tchika-ai-gè-u	thelth-ta-nai-ilkh-thelth	he hews with a hatchet.
Ne-tchika-iggan......	thelth-ta-nai-ilkh-thell	I hew with a hatchet.
Tchi-kè-si-sè-u.......	dzerè-hai-èllè	he plays at draughts.
Tchi-kè-si-sè-ă-wukk	....	they play at draughts.
Tchi-kwa-ha-mè-u ...	belekh-hered-ye	he crumbles the leaves (rubs them to powder).
Ingā-tchi-kwa-hain ...	....	I will crumble the leaves.
Tchi-pè-tŭkk-wow....	tèl-klŭkk...........	it is light blue.
Ni-ghe-tchi-pusti-hà-u	thilk-tas	I put it with my arrow.
Wa-was-ki-shu	tsè-thil	the wapiti.
Oya-peyu-mus-tus....	ettirrè-yā-nè........	bison bull.
Nosia-mustus........	ettirre-su-ta-ha	bison cow.
Wā-pis-tănn	tha	a marten.
Si-kāk.............	nult-si-ai	a skunk.
Si-ku-sew; sigus.....	del-kathlei	an ermine.
Winusk............	tel-leh.............	a marmot or spermophile.
Winustè-key.........	....	a Quebec marmot.
Wapusk	sass-del-gai	white bear.
Apek-ku-sis	tlunnè.............	a mouse.
Shi-shi-pise	elgarrè	a teal.
Key-ask; kai-ask	bess-gai-è...........	a gull.
	Kallei	a plover.
	Bekhu-hulla........	*Salmo mackenzii.*
Okkau; uka.........	ettchu-è	*Dorè.*
Miki-sew	ded-donnè-tcho	an eagle.

Cree, of Carlton House.	Chepewyan, of Athabasca.	English.
Ahà-sew	dadsang	an American crow.
Ottoni-bis	thè-chuthè	*Coregonus artedi* (Tullibee).
Namay-pith	till-tulei	*Catastomus.*
O-wi-pi-tchi-sis	thlu-dathé	*Hiodon.*
Nipe; nipi	tu; to	water.
The-kwus-kwun	....	it is cloudy.
Kŭsku-wŭnŭsk	kothè	clouds.
Kishi-kau; kis-gau; wa-pan	....	day; day-light.
Ki-ki-ship	kambi	morning.
Apikh-tow-kishi-kau (*middle-day*)	tsindéssai:	noon.
Pakkisimu (*sun-set*)	tchilsin	evening.
Tippis-kak; tippiskau	....	night.
Mistiko-tcheman	tetsin-tsi	a boat.
Wini-pègh	tu-tcho	the sea.
Thaka-stimmun-aigan	tsini-ball	a sail.
Paskè-sèggŭn-nis	telgurthe-yaze	a pistol.
Kitchè-kuman	bèss-tcho	a sword (big knife).
Tappis-kā-gan	kothi-ghirrè	a handkerchief.
Mokasin; muskesin	ke	a shoe.
Tippiskā-wi 'peshim	eltsi	the moon.
Kesik; kishik	yaha	the sky.
Pinasi-wuk	edihi	thunder.
Wa-waskhsta-punu; owā-sāmusk	tsinago-thethi	lightning.
Kunu; konă	yath	snow.
Miskwumi	ti-enn	ice.
Piki-sè-u (*it is foggy*)	etzil	fog.
	Hothin	frost.
	Nahalgi	thaw.
Utin; thow-tin	niltsi	wind.
Atchimow	yalthi	to speak.
	Netghin	to sing.
Ni-ku-mun	....	a song.
Mitzu	tchèli	to eat.
Wappamow	etethi	to see.
	Ureltha-nelsi	to hear from you.
	Su-sinnè	a great happiness.

The following words of Dog-rib were collected by myself at Fort Confidence. The want of a good interpreter caused me to discontinue the formation of a vocabulary of this dialect.

Dog-Rib Vocabulary.

English.	Dog-Rib of Fort Confidence.	English.	Dog-Rib, of Fort Confidence.
A kettle	tillè.	Little kettle	tillè-yazè.
Large ditto	tillè-tcho.	Fire	kun.

English.	Dog-Rib, of Fort Confidence.
Fire-wood	sus.
Gunpowder	tel-kithe-kŭn.
Shot	tel-kithé-ka.
Shot-pouch	tel-kètha.
Ball..........	tel-kethi-'tcho.
No meat!	par-ulla!
Dried ribs of reindeer	átcharna; et-chanka.
Water	to.
A tin pan	thai.
A coat, or capot.	i.
A blanket	zidda (tzud-di-e, (Mr. O'Brian).
Indian hose	thelth.
Hair	theo-ya.
The beard	tarra.
A crooked knife.	bèss-ha.
A knife	bèss.
A knife sheath .	bess-thè.
A fork	pakwa.
Snow.........	tzill; tchill.
Smoke	thlet.
A stone	thai.
A brisket......	ana-rāne; ei yid-da.
The shoulder ...	akkànna.
Leg bone or knuckle	ak-kai-tchinna.
A firebrand	halai-kun.
A tent........	nepàlle.
Tent poles.....	thai-è.
Transverse poles to hang meat upon	tanè-ai.
Tent door	ku-latche.
Leathern door for tent	thidai-nepàlle.
Dressed leathern blanket	tel.
A spark from the fire	kantida.
Reindeer tongue	et-thu.
Deer-skin hose..	et-thidda.
Breeches	thlai-i.
Deer head	et-thi.
A shoe........	ku.
Cloth worn by men round the middle	than.
A bag	naltchè; klethè.
A hatchet	thelth.
A spoon.......	thlus, or slus.
A file	kokètha.
Pole for hanging a kettle npon	telle-kaiza.
Buttons	pai-illa.
Mittens	gis.
The head......	ta.
The nose	tinnetze.
The knee	et-thĕtha.
An encampment	zutès.
The encampment is distant	in-tu-è-zutès.
The encampment is near	thi-si-tè-zutès.
A warm woolen collar; a comforter	kow-i-tchitha.
One	'nthlarè.
Two	nakhkè.
Three.........	khtarre.
Four	'tinge.
Five	zazunlarrè.
Six	elkatharrè.
Seven	nthlazintinge.
Eight	alkatingè.
Nine	'nthla-otta.
Ten	'nthla-una.

The vocabularies which follow were made by gentlemen whose system of orthography varies more or less from that adopted in the preceding pages. The dialects of the Dog-ribs who resort to Great Bear Lake, and of those who hunt on Marten Lake and visit Fort Simpson, differ little when spoken, and offer no difficulty to an interpreter who is acquainted with either; but many of the words have a very different aspect when written in English characters; and these tables may serve to illustrate a remark made in a preceding page respecting the difficulty which an English ear experiences in apprehending the sounds of the Tinnè languages. The Kutchin words collected by Mr. M'Murray, though not numerous, show a close affinity between the language spoken by that people and the Tinnè, and will perhaps be considered as a proof of the common origin of the Tinnè and the Kolush tribes down to the 54th parallel of latitude.

English.	Dog-Rib.	English.	Dog-Rib.
Head	bet-thi.	Go!	aga!
Neck	bdi-korh.	Come!	ya-kusi!
Tongue	eth-thadu.	Take!	hi-tcho!
Eyes	mendi.	Cut!	bekan-nèthu!
Ears	bed-ze-gai.	Bring!	si-nekai.
Nose	mi-gou.	Hunt!	no-sai.
Cheeks, chin	mi-ta.	Large	nai-tcha.
Shoulders	ai-kon-nai.	Small	ti-ula.
Thighs	ed-zaddai.	Long	nundeth.
Brisket	a-ethin.	Short	nundeth-helai.
Rump	etchin-nai.	Far	nitha.
Belly	be-tchuki.	Near	whâ-yai.
Hands	mila.	Cry!	azel!
Feet	ak-kai.	Laugh!	mena-thi-ukla!
Fingers	mila-tchinnai.	Speak! or talk!	betha!
Nails	mila-konnai.	How many?	tanna-itai?
Teeth	baighu.	What do you want?	addow-adlis?
Brain	bet-the-ghu.		
Liver	et-hut.	Heavy	tai-it.
Heart	ed-zai.	Light	naikel-helai.
Blood	et-tillai.	High	yu-te-gai.
Skull	et-thi-thu-ine.	Low	u-ai.
Entrails	et-si-si.	Good	naisou.
Udder; milk	et-tuzai.	Bad	tlenai.
Butter	edgiddai-thlissai.	Fat	tlaika.
Flour	hatai-kotliss.	Lean	tlaika-helai.
Sugar	suka.	Eat!	shanai-tai!
Tea	suka-tu (*sugar water*).	Drink!	ath-uluston!
		Smoke!	ustud!
Pepper	tenni-tsi.	Sleep!	notai!
Medicine	na-diddu.	Give!	mi-ne-kai!
Paper	eddithi.	Tell!	adin-dai!

The above vocabulary was formed, I believe, at Fort Simpson, by one of the Hudson's Bay officers for his own use; but, having forgotten to note the circumstances under which it was drawn up, I can give no further information regarding it.

A Vocabulary of Fort Simpson Dog-Rib, by Mr. O'Brian of the Hudson's Bay Company.

Dog-Rib.	English.	Dog-Rib.	English
Edza-zinnè	*Tetrao umbellus.*	Noga	wolverene.
Tih	*Tetrao canadensis*	Kling	dog.
Bet-theu	owl.	Tzus	wood fire.
Thlu-ai	*Coregonus albus.*	Tai-tchin	trees.
Samba	trout.	Tzu	pine-tree.
Kazè	salmon.	Ki	birch.
Tsai-teu	Back's grayling.	Sinnai	I.
Tai-tellai	*Catastomus.*	Tlinnai	thou.
Klogai	squirrel.	Ottinai	he.
Emmu-i-u-ai	*Columba migratoria.*	Ige	it.
		Edetata	yes.
Khun	fire.	Helai; odelis	no.
Tu	water.	Id-zeunai	to-day.
Tchon	rain.	Kambai	to-morrow.
Yah	snow.	Zeunai	day.
Teu	ice.	Tethi	night.
Sa	sun	Yakh-kai	winter.
Tethi-sa	moon.	Klukai	spring.
Thi-u	stars.	Senai	summer.
Kose	clouds.	Ai-tonkai	autumn.
E-tu-ai	girl.	Tai-chin-ala	boat.
Ah	snow-shoes.	Ki-ala	canoe.
Kai	shoes.	Tami	net.
Whoghi	snare.	Tau-ai-on	full.
Thai	sinew.	Tu-tai	empty.
Do	now.	Tlon	plenty.
Ye-won	then.	Hulai	none.
Tau-dezzei	half.	Tzuddi-è	a blanket.
Mal-lionai	rings.	Tai-si-ai	a shirt.
Hai-ai	trowsers.	Ed-geid-dai	a powder-horn.
Męmba-ulai	waistcoat.	Mad-deli	buttons.
Tsi	vermilion.	Thai-on-tithei	thread.
Sat-su-wai	wire snare.	Et-thai-ai	scissors.
Sâs	black bear.	Meni-di-e-dai	looking-glass.
Sa-tai-kuzè	brown bear.	Ai-tchusai	beads.
Tsa	beaver.	Ai-tai	ice-chisel.
Tsa-thu-ai	castoreum.	Bai-huch	crooked knife.
Tai-tchesi	mink.	Bai-chin-ai-i	clasp.
Tzin	musk-rat.	Bed-do-ai-du	pot.
Tèki	wolf.	Tha	pan.

The following vocabulary of the language of a tribe dwelling near the sources of the River of the Mountains, and known to the voyagers by the name of "Mauvais Monde," and of the Dog-rib dialect, was drawn up by Mr. O'Brian, of the Hudson's Bay Company's service.

Mauvais Monde.	Dog-Rib, or Slave.	English
Thèlgai	thlie	one.
Olki-e	olki-e	two.
Ta-dette	ti-e	three.
Tinghi	tinghè	four.
Sazelli	sazelli; lakithe, *the hand.*	five.
Et-seu-ti	et-seu-ti	six.
Thlad za-di-e	han-die	seven.
Et-zan-di-e	et-zan-di-e	eight.
Et-thlei-hu-lai	ethli-e-houlai	nine.
Ken-na-tai	o-nai-u-non	ten.
El-lai-zai*	tel-kithi-kun	gunpowder
Bai-ka	tel-kithi-tcho	ball.
Ni-tai-ton	thai-thi	shot.
Ai-tai-kai	sel-tu-e	tobacco.
Et-ton-nai	tel-kithè	gun.
E-tha-thai-on	hai-ko	gun-flint.
Utha	tiu-ni-e	kettle.
Thei	thei	ax.
Ai-tchut	ai-tchut	awl.
Bèss	bèss	knife.
Ta-chill-ai	et-ley-nai	cloth (strouds).
Kestu-ai	ai	coat (capot).
Theth	theth	leggings; also a belt.
	Edgiddai	powder-horn.
Kothegettai	ko-the-gat	handkerchief.
Set-tsa-tai	tsa	bonnet-cap.
Hai	kun	fire-steel.
A-tai-kai-tenney	seltu-tenne	tobacco-box.
Ta-ti-e	ta-ti	needle.
Thai-ka	ko-kassè	file.
Et-hai-ai	baith-laika	scissors.
E-kadzi	kud-dai	gun-worm.
Ai-kathai-tai	sa-kathai-tai	garters.
Klai-si	sa-tai-kai	gray bear.
U-thai	no-githi	fox.
Ustaidgè	no-ta	lynx.
Ustai	no-thai	marten.
Kasho	nom-be-ai	otter
Wollon	teu-di-e	male moose-deer.
Intsei	teu-di-etse	female moose-deer.
Wod-su-tchu	bed-su	male reindeer.
Wod-su-mon-bed-sai	bed-su-tsi	female reindeer.
Kāg-kalai	kam'ba	ptarmigan.
Ogha-tchai	ogha	goose.
Ea-sai	tai-tonna-tcho	eagle.
U-ta-dja	uti	pike-fish.
Niton	monalla	white man.
Setsa-on	tchikwe	woman.

* This word is used also by the Beaver and Thekannè Indians.

Mauvais Monde.	Dog-Rib, or Slave.	English.
Te-sonnai	tchillawe	boy.
Klu-chu-i-nai	bai-tchinai	sled.
Sunbaddei	dsheth	mittens.

A Vocabulary of Chepewyan and Dog-Rib Words.

The Chepewyan was taken down from the mouth of the interpreter at Great Slave Lake. The Dog-rib from that of the female interpreter (Nanette) at Fort Simpson. The whole in 1844.

J. H. Lefroy.

Toronto, March, 1850.

Broad, ' nasal, ' guttural, " nasal and guttural.

English.	Chepewyan.	Dog-Rib.
Yes!	e-h!	
No!	he-li	he-li.
A man	denne	tchel-a-qui.
A woman	tza-qui-ie	tzek-qui.
A little girl	ett-er-e-ka	tzek-qui-azzè.
A boy	tchilla-qui-ie	tenai-u.
A little boy	tza-qui-the	tenai-u-azzè.
Father	tza-tah	tza-tah.
Mother	en-nè	en-nè.
Brother, elder	tzoon-noi	tzoon-noi.
" younger		tzachilli.
Sister, elder	ssa-ra	sa-rah, or tza-rah.
" younger		sa-tez-zah, or tza-tazze.
To-morrow	kom pee	koume.
Yesterday	ou-ah-ta-tzenke	ye-hho-a.
Tobacco	tza-twe	tza-twe.
A knife	pa-as	pa-as.
A gun	thel-ki-the	qua-ka-he.
An ax	than-thye	qua-a-qui.
Gunpowder	thel-ki-the-"conne	thi-ke-e-"connè.
Ball	thel-ke-the-chou	the-ke-e-chou.
Air	nutze	e-at-ti-ge.
Fire	kkon	kkon.
Water	tto	two.
Earth	wa-kklas	ko-eccla.
A fish	clou-a	clou-a.
A dog	"cling	"cle.
A fox	no-ki-ki-the	e-et-tha-thà.
A buffalo, masculine	et-cherre	et-cherri.
" feminine		" ettzae.
Reindeer	eet-than	et-thun (ettzae, f.)
A moose	den-nee	denne-a.
Snow-shoes	ah-he	a-e.
A sled	betch-chin-ne	ba-chen-ne.
A kettle	tille	than-ne.

English.	Chepewyan.	Dog-Rib.
Evening	eetzson	eya-kka-ezza.
Morning	kompe	sa-tcho.
Colors—black	tel-zonne	ta-zun.
white	tel-ka-ye	tel-ka.
red	tel kkosse	et-tel-kkos.
green	ta-e̋cloze	ta-eck-cles.
blue	not distinguished from black	ten-è-c̋lè.
yellow	tel-thoi	tel-thoi.
The sun	ssa	ssa.
The moon	et-cha-aza	tthe-tha-za.
A star	thun	thun.
(The Great Bear)	(ya-ee-telli)	(ya-tha.)

English.	Dog-Rib.	English.	Dog-Rib.
Snow	yya.	Old	e-e-ranna.
Ice	t-than.	Dear (beloved)	The word unknown to the language.*
Numerals—1	en-c̋lai.		
2	nà-kka.		
3	tta-rgha.	Wise	koo-rac-yon.
4	tting.	Foolish	nà-a-ghal.
5	sa-soo-la.	Strong	na-tz-ap.
6	ut-ke-ttai.	Weak	pa-a-ttha-to-rghelli.
7	kkosing-ting.		
8	etzenting.	Right	nochnesse.
9	kka-hooli.	Left	intzesse.
10	ho-nanna.	My friend	tza-teleg-ga.
20	nou-nanna.	My companion	tza-onenya.
Good	naa-zo.	Head	tzat-the.
Bad	naa-zo-heli; dzoun-de.	Eyes	tzen-nhae.
		Nose	tze-etze.
Beautiful	bur-a-oonde; tzoo-na-e-ti.	Mouth	tze-thä.
		Ears	setz-r-rgha.
Ugly	pa-chi-ri.	Hair	setz-thè-rgha.
Large	natza-konde; in-cha.	Tongue	tze-tthou.
		Teeth	tze-w-who?
Small	tzoo-ta.	Neck	tze-e-e-cottle.
Heavy	net-ta; hinka.	Arm	tze-int-chinne.
Light	hin-ka-he-li.	Hand	ssa-la.
Dark	tel-zen.	Foot	tzè-ka.
Bright	atz-za.	Legs	tze-thunna.
Low	ne-otzin-ik.	Canoe paddle	ola; tho.
High	tan-ne-e-tha.	Here	d-jahn.
Hard	tan-y-eet.	There	a-c-yà.
Soft	taa-yeet-heli.	Where?	djhan-tin?
New	e-e-yes-e.	When	kkonde.

* I endeavored to put this intelligibly to Nanette, by supposing such an expression as ma chère femme, ma chère fille. When at length she understood it, her reply was (with great emphasis): "I' dit jamais ça. I' dit ma femme, ma fille."

English.	Dog-Rib.
Which	mee.
What?	et-cloy?
To me	tzen-ez-etze.
To him	ne-ghon-em-etze.
To you	ne-nin-etze.
To us	e-e-cla-toon-nim-etze.
I don't understand	nè-ad-'hear-des-tha-helili.
I don't speak Cree	(same sentence taken down.)
I won't give it you	na 'rha tchou-heli.
I will give it to you	na 'rha ochou-eze.
What shall I give you for this?	than-etcha-na-rha-ócla-hãze?
Take care	c̋a-re.
Make haste	ag̋a-annitè.
Get out (va t'en)	or-rhink-là.
Where is it?	ye-in-kon-ecla?
Carry this for me	sse-ragh-di-ach.
Don't touch that	perrone-te-sonna.
What do you want?	na-nu-ăt-cloy?
What do you want for this?	etcha-nette-ousa-nousa-ou-sin-ne?
Give me a piece of tobacco	tza-twe-tza-gan-a-two.
I have no tobacco	tza-twe-ta-oo-twe.
Hold this	Ou-net-ton.
Whose is this?	me-etze-hande?

THE END.

The Pictorial Field-Book of the Revolution;

or, Illustrations by Pen and Pencil, of the History, Scenery, Biography, Relics, and Traditions of the War for Independence. By Benson J. Lossing, Esq. Embellished with 600 Engravings on Wood, chiefly from Original Sketches by the Author. In about 20 Numbers, 8vo, Paper, 25 cents each.

Life and Writings of Thomas Chalmers, D.D.,

LL.D. Edited by his Son-in-Law, Rev. William Hanna, LL D 3 vols. 12mo, Paper, 75 cents; Muslin, $1 00 per Volume.

Life of John Calvin.

Compiled from authentic Sources, and particularly from his Correspondence. By T. H. Dyer. Portrait. 12mo, Muslin, $1 00.

Leigh Hunt's Autobiography,

With Reminiscences of Friends and Contemporaries. 2 vols 12mo, Muslin, $1 50.

Southey's Life and Correspondence.

Edited by his Son, Rev. Charles Cuthbert Southey, M.A. In 6 Parts, 8vo, Paper, 25 cents each; one Volume, Muslin, $2 00.

Dr. Johnson: his Religious Life and his Death.

12mo, Muslin, $1 00.

Life and Letters of Thomas Campbell.

Edited by William Beattie, M.D., one of his Executors. With an Introductory Letter by Washington Irving, Esq. Portrait 2 vols. 12mo, Muslin, $2 50.

Hume's History of England,

From the Invasion of Julius Cæsar to the Abdication of James II., 1688. A new Edition, with the Author's last Corrections and Improvements. To which is prefixed a short Account of his Life, written by Himself. With a Portrait of the Author. 6 vols. 12mo, Cloth, $2 40; Sheep, $3 00.

Macaulay's History of England,

From the Accession of James II. With an original Portrait of the Author. Vols. I. and II. Library Edition, 8vo, Muslin, 75 cents per Volume; Sheep extra, 87½ cents per Volume; Calf backs and corners, $1 00 per Volume.—Cheap Edition, 8vo, Paper, 25 cents per Volume.—12mo (uniform with Hume), Cloth, 40 cents per Volume.

Gibbon's History of Rome,

With Notes, by Rev. H. H. Milman and M. Guizot. Maps and Engravings. 4 vols. 8vo, Sheep extra, $5 00.—A new Cheap Edition, with Notes by Rev. H. H. Milman. To which is added a complete index of the whole Work and a Portrait of the Author. 6 vols. 12mo (uniform with Hume), Cloth, $2 40; Sheep, $3 00.

Journal and Memorials of Capt. Obadiah Congar:

for Fifty Years Mariner and Shipmaster from the Port of New York. By Rev. H. T. Cheever. 16mo, Muslin.

Benjamin Franklin's Autobiography.

With a Sketch of his Public Services, by Rev. H. HASTINGS WELD. With numerous exquisite Designs, by JOHN G. CHAPMAN. 8vo, Muslin, $2 50; Sheep, $2 75; half Calf, $3 00.

History of Spanish Literature.

With Criticisms on the particular Works and Biographical Notices of prominent Writers. By GEORGE TICKNOR, Esq. 3 vols. 8vo, half Calf extra, $7 50; Sheep extra, $6 75; Muslin, $6 00.

History of the National Constituent Assembly,

From May, 1848. By J. F. CORKRAN, ESQ. 12mo, Muslin, 90 cents; Paper, 75 cents.

The Recent Progress of Astronomy,

Especially in the United States. By ELIAS LOOMIS, M.A. 12mo, Muslin, $1 00.

The English Language

In its Elements and Forms. With a History of its Origin and Development, and a full Grammar. By W. C. FOWLER, M.A. 8vo, Muslin, $1 50; Sheep, $1 75.

History of Ferdinand and Isabella.

By WILLIAM H. PRESCOTT, Esq. 3 vols. 8vo, half Calf, $7 50; Sheep extra, $6 75; Muslin, $6 00.

History of the Conquest of Mexico.

With the Life of the Conqueror, Hernando Cortez, and a View of the Ancient Mexican Civilization. By WILLIAM H. PRESCOTT, Esq. Portrait and Maps. 3 vols. 8vo, half Calf, $7 50; Sheep extra, $6 75; Muslin, $6 00.

History of the Conquest of Peru.

With a Preliminary view of the Civilization of the Incas. By WILLIAM H. PRESCOTT, Esq. Portraits, Maps, &c. 2 vols. 8vo, half Calf, $5 00; Sheep extra, $4 50; Muslin, $4 00.

Biographical and Critical Miscellanies.

Containing Notices of Charles Brockden Brown, the American Novelist.—Asylum for the Blind.—Irving's Conquest of Grenada.—Cervantes.—Sir W. Scott.—Chauteaubriand's English Literature.—Bancroft's United States.—Madame Calderon's Life in Mexico.—Moliere.—Italian Narrative Poetry.—Poetry and Romance of the Italians.—Scottish Song.—Da Ponte's Observations. By WILLIAM H. PRESCOTT, Esq. Portrait. 8vo, Muslin, $2 00; Sheep extra, $2 25; half Calf, $2 50.

Past, Present, and Future of the Republic.

By ALPHONSE DE LAMARTINE. 12mo, Muslin, 50 cents; Paper, 37½ cents.

The War with Mexico

By R. S. RIPLEY, U.S.A. With Maps, Plans of Battles, &c 2 vols. 12mo, Muslin, $4 00; Sheep, $4 50; half Calf, $5 00

The Conquest of Canada.

By the Author of "Hochelaga." 2 vols. 12mo, Muslin, $1 70.

History of the Confessional.

By John Henry Hopkins, D.D., Bishop of Vermont. 12mo, Muslin, $1 00.

Dark Scenes of History.

By G. P. R. James, Esq. 12mo, Paper, 75 cents; Muslin, $1 00.

Library of American Biography.

Edited by Jared Sparks, LL.D. Portraits, &c. 10 vols. 12mo, Muslin, $7 50. Each volume sold separately, if desired, price 75 cents.

Gieseler's Ecclesiastical History.

From the Fourth Edition, revised and amended. Translated from the German, by Samuel Davidson, LL.D. Vols. I. and II. 8vo, Muslin, $3 00.

History of the American Bible Society,

From its Organization in 1816 to the Present Time. By Rev. W. P. Strickland. With an Introduction, by Rev. N. L. Rice, and a Portrait of Hon. Elias Boudinot, LL.D., first President of the Society. 8vo, Cloth, $1 50; Sheep, $1 75.

Biographical History of Congress:

Comprising Memoirs of Members of the Congress of the United States, together with a History of Internal Improvements from the Foundation of the Government to the Present Time. By Henry G. Wheeler. With Portraits and Fac-simile Autographs. 8vo, Muslin, $3 00 per Volume.

Schmitz's History of Rome,

From the Earliest Times to the Death of Commodus, A.D. 192. With Questions, by J. Robson, B.A. 18mo, Muslin, 75 cents

Louis the Fourteenth,

and the Court of France in the Seventeenth Century. By Miss Pardoe. Illustrated with numerous Engravings, Portraits, &c. 2 vols. 12mo, Muslin, $3 50.

History of the Girondists;

Or, Personal Memoirs of the Patriots of the French Revolution. By A. de Lamartine. From unpublished Sources. 3 vols. 12mo, Muslin, $2 10.

Josephus's Complete Works.

A new Translation, by Rev. Robert Traill, D.D. With Notes, Explanatory Essays, &c., by Rev. Isaac Taylor, of Ongar. Illustrated by numerous Engravings. Publishing in Monthly Numbers, 8vo, Paper, 25 cents each.

History of the French Revolution.

By Thomas Carlyle. Newly Revised by the Author, with Index, &c. 2 vols. 12mo, Muslin, $2 00.

Letters and Speeches of Cromwell.

With Elucidations and connecting Narrative. By T. Carlyle. 2 vols 12mo, Muslin, $2 00.

Life of Madame Guyon.

Life and Religious Opinions of Madame Guyon: together with some Account of the Personal History and Religious Opinions of Archbishop Fenelon. By T. C. Upham. 2 vols. 12mo, Muslin, $2 00.

Life of Madame Catharine Adorna.

Including some leading Facts and Traits in her Religious Experience. Together with Explanations and Remarks, tending to illustrate the Doctrine of Holiness. By T. C. Upham. 12mo, Muslin, 50 cents; Muslin, gilt edges, 60 cents.

Homes and Haunts of the British Poets.

By William Howitt. With numerous Illustrations. 2 vols. 12mo, Muslin, $3 00.

History of Wonderful Inventions.

Illustrated by numerous Engravings. 12mo, Muslin, 75 cents; Paper, 50 cents.

Life and Writings of Cassius M. Clay;

Including Speeches and Addresses. Edited, with a Preface and Memoir, by Horace Greeley. With Portrait. 8vo, Muslin, $1 50.

The Valley of the Mississippi.

History of the Discovery and Settlement of the Valley of the Mississippi, by the three great European Powers, Spain, France, and Great Britain; and the subsequent Occupation, Settlement, and Extension of Civil Government by the United States, until the year 1846. By John W. Monette, Esq. Maps. 2 vols. 8vo, Muslin, $5 00; Sheep, $5 50.

Southey's Life of John Wesley;

And Rise and Progress of Methodism. With Notes by the late Samuel T. Coleridge, Esq., and Remarks on the Life and Character of John Wesley, by the late Alexander Knox, Esq. Edited by the Rev. Charles C. Southey, M.A. Second American Edition, with Notes, &c., by the Rev. Daniel Curry, A.M. 2 vols. 12mo, Muslin, $2 00.

Pictorial History of England.

Being a History of the People as well as a History of the Kingdom, down to the Reign of George III. Profusely Illustrated with many Hundred Engravings on Wood of Monumental Records; Coins; Civil and Military Costume; Domestic Buildings, Furniture, and Ornaments; Cathedrals and other great Works of Architecture; Sports and other Illustrations of Manners; Mechanical Inventions; Portraits of Eminent Persons; and remarkable Historical Scenes. 4 vols. imperial 8vo, Muslin, $14 00; Sheep extra, $15 00; half Calf $16 00.

Diplomatic and Official Papers of Daniel Webster,

while Secretary of State. With Portrait. 8vo, Muslin, $1 75.

Life of the Chevalier Bayard.

By WILLIAM G. SIMMS, Esq. Engravings. 12mo, Muslin, $1 00.

History of Europe,

From the Commencement of the French Revolution in 1789 to the Restoration of the Bourbons in 1815. By ARCHIBALD ALISON, F.R.S. In addition to the Notes on Chapter LXXVI., which correct the errors of the original work concerning the United States, a copious Analytical Index has been appended to this American Edition. 4 vols. 8vo, Muslin, $4 75; Sheep extra, $5 00.

Boswell's Life of Dr. Samuel Johnson.

Including a Journal of a Tour to the Hebrides. With numerous Additions and Notes, by JOHN W. CROKER, LL.D. A new Edition, entirely revised, with much additional Matter. Portrait. 2 vols. 8vo, Muslin, $2 75; Sheep, $3 00.

Life and Speeches of John C. Calhoun.

With Reports and other Writings, subsequent to his Election as Vice-president of the United States; including his leading Speech on the late War, delivered in 1811. 8vo, Muslin, $1 12½.

Life of Charlemagne.

With an Introductory View of the History of France. By G. P. R. JAMES, Esq. 18mo, Muslin, 45 cents.

Life of Henry IV.,

King of France and Navarre. By G. P. R. JAMES, Esq. 2 vols. 12mo, Muslin, $2 50.

History of Chivalry and the Crusades.

By G. P. R. JAMES, Esq. Engravings. 18mo, Muslin, 45 cents.

Neal's History of the Puritans,

Or, Protestant Non-conformists; from the Reformation in 1518 to the Revolution in 1688; comprising an Account of their Principles, Sufferings, and the Lives and Characters of their most considerable Divines. With Notes, by J. O. CHOULES, D.D. With Portraits. 2 vols. 8vo, Muslin, $3 50; Sheep, $4 00.

Neander's Life of Christ;

In its Historical Connections and its Historical Development. Translated from the Fourth German Edition, by Professors M'CLINTOCK and BLUMENTHAL, of Dickinson College. 8vo, Muslin, $2 00; Sheep extra, $2 25.

Lives of Celebrated British Statesmen.

The Statesmen of the Commonwealth of England; with a Treatise on the popular Progress in English History. By JOHN FORSTER. Edited by the Rev. J. O. CHOULES. Portraits. 8vo, Muslin, $1 75; Sheep, $2 00.

The Island World of the Pacific:

Being the Personal Narrative and Results of Travel through the Sandwich or Hawaiian Islands and other parts of Polynesia. By Rev. H. T. CHEEVER. Engravings. 12mo, Muslin, $1 00.

The Whale and his Captors;

Or, the Whaleman's Adventures and the Whale's Biography, as gathered on the Homeward Cruise of the "Commodore Preble." By Rev. H. T. CHEEVER. Engravings. 16mo, Muslin, 60 cents.

Five Years of a Hunter's Life in the Far Interior of South Africa.

With Notices of the Native Tribes, and Anecdotes of the Chase of the Lion, Elephant, Hippopotamus, Giraffe, Rhinoceros, &c. By R. GORDON CUMMING, Esq. With Illustrations. 2 vols. 12mo, Muslin, $1 75.

The Pillars of Hercules;

Or, a Narrative of Travels in Spain and Morocco in 1848. By DAVID URQUHART, M.P. 2 vols. 12mo, Paper, $1 40; Muslin, $1 70.

Glimpses of Spain;

Or, Notes of an Unfinished Tour in 1847. By S. T. WALLIS, Esq. 12mo, Paper, 75 cents; Muslin, $1 00.

Sketches of Minnesota,

The New England of the West. With Incidents of Travel in that Territory during the Summer of 1848. By E. S. SEYMOUR. 12mo, Paper, 50 cents; Muslin, 75 cents.

Life in the Far West.

By GEORGE FREDERICK RUXTON. 12mo, Paper, 37½ cents; Muslin, 60 cents.

Adventures in Mexico and Rocky Mountains.

By G. F. RUXTON. 12mo, Paper, 50 cents; Muslin, 60 cents.

Fresh Gleanings;

Or, a New Sheaf from the Old Fields of Continental Europe. By IK. MARVEL. 12mo, Paper, $1 00; Muslin, $1 25.

A Second Visit to the United States.

By Sir CHARLES LYELL, F.R.S. 2 vols. 12mo, Paper, $1 20, Muslin, $1 50.

Oregon and California in 1848.

With an Appendix, including recent and authentic Information on the Subject of the Gold Mines of California, and other valuable matter of Interest to the Emigrant. By Judge THORNTON. With Illustrations and a Map. 2 vols. 12mo, Muslin, $1 75

The Little Savage.

Being the History of a Boy left alone on an uninhabited Island By Captain MARRYAT, R N. 12mo, Paper, 37½ cents; Muslin, 45 cents.

White-Jacket;

Or, the World in a Man-of-War. By HERMAN MELVILLE, Esq. 12mo, Paper, $1 00; Muslin, $1 25.

Redburn:

His First Voyage. Being the Reminiscences of the Son-of-a-Gentleman in the Merchant Service. By HERMAN MELVILLE, Esq. 12mo, Paper, 75 cents; Muslin, $1 00.

Mardi: and a Voyage Thither.

By HERMAN MELVILLE, Esq. 2 vols. 12mo, Paper, $1 50; Muslin, $1 75.

Omoo;

Or, a Narrative of Adventures in the South Seas. By HERMAN MELVILLE, Esq. 12mo, Paper, $1 00; Muslin, $1 25.

Typee:

A Peep at Polynesian Life, during a Four Months' Residence in the Marquesas. By HERMAN MELVILLE, Esq. The revised Edition, with a Sequel 12mo, Paper, 75 cents; Muslin, 87½ cents.

Loiterings in Europe;

Or, Sketches of Travel in France, Belgium, Switzerland, Italy, Austria, Prussia, Great Britain, and Ireland With an Appendix. containing Observations on European Charities and Medical Institutions. By J. W. CORSON, M.D. 12mo, Paper, 75 cents; Muslin, $1 00.

Narrative of the Texan Santa Fé Expedition.

Comprising a Description of a Tour through Texas, and across the great Southwestern Prairies, the Camanche and Caygua Hunting-grounds, &c. By G. W. KENDALL, Esq. With a Map and Illustrations. 2 vols. 8vo, Muslin, $2 50.

Travels in Egypt, Arabia Petræa, and the Holy

Land. By JOHN L. STEPHENS, Esq. Engravings. 2 vols. 12mo, Muslin, $1 75.

Travels in Greece, Turkey, Russia, and Poland.

By JOHN L. STEPHENS, Esq. Engravings. 2 vols. 12mo, Muslin, $1 75.

Incidents of Travel in Yucatan.

By JOHN L. STEPHENS, Esq. 120 Engravings, from Drawings by F. CATHERWOOD, Esq. 2 vols. 8vo, Muslin, $5 00.

Travels in Central America, Chiapas, and Yuca-

tan. By JOHN L. STEPHENS, Esq. With a Map and 88 Engravings. 2 vols. 8vo, Muslin, $5 00.

Consular Cities of China.

A Narrative of an Exploratory Visit to each of the Consular Cities of China, and to the Islands of Hong Kong and Chusan, in behalf of the Church Missionary Society, in the Years 1844–1846 By Rev. G. SMITH. 12mo, Paper, $1 00; Muslin, $1 25.

Loomis's Recent Progress of Astronomy,
Especially in the United States. 12mo, Muslin, $1 00.

Cheever's (Rev. H. T.) Island World of the Pacific: being the Personal Narrative and Results of Travel through the Sandwich or Hawaiian Islands, and other Parts of Polynesia. With Engravings. 12mo, Muslin, $1 00.

Lossing's Pictorial Field-Book of the Revolution; or, Illustrations, by Pen and Pencil, of the History, Scenery, Biography, Relics, and Traditions of the War for Independence. Embellished with 600 Engravings on Wood, chiefly from Original Sketches by the Author. In about 20 Numbers, 8vo, Paper, 25 cents each.

Abbott's Illustrated Histories:
Comprising, Xerxes the Great, Cyrus the Great, Alexander the Great, Darius the Great, Hannibal the Carthaginian, Julius Cæsar, Cleopatra Queen of Egypt, Constantine, Nero, Romulus, Alfred the Great, William the Conqueror, Queen Elizabeth, Mary Queen of Scots, Charles the First, Charles the Second, Queen Anne, King John, Richard the First, William and Mary, Maria Antoinette, Madame Roland, Josephine. Illuminated Title-pages, and numerous Engravings. 16mo, Muslin, 60 cents each; Muslin, gilt edges, 75 cents each.

Abbott's Kings and Queens;
Or, Life in the Palace: consisting of Historical Sketches of Josephine and Maria Louisa, Louis Philippe, Ferdinand of Austria, Nicholas, Isabella II., Leopold, and Victoria. With numerous Illustrations. 12mo, Muslin, $1 00; Muslin, gilt edges, $1 25.

Abbott's Summer in Scotland.
Engravings. 12mo, Muslin, $1 00.

Southey's Life and Correspondence.
Edited by his Son, Rev. Charles Cuthbert Southey, M.A. In 6 Parts, 8vo, Paper, 25 cents each; one Volume, Muslin, $2 00.

Howitt's Country Year-Book;
Or, the Field, the Forest, and the Fireside. 12mo, Muslin, 87½ cents.

Fowler's Treatise on the English Language
In its Elements and Forms. With a History of its Origin and Development, and a full Grammar. Designed for Use in Colleges and Schools. 8vo, Muslin, $1 50; Sheep, $1 75.

Seymour's Sketches of Minnesota,
The New England of the West. With Incidents of Travel in that Territory during the Summer of 1849. With a Map. 12mo Paper, 50 cents; Muslin, 75 cents.

Dr. Johnson's Religious Life and Death.
12mo, Muslin, $1 00.

Cumming's Five Years of a Hunter's Life

In the Far Interior of South Africa. With Notices of the Native Tribes, and Anecdotes of the Chase of the Lion, Elephant, Hippopotamus, Giraffe, Rhinoceros, &c. Illustrated by numerous Engravings. 2 vols. 12mo, Muslin, $1 75.

Thornton's Oregon and California in 1848:

With an Appendix, including recent and authentic Information on the Subject of the Gold Mines of California, and other valuable Matter of Interest to the Emigrant, &c. With Illustrations and a Map. 2 vols. 12mo, Muslin, $1 75.

Southey's Common-place Book.

Edited by his Son-in-Law, JOHN WOOD WARTER, B.D. 8vo, Paper, $1 00 per Volume; Muslin, $1 25 per Volume.

Gibbon's History of Rome,

With Notes, by Rev. H. H. MILMAN and M. GUIZOT. Maps and Engravings. 4 vols. 8vo, Sheep extra, $5 00.—A new Cheap edition, with Notes by Rev. H. H. MILMAN. To which is added a complete Index of the whole Work and a Portrait of the Author. 6 vols. 12mo (uniform with Hume), Cloth, $2 40; Sheep, $3 00.

Hume's History of England,

From the Invasion of Julius Cæsar to the Abdication of James II., 1688. A new Edition, with the Author's last Corrections and Improvements. To which is prefixed a Short Account of his Life, written by Himself. With a Portrait of the Author. 6 vols. 12mo, Cloth, $2 40; Sheep, $3 00.

Macaulay's History of England,

From the Accession of James II. With an original Portrait of the Author. Vols. I. and II. Library Edition, 8vo, Muslin, 75 cents per Volume; Sheep extra, 87½ cents per Volume; Calf backs and corners, $1 00 per Volume.—Cheap Edition, 8vo, Paper, 25 cents per Volume.—12mo (uniform with Hume), Cloth. 40 cents per Volume.

Leigh Hunt's Autobiography,

With Reminiscences of Friends and Contemporaries. 2 vols 12mo, Muslin, $1 50.

Life and Letters of Thomas Campbell.

Edited by WILLIAM BEATTIE, M.D., one of his Executors. With an Introductory Letter by WASHINGTON IRVING, Esq. Portrait. 2 vols. 12mo, Muslin, $2 50.

Dyer's Life of John Calvin.

Compiled from authentic Sources, and particularly from his Correspondence. Portrait. 12mo, Muslin, $1 00.

Moore's Health, Disease, and Remedy,

Familiarly and practically considered, in a few of their Relations to the Blood. 18mo, Muslin, 60 cents.

www.ingramcontent.com/pod-product-compliance
Lightning Source LLC
LaVergne TN
LVHW021308110826
845150LV00003B/524